Risks of Hazardous Wastes

Risks of Hazardous Wastes

Paul E. Rosenfeld
Lydia G. H. Feng

ELSEVIER

AMSTERDAM • BOSTON • HEIDELBERG • LONDON
NEW YORK • OXFORD • PARIS • SAN DIEGO
SAN FRANCISCO • SINGAPORE • SYDNEY • TOKYO
William Andrew is an imprint of Elsevier

William Andrew is an imprint of Elsevier
The Boulevard, Langford Lane, Kidlington, Oxford OX5 1GB, UK
30 Corporate Drive, Suite 400, Burlington, MA 01803, USA

First edition 2011

British Library Cataloguing-in-Publication Data
A catalogue record for this book is available from the British Library

Library of Congress Cataloging-in-Publication Data
A catalog record for this book is available from the Library of Congress

ISBN: 978-0-323-16565-5

For information on all William Andrew publications
visit our website at books.elsevier.com

Typeset by MPS Limited, a Macmillan Company, Chennai, India
www.macmillansolutions.com

11 12 13 14 15   10 9 8 7 6 5 4 3 2 1

Contents

Preface

In early August 1978, the *New York Times* brought the now infamous toxic waste exposure at Love Canal to the public eye, with a front page article titled unexceptionally: *Upstate Waste Site May Endanger Lives*. The article described a town life with cases of miscarriages, birth defects and family pets not living past the age of three. A town unwittingly sitting on top of an "industrial dump," containing 82 compounds, 11 of which were known carcinogens. It described children and expectant mothers ignorant of the baneful cocktail percolating up through their back yards. Yet, the most possible lasting impact of the article was the quote by the regional director for the USEPA, Eckhardt C. Beck: *"We've been burying these things like ticking time bombs."* According to his estimate, there were thousands of industrial waste landfills throughout the nation, all with the potential to leech harm out onto an unsuspecting American public. In an instant, Beck turned Love Canal from an isolated incident, to one that could occur anywhere or to anyone. Over the next week, front page articles like "Health Chief Calls Waste Site a 'Peril,'" "First Families Leaving Upstate Contamination Site," and "Carter Approves Emergency Help In Niagara Area," reflected the growing severity of the problem at Love Canal, and the national attention it was receiving. Later that year, President Carter described the state of America's numerous industrial waste dumps and the presence of toxins in our environment as "one of the grimmest discoveries of the modern era." The public endeavor to stem the flow of hazardous waste had begun.

President Carter was only partially correct in his statement. What the public was just discovering had been known to the corporations and government agencies for decades. Many of these entities had full knowledge of the health hazards associated with their products, and condoned (even supported) their negligent discharge into the environment. Since 1978, corporations, and government agencies to some degree, have continued to release hazardous waste into our environment in the form of products and production byproducts. As you read this book, agency reports, regulations, lawsuits, case studies and records of hazardous waste mismanagement would depict the struggle of government regulators, public associations, law firms and individuals to fight polluters, and to unearth and remediate their toxic legacies.

The purpose of this book is to provide an overview of the issues surrounding hazardous waste, from its sources to its consequences, focusing on the risks posed to human health and the environment. It explains the legislation and regulations surrounding hazardous waste, as well as describes deficiencies in policy, regulation, and science that have allowed the public to be subjected to a myriad of potentially hazardous agents. This book provides a look into some of the major generators of hazardous wastes, explains the pathways by which humans and wildlife are exposed, and includes discussion of the adverse health effects linked to these pollutants. Finally, it provides a discussion of measures that will be necessary to control society's hazardous waste problem.

Acknowledgements: We would like to extend our most heartfelt thanks to David Molmen, Gabe Arom, Grace Lee, Joyce Chen, Chris Waller, Helen Sok, and all of the students of Environmental Health Sciences at the UCLA School of Public Health, whose help researching and contributing to this book has proved to be indispensable. Without the talents and tireless efforts of these individuals, this book would not have been possible. At last, we must thank the staff at Elsevier, including Susan Li, Frank Hellwig, Matthew Deans, and Melissa Read, for their hard work, patience, and support throughout this process.

Paul E. Rosenfeld
Lydia G. H. Feng

About the Authors

Paul E. Rosenfeld, Ph.D. is an environmental chemist with over twenty years of experience, focusing on fate and transport of environmental contaminants, risk assessment, and ecological restoration. He is an expert in the monitoring and modeling of pollution sources as they relate to human and ecological health. Dr. Rosenfeld has served as a remedial project manager with the U.S. Navy Base Realignment and Closure Program. He has conducted risk assessments and designed cleanup programs for contaminated sites containing pesticides, radioactive waste, PCBs, PAHs, dioxins, furans, volatile organics, semi-volatile organics, chlorinated solvents, perchlorate, heavy metals, asbestos, odorants, petroleum, PFOA, unusual polymers, and fuel oxygenates. His publications have covered pollution management, environmental odorants, environmental sampling methodology, and pollutant levels in communities impacted by industrial emissions. He currently teaches environmental health at the UCLA School of Public Health.

Lydia G. H. Feng is an environmental health scientist at Soil Water Air Protection Enterprise in Santa Monica, CA. Lydia serves as a consultant on major environmental and toxic tort cases, conducting research, exposure and risk analyses on communities impacted by pollution. Her project experience includes evaluation of exposure and risk at numerous industrial sites and fenceline communities, to various pollutants including dioxins, polychlorinated biphenyls, polycyclic aromatic hydrocarbons, volatile organics, benzene, and particulate matter. She studies adverse health effects including respiratory illnesses and cancers associated with environmental pollutants, and has published on pollutant levels in communities impacted by industrial emissions.

1 Definition of Hazardous Waste

Chapter Contents

When one thinks of hazardous waste, images of 55-gallon drums labeled with skull and crossbones and massive factories spewing neon-green sludge probably come to mind. In truth, hazardous wastes take on a variety of guises, sometimes as the mercury in old CRT computer monitors, or lead paint in old demolished buildings, or as vinyl chloride leftover from industrial processes. Hazardous waste can be generated from numerous processes and facilities including construction sites, hospitals, factories, military bases and discarded electronics. In theory, if a material has been discarded and it can cause substantial harm to humans or the environment, it can be considered a hazardous waste.

1.1 The Environmental Protection Agency's Regulatory Definition

According to the US Environmental Protection Agency (2009), waste is hazardous if it poses a substantial threat to human health or the environment. While this definition suffices in a looser, philosophical sense, the words are not specific enough

Risks of Hazardous Wastes. DOI: 10.1016/B978-1-4377-7842-7.00001-5

for regulatory purposes. For a more utilitarian definition, the EPA relies on the specifications set by the Resources Conservation and Recovery Act (RCRA) in 1976. According to the RCRA, a waste first must be considered a 'solid waste.' This term is misleading in that the RCRA defines 'solid waste' as any discarded material, including solids, liquids, and contained gases. Once determined to be a 'solid waste,' the discards are eligible for hazardous status. There are two regulatory definitions the EPA uses to determine if waste is hazardous. The first is the empirical, so-called 'characteristic' definition. If a hazardous waste is *ignitable*, *corrosive*, *reactive*, or *toxic* to the degree specified in the RCRA, then it is considered hazardous. The second method involves cross-referencing the waste in question with a predetermined list of chemicals (found in the RCRA) that pose a substantial threat to human health or the environment in the present or future. The toxins included on this list do not necessarily need to exhibit any of the four aforementioned hazard characteristics to be considered a toxic waste.

1.1.1 Characteristic Hazardous Waste

The first characteristic, *ignitability*, is based on a number of criteria involving the material's tendency to burst into flame (spontaneously or under certain conditions), or a material's flash point, if less than 60°C. Included in these criteria, but not limited to them, are non-liquids capable under STP of igniting and burning through friction, moisture absorption, or spontaneous chemical changes. Tests listed in regulations, such as the Pensky-Martens Closed-Cup Method for Determining Ignitability (Method 101A) and the Ignitability of Solids Method (Method 1030), can also be employed to determine the degree of ignitability. Waste oils and used solvents are two examples of hazardous waste that fulfill the *ignitability* criterion.

If an aqueous waste-solution is a strong acid ($pH \leq 2$) or a strong base ($pH \geq 12.5$) capable of corroding metal containers, then it fulfills the *corrosivity* criterion. If a liquid dissolves metal at a rate greater than 6.35 mm (0.250 inch) per year at a test temperature of 55°C, it too is considered corrosive and hazardous. Hazardous wastes that exhibit *corrosive* properties include sulfuric acid and hydrochloric acid. The testing method known as the Corrosivity Towards Steel (Method 1110A) is the only method to determine corrosivity.

Wastes that are unstable, i.e. those that can cause explosions, release toxic fumes/gases/vapors upon heating, and react upon compression or mixing with water, are considered *reactively* hazardous. There are no existing tests to determine reactivity. Examples of reactive chemicals include lithium sulfur batteries, ammunition,.aerosols, and explosives.

Toxicity is determined by an analysis of the waste's leachate using a testing procedure known as the Toxicity Characteristic Leaching Procedure (EPA Method 1311). If toxins (listed in federal regulations) such as arsenic, trichloroethylene, or mercury are found at levels that surpass regulatory levels, then the material in question is designated as toxic and therefore hazardous.

1.1.2 Listed Hazardous Waste

Hazardous wastes are added to the RCRA list (40 CFR §261.31-33) if the EPA determines they fulfill one of four requirements:

- The waste typically contains harmful chemicals, and other factors indicate that it could pose a threat to human health and the environment in the absence of special regulation. Such wastes are known as *toxic listed wastes.*
- The waste contains such dangerous chemicals that it could pose a threat to human health and the environment even when properly managed. Such wastes are known as *acutely hazardous wastes.*
- The waste typically exhibits one of four characteristics of hazardous waste (see Section 1.1.1).
- When EPA has cause to believe that for some other reason, the waste typically fits within the statutory definition of hazardous waste developed by Congress (EPA, 2005).

These four criteria are used by the US EPA internally for identifying wastes to list in the RCRA. Once posted by the US EPA, manufacturers and other industries can refer to these lists for identification of hazardous waste.

There are four lists maintained by the RCRA for identifying listed hazardous wastes, known individually as F, K, P, and U lists. The F-list is reserved for non-specific source wastes. In English this designation refers to wastes that are produced in general manufacturing and industrial processes that are common throughout industry. Examples of waste generated from these processes include solvents used in the cleaning (degreasing) of machine parts, or wood preserving. As of publication date there are 39 non-specific source waste definitions which can be found in 40 CFR §261.31.

The K-list, or source-specific waste list, describes wastes that are produced in specific manufacturing sectors only, such as pesticide manufacturing or petroleum refining. Within the list, categories pertaining to industry are as specific as 'ink formulation,' or 'veterinary pharmaceuticals,' or 'organic chemical.' The K-list to date contains 148 hazardous waste categories, 111 more than the F-list. This discrepancy is because the K-list deals with wastes in more specificity than its counterpart F-list.

The U and P lists are frequently described in unison, given their similarities. Both refer to discarded chemical products (e.g. pesticides and pharmaceuticals) as opposed to mixed wastes, but differ in designation as either 'toxic' or 'acutely hazardous' waste, respectively. Toxic chemicals, which are less dangerous, present a threat to the public/environment and require regulation and management to reduce the threat. Acutely hazardous chemicals present a threat despite regulation and effective management. In the RCRA, 959 toxic and 487 acutely hazardous wastes are listed.

1.1.3 Excluded Hazardous Waste

Excluded hazardous wastes are defined extensively in 40 CFR §261.4. These excluded materials range from chlorofluorocarbons (CFCs) in old refrigerators, to zinc fertilizers containing limited amounts of arsenic and lead, to secondary materials that are reclaimed and returned to the original process in which they were generated. Excluded wastes are not controlled by the principal hazardous waste regulations found in the RCRA; however, many of these are regulated by other means. Mining and mineral processing wastes, for example, are regulated by mining statutes and regulations. Wastes excluded (even if they exhibit characteristics of hazardous waste) from the RCRA §261.4(b) are as follows (EPA, 2010a):

- Household Hazardous Waste
- Agricultural Waste
- Mining Overburden
- Fossil Fuel Combustion Waste (Bevill)
- Oil, Gas, and Geothermal Wastes (Bentsen Amendment)
- Trivalent Chromium Wastes
- Mining and Mineral Processing Wastes (Bevill)
- Cement Kiln Dust (Bevill)
- Arsenically Treated Wood
- Petroleum Contaminated Media & Debris from Underground Storage Tanks
- Injected Groundwater
- Spent Chlorofluorocarbon Refrigerants
- Used Oil Filters
- Used Oil Distillation Bottoms
- Landfill Leachate or Gas Condensate Derived from Certain Listed Wastes
- Project XL Pilot Project Exclusions.

1.2 Regulatory History of Hazardous Waste in the US

1.2.1 The Resource Conservation and Recovery Act and Amendments

The regulation of hazardous waste began in 1960 when the US Congress passed the Federal Hazardous Substances Act (FHSA). The Act requires the labeling of all hazardous household products to warn consumers of the possible dangers associated with handling (NYCDS, 2010). Disposal and management of hazardous materials, however, was not tackled until 16 years later.

The Resource Conservation and Recovery Act (RCRA) was signed into law by President Gerald Ford in 1976, and is currently the primary federal law regulating the disposal of solid and hazardous waste. The law authorizes the US EPA to regulate hazardous wastes and issue federal guidelines for state-based solid waste disposal. Prior to the RCRA, the US Government had no system of tracking and monitoring solid waste. Arguably the most significant contribution of the Act was to establish so-called 'cradle to grave' monitoring of hazardous wastes. Under the RCRA, generators of hazardous waste are required to identify their material with the EPA upon introduction into the United States (either through production or international import). Once identified, the producer must then apply for a permit and begin documenting all hazardous waste outputs on a monthly basis.

While groundbreaking at the time, the RCRA contained many loopholes soon exploited by manufacturers and other generators of hazardous waste. Uncontrolled incineration and mixing of hazardous waste for 'energy recovery,' and dumping of hazardous wastes in municipal landfills and sewer systems by smaller-quantity generators allowed roughly 40 million metric tons to be emitted annually (Davis, 2001; EPA, 2003). The Reagan administration did little to remedy the situation, supporting a policy of voluntary industry compliance with the RCRA. Furthermore, under the budget-cutting policy of Reaganomics, the EPA was forced to reduce expenditures, leading to a decline in RCRA-based programs. Between 1980 and 1981, litigation against violators submitted to the justice department dropped from 46 to 8 cases. The EPA was not only severely behind in enforcement, but also in its duties to define, classify and remediate hazardous waste as assigned in the RCRA (Durant, 1993). In 1984, a Democrat-dominated Congress crafted a solution known as the Hazardous and Solid Waste Amendments (HSWA). The law was intended to close loopholes by increasing the range of facilities regulated under the RCRA, and jump-start the stalled EPA into legislative action by imposing 70 mandates upon the agency relating to efficiency.

The HSWA was largely successful in its goals. By 1989 the EPA had increased the amount of pages of hazardous waste regulations in the Code of Federal Regulations by 150%. By 1990, the EPA increased the scope of facilities and wastes requiring regulation nine-fold (Durant, 1993). The mandates in the HSWA were so effective in moving the EPA to action that states had difficulty taking the reins of hazardous waste regulation away from the agency. For the states to take over as the RCRA originally intended, they had to exhibit an EPA-comparable monitoring and enforcement system, and this was difficult to do.

1.2.2 Superfund Legislation

In 1967, a Mr. Taylor who owned a 23-acre property in Bullitt County, Kentucky, began haphazardly stocking hazardous wastes. Hazardous waste was deposited on a 13-acre tract from 1967 to 1977 and accumulated to such an extent that, by 1979, 17,051 drums had accumulated on the surface of the tract alone. Of the 140 hazardous compounds identified on-site by the EPA, the chemicals found in the highest concentrations were xylene, acetone, dichloroethylene, vinyl chloride, methyl ethyl ketone, and anthracene. The property was graded in such a way that contaminated water and leachate flowed directly from the barrels into nearby Wilson Creek. Five residences and a country club were less than a mile downstream, not to mention communities further away. Despite being identified as a non-permitted hazardous waste disposal site by the Kentucky Department of Natural Resources and Environmental Protection (KDNREP) as early as 1967, dumping continued until 1977 (EPA, 1986). In 1979, KDNREP requested the assistance of the US EPA to aid in the monumental task of cleanup. The story of the site, dubbed the Valley of Drums, drew national headlines.

Around the same time, residents in a once innocuous-looking community near Niagra Falls, New York, known as Love Canal, drew national attention as it became apparent they were living on a hazardous waste dump site. Hazardous wastes accumulated in Love Canal since the 1920s when municipal and industrial producers began using a disused and dry canal for discarding wastes. The indiscriminate dumping of wastes into the canal continued until 1953, when the Hooker Chemical

Company, then-owners of the dump site, covered the canal with dirt and sold it to the city for one dollar. In the late 1950s, unknowing developers constructed a school and roughly 100 homes on the site where dumping had occurred. The residents of Love Canal lived there in ignorance until 1978, when record rainfalls helped uncover the horrors that existed underneath their homes. The inundation of water leached the hazardous waste out from storage and into the surrounding environment. Trees turned black and died. Corroding 55-gallon drums emerged in people's back yards. A foul 'choking smell' permeated the air. Hazardous waste literally appeared in puddles throughout the community, and children playing outdoors returned home with chemical burns. Upon further investigation, an unusually high amount of miscarriages and birth defects were found in the community, and a large percentage of residents had elevated white blood cell counts (an indicator of future leukemia). On August 1, 1978 the *New York Times* ran a front-page story on the catastrophe at Love Canal. Six days later, President Jimmy Carter approved emergency funding to aid the families affected (Beck, 1979).

Both the incidents at Love Canal and the Valley of the Drums shocked the public and government into action against hazardous waste dump sites and the presence of hazardous wastes in the environment in general. President Carter called the unbridled disposal of hazardous wastes throughout the nation 'one of the grimiest discoveries of the modern era' (Beck, 1979). The very-real specter of unreported and unregulated hazardous dump sites threatening public health prompted Congress to pass the Comprehensive Environmental Response, Compensation, and Liability Act (CERCLA) in 1980. The Act was colloquially known as Superfund due to the $1.6 billion trust fund – a special tax on chemical and petroleum producers established over five years for remediating uncontrolled or abandoned hazardous waste sites. In 1986 Congress passed the Superfund Amendments and Reauthorization Act (SARA), which attempted to increase state and public involvement, reprioritize sites based on potential adverse human health effects, include government sites under Superfund, and increase the trust fund to $8.5 billion. SARA reauthorized the program until 1991. Since then Superfund has been kept afloat by a series of stop-gap measures.

CERCLA provided a regulatory framework for dealing with closed or abandoned sites including a means to establish liability for the original polluters. Superfund provisions for two types of response to a hazardous waste site: short-term removals and long-term remediation. Short-term removals are designed to remove wastes that are immediately hazardous to public health. These usually take several months to complete. Long-term remediation is geared more toward eliminating non-lethal, albeit dangerous hazards permanently from sites. These can span years. The cleanup itself is paid for by Probably Responsible Parties (PRPs) either during the remediation, or afterward in the form of a reimbursement.

Upon CERCLA's inception, the National Priorities List (NPL) – the list of Superfund sites – contained 406 sites. As of August 2010, 1,277 sites across the nation were undergoing some kind of removal/remediation, 343 sites were remediated to the point where they no longer posed a 'significant threat to public health or the environment,' and 61 sites were undergoing assessment as possible future Superfund sites (EPA, 2010a,b). Since the Superfund legislation was enacted, it has

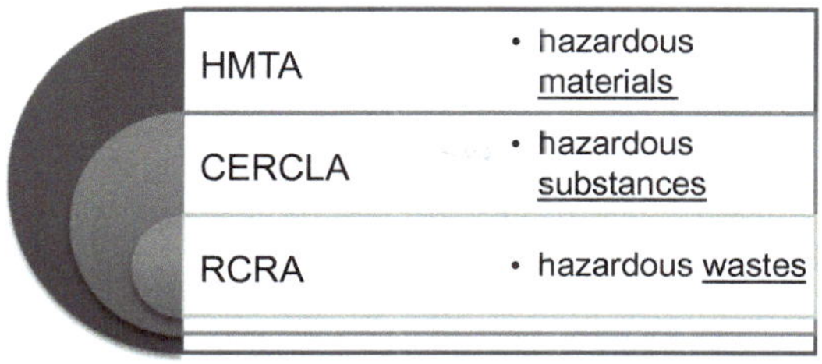

Figure 1.1 Federal Acts and Respective Domains of 'Hazardous' (derived from Gerrard, 1998).

drawn criticism namely from the provisions establishing liability and the allocation of the trust fund. These issues will be discussed in Chapter 4.

1.2.3 *Other Hazardous Waste Regulations*

The Hazardous Material Transportation Act (HMTA) regulates the transportation of hazardous materials and was signed into law in 1975. The law gives the Secretary of Transportation the power to designate a material as hazardous if any 'particular quantity or form ... pose[s] an unreasonable risk to health and safety or property' (HMTA). The Act covers the movement by land, sea, and air, of any hazardous material. The HMTA's definition of hazardous material is much more expansive than the RCRA's and CERCLA's; in fact, hazardous waste, as covered in the RCRA, only makes up about 1% of the hazardous materials covered in the HMTA (Gerrard, 1998).

According to Gerrard (1998), the HMTA is the least known, but most important of the three acts governing hazardous materials (see Figure 1.1). He argues that HMTA-regulated materials pose substantially more of a threat than CERCLA and RCRA materials combined. Over 100 people on average die annually from HMTA-related substances/activities whereas deaths relating to CERCLA and RCRA substances/activities are an 'order of magnitude or two' less. Consider how many lives could be lost if an 18-wheeler carrying chlorine crashes on an interstate highway or if a munitions train derails. It should be noted that Gerrard does not take into account indirect deaths and non-fatal life-altering illnesses caused by hazardous wastes in the air or groundwater.

The HMTA requires that all transported hazardous materials be identified (Magnussen, 1997) and documented. These identifiers are most easily recognized as the diamond-shaped warning placards fastened to the sides and backs of container trucks familiar on interstate highways (Gerrard, 1998).

1.3 Categories and Sources of Hazardous Waste

1.3.1 *Nuclear Waste*

Most nuclear waste comes from nuclear power plants and weapons reprocessing operations and to a lesser extent from natural sources. The amount of man-made nuclear waste in storage on planet earth was estimated at 5,843,399 cubic meters in 2008. This value is only expected to increase as plants currently in operation keep producing, and newer plants are built. Of that 2008 figure, 363,574 cubic meters are designated high-level nuclear waste. Radioactive waste is categorized into three types, aptly named high-level, intermediate, and low-level. These are determined by the amount of radioactivity exhibited per unit volume of waste. Unlike other types of hazardous waste, there is no way to neutralize radioactive waste. Long-term storage on the order of millennia is the only option for mitigation.

1.3.2 Industrial Waste

Industries in the United States generate 265 million metric tons of hazardous waste annually (Ditz, 1988). These industrial hazardous wastes are monitored and regulated by the US EPA under the RCRA; in fact, IHW is the only category of hazardous waste that the EPA monitors. Examples of these wastes are the chemicals in the F, K, U, and P lists, explained in Section 1.1.2.

1.3.3 'Universal' Waste

Universal hazardous wastes, otherwise known as household hazardous waste (HHW), are hazardous wastes that are produced by individuals and homes. Many of these wastes are regulated at the industry level in larger amounts; however, in the smaller, discrete amounts found in consumer items, they are relatively unrestricted. Batteries, household pesticides, and light bulbs are common sources of HHW. According to one estimate, households in the US generate 1.6 million tons of HHW annually (Bonneville, 2010). The actual number may vary widely from this value due to the absence of an official monitoring and disposal system for HHW. Unlike with industrial hazardous waste, no disposal regulations exist for HHW. Many products are put out with non-hazardous waste for curbside pickup. The waste is then transported with regular refuse to municipal landfills incapable of handling hazardous waste. HHW then leaches toxins into underlying aquifers and water systems.

1.3.4 Medical Waste

According to the American Medical Association (AMA, 1998), the United States generates 465,000 tons of hazardous medical waste per year. Medical waste ranges from biological wastes such as human blood, and cultures and reserves of infectious agents, to the man-made, such as contaminated hospital bedding, and used needles. These wastes hold the potential to spread disease or cause injury to the public. The removal and management of hazardous medical waste are regulated at the local state level, and are not currently controlled by federal laws. In 1988, mismanagement of hazardous medical waste was so rampant that it began appearing along the Eastern Seaboard. The presence of syringes and needles on beaches prompted Congress to enact the Medical Waste Tracking Act (MWTA) that fall. The MWTA was a 2-year demonstration program under RCRA Subtitle J designed to track medical waste from cradle to grave. It involved just four states/territories and was not continued at the end of the trial period (http://www.epa.gov/oecaerth/civil/rcra/medwastereq html).

To neutralize the large amount of hazardous medical waste pouring out of hospitals and laboratories, incineration, autoclaving or electro-thermal deactivation methods are applied. Incineration, used for pathological wastes and chemical wastes, heats the material above 700°C, effectively killing all pathogens. This method, however, produces serious point-source emissions of CO_2, odors, and burn byproducts, as well as a concentrated ash containing poly-vinyl chloride and heavy metals. In autoclaving, the hazardous waste is exposed to steam in an enclosed (and therefore high-pressure) environment for under an hour. This method produces contaminated water, which is usually discharged into the local sewage system, and also produces significant amounts of volatile organic carbon compounds. Electro-thermal deactivation operates on essentially the same principles as a microwave. It is achieved by exposing the waste to an electric oven that emits low-frequency waves, heating the waste internally and rapidly, killing all pathogens. While ETD is the cleanest of the three methods, it is also the most cost-intensive (Baldwin, 2003).

1.3.5 Construction Waste

Asbestos insulation, lead paint, and mercury-containing exit signs and thermostats can all present health hazards when demolishing a building, especially an older one. Potentially hazardous wastes in construction projects are regulated by the EPA under the RCRA. Hazardous wastes encountered in demolition projects are difficult to identify and mediate. For example, lead paint applied 30 years prior, and painted over since, can pose a hidden threat. The accidental sanding or crushing of asbestos tiles can release harmful fibers into the air (EPA, 2009).

1.3.6 Electronic Waste

While electronic waste (EW) only makes up 2% of total waste in US landfills, it is responsible for 70% of the heavy metals (including mercury and chromium) found in landfills. The upsurge in home computing and cell phone purchasing over the last two decades, combined with the quick turnover rates such technologies experience, are expected to contribute to a four-fold increase in EW over the next few years. Hazardous materials, such as lead and mercury, can be found at trace levels in consumer electronics including cell phones and CRT monitors. These materials overlap somewhat with household hazardous wastes, in that there is no formal path of safe disposal. Old televisions and computers are either tossed out and deposited in ill-equipped municipal landfills, or shipped out to developing countries for informal salvage where exposures of workers to hazardous materials are all but guaranteed (CAW, 2010).

References

American Medical Association (1998). *Expense of medical and biohazardous waste removal CMS report 2-I-98*. Council on Medical Service. August 13, 2010.

Beck, E. C. (1979). The love canal tragedy. *EPA Journal*. Retrieved August 13, 2010 from <http://www.epa.gov/history/topics/lovecanal/01.htm/>.

Bonneville County (2010). *Solid waste disposal information*. Public Works – Solid Waste Division. Retrieved August 13, 2010 from <http://www.co.bonneville.id.us/pw-transfer.php/>.

Californians Against Waste (CAW) (2010). *Poison PCs and toxic TVs*. Retrieved August 16, 2010 from <http://www.cawrecycles.org/issues/ewaste/poisonpc_exec_summary/>.

Davis, R. C., Ridgeway, H. M., & Swartz, R. E. (2001). *RCRA hazardous wastes handbook* (12th ed.). Rockville, MD: Government Institutes, Inc.

Ditz, D. W. (1988). Hazardous waste incineration at sea: EPA decision making on risk. *Risk Analysis*, *8*(4), 499–508.

Durant, R. F. (1993). Hazardous waste, regulatory reform, and the Reagan revolution: The ironies of an activist approach to deactivating bureaucracy. *Public Administration Review*, *53*(6), 550–560.

Gerrard, M. B. (1998). Demons and angels in hazardous waste regulation: Are justice, efficiency and democracy reconcilable? *Northwestern University Law Review*, *92*(2), 706–749.

Magnussen, N. (1997). *Introduction – Hazardous Materials Transportation Act*. College of Science, Texas A&M University. Retrieved August 24, 2010 from <http://safety.science.tamu.edu/DOT.html/>.

New York City Department of Sanitation (NYCDS). (2010). *Federal Hazardous Substances Act*. Bureau of Waste Prevention, Reuse and Recycling. Retrieved August 13, 2010 from <http://nyc.gov/test/nycwasteless/html/laws/fed_hazardousproducts.shtml/>.

US Environmental Protection Agency (1986). EPA superfund record of decision: A.L. Taylor (Valley of Drums), EPA/ROD/R04-86/009. <http://www.epa.gov/superfund/sites/rods/fulltext/r0486009.pdf/>.

US Environmental Protection Agency (2003). *25 years of RCRA: Building on our past to protect our future*. Retrieved August 26 from <http://www.answers.com/topic/hazardous-and-solid-waste-amendments-of-1984/>.

US Environmental Protection Agency (2005). Introduction to hazardous waste identification, EPA530-K-05-012. Solid Waste and Emergency Response. Retrieved April 13, 2010 from <http://www.epa.gov/osw/inforesources/pubs/hotline/training/hwid05.pdf/>.

US Environmental Protection Agency (2009). *Hazardous waste characteristics: A user-friendly reference document*. <http://www.epa.gov/osw/hazard/wastetypes/wasteid/char/hw-char.pdf/>.

US Environmental Protection Agency (2010a). Wastes specificially excluded from RCRA. *EPA Hazardous Wastes*. Retrieved August 13, 2010 from <http://www.epa.gov/osw/hazard/wastetypes/wasteid/exclude.htm/>.

US Environmental Protection Agency (2010b). *National priorities list*. Retrieved August 13, 2010 from <http://www.epa.gov/superfund/sites/npl/index.htm/>.

www.epasuperfund.com. Retrieved August 13, 2010 from <http://epasuperfund.com/images/toxic_waste_barrels.jpg/>.

2 The Biggest Generators of Hazardous Waste in the US

Chapter Contents

Of the industries regulated by the EPA, chemical manufacturing and petroleum/coal products manufacturing together are responsible for 84% of the hazardous waste generated. Some of the largest of these manufacturing facilities can generate over 8 million tons of hazardous waste annually, much of which is stored on-site. The creation and storage of hazardous waste at such quantities inevitably results in environmental contamination, causing harm to humans and other organisms. Federal facilities including the military, Department of Energy, and Department of the Interior generate millions of tons of hazardous waste annually; 178 sites on the National Priorities List are federally owned. Additionally, a small but considerable amount of unregulated household hazardous waste is dumped annually into landfills, down sinks, and into street gutters by ordinary citizens. This unregulated and largely unmonitored waste stream also presents cause for concern.

2.1 Federal Criteria for Regulation of Waste Generators

Hazardous waste originates from various sources, ranging from large extensive manufacturing operations, universities, hospitals, to the small: small businesses,

Risks of Hazardous Wastes. DOI: 10.1016/B978-1-4377-7842-7.00002-7

laboratories, dry cleaners, and auto mechanics. To facilitate federal management under the RCRA and HSWA (see Section 1.2.1), hazardous waste emitters are sorted into categories in accordance with the monthly amount of hazardous waste produced. There are three categories: large-quantity generators (LQGs), small-quantity generators (SQGs) and conditionally exempt small-quantity generators (CESQGs).

To be designated as a LQG, facilities must produce greater than 1,000 kg of hazardous waste per month, or greater than or equal to 1 kg of acutely hazardous waste. Facilities are also considered LQGs if they generate or accumulate more than 100 kg of contaminated spill cleanup material in a month. Examples of LQGs include Dow Chemical and Monsanto. A majority of domestic hazardous waste is generated by the roughly 16,000 LQGs operating within the United States (EPA, 2010a). Small-quantity generators produce between 100 kg and 1,000 kg per month, and can accumulate less than 6,000 kg of hazardous waste at any time. This latter requirement, designed to ease the burden of transport for smaller producers, allows SQGs to hold waste until it can be shipped out in bulk. SQGs were largely ignored by regulators at first; however, due to exploitation the HSWA was passed to include SQGs in the RCRA's regulatory scope (see Section 1.2.1). As of 2001 there were roughly 200,000 SQGs in the United States. The final category, CESQGs, was established under the HSWA in 1984 and only applies in some states. These produce less than 100 kg of hazardous waste each month, or less than 1 kg of acutely hazardous waste per month. In 1997, the EPA estimated that there were between 400,000 and 700,000 CESQGs in the US. Because the categorization system is based on monthly hazardous waste output, producers can move freely between small, large, and conditional levels. This fluctuating is known as episodic generation. Table 2.1 is a summary of this information.

As maintained in Federal Regulations (40 CFR Part 262), LQGs (and to a lesser extent SQGs) are required to document, register, and properly manage on-site hazardous wastes. Large-quantity generators must first identify and count the type and amount of hazardous waste. Once the proper paperwork is submitted, the EPA assigns the generators a numerical ID. To interact with other purveyors of

Table 2.1 Criteria for LQG, SQG, CESQG Designations (EPA, 2010)

	Large-Quantity Generators	**Small-Quantity Generators**	**Conditionally Exempt Generators**
Amount of HW generated (kg/month)	>1,000; >1 acutely hazardous waste	>100, <1,000	<100; <1 acutely hazardous waste
Amount of HW accumulated (kg/month)	>1,000	<6,000	<1,000
Spill residue allowed (kg/month)	>100	–	<100
No. operating in US	16,000	200,000	>400 k, <700 k

hazardous waste, one must have an EPA-assigned ID number. Waste stored on-site must be shipped out after 90 days, but can be stored longer in special cases approved by the EPA. Hazardous waste must be labeled as such, and official, written emergency arrangements must be kept at the LQG facility in case of accidental release. SQGs on the other hand are allowed to store waste on-site for up to 180 days, and can do so up to 270 days if the SQG is transporting hazardous waste over 200 miles. Only LQGs are required to submit National Biennial RCRA Hazardous Waste Reports detailing their activities surrounding hazardous waste generation and disposal.

2.2 National Biennial RCRA Hazardous Waste Reports

In March on every even year, LQGs are required to submit a National Biennial RCRA Hazardous Waste Report to the US EPA. This report contains information on hazardous waste type and amount produced by each facility. After an analysis by the EPA, the reports are released to the public. The most recently available report of this type was synthesized from data in 2007 (EPA, 2010b).

According to the EPA, 46.7 million tons of hazardous waste were generated in the United States (excluding the military) in the year 2007 (EPA biennial report). The EPA categorizes 50 types of producers of hazardous wastes, with the top five accounting for 93% of the wastes produced. These include chemical manufacturing (72%), petrol and coal product manufacturing (12%), waste treatment and disposal (4%), iron and steel mills (3%), and semiconductor and electronic manufacturing (2%); 45 other types of producers account for the remaining 11% (Table 2.4, Figure 2.2).

The production of hazardous wastes is not uniformly distributed across the nation. More than half of all production occurs in two states, with Louisiana accounting for 34% and Texas contributing 28.4%. Other top producing states

Table 2.2 Worst 10 States for Hazardous Waste Production (EPA, 2010b)

Rank	State	Tons of HW Generated	% of National Production	% of Nationwide Generators	#LQG
1	LA	15,892,592	34	2.1	324
2	TX	13,272,307	28.4	5.6	918
3	MI	2,397,357	5.1	4.2	536
4	MS	2,239,718	4.8	0.8	133
5	OH	1,608,186	3.4	5.8	794
6	NY	1,267,648	2.7	7.2	896
7	IL	1,122,937	2.4	4.9	697
8	TN	1,079,070	2.3	2.2	358
9	ID	958,019	2.1	3.2	427
10	NM	944,581	2.0	0.3	37
	Others	n/a	12.8		

Table 2.3 Top 10 Companies/Facilities with Most Generated Hazardous Waste (EPA, 2010b)

Rank	Company	City	Tons of HW Generated
1	The Dow Chemical Company	Plaquemine, LA	8,125,031
2	Solutia Inc.	Alvin, TX	3,172,178
3	Occidental Chemical Corp. – Taft Plant	Hahnville, LA	2,804,999
4	Rubicon LLC	Geismar, LA	2,529,760
5	The Dow Chemical Company	Midland, MI	2,037,721
6	The Shamrock Pipe Line Corp.	Sunray, TX	1,776,386
7	E.I. Du Pont de Nemours and Co.	Pass Christian, MS	1,735,698
8	Cytec Industries Inc.	Waggaman, LA	1,603,618
9	BP Products North America Inc.	Texas City, TX	1,531,604
10	Ineos USA LLC.	Port Lavaca, TX	1,072,849

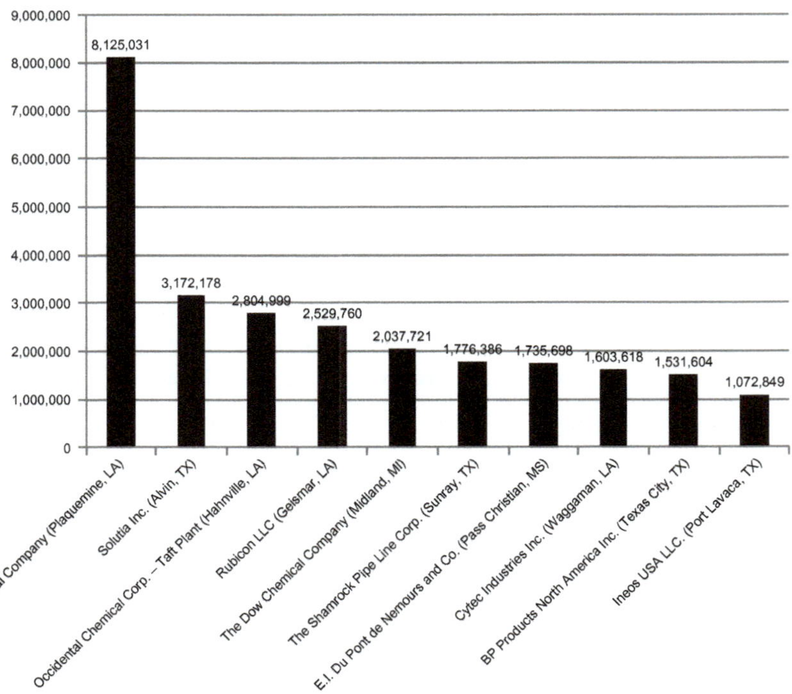

Figure 2.1 Top 10 companies/facilities with most generated hazardous waste (EPA, 2010b).

Table 2.4 Top 10 Waste Producing Industries (EPA, 2010b)

Rank	Description	Tons of HW Generated
1	Basic chemical manufacturing	31,666,943
2	Petroleum and coal products manufacturing	5,130,594
3	Waste treatment and disposal	1,963,662
4	Iron and steel mills and ferroallow manufacturing	1,557,174
5	Semiconductor and other electronic component manufacturing	1,070,605
6	Remediation and other waste management services	691,782
7	Coating, engraving, heat treating, and allied activities	689,847
8	Nonferrous metal (except aluminum) production and processing	447,961
9	Pharmaceutical and medicine manufacturing	324,826
10	Resin, synthetic rubber, and artificial synthetic fibers and filaments manufacturing	278,763

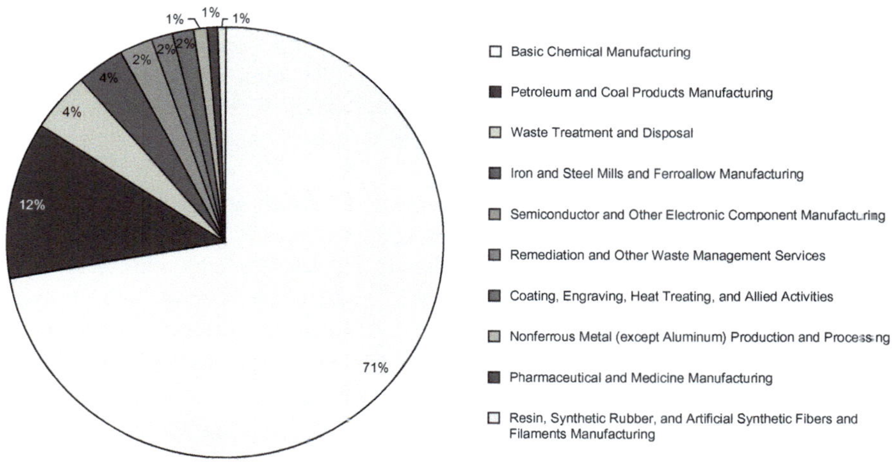

Figure 2.2 Top 10 waste producing industries (EPA, 2010b).

account for a much smaller amount with Michigan producing 5.1% and Mississippi producing 4.8% of the total annual hazardous wastes generated in the United States (Table 2.2).

2.3 The Chemical Industry

As mentioned before, the chemical industry, specifically basic chemical manufacturing, is responsible for the largest portion (72%) of hazardous waste produced by LQGs in the country. These basic chemicals are sold as intermediates for use as ingredients in manufacturing other more complex chemicals, and as simple product chemicals used in everything from everyday household cleaning agents to explosives. These basic chemicals include chlorine and arsenic, which appear 6 and 16 times (respectively) as listed, or components of listed hazardous wastes. Examples of other basic chemicals also listed as hazardous wastes are chromium, fluorine, hydrazine, and mercury compounds. (For more on the chemical industry, see Chapter 3.)

2.4 Local Effects of Hazardous Waste Production: Case Studies of the Top 3 Hazardous Waste Generators and the Communities that House Them

2.4.1 The Dow Chemical Company, Plaquemine Facility, Plaquemine, Louisiana

According to the US EPA's most recent report (2007), the nation's largest hazardous waste producer is a facility owned and operated by The Dow Chemical Company (Dow) in the small town of Plaquemine just south of Baton Rouge, Louisiana. In 2007, the facility produced 8,125,031 tons of hazardous waste, including vinyl chloride and formaldehyde. Dow, which first came to Plaquemine in 1958, has been disposing waste on-site and in nearby landfills since the 1960s (COP, 2010).

In 1997 and 1998 the Louisiana Department of Health and Hospitals sampled the aquifer beneath Myrtle Grove Trailer Park, approximately one mile south of the Dow facility, and found levels of vinyl chloride and cis-1,2-dichloroethylene to be above the Maximum Contaminant Level (MCL) set in the Federal Safe Drinking Water Act (LDEQ, 2008). The presence of vinyl chloride in drinking water, which has been associated with liver cancer, nerve damage, circulatory problems, skin

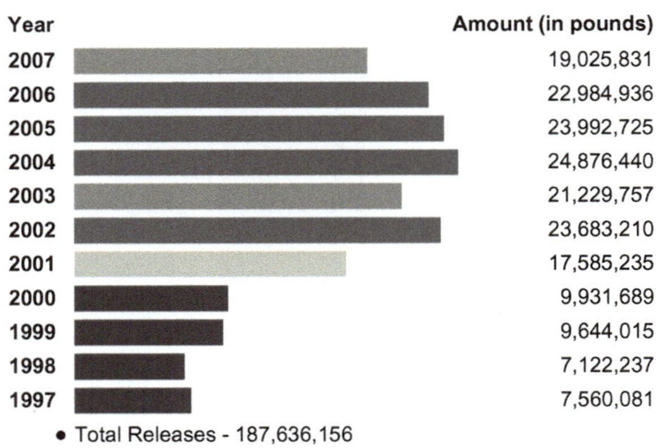

Figure 2.3 Total aggregate releases of TRI chemicals to the environment from 1997 to 2007 by Solutia Inc.'s plant in Alvin, TX (EPA, 2010b).

lesions, and teratogenic effects, concerned local regulators and in 2001 the US EPA was notified (Bragg, 2003). In a *New York Times* report covering the contamination at Plaquemine, residents complained of a high rate of miscarriages (13 on one street alone). The trailer park was shut down and the residents forced to relocate.

Several parties, including Dow and the US EPA, drilled over 120 wells to a depth of 190 ft in order to determine the origin and extent of the toxic plume. The findings differed greatly. Dow reported that no contamination existed in the aquifer directly below their facility and that flow was to the west, not south toward the trailer park. The US EPA, however, determined that the aquifer traveled in a southwest direction, directly toward Myrtle Grove Trailer Park. According to US EPA Region 6, the likely source is a now-defunct disposal site used by Dow in the 1960s and 1970s.

In 2003, the Dow Chemical Company in Plaquemine reached a settlement with the state for hazardous waste violations dating back to 1996. The settlement for $2.4 million dollars channeled $1 million to the state and $1.4 million to local remediation projects (DOW, 2003). In 2009, Dow Chemical announced plans to shut down the ethylene dichloride and vinyl chloride production facilities in Plaquemine by 2011.

2.4.2 Solutia Inc. and Alvin, Texas

Solutia Inc.'s Chocolate Bayou Facility in Alvin, Texas, is the second largest generator of hazardous waste in the nation. The facility, as stated on Solutia's investor outreach website, was the world's largest acrylonitrile production unit when it began production in November 2000 (Solutia, 2002). The Chocolate Bayou Facility has also earned another superlative: Texas' most polluting facility monitored in the EPA's Toxic Release Inventory. In 2007 alone, the facility reported the release of 19,025,831 pounds of toxic chemicals into the environment (see Figure 2.3). The value self-reported by corporations is almost always an underestimate. In meta-studies conducted by the EPA, discrepancies were found in fugitive emission estimates and in the estimation methods themselves. Also, corporations only have to report on listed toxic chemicals, i.e. those required by the EPA to be reported. Other chemicals, such as fluorene, pyrene, 1,2,4,5-tetrachlorobenzene, and acenaphthylene, are not reported (Gerde, 2001; EPA, 2006).

According to the most recent data available (2008 TRI), the Chocolate Bayou production facility released 18,517 lbs of hydrogen cyanide, 6,839 lbs of ammonia, and 1,984 lbs of xylene (mixed isomers) into the air (EPA, 2010c). That same year, the plant injected 1,600,000 lbs of acrolein, 81,865 lbs of formaldehyde, and 100,000 lbs of pyridine into the rock beneath the Chocolate Bayou facility. Both acrolein and formaldehyde are listed hazardous wastes. Acrolein is so harmful that it was employed in World War I as an irritant gas to be used against enemy forces (Romano et al., 2007). Formaldehyde is considered a hazardous air pollutant by the EPA and is known to cause pulmonary edema and pneumonitis. It is also a possible carcinogen.

Pyridine has been shown to cause damage to the liver, and may be linked to neurological and renal effects, as well as skin and eye irritation (MSDS, 2009).

2.4.3 Occidental Chemical Corporation and Hahnville, Louisiana

Ranked as the third largest producer of hazardous waste in the nation, Occidental Chemical's Taft Plant in Hahnville, Louisiana, generated 2,804,999 lbs of hazardous waste in 2007. The next year, the *New York Times* reported that the facility accumulated seven effluent violations and was out of compliance with the EPA in one out of the past 12 quarters (NYT, 2009). In the 5 years prior to 2009, the plant reported a 68-pound release of chlorine that resulted in the hospitalization of two on-site contractors and a 54-pound release that forced employees at a nearby industrial facility to 'shelter-in-place.'

In September 2009, the US EPA recommended to the Louisiana Department of Environmental Quality that it establish quantitative water quality criteria for toxic substances. This recommendation was prompted by the large and unmonitored amounts of priority, and non-priority pollutants that were being discharged by industries (including the Occidental plant) into the state's waterways. According to the EPA, Occidental Chemical Corp.'s plants at Geismar, LA and Convent, LA are responsible for the discharge of the priority pollutants bis(2-chloroethyl)ether and carbon tetrachloride into the Mississippi River. The Occidental Taft Plant in Hahnville is responsible for the discharge of chloride into the Mississippi River.

The Dow Chemical Company alone manufactures over 5,000 separate products at 214 sites in 37 countries. Of the top 10 facilities that generate hazardous waste, eight are chemical companies (Table 2.3, Figure 2.1). Of the hazardous waste generated by those top ten, 38.5% (10,162,752 tons) comes from Dow Chemical facilities.

2.5 The United States Military and Other Federal Facilities

2.5.1 The Department of Defense as a Major Contributor

While years of immunity from hazardous waste regulation and accounting prevent researchers from accurately estimating the amount of hazardous waste created by the military, there is little doubt that the Department of Defense is our nation's largest polluter (Layton, 2008). As of 2010, the DOD was, or has been, responsible for 141 of the Superfund sites listed on the National Priority List, the most, by far, attributed to any single organization. This astounding figure is not as remarkable when one considers the vast amount and variety of hazardous wastes typically generated by and at military bases.

Common hazardous waste-generating activities typical at DOD facilities include (EPA, 1992):

- Manufacturing, testing, loading, and packaging weapons
- Maintaining and repairing aircraft and vehicles

- Plating metal
- Producing, processing, and recovering nuclear materials (more typical of Dept. of Energy facilities).

These processes produce wastes in the form of explosives, solvents, cleaning agents, paints, heavy metals, pesticides, and waste oil. For example, in one study by Lemasters et al. (1999) conducted at Hill Air Force Base, UT, the authors found that 8,200 gallons of solvents were used on an annual basis. These degreasing solvents, used to clean vehicle components, included the agents 1,1,1-trichloroethane (TCA), xylene, toluene, methyl ethyl ketone and methylene chloride (all are listed hazardous wastes). It was common practice for these solvents, after use, to be dumped down the nearest drain or into nearby soil (Shulman et al., 1989). TCA and these other chemicals are harmful enough; however, military bases also suffer from abundance of other hazardous wastes. At the now decommissioned Fort Ord, in Monterey, CA, lead, copper, and antimony from firing ranges contaminate the soil. In some areas of the fort, shell casings make up 10% of the ground cover. Unexploded ordinance such as claymore mines and mortar shells also litter the area. While these are specific examples, they can be applied to many if not all DOD bases to varying degrees. As of 1993, the US Military had 440 army, navy, and air force bases established throughout the continental US (DOI, 1993). (For more on the Department of Defense, see Chapter 4.)

2.5.2 *The Department of the Interior*

The Department of the Interior (DOI) is responsible for a rich mix of agencies including the bureaus of: Indian Affairs, Land Management, Reclamation, Fish and Wildlife Service, National Park Service and the US Geological Survey. The DOI established the Central Hazardous Materials Fund (CHMF) to provide financial backing to remediate the hazardous mining, landfill, agricultural, and other sites under its jurisdiction. As of 2010, 23 sites had completed, or were scheduled to complete their remediation program by the end of the year. Fifty sites still were undergoing CHMF remediation (DOI, 2010). In the DOI's most recent Principal Financial Statement (2008), the Department reported $155 million in environmental liabilities, mostly originating from defunct sites. Since 2004, the environmental liability burden has increased by over $50 million.

2.5.3 *The Department of Energy*

At the closing of the Cold War, and the subsequent halt of nuclear weapons manufacturing, the United States, specifically the Department of Energy, was left with an immense cache of radioactive waste. Forty-five years of spent reactor fuel, nuclear material, contaminated soil and water had accumulated at countless sites around the nation. In Fernald, OH, for example, the weapons-grade uranium-processing facility had accumulated a staggering '31 million pounds of uranium product present on site, 2.5 billion pounds of waste, and 2.75 million cubic yards of contaminated soil and debris' (DOE, 2006). While the DOE may not create hazardous waste at the rate of the chemical industry or the DOD, the longevity of radioactive materials (up to billions of years) will ensure that the DOE's waste will persist and accumulate to the same if not a greater extent.

2.6 Unregulated Household Hazardous Waste

Household hazardous wastes (HHW) are as varied as the consumer products that contain them. Some, such as garden pesticides and paint thinner, are overtly hazardous, while others like suntan oil and deodorants are more discreet, but hardly innocuous (Reinhart, 1993). School glue contains acetaldehyde, adhesive remover contains chloroform, and toluene appears in rust preventers, lacquer, and metallic paints, to name just a few (HPD, 2010). Hazardous waste chemicals, as defined in the RCRA, are present in household products. They are not, however, regulated under the RCRA (Slack, 2007). Therefore, formal monitoring and disposal of HHW is nonexistent, and we do not have a firm idea of how much HHWs are deposited in municipal landfills. According to one estimate, Americans generate 1.6 million tons of HHW annually (Bonneville, 2010). In a 2002 report on HHW disposal in Europe, the amount of HHW discarded annually in the European Union was estimated at 1.76 million tons (ECDGE, 2002). Municipal solid waste landfills are not designed to handle hazardous waste materials (see Chapter 12), and some water-soluble wastes have potential to leach out of containment and drain into adjacent aquifers.

While HHW is not considered 'hazardous waste' under the RCRA, the Superfund Act considers HHW to be a 'hazardous waste' and polluters (municipalities in this case) liable for damages. As of 2009, there were roughly 250 municipal solid waste landfills listed for cleanup under Superfund (23% of all sites listed on the National Priority List as requiring Superfund action). The EPA has offered municipalities exemption from legal action under Superfund; however, only for landfills that do not contain large amounts of hazardous wastes.

The liability of municipalities under Superfund has played a part in the creation of local and state-sponsored hazardous waste disposal. Many cities now have HHW pickup facilities at waste management plants, universities, and other publically accessible locations. Some communities also provide special collection days for HHW pickups. Ultimately, however, it is the individual's task to identify and deposit HHW safely.

References

Bragg, R. (2003, May 5). Toxic water numbers days of a trailer park. *New York Times*. <http://www.nytimes.com/2003/05/05/us/toxic-water-numbers-days-of-a-trailer-park.html?pagewanted=all/>.

Bonneville County (2010). *Solid waste disposal information*. Public Works – Solid Waste Division. Retrieved August 13, 2010 from <http://www.co.bonneville.id.us/pw-transfer.php/>.

City of Plaquemine (COP) (2010). *A brief history*. City of Plaquemine. Retrieved August 13, 2010 from <http://www.plaquemine.org/PageDisplay.asp?p1=3554/>.

Dow to Pay $2.4 million, settle case (DOW). *The advocate*. Retrieved April 16, 2003 from <http://nl.newsbank.com/nl-search/we/Archives?p_product=AD&p_theme=ad&p_action=search&p_maxdocs=200&p_topdoc=1&p_text_direct-0=0FA7C0BED49A46B3&p_field_direct-0=document_id&p_perpage=10&p_sort=YMD_date:D&s_trackval=GooglePM/>.

European Commission – Directorate – General Environment (ECDGE) (2002). Study on Hazardous Household Waste (HHW) with a main emphasis on Hazardous Household Chemicals (HHC) – final report, CO 5089-2. <http://ec.europa.eu/environment/waste/studies/pdf/household_report.pdf/>.

Gerde, V. W., & Logsdon, J. M. (2001). Measuring environmental performance: Use of the toxics release inventory (TRI) and other US environmental databases. *Business Strategy and the Environment*, *10*, 269–285.

Household Products Database (HPD) (2010). *Health & safety information on household products*. Retrieved August 13, 2010 from <http://householdproducts.nlm.nih.gov/cgi-bin/household/brands?tbl=chem&id=1923/>.

Layton, L. (2008, June 30). Pentagon fights EPA on pollution cleanup. *Washington Post*. <http://www.washingtonpost.com/wp-dyn/content/article/2008/06/29/AR2008062901977.html?sid=ST2008063001979/>.

Lemasters, G., Kawas, L., Olsen, J. E., Selevan, D. M., Tabor, S. G., & Wilson, M., et al. (1999). Comparison of internal dose measures of solvents in breath, blood and urine and genotoxic changes in aircraft maintenance personnel. *Drug and Chemical Toxicology*, *22*(1), 181–200.

Louisiana Department of Environmental Quality (LDEQ) (2008). *AI 81438 – A. Wilbert and Sons trailer park: Site background and activities*. Remediation Services. Retrieved August 18, 2010 from <http://www.deq.louisiana.gov/portal/DIVISIONS/RemediationServices/RemediationHighProfileSites/AWilbertandSonsTrailerPark.aspx/>.

Mallinckrodt Baker, Inc. (MSDS) (2009). *Formaldehyde*. Mallinckrodt Chemicals, J.T. Baker. Retrieved August 13, 2010 from <http://www.jtbaker.com/msds/englishhtml/f5522.htm/>.

New York Times (*NYT*) (2009). Find water polluters near you. *New York Times*, US. Retrieved August 13, 2010 from <http://projects.nytimes.com/toxic-waters/polluters/louisiana/hahnville/>.

Reinhart, D. R. (1993). A review of recent studies on the sources of hazardous compounds emitted from solid waste landfills: A U.S. experience. *Waste Management & Research*, *11*, 257–268.

Romano, J. A., Jr., Lukey, B. J., & Salem, H. (2007). *Chemical warfare agents: Chemistry, pharmacology, toxicology, and therapeutics* (2nd ed.). <http://www.aeonia.com/files/Chemical%20Warfare%20Agents.pdf#page=13/>.

Shulman, S., & Pollack, S. (1989, March). Toxic responsibility: The U.S. military's problems of toxic waste disposal are as great as those of industry. *The Atlantic*.

Slack, R. J., Bonin, M., Gronow, J. R., Van Santen, A., & Voulvoulis, N. (2007). Household hazardous waste data for the UK by direct sampling. *Environmental Science and Technology*, *41*(7), 2566–2571. <http://pubs.acs.org/doi/abs/10.1021/es061775w?prevSearch=[abstract%3A/>.

Solutia Incorporated (Solutia) (2002). Solutia milestones: The first five years. *News Release*. Retrieved August 13, 2010 from <http://investor.solutia.com/phoenix.zhtml?c=88803&p=irol-newsArticle&ID=329250&highlight=/>.

US Department of Energy (DOE) (2006). *Five year plan FY 2007-FY 2011* (Vol. II). Office of the Chief Financial Officer. Retrieved September 15, 2006 from <http://www.em.doe.gov/PDFs/170016EM_FYP_Final_3-6-06.pdf/>.

US Department of the Interior (DOI) (1993). *Military bases in the contiguous United States*. National Park Service. Retrieved September, 15 2010 from <http://www.nps.gov/history/nagpra/documents/basesmilitarymap.htm/>.

US Department of the Interior (DOI) (2010). *Environmental cleanup and liability management team*. Office of Environmental Policy and Compliance. Retrieved September 15, 2010 from <http://www.doi.gov/oepc/eclm.html/>.

US Environmental Protection Agency (EPA) (1992). *CERCLA/SUPERFUND orientation manual*. Office of Solid Waste and Emergency Response. Retrieved September 15, 2010 from <http://www.epa.gov/superfund/policy/remedy/pdfs/542r-92005-s.pdf/>.

US Environmental Protection Agency (EPA) (2006). *National priority chemicals trends report (2004–2006)*. Hazardous Waste Minimization and Management Division Office of Solid Waste. Retrieved August 13, 2010 from <http://www.epa.gov/osw/hazard/wastemin/minimize/trend09/sec4/sec4_01.pdf/>.

US Environmental Protection Agency (EPA) (2010a). *Regulations governing hazardous waste generators*. <http://www.epa.gov/osw/inforesources/data/br07/national07.pdf/>.

US Environmental Protection Agency (EPA) (2010b). The national biennial RCRA hazardous waste report (based on 2007 data). <http://www.epa.gov/osw/inforesources/data/br07/national07.pdf/>.

US Environmental Protection Agency (EPA) (2010c). Chocolate Bayou Plant, FM 2917, Alvin, TX 775129888. EPA Toxics Release Inventory. Retrieved August 13, 2010 from <http://www.oaspub.epa.gov/enviro/multisys2.get_list_tri?tri_fac_id=77511SLTNCFM291/>.

3 The Chemical Industry

Chapter Contents

Behind the federal government, the chemical industry is the largest producer of hazardous waste in the United States. According to the United States Environmental Protection Agency, chemical manufacturing was responsible for 72% of all hazardous waste produced by private industry (EPA, 2010e). Of those chemical industries, The Dow Chemical Company (Dow), E.I. DuPont de Nemours and Company (DuPont), and The Monsanto Company (Monsanto) are some of the largest. They also exemplify the profit over public business model that has characterized the chemical industry in the 20th century. Together these three companies are responsible for a total of 157 Superfund sites on the National Priority List (PIRG, 1999).

3.1 The Dow Chemical Company

Herbert Henry Dow founded the Dow Chemical Company in 1897 to manufacture and sell bleach on a commercial scale. After founding, Dow grew to manufacture many chemical-related products. Within about 100 years, the company had dabbled in agricultural chemicals, food preservatives, plastics, automotive applications, and household products. Additionally, Dow has invented and produced many chemical compounds. Companies that have been and/or are currently associated with Dow

Risks of Hazardous Wastes. DOI: 10.1016/B978-1-4377-7842-7.00003-9

include: Styron, Merrell Dow, Marion Merrell Dow, DowElanco, DuPont Dow Elastomers, Dow Agrosciences, Isopol, Mycogen Corporation, Sentrachem, SAL Petrochemical Company, Shenzen Sino-American OCT International Trading Company, CanStates Holdings, ANGUS Chemical Company, Union Carbide, Elemica, BSL, Innovase, Enichem S.p.A., Rohm and Haas Company, and Dow Corning (The Dow Chemical Company, 2010).

3.1.1 Superfund Sites

The Dow Chemical Company is currently liable for 96 Superfund sites throughout the US. Of those, Dow shares liability at 95 of the sites and complete liability for one (The Center for Public Integrity, 2008).

Rocky Flats, Colorado, is the site of one particular Superfund site. From 1952 to 1975, the site was operated by Dow as a nuclear weapons facility. It was subsequently operated by Rockwell from 1975 to 1989 (USEPA, 1989). In 1969, while operated by Dow, a fire, considered the most costly industrial accident in United States history at that time, released radioactive plutonium (CDPHE, 1999). The incident prompted concern from the public over contaminant releases from Rocky Flats. In 1989, an FBI and EPA raid discovered that there had been many releases of solvents, pesticides, plutonium, and tritium into the surrounding soil and water, and that there were radioactive soils and sediments located off-site (USEPA, 1989; CDPHE, 1999). Afterwards, production of nuclear weapons was put on hiatus due to these and other safety violations, such as uncontrolled carbon tetrachloride releases into the atmosphere. A subsequent lawsuit filed by homeowners in the surrounding community resulted in $925 million in damages in 2006 (Burdick, 2008). After abandonment of the site as a nuclear weapons facility, the EPA, Department of Energy, and Colorado Department of Health have been working together in remediation of the site, which has so far cost $7 billion to turn it into a wildlife refuge (USEPA, 1989; Burdick, 2008).

The Dow plant in Midland, Michigan, is another notorious Superfund site. Significant contamination in and around the facility, including contamination over 50 miles downstream in the Saginaw Bay on Lake Huron, have resulted from years of fugitive emissions, waste disposal, and incineration (USEPA, 2010d). Additionally, dioxins, furans, and other contaminants have been identified in the rivers that carried outflow from the site; dioxins and furans are byproducts formed from the manufacture of chlorine-based products such as chlorophenols. Dioxin levels of 110,000 parts per trillion toxic equivalence concentration (TEQ) and 1,600,000 parts per trillion TEQ have been detected in the Tittabawassee and Saginaw Rivers, respectively, prompting a required cleanup by Dow in 2007 (USEPA, 2010d). Additional cleanups have been scheduled since then, with the EPA overseeing the current cleanup in the Saginaw Bay watershed. EPA has promised to guarantee that Dow 'upholds its responsibility to clean up. . .[and] if Dow fails to meet its responsibilities, EPA will conduct the cleanup at the company's expense' (USEPA, 2009). Hopefully, significant prodding by the EPA will encourage a prompt cleanup effort for contamination of the ecosystem surrounding the facility.

3.1.2 Disaster in Bhopal, India

On the night of December 2, 1984, the Union Carbide factory in Bhopal, India, experienced a runaway chemical reaction that caused the leakage of massive amounts of a gas identified as methyl isocyanate (MIC). MIC is used as an intermediate chemical in pesticide, plastics, paint, and drug manufacture, and produces toxic effects in humans including ocular damage, respiratory tract irritation, and asthma attack (Raina, 1985). Approximately 40 tons of MIC were released from the facility that night, engulfing the low-lying regions of the city (Raina, 1985). During the release, panicked residents of Bhopal attempted to flee the city and trampled other residents in their haste (Government of MP, 2010). The accident killed an estimated 15,000 people and injured an estimated 500,000 in and around Bhopal (Dhar, 2010). The Bhopal leak has been called 'the worst industrial disaster the world has ever seen' due to the company mishandling and lack of prevention of the situation (AP, 2009).

After the incident, Union Carbide did little more than shut the plant down, leading to litigation and continuation of hazardous conditions for nearby residents. Union Carbide has ignored the continuous leakage of hazardous waste into local groundwater from the site (Broughton, 2005). To this day, the toxic legacy of Union Carbide continues to plague survivors. Birth defects, brain damage, and reproductive defects are all present in Bhopal residents 25 years after the gas leak (AP, 2009).

In 1989, a $470 million settlement was reached between Union Carbide and the Indian government (AP, 2009). Surprisingly, the Indian Supreme Court ruled that Union Carbide was not responsible for its subsidiary branch's actions in the leak; thus 'charges in the case were reduced from culpable homicide to negligence' (Dhar, 2010). These reduced charges allowed Union Carbide executives to get off on a serious breach of safety. In 2001, Union Carbide was purchased by Dow, which refuses to take responsibility for the site, claiming that the case was resolved in 1989 (AP, 2009). Another trial against Union Carbide CEO at the time, Warren Anderson, remains on hiatus after Anderson fled to the United States (Kuncheria, 2010). In 2010, the Indian Supreme Court reopened the case after years of public opposition to the light sentences seven Union Carbide executives received. The government is now seeking an additional $325 million from Dow regarding the Bhopal disaster (Dhar, 2010).

3.1.3 Agent Orange Liabilities

During the Vietnam War, Dow and several other companies including Monsanto and Diamond Shamrock were tasked with supplying the government with Agent Orange, a herbicide mixture used as a defoliant to destroy food and cover for the Viet Cong. During the campaign, 42.6 million liters of Agent Orange were sprayed over much of southern Vietnam (Schecter et al., 2003). While meant for destroying vegetation, Agent Orange also contaminated livestock, fish, and other animals, as well as the Vietnamese people who lived in the spray area.

The toxic compound present in Agent Orange as an impurity is considered responsible for most of the health effects associated with Agent Orange. Dow has admitted that an 'unavoidable trace by-product was the dioxin compound 2,3,7,8-TCDD' (The Dow Chemical Company, 2007). TCDD is the most toxic of all dioxins, affecting the skin and causing cancer in humans (ATSDR, 1999). In 2003, a dioxin contamination expert from the United States tested soil around the former US military base at Bien Hoa that stored Agent Orange and discovered TCCD levels 180 million times above the EPA screening level (Fawthrop, 2004; Schecter et al., 2003), producing controversy about how much toxic TCDD was actually present in Agent Orange. In general domestic production, Agent Orange typically contains about 0.05 parts per million TCDD, while the product used in the Vietnam War reportedly contained up to 50 parts per million TCDD (Warwick, 1998). It has been claimed that since the 1960s, Dow 'had information that Agent Orange supplied to the Government contained large levels of dioxin, far in excess of anything Dow considered safe or necessary' (Burnham, 1983).

Agent Orange has produced deleterious health effects in Vietnam veterans and Vietnamese people exposed to the material. The United States Department of Veterans Affairs lists many veterans' diseases that are associated with exposure to the herbicide, including neuropathy, chloracne, chronic B-cell leukemias, type 2 diabetes, Hodgkin's disease, heart disease, multiple myeloma, non-Hodgkin's lymphoma, Parkinson's disease, prostate cancer, and respiratory cancer (US Department of Veterans Affairs, 2010). Additionally, symptoms of exposure to Agent Orange include cancer, nervous disorders, cardiovascular disease, and diabetes (Koo and Sim, 2006). In Vietnam, there are about 150,000 children suffering from crippling birth defects, such as twisted and deformed limbs, whose defects can be traced back to parental exposure to Agent Orange or consumption of contaminated food and water since 1975 (Fawthrop, 2004). Health problems are not only limited to humans; animals can also be adversely affected by Agent Orange. Animals such as duck, chicken, and fish exposed to Agent Orange contain elevated levels of TCDD and other dioxins (Schecter et al., 2003). The lingering legacy of Agent Orange will not disappear easily or quickly, despite Dow's insistence that there is no association with veterans' diseases (The Dow Chemical Company, 2007).

In 1984, a settlement between Vietnam veterans and the chemical companies that manufactured Agent Orange resulted in a $180 million settlement (Elkin, 1984). As part of the settlement, the chemical companies denied any liability in injuries associated with Agent Orange (Elkin, 1984). In March 2006, a US District Court ruled that chemical manufacturers were not liable for damages caused by Agent Orange after dismissing a lawsuit filed for four million Vietnamese citizens (Koo and Sim, 2006). Another case in 2008 concerning the chemical manufacturers ruled that they were shielded from liability because they were government contractors (Graybow, 2008).

3.1.4 Chlorpyrifos

Chlorpyrifos is a chlorinated organophosphate insecticide, acaricide, and nematicide. It was developed by Dow in 1962 and commercialized in 1965 to control insects (PAN North America, 2004). It is used for agricultural and residential applications such as for food crops, indoor pest control, termite control, pet collars, school use, restaurant use, hospital use, and food manufacturing plants (ATSDR, 2006). Chlorpyrifos, from the same family of chemicals as the sarin nerve gas agent, functions by attacking the nervous system (Herbert, 2000).

The consequences of exposure to chlorpyrifos are known. Inhalation or ingestion of chlorpyrifos can produce headaches, blurred vision, lacrimation, runny nose, dizziness, confusion, muscle weakness or tremors, nausea, diarrhea, and sudden changes in heart rate (ATSDR, 1997). Exposure to a high amount of chlorpyrifos can cause severe sweating, loss of bowel control, severe muscle tremors, seizures, loss of consciousness, or death (ATSDR, 1997). Chlorpyrifos is also very toxic to the surrounding ecosystem, with risks to mammals, birds, fish, and aquatic invertebrate species from most registered outdoor uses (PAN North America, 2004). Due to these and other serious health effects, chlorpyrifos was banned for indoor use and phased out of outdoor lawn and garden applications from 2000 to 2004 under the EPA Food Quality Protection Act (Herbert, 2000).

Although chlorpyrifos has been seriously limited in use in the United States, Dow has continued to manufacture the pesticide to sell to countries that have not yet banned it (PAN North America, 2004). Recently, Dow has been strongly pushing its pesticides in India. In 2007, Dow was fined by the US Securities and Exchange Commission for allowing its subsidiary to bribe Indian officials to fast-track pesticide sales in the country (Kumar, 2010). Dow has been convicted of bribing government officials to ensure that its pesticides, including chlorpyrifos, will be registered in the country The company has subsequently been blacklisted until 2015 (Kumar, 2010).

In 2003, the Attorney General of New York charged a subsidiary of Dow $2 million for falsely advertising the safety of Dursban, a trade name of chlorpyrifos, explaining that 'advertising claims about its herbicide products are contradicted by the product labels, other Dow documents (such as Material Data Safety Sheets), and government findings' (State of New York, 2003). In advertisements, Dow had claimed 'there is simply no credible scientific evidence that Dursban products harm people or the environment when used properly' (State of New York, 2003). Dow agreed to reform its advertising and marketing practices in 1994 after another Attorney General investigation found that false advertising had occurred (State of New York, 2003). In 1995, Dow was fined $732,000 for failing to quickly report user poisoning (PAN North America, 2004).

3.2 E.I. DuPont de Nemours and Company

DuPont was founded in 1802 by the French expatriate E.I. du Pont. Du Pont learned powder-making in France before he was uprooted during the French Revolution and fled to the United States. In 1802 he established a powder mill on the Brandywine River, Delaware. DuPont soon earned a government contract for making black powder and in 1811 became the largest manufacturer of black powder in the US (ASME, 2010; DuPont, 2002). In 1902 DuPont began manufacturing industrial chemicals in addition to explosives, and soon thereafter opened lacquer facilities in 1904 and industrial dye facilities in 1917. By the end of World War I, DuPont had produced and delivered 1.5 billion pounds of military explosives to the Allied forces – roughly 40% of all explosives used by the Allies (DuPont, 2002).

Since that time, DuPont has established an extensive toxic legacy. In 1999 DuPont was included in the US Public Interest Research Group's 'Dirty Five'; a study of the five largest polluters in the United States. DuPont was ranked number 2, with 81 sites on the EPA Superfund's National Priority List. One of the more striking traits of the 'Dirty Five' was the willingness of the involved companies to place profits before public health. Between 1991 and 1998, the 'Five' spent $6,523,677 lobbying Congress, the House of Representatives, and CERCLA-related groups to maintain lax regulatory levels for hazardous waste (CW, 2002). DuPont exemplified this focus on profits in several instances over the past 100 years.

3.2.1 Tetraethyl Lead

DuPont, Standard Oil, and General Motors spearheaded the development and marketing of tetraethyl lead (TEL) in 1921 for use as a fuel additive to improve engine efficiency. Of the three companies, DuPont was the largest producer of TEL. The three formed a joint venture known at Ethyl Gasoline, and began selling it by the barrel to interested buyers, avoiding the word 'lead' in the brand name. In the 1920s, the dangers of lead were well known. In a *Time Magazine* article from 1924, three sources of danger from tetraethyl lead were described: harm incurred when manufacturing and handling concentrated TEL, the possible harm associated with handling Ethyl Gasoline, and the possible hazard to the public from the exhaust of automobiles using the TEL-laced Ethyl Gasoline (Time, 1924). Contemporary newspapers at the time, reporting cases of DuPont worker hospitalizations and deaths occurring at TEL manufacturing plants, colloquially referred to TEL as 'looney gas.' This nickname referred to the bouts of insanity that some workers at DuPont plants exhibited after exposure to the lead-containing substance.

Experts at the time, including those at the US Public Health Service, voiced concerns about the substance. One lab director wrote to the Assistant Surgeon General A.M. Stimson that TEL posed a 'serious menace to public health,' citing reports from the DuPont manufacturing plant of 'very serious lead poisoning'; the letter also voiced concerns over the public health dangers associated with exhaust stating that 'on busy thoroughfares it is highly probably that the lead oxide dust will remain in the lower stratum' (Kitman, 2000). Stimson, understandably concerned, requested more research from the PHS laboratories, but was denied and given the explanation that the government would rely on industry testing to determine the dangers. Eventually, the lab director and Stimson's correspondence prompted a concerned US Surgeon General H.S. Cumming to write a letter addressed to DuPont in 1922 requesting information on the health effects of DuPont's TEL. The company responded with a letter drafted by the compound's creator, Thomas Midgely Jr., stating that 'the average street will probably be so free from lead that it will be impossible to detect it or its absorption' (GLN, 2010). Despite Midgely's assurances, no data had been collected by DuPont regarding the health effects of TEL at that time.

After a disaster at a Standard Oil TEL plant in 1924, tetraethyl lead was barred from New York City and parts of New Jersey. Forty-four workers were injured and five died from exposure. The New York Board of Health dubbed TEL a 'public menace' and a *New York Times* article reporting the release referred to the additive as a perilous 'insanity producing substance' (Bent, 1925). To say that DuPont and other manufacturers were slow to acknowledge the danger associated with concentrated TEL is an understatement. DuPont, in fact, hid a huge number of poisonings from the press and public. At the DuPont factory at Deepwater, NJ, over 300 cases of lead poisoning were uncovered by the *New York Times* (1925). Accounts of medical staff at the DuPont-run hospital adjoining Deepwater refusing to speak to reporters or giving them false information have been recorded. Some even omitted the cause of death on certificates of workers who, as it was later found, clearly died of lead poisoning. In a contemporary newspaper, it was estimated that 80% of people who worked or visited the factory suffered from organic lead poisoning associated with TEL (Bent, 1925). The problem was so regnant that the factory came to be called the 'House of Butterflies' by the workers, the name derived from the workers' hallucinations of winged insects that would appear at the onset of lead poisoning.

After a while, it became impossible for DuPont and the other manufacturers of TEL to deny the negative effects of tetraethyl lead; they, however, still claimed that the lower concentrations found in exhaust were not harmful to the public. To enforce their claim, the manufacturers requested the opinion of expert Dr. Yandel Henderson, Professor of Physiology at Yale, who, much to their chagrin, disagreed with their assertion entirely. Dr. Henderson was paraphrased in a *New York Times* article as saying that 'this poison (TEL) was so insidious that the breakdown in health might occur long after the breathing of the poison,' and that 'the absence of reports of injury due to breathing this gas in automobile exhaust fumes was no indication that those fumes were not dangerous.' In a later article in 1925, Dr. Henderson is paraphrased as stating that the widespread use of TEL as an additive 'would cause widespread lead poisoning.' Needless to say, the manufacturers rejected his opinion, continually saying that levels of TEL in gasoline were not high enough to cause harm. Industrial hygienists were up in arms, including Dr. Alice Hamilton, a prominent Harvard lead specialist, Cecil K. Drinker, editor of the *Journal of Industrial Hygiene* and Professor of Public Health at Harvard, and Dr. David Edsall, Dean of Harvard Medical School (NYT, 1925).

Despite the protests of experts, the public and government agreed to permit this 'poison' and 'looney gas' to continue being produced and distributed. This would not have been possible if Ethyl Gasoline's wholly unethical dealings with the US Bureau of Mines regarding 'testing' of tetraethylene lead provided the public and government with a false bill of safety for TEL. DuPont and General Motors, realizing that after the highly publicized lead poisoning incidents encountered during the manufacturing of TEL, their additive would need a federally endorsed testing process to garner public approval. Therefore, Ethyl Gasoline formed a contract with the pliable US Bureau of Mines (BM) to fund the testing of TEL while the Bureau provided the facilities and seal of approval. As part of the contract, Ethyl Gasoline stipulated that BM must submit all documents intended for publication to Ethyl Gasoline for approval and that the BM substitute 'ethyl' for lead in all documents. Ethyl Gasoline was in control of the relationship. The BM released its findings and, unsurprisingly, exonerated TEL. The headline of an article reporting the release effectively sums up the BM's testing, 'No Peril to Public Seen in Ethyl Gas/Bureau of Mines Reports after Long Experiments with Motor Exhausts/More Deaths Unlikely' (*NYT*, 1924).

When the report by the BM emerged, the public was skeptical. The US Surgeon General Cumming held hearings that were inconsequential and established a blue-ribbon group of experts to assess the toxicity of TEL. The scientists participating in this rushed and short-term study could not find convincing evidence that lead would be present to such a degree that it would be a critical threat (Kovarik, 1999). In retrospect, the study period was too short to fully capture the persistent and bioaccumulative properties of lead. The blue-ribbon group gave TEL approval, and unfortunately TEL and Ethyl Gasoline were permitted to continue being produced. In 1936, tetraethyl lead was present in 90% of gasoline used in the United States (Agin, 2006).

In the 1970s, tetraethyl lead came under duly harsh criticism as more studies emerged pointing to the persistent and bioaccumulative nature of lead from gasoline exhaust. Clair Patterson, for example, examining ice cores in Greenland, found that lead contaminants appeared in the ice around the same time that TEL was introduced (Patterson, 1965). In 1972, the US EPA began a campaign to eliminate leaded gasoline, and was subsequently sued by DuPont and Ethyl Corp. but eventually won the case in 1976 (*NYT*, 1976). Even after leaded gasoline was beginning to be phased out in the 1980s, DuPont continued to add illegally high levels of lead into gas. The EPA filed a complaint that DuPont and three other companies illegally added lead to gasoline between 1983 and 1985. The EPA sought an $8.23 million fine on the grounds that DuPont had added 2,200 tons of excess lead to gasoline (AP, 1988).

A 1993 report released by the Institute for Policy Studies accused DuPont, a proponent of the North American Free Trade Agreement (NAFTA), of exploiting the agreement (DePalma, 1993). The report states that after TEL was phased out in the United States, DuPont relocated its manufacturing plant from Deepwater, NJ, to Coatzacoalcos, Mexico, and restarted production.

The effects of leaded gasoline on our environment have been substantial. In a famous study by Reyes (2007), a strong and significant correlation was found between a national drop in violent crime and a drop in ambient lead levels after TEL was banned in the 1980s. Reyes argues in her paper that the decrease in childhood lead exposure resulted in a 56% decrease in violent crime in the 1990s. The author argues that lead exposure to children increases the likelihood 'of behavioral and cognitive traits such as impulsivity, aggressivity, and low IQ that are strongly associated with criminal behavior.' Reyes also concluded that between 1990 and 2020, the switch from leaded to unleaded in the 1980s saved $75 billion in prevented violent crimes.

3.2.2 Pompton Lakes, New Jersey

The borough of Pompton Lakes, New Jersey, is a 3-square-mile town, with roughly 10,500 residents. This 'pleasant downtown business community,' as described on its website, is also home to 200,000 tons of soil contaminated with hazardous waste (Applebome, 2010). In 1902, DuPont opened an explosives manufacturing facility in Pompton Lakes, New Jersey. By 1998, 5 years after DuPont stopped production of explosives, Pompton Lake contained a maximum of 367 mg/kg of mercury in the top 6 inches of sediment, and an average of 9.2 mg/kg (DuPont, 2008).

The 570-acre site is located adjacent to the aptly named Acid Brook. The waterway received its name from the various colors it would change depending on the particular chemical cocktail in the stream at the time (Applebome, 2010). The site today is contaminated with lead salts and mercury compounds, explosive powders, chlorinated solvents, detonated blasting caps, and waste wire drawing solution (EPA, 2010c). Lead and mercury contaminated 140 nearby homes that the US EPA has since remediated. Volatile organic compounds (VOCs) such as tetrachloroethylene and perchloroethylene found in the water have been blamed by the New Jersey Department of Health and Senior Services for causing 'significantly elevated' rates of kidney cancer in women and elevated rates of non-Hodgkin's lymphoma in men (Applebome, 2010).

People first began to worry about the polluting plant in the 1970s, the local government became opposed in the 1980s, and DuPont settled a lawsuit with residents for $38.5 million in the 1990s. In 2002, DuPont lost another smaller lawsuit (Applebome, 2010). John Sinsimer, the mayor of Pompton Lakes between 1988 and 1992, described his dealings with DuPont as such: 'DuPont will try to get away with as much as they can get away with anytime they can. First they try to buy you. Then they try to bully you. Then they try and bury you' (Applebome, 2010). DuPont would certainly have enough soil to bury opponents if they were to take Sinsimer's comments literally. Over 200,000 tons of contaminated soil and sediment have been removed, or are cited for removal at and around the Pompton Lakes facility (Applebome, 2010).

3.2.3 Other Recent DuPont Transgressions: Teflon, Dioxins, and Safety Violations

2005	DuPont is fined for withholding information on Teflon (perfluoroctanoic acid – PFOA). A now designated 'known carcinogen' by the EPA, the dangers of Teflon were exposed by whistle-blower Glenn Evers. According to Evers, a DuPont employee from 1988–2002, DuPont knew about the dangers

	of PFOA, but did not disclose them to the public. As early as 1973 DuPont had internally conducted animal studies that demonstrated PFOA caused kidney damage, liver damage, anemia, and higher cholesterol (AP, 2005).
2005	DuPont pays \$107 million to settle a class action lawsuit in Ohio and West Virginia for intentionally withholding information about the human health threat associated with PFOA. DuPont was also fined \$10–\$15 million for failure to report risks to the public about PFOA after discharges at a Pascagoula plant (Janofsky, 2005).
2007	The DuPont titanium oxide (TiO_2) facility in Edge Moor, DE, contained an estimated 500,000 tons of soil contaminated with arsenic, heavy metals, dioxin, dioxin-like compounds, PCBs, and benzene compounds (Evers, 2007). The soil began accumulating significantly after the mid-1900s when DuPont introduced a new product called Iron Rich™. This product was marketed as a soil substitute for use in landfills, roadbeds, and other construction projects. The 'scathing truth,' as DuPont whistleblower Glenn Evers puts it in his 2007 report on the site, is that Iron Rich™ was really soil mixed with waste from the TiO_2 production process. By selling the plant's toxic waste as a product, DuPont enjoyed a 'tremendous economic benefit' than if it were classified as waste, according to Evers. Allegedly, three DuPont researchers warned the Edge Moor facility producing Iron Rich™ that the soil substitute was too contaminated to be sold as a product, but to no effect. It took until 2001 for DuPont to halt production, when a Consent Order was lodged with the Delaware Superior Court (DNREC, 2004). Currently DuPont has placed a plastic sheet over the earth and is working to effectively cap the soil (DNREC, 2010). However, the poorly sealed underside of the parcel of soil is prone to leaching, according to Evers, and local residents are skeptical if the cap will last the purported amount of time (Montgomery, 2005).
2008	Invista, a corporation that bought out DuPont's fiber and resin business in 2004, sued the chemical giant in 2008 for \$800 million after finding that two plants in Texas were operating in severe non-compliance with safety standards. According to Invista, DuPont was well aware of the most dangerous violations, including one plant that had been releasing benzene directly into the ambient air since 1992 (TCE, 2008).

3.3 Monsanto Company

Monsanto was founded in 1901 by John F. Queeny, who opened the Monsanto Chemical Works in St. Louis, MO. The first factory produced saccharine, an artificial sweetener and sold it to Coca-Cola. During World War I, the company began to expand production to include phenol and acetylsalicyclic acid. In 1928, the time of Queeny's retirement, the company had 1,000 or so employees and had gone public. By the early 1930s, Monsanto's PCB plants at Sauget, Illinois, and Anniston, Alabama, had opened, as well as a herbicide production facility in Nitro, West Virginia. In the first years of the 21st century, all three of these plants were involved in litigation on the order of millions of dollars, for the PCBs (Sauget,

Anniston) and dioxins (Nitro) they released into the environment (AP, 2003; Dickerson, 2007). In total, Monsanto is considered a 'Potentially Responsible Party' at 98 polluted sites, meaning that Monsanto is liable for the hazardous waste deposited at those sites (EPA, 2001d).

3.3.1 Monsanto's Toxic PCB Legacy at Sauget, Illinois

Two industrial facilities, Monsanto Corporation's W.G. Krummrich facility and Cerro Copper Products Company facility, have operated in Sauget, Illinois, since the early 1900s. Sauget is a primarily industrial area that was originally incorporated as the Village of Monsanto in 1926. Sauget is immediately surrounded by Cahokia to the south, East St. Louis to the north and east, and Centerville to the east. These cities and other villages in the area, such as Alorton and Washington Park, all contain dense residential areas located within close proximity to the Monsanto and Cerro Copper facilities.

Monsanto was the sole domestic producer of polychlorinated biphenyls (PCB) in the United States, with plants in Sauget and Anniston. Manufacturing of PCBs was conducted at the Sauget facility from approximately 1929–30 to 1975, and PCB production peaked in 1970 with approximately 85 million pounds (US EPA, 1976). Monsanto produced extensive lines of other chemical products at its W.G. Krummrich plant in Sauget, including chlorinated pesticides and other chemicals. Several chlorinated chemical products manufactured at and/or used at the Monsanto facility are known or suspected precursor chemicals that can lead to the formation of dioxins/furans during chemical manufacturing (EPA, 1980).

As a result of decades of manufacturing activities conducted by Monsanto at these facilities, hazardous wastes, effluents, and emissions containing PCBs and dioxins/furans were released from the facilities. Unsystematic dumping of hazardous wastes into landfills and dump sites within the vicinity of the Monsanto facility have occurred since before the 1940s. The US EPA has named Monsanto and Cerro Copper as the primary responsible parties involving these discharges. In 2001, US EPA placed the many hazardous dump sites surrounding the Monsanto and Cerro Copper facilities on the National Priorities List (NPL) of Superfund Sites (EPA, 2001a; EPA, 2001b; EPA, 2001c).

The decades of industrial activity and discharges in the area have raised concerns regarding potential health effects to residents of the nearby communities. PCBs and dioxins/furans are toxic substances which have been shown to cause adverse health effects, including cancer in humans and animals, neurotoxicity, reproductive and developmental toxicity, elevated blood pressure, changes in serum chemistry, immune system suppression, liver damage, skin irritation, and endocrine disruption (ATSDR, 1994; ATSDR, 1998; ATSDR, 2000).

PCBs and dioxins have been found in the environment at various locations in Sauget, such as hazardous waste disposal areas, surrounding the Monsanto and Cerro Copper facilities according to numerous extensive investigations and remedial actions conducted since the 1980s by Monsanto, Cerro Copper, Illinois Environmental Protection Agency (IEPA) and the US EPA. Dioxins have been detected within the vicinity of the Monsanto and Cerro Copper facilities at concentrations that are among the highest reported in the United States. Solutia (2000) reported the detection of dioxins in samples of soil collected from the Monsanto facility in their 2000 *Description of Current Conditions Report*. One sample was reported to have a dioxin concentration of over 2,800,000 parts per trillion TCDD TEQs. Investigations have reported highly elevated concentrations of dioxins in many landfills and dumps throughout the Sauget area. Site O, the Village of Monsanto/Sauget WWTP sludge spreading lagoons, most notably received wastewater from both Monsanto and Cerro Copper. Dioxins have been detected at Site O in surface soil at concentrations as high as 427,000 parts per trillion TCDD TEQs (URS, 2004).

Monsanto has determined that a four-acre area of soil at the Sauget location is contaminated with PCB concentrations greater than 25 parts per million. PCB contamination has been discovered 15 feet down near groundwater aquifers. Monsanto/Solutia estimates that about 60,000 pounds of PCBs are spread throughout the upper 15 feet of soil in this area. PCBs have also been found 60 feet below the groundwater table. High levels of PCBs in groundwater have additionally migrated over 1,600 feet west of the former PCB manufacturing area. Despite the best cleanup efforts, many PCBs are still contaminating the environment at Monsanto's Sauget plant (EPA, 2007).

3.3.2 Monsanto's (Solutia Inc.) Plant at Anniston, Alabama

In 1935, Monsanto/Solutia purchased the PCB producing plant in Anniston from Theodore Swan Company. Over the course of roughly 40 years, Monsanto discharged millions of tons of PCBs and amounts of mercury into the neighboring creek (Grunwald, 2002).

Well before 1967, when a study by Swedish scientists brought the dangers of PCBs to the public's eye, Monsanto knew about the toxic nature of the chemical it was manufacturing and discharging as effluent.

1935	A company memo states that PCBs 'cannot be considered non-toxic.'
1937	Researchers at Harvard University found that long-term exposure to PCBs can cause liver damage and chloroacne. The lead researcher was soon hired by Monsanto as a consultant and company memos began to refer to the 'systemic toxic effects' of PCBs.
1950	Aroclor (Monsanto's brand-name PCB) was attributed by Monsanto's medical director to have caused worker liver problems at a customer's factory in Indiana.
1952	Monsanto, in directions to customers, warned them that if Aroclor is 'discharged in large concentrations it will adversely affect . . . aquatic life in the stream.'
1956	Monsanto attempted to sell hydraulic fluid containing PCBs to the Navy. After running tests of their own, the Navy determined the fluid was 'too toxic' for use and caused 'definite liver damage.'

1963	A Monsanto employee re-examined areas where Aroclor was injected into the soil as part of a study in 1939 and found there 'was still visual evidence of a presence of Aroclor,' clearly demonstrating the chemical's persistence in the environment some time before 1967.
1966	A study conducted by a Mississippi State University biologist hired by Monsanto of Snow Creek attributed the horrendous environmental damage to PCBs discharged by the Anniston plant, and called PCB levels in the creek 'extremely toxic.' He also stated that 'since [Snow Creek] is a surface stream that passes through residential areas, it may represent a potential source of danger to children, domestic animals, etc.'

(Grunwald, 2002; EWG, 2009)

Probably the most telling study was the one conducted by the MSU biologist Denzel Ferguson. In 1966 he was hired by Monsanto to examine the water quality around the Anniston plant. Ferguson's experiment was simple. He submerged cloth containers of 25 Bluegill fish into Snow Creek and the surrounding waterways, and recorded the effect of the water on the fish. One of his graduate students working with him at the time best described the results: 'It was like dunking the fish in battery acid.' All 25 of the fish submerged in Snow Creek 'lost equilibrium and turned on their sides in 10 seconds and all were dead in 3 minutes,' reported Ferguson. 'Their skin would literally slough off, like a blood blister on the bottom of your foot,' recalled another graduate student. Ferguson estimated that even if the water from Snow Creek were diluted 1,000 times, it would kill the fish. He concluded 'the outflow to Snow Creek from the east side of the Monsanto Plant . . . contains some extremely toxic materials and kills fish in less than 24 hours when diluted 300 times' (EWG, 2009).

Despite this evidence of toxicity that was clearly known to Monsanto, management still managed to annually discharge 50,000 pounds of polychlorinated biphenyls into nearby Snow Creek. In 1969 alone, 45 tons of PCBs were discharged into Snow Creek (Sack, 2002). On the property itself, Monsanto buried over 1 million pounds of PCB-contaminated waste in inadequate landfills. One outside consultant in a later lawsuit against Monsanto stated that the Anniston plant was not compliant with even the most basic industry practices. Spills of hazardous chemicals, such as PCBs and mercury, a by-product of PCB production, were washed down drains and into sewers. The plant did not use catch basins, settling ponds or carbon filters to clean wastewater before releasing it (Grunwald, 2002). Only in 1970 did the company begin to improve pollution abatement – 12 months before shutting down the plant permanently.

The effects of the Anniston facility effluent on the neighboring town and streams were significant. A study of a larger river downstream from Snow Creek found one fish with PCB concentrations as high as 37,800 parts per million; 5 parts per million was the legal maximum at the time. Fish in the same creek were found to be mutated and deformed. Efforts by Monsanto to conceal the contamination (revealed in subpoenaed 'CLASSIFIED' memos) and incompetent regulatory

Table 3.1 RCRA K-listed Hazardous Wastes – Chemical Industry

Industry and Hazardous Waste Number	Hazardous Waste	Haz. Waste Code
Inorganic Pigments		
K002	Wastewater treatment sludge from the production of chrome yellow and orange pigments	(T)
K003	Wastewater treatment sludge from the production of molybdate orange pigments	(T)
K004	Wastewater treatment sludge from the production of zinc yellow pigments	(T)
K005	Wastewater treatment sludge from the production of chrome green pigments	(T)
K006	Wastewater treatment sludge from the production of chrome oxide green pigments (anhydrous and hydrated)	(T)
K007	Wastewater treatment sludge from the production of iron blue pigments	(T)
K008	Oven residue from the production of chrome oxide green pigments	(T)
Organic Chemicals		
K009	Distillation bottoms from the production of acetaldehyde from ethylene	(T)
K010	Distillation side cuts from the production of acetaldehyde from ethylene	(T)
K011	Bottom stream from the wastewater stripper in the production of acrylonitrile	(R,T)
K013	Bottom stream from the acetonitrile column in the production of acrylonitrile	(R,T)
K014	Bottoms from the acetonitrile purification column in the production of acrylonitrile	(T)
K015	Still bottoms from the distillation of benzyl chloride	(T)
K016	Heavy ends or distillation residues from the production of carbon tetrachloride	(T)
K017	Heavy ends (still bottoms) from the purification column in the production of epichlorohydrin	(T)
K018	Heavy ends from the fractionation column in ethyl chloride production	(T)
K019	Heavy ends from the distillation of ethylene dichloride in ethylene dichloride production	(T)
K020	Heavy ends from the distillation of vinyl chloride in vinyl chloride monomer production	(T)
K021	Aqueous spent antimony catalyst waste from fluoromethanes production	(T)
K022	Distillation bottom tars from the production of phenol/acetone from cumene	(T)

(*Continued*)

Table 3.1 (Continued)

Industry and Hazardous Waste Number	**Hazardous Waste**	**Haz. Waste Code**
K023	Distillation light ends from the production of phthalic anhydride from naphthalene	(T)
K024	Distillation bottoms from the production of phthalic anhydride from naphthalene	(T)
K025	Distillation bottoms from the production of nitrobenzene by the nitration of benzene	(T)
K026	Stripping still tails from the production of methyl ethyl pyridines	(T)
K027	Centrifuge and distillation residues from toluene diisocyanate production	(R,T)
K028	Spent catalyst from the hydrochlorinator reactor in the production of 1,1,1-trichloroethane	(T)
K029	Waste from the product steam stripper in the production of 1,1,1-trichloroethane	(T)
K030	Column bottoms or heavy ends from the combined production of trichloroethylene and perchloroethylene	(T)
K083	Distillation bottoms from aniline production	(T)
K085	Distillation or fractionation column bottoms from the production of chlorobenzenes	(T)
K093	Distillation light ends from the production of phthalic anhydride from ortho-xylene	(T)
K094	Distillation bottoms from the production of phthalic anhydride from ortho-xylene	(T)
K095	Distillation bottoms from the production of 1,1,1-trichloroethane	(T)
K096	Heavy ends from the heavy ends column from the production of 1,1,1-trichloroethane	(T)
K103	Process residues from aniline extraction from the production of aniline	(T)
K104	Combined wastewater streams generated from nitrobenzene/aniline production	(T)
K105	Separated aqueous stream from the reactor product washing step in the production of chlorobenzenes	(T)
K107	Column bottoms from product separation from the production of 1,1-dimethylhydrazine (UDMH) from carboxylic acid hydrazines	(C,T)
K108	Condensed column overheads from product separation and condensed reactor vent gases from the production of 1,1-dimethylhydrazine (UDMH) from carboxylic acid hydrazides	(I,T)
K109	Spent filter cartridges from product purification from the production of 1,1-dimethylhydrazine (UDMH) from carboxylic acid hydrazides	(T)

(*Continued*)

Table 3.1 (Continued)

Industry and Hazardous Waste Number	Hazardous Waste	Haz. Waste Code
K110	Condensed column overheads from intermediate separation from the production of 1,1-dimethylhydrazine (UDMH) from carboxylic acid hydrazides	(T)
K111	Product washwaters from the production of dinitrotoluene via nitration of toluene	(C,T)
K112	Reaction by-product water from the drying column in the production of toluenediamine via hydrogenation of dinitrotoluene	(T)
K113	Condensed liquid light ends from the purification of toluenediamine in the production of toluenediamine via hydrogenation of dinitrotoluene	(T)
K114	Vicinals from the purification of toluenediamine in the production of toluenediamine via hydrogenation of dinitrotoluene	(T)
K115	Heavy ends from the purification of toluenediamine in the production of toluenediamine via hydrogenation of dinitrotoluene	(T)
K116	Organic condensate from the solvent recovery column in the production of toluene diisocyanate via phosgenation of toluenediamine	(T)
K117	Wastewater from the reactor vent gas scrubber in the production of ethylene dibromide via bromination of ethene	(T)
K118	Spent adsorbent solids from purification of ethylene dibromide in the production of ethylene dibromide via bromination of ethene	(T)
K136	Still bottoms from the purification of ethylene dibromide in the production of ethylene dibromide via bromination of ethene	(T)
K149	Distillation bottoms from the production of alpha- (or methyl-) chlorinated toluenes, ring-chlorinated toluenes, benzoyl chlorides, and compounds with mixtures of these functional groups. (This waste does not include still bottoms from the distillation of benzyl chloride)	(T)
K150	Organic residuals, excluding spent carbon adsorbent, from the spent chlorine gas and hydrochloric acid recovery processes associated with the production of alpha- (or methyl-) chlorinated toluenes, ring-chlorinated toluenes, benzoyl chlorides, and compounds with mixtures of these functional groups	(T)
K151	Wastewater treatment sludges, excluding neutralization and biological sludges, generated during the treatment of wastewaters from the production of alpha- (or methyl-) chlorinated toluenes, ring-chlorinated toluenes, benzoyl chlorides, and compounds with mixtures of these functional groups	(T)

(Continued)

Table 3.1 (Continued)

Industry and Hazardous Waste Number	Hazardous Waste	Haz. Waste Code
K156	Organic waste (including heavy ends, still bottoms, light ends, spent solvents, filtrates, and decantates) from the production of carbamates and carbamoyl oximes. (This listing does not apply to wastes generated from the manufacture of 3-iodo-2-propynyl n-butylcarbamate)	(T)
K157	Wastewaters (including scrubber waters, condenser waters, washwaters, and separation waters) from the production of carbamates and carbamoyl oximes. (This listing does not apply to wastes generated from the manufacture of 3-iodo-2-propynyl n-butylcarbamate)	(T)
K158	Bag house dusts and filter/separation solids from the production of carbamates and carbamoyl oximes. (This listing does not apply to wastes generated from the manufacture of 3-iodo-2-propynyl n-butylcarbamate)	(T)
K159	Organics from the treatment of thiocarbamate wastes	(T)
K161	Purification solids (including filtration, evaporation, and centrifugation solids), bag house dust and floor sweepings from the production of dithiocarbamate acids and their salts (This listing does not include K125 or K126)	(R,T)
K174	Wastewater treatment sludges from the production of ethylene dichloride or vinyl chloride monomer (including sludges that result from commingled ethylene dichloride or vinyl chloride monomer wastewater and other wastewater), unless the sludges meet the following conditions: (i) they are disposed of in a subtitle C or non-hazardous landfill licensed or permitted by the state or federal government; (ii) they are not otherwise placed on the land prior to final disposal; and (iii) the generator maintains documentation demonstrating that the waste was either disposed of in an on-site landfill or consigned to a transporter or disposal facility that provided a written commitment to dispose of the waste in an off-site landfill. Respondents in any action brought to enforce the requirements of subtitle C must, upon a showing by the government that the respondent managed wastewater treatment sludges from the production of vinyl chloride monomer or ethylene dichloride, demonstrate that they meet the terms of the exclusion set forth above. In doing so, they must provide appropriate documentation (e.g., contracts between the generator and the landfill owner/operator, invoices documenting delivery of waste to landfill, etc.) that the terms of the exclusion were met	(T)

(Continued)

Table 3.1 (Continued)

Industry and Hazardous Waste Number	Hazardous Waste	Haz. Waste Code
K175	Wastewater treatment sludges from the production of vinyl chloride monomer using mercuric chloride catalyst in an acetylene-based process	(T)
K181	Nonwastewaters from the production of dyes and/or pigments (including nonwastewaters commingled at the point of generation with nonwastewaters from other processes) that, at the point of generation, contain mass loadings of any of the constituents identified in paragraph (c) of this section that are equal to or greater than the corresponding paragraph (c) levels, as determined on a calendar year basis. These wastes will not be hazardous if the nonwastewaters are: (i) disposed in a Subtitle D landfill unit subject to the design criteria in §258.40, (ii) disposed in a Subtitle C landfill unit subject to either §264.301 or §265.301, (iii) disposed in other Subtitle D landfill units that meet the design criteria in §258.40, §264.301, or §265.301, or (iv) treated in a combustion unit that is permitted under Subtitle C, or an on-site combustion unit that is permitted under the Clean Air Act. For the purposes of this listing, dyes and/or pigments production is defined in paragraph (b)(1) of this section. Paragraph (d) of this section describes the process for demonstrating that a facility's nonwastewaters are not K181. This listing does not apply to wastes that are otherwise identified as hazardous under §§261.21–261.24 and 261.31–261.33 at the point of generation. Also, the listing does not apply to wastes generated before any annual mass loading limit is met	(T)
Inorganic Chemicals		
K071	Brine purification muds from the mercury cell process in chlorine production, where separately prepurified brine is not used	(T)
K073	Chlorinated hydrocarbon waste from the purification step of the diaphragm cell process using graphite anodes in chlorine production	(T)
K106	Wastewater treatment sludge from the mercury cell process in chlorine production	(T)
K176	Baghouse filters from the production of antimony oxide, including filters from the production of intermediates (e.g., antimony metal or crude antimony oxide)	(E)
K177	Slag from the production of antimony oxide that is speculatively accumulated or disposed, including slag from	(T)

(*Continued*)

Table 3.1 (Continued)

Industry and Hazardous Waste Number	**Hazardous Waste**	**Haz. Waste Code**
	the production of intermediates (e.g., antimony metal or crude antimony oxide)	
K178	Residues from manufacturing and manufacturing-site storage of ferric chloride from acids formed during the production of titanium dioxide using the chloride-ilmenite process	(T)
Pesticides		
K031	By-product salts generated in the production of MSMA and cacodylic acid	(T)
K032	Wastewater treatment sludge from the production of chlordane	(T)
K033	Wastewater and scrub water from the chlorination of cyclopentadiene in the production of chlordane	(T)
K034	Filter solids from the filtration of hexachlorocyclopentadiene in the production of chlordane	(T)
K035	Wastewater treatment sludges generated in the production of creosote	(T)
K036	Still bottoms from toluene reclamation distillation in the production of disulfoton	(T)
K037	Wastewater treatment sludges from the production of disulfoton	(T)
K038	Wastewater from the washing and stripping of phorate production	(T)
K039	Filter cake from the filtration of diethylphosphorodithioic acid in the production of phorate	(T)
K040	Wastewater treatment sludge from the production of phorate	(T)
K041	Wastewater treatment sludge from the production of toxaphene	(T)
K042	Heavy ends or distillation residues from the distillation of tetrachlorobenzene in the production of 2,4,5-T	(T)
K043	2,6-Dichlorophenol waste from the production of 2,4-D	(T)
K097	Vacuum stripper discharge from the chlordane chlorinator in the production of chlordane	(T)
K098	Untreated process wastewater from the production of toxaphene	(T)
K099	Untreated wastewater from the production of 2,4-D	(T)
K123	Process wastewater (including supernates, filtrates, and washwaters) from the production of ethylenebisdithiocarbamic acid and its salt	(T)
K124	Reactor vent scrubber water from the production of ethylenebisdithiocarbamic acid and its salts	(C,T)
K125	Filtration, evaporation, and centrifugation solids from the production of ethylenebisdithiocarbamic acid and its salts	(T)
K126	Baghouse dust and floor sweepings in milling and packaging operations from the production or formulation of ethylenebisdithiocarbamic acid and its salts	(T)

(*Continued*)

Table 3.1 (Continued)

Industry and Hazardous Waste Number	Hazardous Waste	Haz. Waste Code
K131	Wastewater from the reactor and spent sulfuric acid from the acid dryer from the production of methyl bromide	(C,T)
K132	Spent absorbent and wastewater separator solids from the production of methyl bromide	(T)
Explosives		
K044	Wastewater treatment sludges from the manufacturing and processing of explosives	(R)
K045	Spent carbon from the treatment of wastewater containing explosives	(R)
K046	Wastewater treatment sludges from the manufacturing, formulation and loading of lead-based initiating compounds	(T)
K047	Pink/red water from TNT operations	

agencies kept the contamination from the public at large until 1993 when the discovery of a largemouth bass with blistering scales downstream of Snow Creek made headlines. After testing for PCBs, residents were finally warned against eating fish caught in the area. In 1996, testing of soil, homes, and Monsanto's drainage ditches found PCB levels 940, 200, and 2,000 times the federal limits (Grunwald, 2002). Roughly one-third of nearby residents had elevated levels of PCBs in their blood, and the state warned the increased cancer risk was high among people living near Snow Creek. Liver problems, acute toxic syndrome, learning impairment, and chloroacne had also been reported by residents living around the facility. The communities of Sweet Valley and Cobbtown near the plant were considered public health hazards (EPA, 2010).

In a 2003 settlement, Solutia/Monsanto paid $700 million to Anniston residents affected by the plant's PCB effluent. Two earlier lawsuits relating to PCBs released by the Anniston plant were settled for $83.7 million. Partly due to these and other lawsuits pertaining to environmental damages caused by corporate negligence, Solutia Inc. filed for bankruptcy in 2003. In 2008, the company emerged from bankruptcy and is now fully operational again (AP, 2003; Solutia, 2008).

(A discussion of PCBs, their sources, and health effects can be found in Chapter 15.)

3.4 Chemical Industry Hazardous Wastes

The chemicals and hazardous waste producing processes generated by the chemical industry comprise the K-listed hazardous wastes found in the Federal Resource Conservation and Recovery Act (Table 3.1).

The hazard codes in the third column are employed by administrators to signify the type of hazard that is associated with each hazardous waste: (T) Toxic, (C) Corrosive, and (I) Ignitable (EPA, 2010d).

Acknowledgment

The section 'Monsanto's Toxic PCB Legacy at Sauget, Illinois' was contributed by Jess Strange. Jess Strange is a Clinical Health Educator for La Clinica de la Raza, a Federally Qualified Health Center. She lives in Oakland, CA.

References

Agin, D. (2006). *Junk science: How politicians, corporations, and other hucksters betray us*. New York: St. Martin's Press.

Agency for Toxic Substances and Disease Registry (ATSDR) (1994, May). *Toxicological profile for chlorodibenzofurans*. US Department of Health and Human Services.

Agency for Toxic Substances & Disease Registry (ATSDR) (1997, September). *ToxFAQs for chlorpyrifos*. US Department of Health and Human Services.

Agency for Toxic Substances and Disease Registry (ATSDR) (1998, December). *Toxicological profile for chlorinated dibenzo*-p-*dioxins*. US Department of Health and Human Services.

Agency for Toxic Substances and Disease Registry (ATSDR) (1999, February). *ToxFAQs for chlorinated dibenzo-p-dioxins (CDDs)*. US Department of Health and Human Services.

Agency for Toxic Substances and Disease Registry (ATSDR) (2000, November). *Toxicological profile for polychlorinated byphenyls (PCBs)*. US Department of Health and Human Services.

Agency for Toxic Substances and Disease Registry (ATSDR). (2006). *Interaction profiles for: chlorpyrifos, lead, mercury, and methylmercury*. Public Health Service. Retrieved August from <http://www.atsdr.cdc.gov/interactionprofiles/IP-11/ip11.pdf>.

American Society of Mechanical Engineers (ASME) (2010). *#221 Brandywine river powder mills (1803-1921)*. ASME. Retrieved September 9, 2010 from <http://www.asme.org/Communities/History/Landmarks/Brandywine_River_Powder_Mills.cfm>.

Applebome, P. (2010). Old story of pollution; new urgency this time. *New York Times*. Retrieved January 31 from <http://www.nytimes.com/2010/02/01/nyregion/01towns.html>.

Associated Press (AP) (1988). U.S. is seeking to fine Du Pont $8.23 million. *New York Times*. Retrieved December 29 from <http://ucelinks.cdlib.org:8888/sfx_local?genre=article&issn=03624331&title=New+York+Times&volume=&issue=&date=19881229&atitle=U.S.+Is+Seeking+to+Fine+Du+Pont+$8.23+Million.&spage=14&sid=EBSCO:nfh&pid=>.

Associated Press (AP) (2003). $700 million settlement in Alabama PCB lawsuit. *New York Times*. Retrieved August 21 from <http://www.nytimes.com/2003/08/21/business/21PCB.html>.

Associated Press (AP) (2005). *DuPont data indicates it hid risks of chemical*. Retrieved September 9 from <http://www.msnbc.msn.com/id/10082924/>.

Associated Press (AP) (2009). *Worst industrial disaster still haunts India – 25 years after lethal Bhopal gas leak, injuries and birth defects linger*. Associated Press. Retrieved December 2 from <http://www.msnbc.msn.com/id/34247132/ns/world_news-south_and_central_asia/page/2/>.

Bent, S. (1925). Tetraethyl lead fatal to makers. *New York Times*. Retrieved June 22 from <http://proquest.umi.com/pqdweb?index=1&did=98833582&SrchMode=1&sid=7&Fmt=10&VInst=PROD&VType=PQD&RQT=309&VName=HNP&TS=1285283475&clientId=48051>.

Broughton, E. (2005). The Bhopal disaster and its aftermath: A review. *Environmental Health*, *4*(6).

Burdick, B. (2008). *Rocky flats: Dow Chemical and Rockwell International billed $925M for contamination at nuclear site*. The Huffington Post. Retrieved September 17 from <http://www.huffingtonpost.com/2008/06/03/rocky-flats-dow-chemical_n_104974.html#>.

Burnham, D. (1983). Dow says U.S. knew dioxin peril of Agent Orange. *New York Times*. Retrieved May 5 from <http://www.nytimes.com/1983/05/05/us/dow-says-us-knew-dioxin-peril-of-agent-orange.html?sec=health&spon=&pagewanted=all>.

The Center for Public Integrity (2008). *Dow chemical company profile*. <http://projects.publicintegrity.org/Superfund/Company.aspx?act=10312>.

The Chemical Engineer (TCE) (2008). Invista sues DuPont over pollution. *803*(5), 5. September 19, 2008.

Colorado Department of Public Health and Environment (CDPE) (1999). *Rocky flats public exposure studies summary of findings*. <http://www.cdphe.state.co.us/rf/historyofrf.htm>.

Delaware Department of Natural Resources and Environmental Control (DNREC) (2004). *Hay Street sludge drying site/Cherry Island landfill – iron rich staging area*. Wilmington, DE: Site Investigation and Restoration Branch. <http://www.dnrec.state.de.us/dnrec2000/Divisions/AWM/sirb/DuPontIronRichPile_files/Proposed%20Plan%20of%20Remedial%20Action.pdf>.

Delaware Department of Natural Resources and Environmental Control (DNREC) (2010). *DuPont Hay Road "Iron Rich" site*. Divisions of Air and Waste Management. Retrieved August from <http://www.awm.delaware.gov/Pages/SIRBIronRichSite.aspx>.

Dhar, S. (2010). Indian court reopens Bhopal case – will review 1996 decision. *The Washington Times*. Retrieved August 31 from <http://www.washingtontimes.com/news/2010/aug/31/indias-highest-court-reopens-bhopal-disaster-case/>.

DePalma, A. (1993). Report on trade treaty is critical of companies. *New York Times*. Retrieved September 19, 2010 from <http://select.nytimes.com/gst/abstract.html?res=F00617FE35580C7B8EDDAE0894DB494D81>.

The Dow Chemical Company (2007). *Dow sustainability – Agent Orange*. <http://www.dow.com/commitments/debates/agentorange/>.

The Dow Chemical Company (2010). *Our company – history*. <http://www.dow.com/about/aboutdow/history/1890s.htm>.

DuPont. (2002). DuPont overview. Wayback machine, internet archive. Retrieved February 22, 2002 from <http://web.archive.org/web/20011221230858/www.dupont.com/corp/overview/history/>.

DuPont Corporate Remediation Group (DuPont) (2008). *Acid Brook delta remedial investigation report DuPont Pompton Lakes works, Pompton lakes*. New Jersey. Retrieved September 16, 2010 from <http://www.epa.gov/region02/waste/dupont_pompton/7827RIRFinal-June-2008-noapps.pdf>.

Elkin, L. (1984, May 8). Agent Orange plaintiffs to receive $180 million. *Pittsburgh Post-Gazette*.

Environmental Working Group (EWG) (2009). Monsanto responds to PCBs in wildlife. *Chemical Industry Archives*. Retrieved March 27 from <http://www.chemicalindustryarchives.com/dirtysecrets/annistonindepth/wildlife.asp>.

Evers, G. R. (2007). *Public comment to Hay Road sludge drying site (AKA iron rich dioxin pile)*. Cherry Island Study. Retrieved March 2 from <http://www.awm.delaware.gov/Documents/7af20aebd7e94149920da2e72e0b8439GlennEvers.pdf>.

Fawthrop, T. (2004). Vietnam's war against Agent Orange. *BBC News*. Retrieved June 14 from <http://news.bbc.co.uk/2/hi/health/3798581.stm>.

Government of Madhya Pradesh (MP) (2010). *Bhopal gas tragedy relief and rehabilitation department, Bhopal*. Retrieved May from <http://www.mp.gov.in/bgtrrdmp/profile.htm>.

Graybow, M. (2008). Corrected – Update 3 – US Court Upholds Dismissal of Agent Orange Suit. *Reuters*. Retrieved February 25 from <http://uk.reuters.com/article/idUKN2258148920080225>.

Grunwald, M. (2002). Monsanto hid decades of pollution. *The Washington Post*. Retrieved January 1 from <http://www.washingtonpost.com/ac2/wp-dyn?pagename=article&contentId=A46648-2001Dec31>.

Global Lead Network (GLN) (2010). *Current initiatives*. Retrieved September 19 from <http://globalleadnet.com/18/current-initiatives>.

IEPA (2002, September 3). *Clean Air Act Permit Program (CAAPP) permit and title I permit*. Illinois Environmental Protection Agency, division of Air Pollution Control.

Janofsky, M. (2005). DuPont to pay $16.5 million for unreported risks. *New York Times*. Retrieved December 15 from <http://query.nytimes.com/gst/fullpage.html?res=9903E5DB1F31F936A25751C1A9639C8B63>.

Kovarik, B. (1999). *Charles F. Kettering and the 1921 Discovery of Tetraethyl Lead in the Context of Technological Alternatives*. Working Paper. <http://www4.hmc.edu:8001/Chemistry/Pb/resources/Kovarik.pdf>.

Koo, H., & Sim, W. (2006). Dow, Monsanto lose lawsuit in Korea over Agent Orange. *Bloomberg*. Retrieved January 26 from <http://www.bloomberg.com/apps/news?pid=newsarchive&sid=aqYCjwCCjIPE>.

Kumar, D. (2010). Dow AgroSciences blacklisted for bribing. *The Economic Times*. Retrieved September 18 from <http://economictimes.indiatimes.com/news/news-by-industry/indl-goods-/-svs/chem-/-fertilisers/Dow-AgroSciences-blacklisted-for-bribing/articleshow/6575870.cms>.

Kuncheria, C. J. (2010). India top court begins Bhopal gas verdict review. *Reuters*. Retrieved August 31 from <http://www.reuters.com/article/idUSSGE67U0H420100831>.

Monsanto (2010). *Company history*. Retrieved September 19 from <http://www.monsanto.com/whoweare/Pages/monsanto-history.aspx>.

Montgomery, J. (2005). It's toxic, but DuPont wants it near homes. *The News Journal*. Retrieved July 17 from <http://atkeison.org/home/issues/hazmat/Its_toxic_DuPont_IronyRich_pile_NJ_20050717.html>.

New York Times (NYT) (1924). *No peril to public seen in ethyl gas*. Retrieved November 1 from <http://proquest.umi.com/pqdweb?index=0&did=119045345& SrchMode=1&sid=8&Fmt=10&VInst=PROD&VType=PQD&RQT=309&VName=HNP&TS=1285283934&clientId=48051>.

New York Times (NYT) (1925). Scientists to pass on tetra-ethyl gas. *ProQuest Historical Newspapers*. Retrieved May 20 from <http://proquest.umi.com/pqdweb?index=0&did=98829569&SrchMode=1&sid=6&Fmt=10&VInst=PROD&VType=PQD&RQT=309&VName=HNP&TS=1284666951&clientId=48051>.

New York Times (NYT) (1976). Summary of actions taken by the United States supreme court yesterday on a wide variety of matters. *ProQuest Historical Newspapers*. Retrieved June 15 from <http://proquest.umi.com/pqdweb?index=0&did=79772719&SrchMode=1&sid=7&Fmt=10&VInst=PROD&VType=PQD&RQT=309&VName=HNP&TS=1285112016&clientId=1564>.

Patterson, C. C. (1965). Contaminated and natural lead environments of man. *Archives of Environmental Health, 11*(3), 344–360.

Pesticide Action Network (PAN) North America (2004). *Chemical trespass – pesticides in our bodies and corporate accountability*. Retrieved May from <http://pulse.pharmacy.arizona.edu/resources/fertilizers/ChemTres.pdf>.

Raina, R. K. (1985). Gas bomb at Bhopal. *Indian Journal of Pharmacology, 17*(2), 1–3.

Reyes, J. (2007). *Environmental policy as social policy? The impact of childhood lead exposure on crime*. National Bureau of Economic Research. Retrieved May from <http://www3.amherst.edu/~jwreyes/papers/LeadCrime.pdf>.

Sack, K. (2002). PCB pollution suits have day in court in Alabama. *New York Times*. Retrieved January 25 from <http://query.nytimes.com/gst/fullpage.html?res=9C04EED7143AF934A15752C0A9649C8B63>.

Schecter, A., Quynh, H. T., Pavuk, M., Papke, O., Malisch, R., & Constable, J. D. (2003). Food as a source of dioxin exposure in the residents of Bien Hoa City, Vietnam. *Journal of Occupational and Environmental Medicine, 45*(8), 781–788.

Solutia (2000, August 1). Description of current conditions – W.G. Krummrich plant, Sauget, Illinois – Draft. Solutia, Inc.

Solutia (2008). Solutia Inc. Emerges from chapter 11. Solutia Inc. Press Release. Retrieved February 28 from <http://investor.solutia.com/phoenix.zhtml?c=88803&p=irol-newsArticle&ID=1113553&>.

State of New York (2003). *Dow subsidiary to pay $2 million for making false safety claims in pesticide ads*. Office of the Attorney General. Retrieved December 15 from <http://www.ag.ny.gov/media_center/2003/dec/dec15a_03.html>.

Time (1924). Science: Tetraethyl lead. *Time Magazine*. Retrieved November 10 from <http://www.time.com/time/magazine/article/0,9171,728055,00.html>.

URS (2004, January 30). *Remedial investigation/feasibility study report, volume 1 – text, volume 2 – tables and figures*. URS Corporation.

US Public Interest Research Group (PIRG) (1999). *Top 25 'potentially responsible parties' at superfund national priority list (NPL) sites*. <http://www.pirg.org/reports/enviro/super25/page6.htm>.

US Department of Veterans Affairs (VA) (2010). *Agent Orange: Diseases associated with Agent Orange exposure*. Office of Public Health and Environmental Hazards. Retrieved September 13 from <http://www.publichealth.va.gov/exposures/agentorange/diseases.asp>.

US Environmental Protection Agency (EPA) (1976, February 25). *PCBs in the United States industrial use and environmental distribution – task 1 – final report. EPA 560/6/76-005*. United States Environmental Protection Agency, Office of Toxic Substances.

US Environmental Protection Agency (EPA) (1980, November). *Dioxins – EPA-600/2-80-197*. United States Environmental Protection Agency, Industrial Environmental Research Laboratory, Office of Research and Development.

US Environmental Protection Agency (EPA) (1989). *NPL site narrative for Rocky Flats plant (USDOE)*. Retrieved October 4 from <http://www.epa.gov/superfund/sites/npl/nar865.htm>.

US Environmental Protection Agency (EPA) (2001c). Federal Register Notice – 47612 – 47618 *Federal Register*/Vol. 66, No. 178/Thursday, September 13, 2001/Proposed Rules. <http://www.epa.gov/superfund/sites/npl/p010913.htm>. 2001.

US Environmental Protection Agency (EPA) (2001d). *Top 100 PRPs based on number of sites at which they are involved*.

US Environmental Protection Agency (EPA) (2007, July). *Statement of basis for Solutia, Inc.*, Sauget, Illinois: United States Environmental Protection Agency.

US Environmental Protection Agency (EPA) (2009). *EPA Administrator pledges strong federal cleanup presence at Dow dioxin site in Michigan and accelerated assessment of dioxins' human health impacts*. Retrieved May 26 from <http://yosemite.epa.gov/opa/admpress.nsf/d0cf6618525a9efb85257359003fb69d/3ffa6e8e70763f28852575c20064b26b!OpenDocument>.

US Environmental Protection Agency (EPA) (2010a). *The national biennial RCRA hazardous waste report (based on 2007 data)*. <http://www.epa.gov/osw/inforesources/data/br07/national07.pdf>.

US Environmental Protection Agency (EPA) (2010b). *Anniston PCB site. Region 4 superfund*. Retrieved September 13 from <http://www.epa.gov/region4/waste/npl/npl al/annpcbal.htm>.

US Environmental Protection Agency (EPA) (2010c). *DuPont/Pompton lakes works. Region 2 hazardous waste cleanups*. Retrieved August 5 from <http://www.epa.gov/region02/waste/dupont_pompton/index.html>.

US Environmental Protection Agency (EPA) (2010d). *Hazardous waste generator regulations. Version 4*. Retrieved August 2010 from <http://ecfr.gpoaccess.gov/cgi/t/text/text-idx?c=ecfr;sid=4990e762d7b81851bef18f82dc851826;rgn=div5;view=text;node=40:25.0.1.1.2;idno=40;cc=ecfr#40:25.0.1.1.2.4.1.3>.

US Environmental Protection Agency (EPA) (2010e). *Region 5 cleanup sites – background*. Retrieved June 10 from <http://www.epa.gov/region5/sites/dowchemical/background.htm>.

Warwick, H. (1998). Agent Orange: the poisoning of Vietnam. *The Ecologist, 28*(5), 264–265.

4 The United States Military

Chapter Contents

The United States military generates as much if not more hazardous waste than some of the largest industries in the country, yet has historically received immunity from hazardous waste regulation and accounting. Sites of military bases, battlefields, and coastal waters at both foreign and domestic settings have been severely contaminated due to the Department of Defense.

4.1 Range and Scope of Military Hazardous Waste

The immensity of the US military coupled with its lack of transparency makes it difficult to ascertain the exact amount of hazardous waste generated; however, we do know that it is a lot. The Defense Department is the nation's biggest polluter (Layton, 2008), releasing both hazardous wastes typical to industry as well as those unique to the military. 'Military bases are really very, very large industrial facilities or manufacturing plants that use lots of solvents and paints,' said J. Winston Porter, assistant administrator at the EPA in a 1989 article (Shulman et al., 1989). Tanks, airplanes, and weapons, for example, were cleaned with the solvent (and severe environmental pollutant) trichloroethylene (TCE) until 2005. TCE was frequently washed down the nearest drain or let seep into adjacent soils. Aviation fuel and gasoline spilled while refueling vehicles further contaminated sites with polycyclic aromatic hydrocarbons, benzene, and other toxic petroleum-based chemicals. Other not-so-typical hazardous wastes are also common to ex-military sites. Ordinance such as mortar shells, landmines, grenades, and air-dropped bombs pepper the landscape of old training grounds, and leach chemicals into the surrounding landscape. Even shell casings, containing lead, copper, and antimony, contaminate soils and

Risks of Hazardous Wastes. DOI: 10.1016/B978-1-4377-7842-7.00004-0

groundwater (Stone, 2006). Chemicals unique to military operations, such as GB nerve gas and the high-explosive RDX, also can contaminate sites and spread to aquifers. According to the military in a 1988 report, the army practiced inappropriate disposal techniques for hazardous wastes including: discharge on the ground into unlined pits ... or local creeks, pouring and spraying on the ground, drainage to industrial sewers, burning during fire protection training and storage in leaking underground tanks (Shulman et al., 1989).

4.2 Hazardous Waste from Domestic Manufacturing and Bases

The Rocky Mountain Arsenal (RMA) was established by the Army in 1942 to produce chemical weapons. The military used this site for the manufacture of mustard gas and white phosphorous during World War II, before encouraging private industries, such as Shell Chemical Company, to lease land. Shell produced agricultural pesticides at this location from 1946 to 1982, alongside the military. Due to the 'common industrial and waste disposal practices used by the Army and Shell during those years,' a vast amount of soil was contaminated (EPA, 2010). The most toxic of these disposal sites was a disposal lagoon called Basin F. When cleanup of Basin F began in 1988, 4 million gallons of heavily contaminated sludge was removed. The entire 26 square mile RMA contained an estimated 16 million cubic yards of contaminated soil (Shulman et al., 1989).

The 40 years of accumulated manufacturing by-products, spill residue, and other contaminants associated with pesticide and nerve gas production have left their mark. The most dangerous chemicals on site, as determined by the EPA, are aldrin, dieldrin, dibromochloro-propane (DBCP), and arsenic. Aldrin (parent pesticide) and dieldrin (breakdown product) are bioaccumulative and affect the CNS and liver in humans. DBCP affects the testes, kidneys, liver, respiratory system, central nervous system, and blood cells. Arsenic is a known carcinogen in humans. Other sources, such as leaking fuel storage tanks, have added BTX (benzene, toluene, xylene), bicycloheptadiene, and dicyclopentadiene to the toxic cocktail. By 1989 the contaminants had infiltrated the groundwater and spread to an area roughly the same size as the RMA itself (Shulman et al., 1989). The RMA is just one out of 141 Superfund sites owned by the Pentagon – the most sites owned by any single entity (Layton, 2008; Swanson, 2010).

Despite the staggering number of Department of Defense (DOD) sites on the Superfund list, the Pentagon still fights the EPA's orders to clean up sites under the DOD's control. Under the 1986 SARA amendments to CERCLA, federally owned facilities, including those managed by the DOD, became eligible for Superfund status and EPA mandated cleanups. Some branches of the military have responded to these remediation orders better than others. The US Navy has signed cleanup agreements for all sites under its control, whereas the US Air Force has not signed a single agreement in 14 years (Layton, 2008). As of 2009, the DOD has refused to sign cleanup authorizations for 11 contaminated sites (GAO, 2010). According to the GAO, the DOD has obstructed CERCLA implementation by the EPA in a number of ways, specifically at three of these 11 sites: Fort Meade Army Base, McGuire Air Force Base, and Tyndall Air Force Base. First, the DOD has not signed inter-agency agreements with the EPA. These agreements stipulate that the DOD would have to allot funding specifically toward remediation of the three sites. Second, the DOD did not produce contamination information in a timely fashion, leading to an increase in human health risk. For example, an area littered with lead shot was turned into a child's playground, despite DOD knowledge of contamination. Last, the DOD imposes price caps and deadlines on remediation contracts, leading to rushed and less than ideal solutions to contamination problems.

4.3 Dumping of Munitions in the Ocean

4.3.1 History and Scope of Marine Munitions Disposal by the US Military

The US military is one of the largest producers of hazardous waste in the nation, if not the world. From the chemical warfare agents produced in World War I, and (unused but produced) in World War II, to the Vietnam-era Agent Orange and radioactive waste, the military has dumped its surplus explosives, chemical warfare agents, and even radioactive wastes into the world's oceans. The DOD maintained few records of these disposals and even those were kept classified from the public until 2005. No one, not even the military, has a firm idea of where many of these dumpsites are.

The Army began dumping hazardous materials in World War I. Official policy at that time was to toss military hazardous waste, such as old munitions and poison gas canisters, over the side. Few, if any records were kept of these haphazard and unofficial disposals. As such, no one really has any idea of the extent to which hazardous materials were jettisoned out to sea during this pre-1944 timeframe. In 1944 the Army formalized the disposal process and began its secretive dumping program that it continued until 1970. It was between 1944 and 1970 that the largest amount of hazardous waste was fed into the ocean. In those mere 26 years, 64 million pounds of nerve/mustard agents, 400,000 chemical-filled bombs, landmines, rockets, and over 500 tons of radioactive waste were dumped off US coastlines alone. Hazardous waste dumpsites spanned the coasts of 11 states, including Hawaii, California, Alaska, the Gulf Coast, and those on the Eastern Seaboard. Of the 26 dumpsites recorded by the Army (far more were not recorded), only a few have ever been inspected post-'disposal' by the military. None have been inspected in the past 30 years.

One of the largest and first documented sites began operation in March 1946. The ammunition ship USS *Diamond Head* released four railroad cars of mustard gas, bombs, and mines into the waters off the coast of Charleston, SC, at a location designated 'Disposal Site Baker.' In the months to follow, 23 barges, each carrying up to 350 tons of German nerve gas bombs and American lewisite (similar to mustard gas), are reported to have tossed their cargo in the surrounding area. There is strong evidence, however, that much of this cargo was released by barges prior to reaching Disposal Site Baker, which was a 'safe' depth and distance away from the mainland. Army records show that the barges only took one day to leave port, unload their cargo, and return, indicating a discharge point quite close to the coastline and in shallow water.

Official military regulations in WWI required dumping in at least 600 ft of water. In 1944, those regulations had changed to 6,000 feet; however, evidence in addition to that associated with Disposal Site Baker exists that indicates ship captains frequently flouted these requirements. Sometimes, instances of leaking ordinance forced captains to prematurely discharge hazardous waste en-route to the disposal site. Other times simply the danger of having such waste aboard prompted wary captains to order dumping far before they reached water of 'adequate' depth. Investigators working for Sea Watch International found that munitions were dumped in as little as 130 feet of water. The dumping of hazardous waste in shallow waters greatly increases the chances of human exposure and harm.

4.3.2 The Effects of Dumping: Past Harms and Future Threats

The effects of the dumping have rarely been studied, largely due to the fact that the military kept disposal information classified until 2005. Since declassification scientists have begun to link once inexplicable disturbances in marine ecosystems to the effects of underwater hazardous waste dumpsites. In 1987, several hundred bottlenose dolphins washed ashore in Virginia and New Jersey with large blisters and lesions, similar to those found on humans exposed to mustard gas. Marine biologists at the time could not sufficiently explain the cause of the wounds. After examining the declassified information released by the military, biologists now say that 'it is a good possibility' that the burns were caused by mustard gas from offshore dumpsites. Some have also attributed the decline in benthic sea life on the Eastern Seaboard to the military's wanton dumping of hazardous waste. Inside the once famed fishing ground known as the Grand Banks, crabs with inexplicable mutations and the large-scale deaths of cod larvae have befuddled scientists for some time. Since 2005, however, researchers have discovered the location of a large Canadian/American military dumpsite in the area, and have labeled it as the possible cause of the mutations and mortality.

The military's failure to accurately record the location of submerged dumpsites can also pose unexpected risks to humans. Some residents along the Eastern Seaboard, where a majority of dumping took place, have reported decades-old munitions appearing in their driveways. While the connection between driveways and hazardous ordinance at first seems vague, a clear link can be made when it is understood that clamshells dredged from the sea floor are commonly used in the region as cheap filler for driveways. Occasionally clamshell dredging operations in the area mine clandestine underwater dumpsites, and unknowingly deliver filler studded with antique ordinance to clients. In 2005 alone, homeowners in the

Delmarva Peninsula, NJ, reported finding a total of 318 pieces of ordinance in their driveways. Live hand grenades and shells filled with volatile mustard gas were among the most hazardous objects recovered. Despite spending the better part of a century underwater, these driveway munitions still pose a threat if the canister is still sealed. In 2005, three members of an Air Force bomb disposal squad were burned when they opened an unmarked 75 mm shell containing mustard agent that was partially embedded in a driveway.

Nerve and mustard gas, some of the chemicals common to the older dumpsites, were designed to kill in the trenches of France in WWI, but now haunt the ocean floor adjacent to the East Coast. Mustard agent when released upon the ocean floor is insoluble and congeals, forming a gel that traverses the sea floor for up to 5 years, killing all sea life it encounters. Nerve gas, a drop of which can kill in under a minute, can last underwater for up to 6 weeks. If the agent is contained within a sealed environment, such as in an artillery shell or storage canister, it will remain in its original hazardous state. Because the canisters containing mustard and nerve gas corrode at different rates depending on water temperature, pressure, condition of the container etc., the agent is released intermittently, thus posing a continuing threat to the benthic environment. Chlorine, phosgene, arsenic, sarin, tabun, lewisite (similar to mustard gas), and other deadly chemical agents were also dumped during this time and have contaminated the seabed. Conventional ordinance containing high amounts of trinitrotoluene (TNT) present an 'extremely toxic' threat to marine organisms (Beddington et al., 2005). These conventional munitions have been shown to spontaneously combust and bioaccumulate – presenting another type of threat to marine life.

Beddington et al. describe three ways disposed ordinance impacts marine life: direct contact, bioaccumulation, and spontaneous combustion. Direct contact with lethal agents such as nerve and mustard agents maim and kill marine life in the same way that the weapons were originally intended to kill humans. If a mustard agent touches an animal's skin, it will cause horrendous blistering and cancers. Nerve agents, as said before, can kill within a minute upon dermal absorption. Bioaccumulation, the phenomenon where a persistent toxin builds up in individuals as it moves up the food chain, is a hazard that only increases with time. A Norwegian study in 2002, which examined a British/American dumpsite from 1944, found that soil sediments contained high levels of arsenic – an element that bioaccumulates, posing a possible threat to humans at the top of the food chain. TNT exhibits a bioaccululative effect and is incredibly common in conventional munitions and therefore marine dumpsites. Probably the most directly deleterious problem with ordinance disposal is that of spontaneous combustion. In 1964 the Navy implemented the munitions disposal operation CHASE, short for *Cut Holes And Sink-Em*. As part of the operation, 15 ships between 1964 and 1970 were sunk off the coast of the US, and with them over 31 million pounds of high explosives. Of the 15 ships, seven were blown to pieces on the surface of the water and seven were flooded, sank, and on their way to the ocean floor exploded due to the pressure exerted upon the munitions by the surrounding water. These immense explosions created pressure waves so extreme that seismic sensors, typically used to pick up earthquakes, registered the vibrations far inland. These underwater pressure waves easily killed marine life in the immediate vicinity, but also scattered unexploded ordinance far from the ship. This scattering spread the munitions over a large area of the sea floor and further increased the chances of marine life encountering the harmful compounds.

Despite the ban in 1975, chemical and conventional ordinance still pose hazards in the future. For example, the SS *William Ralston* was scuttled with radioactive waste aboard 117 miles off the coast of San Francisco in 1957. Since then, scientists have been lackadaisically monitoring the site, and have reported no leaks. However, the drums containing the waste have corroded so over the past 50 years that they are now 'paper thin,' some even with holes on the sides. The Norwegian researchers also reported varying degrees of container corrosion by seawater. Some of the canisters had gaping holes, and had released their cargo decades ago; others had dangerously thin shells, while still others remained undamaged. These undamaged and partly corroded shells pose a future threat of release.

The US Army has reported 30 chemical weapons dumpsites outside of United States waters that were created before it signed a treaty in 1975 prohibiting the oceanic dumping of conventional munitions and chemical ordinance. Despite the ban, dumpsites have still caused injury to mariners around the world. Since World War II, encounters with chemical dumpsites have caused the injury of 150 Danish, 52 Japanese and 232 Italian fishermen, among other cases. The most notorious overseas dumping program was titled *Operation Davy Jones Locker*. In the operation, British and US governments loaded numerous German ships with nerve and mustard agents and scuttled them in the shallow waters of the Baltic Sea. Roughly 170,000 tons of Nazi chemical weapons went down with the ships. According to Danish authorities, 150 fishermen have been 'hurt,' by discarded American or British chemical agents in the Baltic. Discovery of chemical ordinance is common enough that Danish fishermen who know they will be trawling in certain waters near old dumpsites don chemical protection suits. Gas masks are a standard safety measure for fishing ships in the Baltic.

4.4 Improper Disposal of Hazardous Waste in US Military Operations Abroad

The United States military entered Iraq in 2003 with a massive force of troops, tanks, ships, and planes. Seven years later, the last of the combat troops have left; however, the seven years of accumulated waste will stay in Iraq. Unexploded ordinance, cleaning solvents, oil and petroleum residue, and radioactive depleted uranium projectiles will haunt the 300 or so former US military bases for some time. There is no official policy regarding environmental cleanup for military operations. If fact, there are no existing laws, US law, international law, or otherwise, which govern the management of hazardous wastes in contingency operations such as the Iraq war. In war, environmental considerations are hardly the military's top priority. According to US Central Command guidelines, 'during combat operations, environmental considerations will be subordinate to mission accomplishment and preservation of human life' (Maron, 2010).

Disposal practices by the military have hardly been up to domestic standards. On the main roads connecting Baghdad, Fallujah, and Mosul, where fighting was most intense, hazardous waste barrels, half-spent aerosol cans, oil filters, and unmarked drums of liquid litter the shoulder. Reports have also surfaced of contractors, hired by the military, mixing hazardous waste with scrap metal for distribution to Iraqi dealers. Consequently, Iraqi scrap yard workers have complained of coughing, rashes, and illness after handling US scrap contaminated with hazardous waste (August, 2010).

The amount of hazardous waste generated by the combat and occupation forces is not trivial. Estimates vary: in an article by *The Times* in 2010, a private contractor working in Iraq estimates US forces generated 11 million pounds of hazardous waste during their occupation of Iraq; according to Brigadier General Kendall Cox – the man responsible for creating and maintaining Iraq's infrastructure – over 30 million pounds of soil alone has been contaminated with oil (August, 2010). Both of these values fail to include unexploded ordinance, or spent munitions containing depleted uranium.

According to US military officials, 3–5% (up to 15% in soft sand) of bombs, rockets, and shells used do not detonate upon impact (Maron, 2010). These undetonated munitions may detonate if disturbed, creating an obvious hazard. Depleted uranium (DU), an extremely dense material (1.7 times more dense than lead), is used by the US military for armor plating and armor-piercing projectiles. When a DU-tipped projectile penetrates the armor of a tank, for example, it enters the interior of the vehicle and explodes in a flurry of conflagrating vapors. Once extinguished, the DU fragments and residue settle in a toxic and mildly radioactive dust. Numerous scientific studies have linked DU to ill-health effects such as genotoxic effects (Miller et al., 2004), 'neurophysiological perturbations' (Lestaevel et al., 2005; McDiarmid et al., 1999; Scuidi et al., 2009), and birth defects (Kundt et al., 2009). During the first three weeks of the Iraq war in 2003, the US military is estimated to have used 2 million to 4.5 million pounds of DU munitions.

References

August, O. (2010). America leaves Iraq a toxic legacy of dumped hazardous materials. *The Times*. Retrieved June 14 from <http://www.timesonline.co.uk/tol/news/world/iraq/article7149611.ece/>.

Beddington, J., & Kinloch, A. J. (2005, June). *Munitions dumped at sea: A literature review* (June). London: Imperial College, Consultants.

Bull, M. R. J. (2005a). *Special report, Part 1: The deadliness below. Daily Press*. Retrieved October 30 from <http://www.dailypress.com/news/dp-02761sy0oct30,0,2199000.story/>.

Bull, M. R. J. (2005b). *Special Report, Part 2: The deadliness below. Daily Press*. Retrieved October 31 from <http://www.dailypress.com/news/dp-02761sy0oct30,0,2199000.story/>.

Bull, M. R. J. (2005c). *What's Off Virginia? 'There's No Guarantee'. Daily Press*. Retrieved October 30 from <dailypress.com/news/dp-09106sy0oct30,0,2395608.story/>.

Kundt, M. S., Martinez-Taibo, C., Muhlmann, M. C., & Furnari, J. C. (2009). *Uranium in drinking water: Effects on mouse oocyte quality*. Health Physics Society.

Layton, L. (2008). Pentagon fights EPA on pollution cleanup. *Washington Post*. Retrieved June 30 from <http://www.washingtonpost.com/wp-dyn/content/article/2008/06/29/AR2008062901977.html?sid=ST2008063001979/>.

Lestaevel, P., Houpert, P., Bussy, C., Dhieux, B., Gourmelon, P., & Paquet, F. (2005). The brain is a target organ after acute exposure to depleted uranium. *Toxicology*, *212*, 219–226.

Maron, D. F. (2010). Pentagon weighs cleanups as it plans Iraq exit. *New York Times*. Retrieved January 13 from <http://www.nytimes.com/gwire/2010/01/13/13greenwire-pentagon-weighs-cleanups-as-it-plans-iraq-exit-21915.html?pagewanted=1&sq=toxic%20waste%20iraq&st=cse&scp=1/>.

Miller, A. C., Brooks, K., Smith, J., & Page, N. (2004). Effect of the militarily-relevant heavy metals, depleted uranium and heavy metal tungsten-alloy on gene expression in human liver carcinoma cells (HepG2). *Molecular and Cellular Biochemistry*, *255*, 247–256.

Shulman, S., & Pollack, S. (1989, March). Toxic responsibility: The U.S. military's problems of toxic waste disposal are as great as those of industry. The Atlantic.

Souidi, M., Tissandie, E., Racine, R., Ben Soussan, H., Rouas, C., & Grignard, E., et al. (2009). Uranium: Properties and biological effects after internal contamination. *Annales de Biologie Clinique*, *67*(1), 23–28.

Stone, L. (2006). *Environmental justice at military superfund sites*. The Fort Ord Environmental Justice Network. <http://www.docstoc.com/docs/610840/Environmental-Justice-at-Military-Superfund-Sites/>.

Swanson, N. (2010). *GAO: Measuring success at military superfund sites is super confusing*. <http://www.govexec.com/dailyfed/0810/081710n1.htm/>.

US Environmental Protection Agency (EPA) (2010, July). Rocky Mountain Arsenal. Superfund Program. <http://www.epa.gov/region8/superfund/co/rkymtnarsenal/>.

US Government Accountability Office (GAO) (2010, July). SUPERFUND: Interagency Agreements and Improved Project Management Needed to Achieve Cleanup Progress at Key Defense Installations, GAO-10-348. Report to Congressional Requesters.

5 The Petroleum Industry

Chapter Contents

5.1 Overview, Emissions and Waste

5.1.1 Overview

The petroleum industry refines crude petroleum and processes natural gas into a multitude of products. It is also involved in the distribution and marketing of petroleum-derived products. Processes include: oil- and gas-field operations, gas plant processing operations, refining and refinery operations, and refining technologies. Due to these activities, pollutants can be produced in the form of toxic emissions and hazardous waste that are released into the environment. Many environmental factors depend largely on the conditions and operations of the facilities themselves. Pollution caused by this industry relies on design and operating practices; frequency of maintenance and inspection; type, age, and quality of equipment; operating

Risks of Hazardous Wastes. DOI: 10.1016/B978-1-4377-7842-7.00005-2

conditions; treatment and processing equipments; and applicable environmental and conservation regulations. Additionally, various accident and equipment failures can further contaminate the environment; possibilities include well blowouts, pipeline breaks, tanker accidents, and tank explosions.

Historically the industry sector has not acted responsibly towards environmental management. Later sections document poor practices that have stemmed from both unintentional as well as intentional actions. As a result, such actions have placed the public at risk from both chronic and acute exposures to various toxic chemicals, including significant amounts of carcinogens like benzene.

5.1.2 Emissions

Within the petroleum industry, the primary family of pollutants emitted from these activities is volatile organic compounds (VOCs) arising from leakage, venting, and the evaporation of raw materials and finished products. However, other significant air pollutants include sulfur oxides, hydrogen sulfide, particulate matter, and various toxic chemicals. The operations within a typical refinery emit a wide variety of criteria pollutants and toxic chemicals from fuel combustion devices. Oil and gas field operations as well as gas processing plants are also significant sources of emissions. As such, air emissions consist of point, fugitive, and area sources. Fugitive emissions are the emissions from equipment leaks, such as from valves, storage tanks, and various support equipment. According to the US Environmental Protection Agency (EPA), over 50% of all VOC and toxic air emissions from refineries are fugitive emissions.

A problem with the industry sector is the lack of a systematic and transparent approach to the quantification and reporting of air emissions. A study by the Environmental Integrity Group (EIG) highlights that under-reporting of emissions occurs because most air pollution is now estimated and not actually monitored in the US. To make matters worse, the estimates are prepared by the facilities that generate pollution. This is a conflict of interest because such facilities have financial incentives to keep emissions reporting figures low. In fact, a report to Congress (Waxman Report, 1999) showed that for valves alone the average oil refinery under-reports fugitive emissions by nearly a factor of five, and that unreported fugitive emissions add 'over 80 million pounds of VOCs and over 15 million pounds of toxic pollutants' every year.

Emissions from the petroleum industry consist of a multitude of different compounds that pose a risk to human health, including known carcinogens like benzene and butadiene. During heat process streams employed by refineries, for instance, incomplete combustion can lead to significant emissions of sulfur oxides (SO_x), nitrogen oxides (NO_x), carbon monoxide (CO), particulates, and hydrocarbon emissions. Air emissions are the largest source of untreated wastes released into the environment. In order to reduce the amount of pollution emitted, and its impact on public health, the San Luis Obispo County Air Pollution Control District guidelines recommend higher inspection frequencies, equipment repairs, the use of accurate test methods, and up-to-date recordkeeping.

5.1.3 Waste

Solid wastes are generated from the refining process, petroleum handling operations, and wastewater treatment. Both hazardous and non-hazardous wastes are generated. Practices that aim to reduce the loadings to processing works are preferable for cleaner production because they reduce volumes of sludge and other wastes that require containment and costly disposal.

Refineries generate solid wastes and sludges (ranging from 3–5 kg per ton of crude processed), 80% of which may be considered hazardous because of the presence of toxic organics and heavy metals. Accidental discharges of large quantities of pollutants can occur as a result of abnormal operation in a refinery and potentially pose a major local environmental hazard. Treatment of these wastes includes incineration, land treating off-site, land filling on-site, land filling off-site, chemical fixation, neutralization, and other treatment methods.

Wastewater treatment processes produce acid sludge, which varies with the stock treated and the conditions of treatment. The sludge may vary from a low-viscosity liquid to a solid. Methods of disposal of this sludge are varied but can

Table 5.1 Solid Wastes Generated in Petroleum Refining Listed as Hazardous Waste by the EPA (EPA, 2010c)

Industry and EPA Hazardous Waste No.	Hazardous Waste	Hazard Code
Petroleum Refining		
K048	Dissolved air flotation (DAF) float from the petroleum refining industry	(T)
K049	Slop oil emulsion solids from the petroleum refining industry	(T)
K050	Heat exchanger bundle cleaning sludge from the petroleum refining industry	(T)
K051	API separator sludge from the petroleum refining industry	(T)
K052	Tank bottoms (leaded) from the petroleum refining industry	(T)
K169	Crude oil storage tank sediment from petroleum refining operations	(T)
K170	Clarified slurry oil tank sediment and/or in-line filter/separation solids from petroleum refining operations	(T)
K171	Spent hydrotreating catalyst from petroleum refining operations, including guard beds used to desulfurize feeds to other catalytic reactors (this listing does not include inert support media)	(I,T)
K172	Spent hydrorefining catalyst from petroleum refining operations, including guard beds used to desulfurize feeds to other catalytic reactors (this listing does not include inert support media)	(I,T)

include: disposal by burning as fuel; dumping in the ground; processing to product by-products such as ammonium sulfate, metallic sulfates, oils, tars, and other materials; and processing for recovery of acid. However, the burning of sludge results in discharge to the atmosphere of excessive amounts of sulfur dioxide and sulfur trioxide from furnace stacks. If the sludge is solid or semi-solid it may be buried in specially constructed pits, but this method risks the problem of acid leaching out to adjacent waters.

EPA's so-called K-list details source-specific wastes from particular industries, such as petroleum refining, iron and steel, and coking. Sludges and wastewaters from treatment and production processes in these sectors are examples of source-specific hazardous wastes. Table 5.1 lists EPA-recognized solid hazardous wastes from the petroleum refining industry.

The hazard codes in the third column are employed by administrators to signify the type of hazard that is associated with each hazardous waste: (T) Toxic, (C) Corrosive, and (I) Ignitable. From the table, you can see that most hazardous wastes from the industry are considered Toxic Wastes.

Various practices can be used to reduce the amount of waste solids produced, including: segregating rainwater runoff in storm sewers from refinery process streams; identifying benzene sources and installing upstream water treatment; controlling the amount of solids entering sewers (e.g. by street sweeping or relining sewers); and training programs to prevent leaks and spills.

5.2 Refinery Workers Studies

Modern-day refineries produce a variety of products including many required as feedstock for the petrochemical industry. The evolution of petroleum refining from simple distillation to today's complex processes has been one of many remarkable technological achievements, but it has brought along devastating effects on the environment and communities that live near and work within the industry. Petroleum workers involved in producing crude oil are exposed to a wide range of carcinogenic agents, including benzene, asbestos, silica dust, various types of organic solvents including chlorinated hydrocarbons, volatile sulfur compounds, and combustion by-products containing polycyclic aromatic hydrocarbons (PAH) (Kirkeleit et al., 2009).

One study (Sorahan, 2007) investigated the correlation between oil refinery workers and cancer risk. Standardized mortality rates were significantly elevated in oil refinery workers for cancer of the pleura (mesothelioma) and melanoma. High levels of carcinogens have been detected from petroleum and chemical refinery air emissions, which can be caused by improper production processes, poor maintenance practices, and internal operational problems (Iyer et al., 2009).

Particularly during the course of gas leaks, petroleum emissions can be very lethal. In Iyer's (2009) assessment on refinery air pollution, petrochemical air sampling measurements identified 30 toxic chemicals that were discharged into the atmosphere. Chemicals like hydrogen sulfide, bromo methane, benzene, toluene, h-hexane, and methylbenzene were found above safe limits. These emissions pose severe health problems to both refinery workers and the public. Chronic and acute diseases can include ulcer, lung cancer, brain damage, premature death, and liver and kidney problems. Toxic effects are produced by prolonged contact with air-borne or liquid petrochemical carcinogenic compounds even in small quantities.

5.3 The Baton Rouge Refinery: Cancer Alley

Baton Rouge is connected to New Orleans by an industrial corridor along the Mississippi River dotted with chemical plants notorious for their emissions. The term 'Cancer Alley' was coined in 1987 after the elevated incidence of cancers in neighboring communities along the strip. Today, there are 138 different companies located in Cancer Alley that consist of all the major corporations in the petrochemical industry including Texaco, Chevron, Dupont, Borden, and Dow.

In 1997, Louisiana ranked second in the nation, behind Texas, in the amount of toxic substances released into the air and water: 183 million tons of toxic chemicals were emitted that year alone (Berry, 2001). Between 1958 and 1973, Dow Corporation buried 46,000 tons of toxic waste in unlined pits that now cover more than 30 underground acres (Berry, 2001). While the company was attempting to pump the waste back before it reached aquifers used for drinking water, every time the Mississippi River flooded the waste site toxins were carried into the river and poisoned the water supply.

Consequently, residents who are exposed to such pollutants are at great risk for health problems, including cancer and respiratory disease. High levels of vinyl chloride were detected in 2001 in the community's water supply (Sissell, 2001), a substance that is highly toxic and causes a rare form of liver cancer. In 2002 Louisiana had the second highest death rate from cancer in the US.

Governor Mike Foster's administration has supported the development of these corporations as a means of perpetuating Louisiana's chemically dependent development. However, lower-income families who reside in the region are essentially forced to work in these areas as well. Due to the large population of African Americans, many believe this area faces environmental racism, the targeting of minority and low-income communities for undesirable facilities.

In Convent, Louisiana, the population of residents within a 4-mile radius of the plant site is 84% African American. The per capita income in Convent is $7,635 and 40% of the population is below the poverty level. In Mossville, residents were found to have three times more dioxin in their blood than the average US citizen (Garchar, 2004). Dioxins have been characterized by the EPA as likely to be human carcinogens and are dangerous to human health because they are persistent and bioaccumulated (EPA, 2010b).

Cancer Alley has been an issue of major controversy. It is an example of how the sustainability of small communities, especially minority communities, can be seriously threatened by the current manufacturing practices of large industrial facilities. Toxic emissions into the air, water, and land often curtail the quality and quantity of life in such communities (Blodgett, 2006). Citizens have begun to take action to stop additional pollution from affecting their community. In 1996, residents were able to stop Shintech Corporation from building a poly-vinyl chloride (PVC) manufacturing plant in Convent.

5.4 Methyl Tertiary Butyl Ether (MTBE)

Methyl tertiary butyl ether (MTBE) is a synthetic chemical compound that is manufactured through the chemical reaction of methanol and isobutylene. It is one of a group of chemicals commonly known as 'oxygenates' because it raises the oxygen content of gasoline, which helps prevent the engine from 'knocking' (ATSDR, 2007). At room temperature, MTBE is a volatile, flammable, and colorless liquid that dissolves easily in water. It has been produced in very large quantities and is almost exclusively used as a fuel additive in unleaded gasoline in the US and throughout the world.

There has been significant controversy over the environmental threat from MTBE. MTBE may be released into the environment wherever gasoline is stored; however, leakage of underground fuel tanks and pipelines, and fuel spillage, account for the largest amount. Opportunities for spillage occur whenever fuel is transported or transferred as well. Because MTBE has a low octanol–water partition coefficient (meaning it is attracted to water) compared to other organic gasoline components (e.g. benzene, toluene, ethylbenzene, and xylene – known collectively as BTEX compounds), MTBE will travel rapidly through soil into groundwater (Hinck, 2001). Likewise, gasoline containing MTBE that leaks, spills, or is otherwise released into the environment will quickly reach the water table and can contaminate wells that draw from the affected underground aquifers. In contrast, BTEX compounds tend to bind to soil and migrate slowly to the water.

MTBE has been detected in water as well as the air since it quickly evaporates from open containers and surface water. The ATSDR (2007) has reported that small amounts of MTBE may dissolve in water when MTBE levels in the water are below those found in the air, and this may travel to underground aquifers. In the Santa Monica area, groundwater used for drinking has been reported to contain up to 610 parts per billion (ppb) MTBE concentrations (EPA, 2008).

Not readily biodegradable underground, MTBE released into groundwater has been shown to linger for many decades. As such, in 1999, the US Geological Survey detected MTBE in 21% of groundwater samples taken from areas chosen specifically because there were no records that gasoline had ever leaked or been spilled there (Hinck, 2001).

5.5 Oil Fields Across America and Damage Done

Schlumberger World Energy Atlas lists more than 40,000 oil and gas fields of varying sizes throughout the world. Approximately 94% of known oil is concentrated in fewer than 1,500 giant and major fields (Ivanhoe and Leckie, 1993). In fact, the largest discovered conventional oil field is the Ghawar Field in Saudi Arabia, accounting for 65% of all Saudi oil produced between 1948 and 2000 (Porter, 2005).

Oil and gas fields are characterized by the geological structure of the field, as well as by the quality and composition of the production streams. Depending on the set of conditions, different and sometimes unique recovery processes are employed. Discoveries of new oil and gas reserves generally require drilling of very deep wells. In many oil reservoirs the naturally occurring pressure is sufficient to force the crude oil to the surface of the well. In situations where the oil reservoir pressure is not sufficient to ensure the desired level of production, pumping systems may be employed. However, many environmental problems can be associated with oil fields.

5.5.1 The Santa Maria Oil Sumps

One significant case study occurred at an oil field in Santa Maria, a coastal city in Santa Barbara County. During the 1950s, large oil-well sumps were built to collect by-products of drilling, including water, drilling mud, and oil. When the oil field started being decommissioned, the oil company removed the oil and covered the sumps with 1–4 feet of clean soil (Santa Barbara County Fire Department, 2006). Without first being decontaminated, the land was taken over by houses, agriculture, and industry. As a result, many residents in Santa Maria lived on top of the oil sumps and were exposed to petroleum waste chemicals. It was not until the turn of the century that cleanup of the sumps was instigated.

Dahlgren et al. (2007) show that homes built on decommissioned oil sumps have higher ambient air and dust concentrations of benzene, xylene, toluene, mercury, and polycyclic aromatic hydrocarbons. The study also shows that people exposed to oil-contaminated soil are ten times more likely to develop systemic lupus erythematosus. Lupus is an autoimmune disease that affects the joints, skin, heart, lungs, brain, and kidneys. Furthermore, Dahlgren found that people living near oil sumps had higher levels of serum calcium, indicating endocrine malfunction and higher levels of mercury, which negatively affects the immune system.

Residents of Santa Maria have suffered from negative health effects from the chemicals in the contaminated soil, homes have been destroyed as the contaminated soil is finally being remediated, and agriculture continues to be grown on the decommissioned oil sumps. In June 2006, 17 residents of Sunrise Hills, a neighborhood in Santa Maria, filed civil lawsuits against the parties responsible – Unocal Corporation (now Chevron Corporation), Kerr-McGee Corporation, and Conoco Phillips. The plaintiffs contend that the corporations contaminated the soil and did not clean it properly before selling the property for the Sunrise Hills residential development (Yale, 2008).

5.5.2 The Santa Barbara Oil Spill of 1969

In 1969, a Union Oil of California drilling rig called Platform A (or alpha), located offshore of Santa Barbara, was extracting pipe from a 3,500 foot deep well. Due to a pressure increase from pumping that was not sufficiently accounted for, extreme pressure built up below the ocean floor resulting in a burst of natural gas from the hole of the pipe. The intensity of the release caused five cracks in the sea floor around the drill casings along an east–west fault, releasing a large volume of oil and natural gas from deep beneath the earth (Clarke and Hemphill, 2002; SBWCN, 2008). A casing is a safety device which reinforces the well to prevent blowouts. Over the next 11 days, there was anywhere from 200,000 (SBWCN, 2008) to 3 million gallons (Clarke and Hemphill, 2002) of crude oil spread over 800 square miles. Thick tar coated beaches from Rincon Point to Goleta, damaging 35 miles of coastline.

Two main players were acknowledged to be responsible for the 1969 oil spill: Union Oil of California and the United States Department of the Interior, specifically the US Geological Survey (USGS). The USGS specified a particular length of casing on the pipe used by Union Oil on Platform A. However, Union Oil was granted a waiver on the length of the casing for this well and used a shorter casing that was below federal and State of California guidelines, which resulted in fragmentation of the wellhead (Clarke and Hemphill, 2002).

Oil along the coast of Santa Barbara was at some points 6 inches thick, muting the waves on the beach and producing a stench of petroleum described as 'inescapable' (SBWCN, 2008). Marine animals were coated with oil, while others ingested it, resulting in poisoning and suffocation. The 1969 oil spill spurred many changes in legislation. For instance, the National Environmental Policy Act (NEPA) was enacted in 1970, requiring all federal agencies to prepare Environmental Assessments and Environmental Impact Statements to consider the environmental implications of their proposed actions (CSBPD, 2005). That same year the Environmental Protection Agency was created, followed by the establishment of the California Coastal Commission in 1972.

5.5.3 Exxon Valdez *Oil Spill*

On March 24, 1989, the *Exxon Valdez* ran aground on a large but newly formed ice shelf in the Prince William Sound. The accident spilled 11 million gallons of oil into the ocean. Over the next 3 days, oil spread across 1,300 miles from Alaska to northern Washington, resulting in widespread damage to the ecosystem. The world's largest estuary would remain permanently damaged by the technological disaster (Paine et al., 1996).

A number of different factors led to this incident. According to testimony before the National Transportation Safety Board, the captain of the *Exxon Valdez* had been drinking alcoholic beverages in town on the day of the accident as well as on the ship (State of Alaska, 1990). His alcohol intake suggested impairment of his ability to react to changing conditions in the volatile Arctic climate (Paine et al., 1996). However, because there was no monitoring system of the captain's behaviors, this mistake also represents the gradual degradation of oversight and safety practices aboard the *Exxon Valdez* (State of Alaska, 1990).

Despite the obvious error, Exxon withheld admission of guilt. For 3 days, the oil spilled while the government, local industry, and Exxon battled over responsibility.

The heightened awareness of future fines and liability lessened cooperation and postponed intervention until government order initiated the process. The delay proved costly. The resultant damage cost $5 billion in lost passive use, over $3 billion in cleanup, and $287 million in lost wages to the local fishing industry (Paine et al., 1996; Duffield, 1997; Hayes and Michel, 1999; Picou, 2000; Carson et al., 2008).

Upon exposure to air, water, and bacteria, the crude oil rapidly began to degrade into harmful by-products. Toxic chemicals found in the water and soil included sulfides, sulfur, benzene, phenanthrene, naphthalene, dioxygenase, and polycyclic aromatic compounds. These chemicals may decrease tissue oxygenation, depress central nervous system function and induce liver, lung, eye, and reproductive abnormalities.

Human exposure was limited by rapid response to cease ingestion of marred animals or polluted water as well as limitations on exposure to the chemicals during cleaning. However, documented reports indicate that depression and social disruption also increased significantly in wake of the Exxon tragedy. A review of Alaska's health problems for the 10-year period following the spill indicates an increase in many health problems, including a 25% increase in cancer rates (Palinkas et al., 1993; Sullivan and Krieger, 2001).

5.6 Citgo's Spill in Lake Charles and Criminal Charges

5.6.1 The Event

On June 19, 2006, several problems at Citgo Petroleum's Lake Charles Manufacturing Complex resulted in air emissions and an oil spill into the Calcasieu River in Louisiana. During a large storm event, lack of adequate preparation resulted in toxic sulfur dioxide emissions and the release of large quantities of slop oil.

Oil skimmers in Citgo's storm water tanks had been non-operational after 1996 and a project to clean the tanks in 2003 was unsuccessful. Failure to repair the skimming equipment allowed a significant amount of oil to build up in the storm water tanks which contributed to the overflow. Furthermore, storm water tanks are required to be kept at a minimal level in anticipation of heavy rain. Citgo allowed the storm water tanks to be above the minimal possible level and did not repair skimmers to prevent the accumulation of slop oil in the tanks. As such, the heavy rainstorm overwhelmed the capacity of the two existing tanks and forced oil from the tanks into the river.

During the event, storm water tanks at Citgo refinery's waste treatment plant overflowed and released nearly 99,000 barrels (4.16 million gallons) of slop oil and 441,239 barrels (18.5 million gallons) of wastewater. The slop oil consisted of oil to be recovered for future use, water, and waste from other refinery processes. Over the next 2 days, the refinery released an additional 53,000 barrels (2.23 million gallons) of slop oil and 259,524 barrels (10.9 million gallons) of wastewater into the Indian Marais, an on-site area designed for drainage. About 25,000 barrels (1 million gallons) of the slop oil in the Indian Marais migrated into the nearby Calcasieu River, affecting the nearby community.

Moreover, the intense rain also disrupted the refinery's Central Amine Unit and initiated the release of excess sulfur dioxide into the atmosphere. The steam lines that ran through trenches in the refinery became submerged in rainwater, resulting in the eventual shutdown of the sulfur recovery units (Citgo, 2006b). Various heaters and boilers exceeded permitted sulfur dioxide limits by a total of 191,205 lbs of the gas between 4:00 am and 1:00 pm on June 19.

5.6.2 The Damage

The 2006 Citgo oil leak was one of the largest chemical spills the Lake Charles region had experienced. After the initial leakage, the oil further permeated the surrounding ecosystem via spreading, advection, turbulent diffusion, evaporation, dissolution, vertical dispersion, shoreline deposition, and absorption by sediment. Due to the ease of oil dispersion, the spill event was a huge devastation to the people, wildlife, and resources of the Lake Charles area.

The spilled slop oil, a mixture consisting of effluent from waste streams and spills, contained many hazardous chemicals including hydrogen sulfide, benzene, toluene, xylene, ethylbenzene, trimethylbenzene, and alkanes. All of these chemicals can provoke various adverse health effects. Specifically, exposure to hydrogen sulfide at low concentrations can cause certain neurological and respiratory effects such as incoordination or poor memory. At high concentrations, hydrogen sulfide can result in respiratory arrest or pulmonary edema (ATSDR, 2006). As an aromatic VOC, benzene has been shown to cause various neurological effects as well. Most consistently, benzene has been linked to acute myelogenous leukemia, a cancer of the blood and bone marrow. Toluene, xylene, and other chemical solvents are known for their synergistically adverse effect on human health, hindering the clearance rate of benzene from the human body. As such, these chemicals severely threaten the health of surrounding communities.

Between June 19 and July 1, many people around Calcasieu River were exposed to doses of benzene and hydrogen sulfide that exceeded chemical threshold values set by the American Conference of Governmental Industrial Hygienists (ACGIH), the US Environmental Protection Agency (EPA), and the National Institute for Occupational Safety and Health (NIOSH). Chemical exposures above regulatory limits extended to individuals many miles away from the spill site.

Workers employed by Ron Williams Construction working downstream from the Calcasieu Refinery reported numerous adverse health effects from the spilled slop oil. Immediate symptoms included lightheadedness; burning and irritation of the eyes, nose, and throat; dripping sinuses; headache; dizziness; blurry vision; nausea; vomiting; diarrhea; fatigue; skin rash; and loss of appetite. Not only did all of these maladies persist in the weeks and months following the spill, the employees also suffered from additional health problems that arose later. Their post-June-19 symptoms included coughing, muscle weakness, sharp chest pains, shortness of breath, acid reflux, peeling and blistering skin, and sleeplessness.

Citgo failed to adequately address how slop oil contaminants in water and air impacted residents and workers in the Lake Charles area. The only chemicals out of the 11 chemicals listed on Citgo's slop oil MSDS analyzed in oil samples were BTEX and hydrogen sulfide. Citgo attempted to calculate the amount of contaminants that volatilized from the slop oil, but their emission rates were calculated based on analytical results from samples collected after the start of the event.

Ships could not access Prien Lake for over 2 weeks and local fisheries were shut down temporarily. Residents were warned by the state health department against consuming fish in oil-affected regions. Furthermore the discharged oil adversely affected over 80 miles of shoreline, including marshes and other terrestrial habitats.

5.6.3 The Consequences

The Citgo Refinery spill was the largest spill in Louisiana state history. The federal Environmental Protection Agency and the Louisiana Department of Environmental Quality sued Citgo in 2008 for negligently failing to maintain storm water tanks and failing to maintain adequate storm water storage capacity at its refinery. In September 2008, Citgo pleaded guilty to the charges and was sentenced to pay \$13 million, the largest fine ever for a criminal misdemeanor violation of the Clean Water Act. This suit has generated more than 200 additional lawsuits against Citgo for the spill.

In January 2009, the Department of Justice and the US Environmental Protection Agency announced a comprehensive Clean Air Act settlement with Citgo as part of the EPA's national Petroleum Refinery Priority. A consent decree will require Citgo to spend an estimated \$320 million to install and implement new control technologies to reduce refinery emissions. Under this agreement, Citgo's actions are expected to reduce annual emissions of nitrogen oxide by more than 7,184 tons and sulfur dioxide by more than 23,250 tons (EPA, 2009). The settlement will also require reductions of VOCs and other hazardous air pollutants.

5.7 BP Oil Spill

5.7.1 The Incident

On April 20, 2010, a drilling rig explosion aboard the Deepwater Horizon – an ultra-deepwater drilling platform leased by the oil company British Petroleum (BP) – resulted in a sea-floor oil gusher that released nearly 5 million barrels of crude oil into the Gulf of Mexico. In the Mississippi Canyon area, Deepwater Horizon was completing a new well when a giant bubble of methane caused an explosion and fire, which sank the rig and killed 11 workers (Joye and MacDonald, 2010). The explosion occurred one mile below the water's surface and was followed by a series of unsuccessful attempts to plug the leak (*NY Times*, 2010). An alarming 86 days later, BP was finally able to cap the well on July 15, 2010. Engineers have estimated that approximately 53,000 barrels of oil were being released each day from the well, with 4.9 million barrels of oil released in total (Robertson, 2010). Although the well was plugged with cement, research is currently under way to find a more effective means of permanent cover. The resulting fine against BP could potentially add up to $21 billion (Robertson, 2010). Deemed the largest accidental oil spill in history, the BP incident released about 19 times more oil than the 1989 Exxon Valdez event (Khan, 2010).

The spill has introduced vast quantities of oil and methane gas into the deep waters of the Gulf of Mexico, spurring major environmental controversy and raising serious concerns for oceanic health. As was seen in the Exxon Valdez case, the aftermath of this incident will likely exhibit negative effects in marine organisms, wildlife, and surrounding communities that may persist for decades. In fact the extent of long-term damage is expected to be increasingly disastrous because large quantities of oil are spreading underwater, rather than rising to the surface. And while methane released at the water's surface would likely evaporate into the atmosphere, methane released underwater will be consumed by microbial bodies, thus resulting in severe oxygen depletion. The depletion of oxygen in bottom waters can introduce a multitude of environmental consequences, affecting oxygen-dependent components of the food web (such as zooplankton, fish, squid, etc.), seafloor benthic communities, cold-water corals, and the organisms that depend on them (Joye and MacDonald, 2010). After air quality monitoring tests, the EPA has furthermore discovered ozone and particulate matter at levels harmful to sensitive organisms (EPA, 2010a).

5.7.2 Oil Dispersants

Throughout the span of the incident, there has been a great deal of debate over the use of oil dispersants, toxic chemicals that can be applied to an oil slick to transform the mass into droplets. In droplet form, oil can be more easily consumed by microbes and concealed from coastal wildlife under ocean currents (Khan, 2010). To date, almost 2 million gallons of the oil dispersant known as Corexit 9500A have been applied over the BP oil spill (Khan, 2010). However, the use of dispersant risks severe ecosystem contamination, which is dangerous especially in areas that are densely populated by marine life. In various studies, Corexit exhibited irreversible effects on aquatic organisms, and at certain concentrations it impeded fertilization and larval development of sea urchins and fish, altered mussel immune functions, and inhibited siphon activities in adult clams (Albers, 1979; Hartwick et al., 1981; Hamoutene et al., 2004). Additionally, it was found that time reactions in response to oil dispersants differed from one organism to another, thus suggesting that organisms with the greatest sensitivity would be at the greatest risk of dispersant toxicity (Belkhir and Hadj, 1986). Of the nearly 5 million gallons of oil that have gushed from the well into the ocean, only about 800,000 gallons have been captured by containment efforts such as this (Robertson, 2010). Thus the environmental risk remains extremely high and additional measures must be employed to mitigate further contamination effects.

Acknowledgment

The section on oil dispersants was contributed by Ernest J Vasti, MD. Ernie is an MPH candidate at the University of Los Angeles School of Public Health.

References

Agency for Toxic Substances and Diseases Registry. (2006). *Toxicological profile for hydrogen sulfide*. US Department of Health and Human Services, Agency for Toxic Substances and Disease Registry.

Agency for Toxic Substances and Diseases Registry. (2007). *ToxFAQs for methyl tert-butyl ether (MTBE)*. Retrieved August 16, 2008 from <http://www.atsdr.cdc.gov/tfacts91.html/>.

Albers, P. (1979). Effects of Corexit 9527 on the hatchability of mallard eggs. *Bulletin of Environmental Contamination and Toxicology*, *23*(4–5), 661–668.

Belkhir, M., & Hadj, A. (1986). Oil spill dispersant toxicity on fish and mollusk. *Bulletin de l'Institut National Scientifique et Technique d'Oceanographie et de Peche de Salammbo*, *13*, 13–18.

Berry, J. (2001). Cancer alley – the poisoning of the American South. *Aperture*, *162*, 30–43.

Blodgett, A. (2006). An analysis of pollution and community advocacy in 'Cancer Alley': Setting an example for the environmental justice movement in St. James Parish, Louisiana. *Local Environment*, *11*(6), 647–661.

Carson, R., et al., (2008, February). *A contingent valuation study of los passive use values resulting from the Exxon Valdez oil spill*. MPRE Paper 6984. Retrieved August 18, 2008 from <http://mpra.ub.uni-muenchen.de/6984/>.

Citgo (2006b). *Citgo oil spill*. Louisiana Department of Environmental Quality. Retrieved November 16, 2008 from <http://kplc.images.worldnow.com/images/incoming/Citgo/CitgoIncident.pdf/>.

Clarke, K., & Hemphill, J. (2002). *The Santa Barbara oil spill, a retrospective*. In D. Danta (Ed.), *Yearbook of the association of the pacific coast geographers*. University of Hawaii Press, 64, 157–162.

County of Santa Barbara Planning and Development (CSBPD) (2005). *Blowout at Union Oil's platform A*. Energy Division. Retrieved August 16, 2008 from <http://www.countyofsb.org/ENERGY/information/1969blowout.asp/>.

Dahlgren, J., Takhar, H., Anderson-Mahoney, P., Kotlerman, J., Tarr, J., & Warshaw, R. (2007). Cluster of systemic lupus erythematosus (SLE) associated with an oil field waste site: A cross sectional study. *Environmental Health*, *8*(6), 1–15.

Duffield, J. (1997). Nonmarket valuation and the courts: The case of the Exxon Valdez. *Contemporary Economic Policy*, *15*, 98–110.

Garchar, R. (2004, Spring). *Urban studies and environmental studies program 515: Race, poverty, and the environment*. San Francisco State University.

Grabowski, M., & Robers, K. H. (1996). Human and organizational error in large scale systems. *IEEE Transactions on Systems, Man, and Cybernetics – Part A: Systems and Humans*, *26*(1), 2–15.

Hamoutene, D., Payne, J., Rahimtula, A., & Lee, K. (2004). Effect of water soluble fractions of diesel and an oil spill dispersant (Corexit 9527) on immune responses in mussels. *Bulletin of Environmental Contamination and Toxicology*, *72*(6), 1260–1267.

Hartwick, E., Wu, R., & Parker, D. (1982). Effects of crude-oil and an oil dispersant (Corexit 9527) on populations of the littleneck clam. *Marine Environmental Research*, *6*(4), 291–306.

Hayes, M., & Michel, J. (1999). Factors determining the long-term persistence of the Exxon Valdez oil spill. *Marine Pollution Bulletin*, *38*(10), 855–863.

Hinck, J. (2001, July 1). *Testing the waters in MTBE litigation*. Retrieved August 18, 2008 from <http://www.Accessmylibrary.com/coms2/summary_0286-10494133_ITM/>.

Ivanhoe, L. F., & Leckie, G. G. (1993). Global oil, gas fields, sizes tallied, analyzed. *Oil and Gas Journal*, *91*(7), 87–91.

Iyer, V., & Mastorakis, N. E. (2009). Unsafe petrochemical refinery air pollution and its environmental impact assessment. *WSEAS Proceedings of the 3rd Energy Problems and Environmental Engineering*, 525–532.

Joye, S., & MacDonald, I. (2010). Offshore oceanic impacts from the BP oil spill. *Nature Geoscience*, *3*(7), 446.

Khan, A. (2010). Oil dispersant effects remain a mystery. *LA Times*. Retrieved Septemper 4 from <http://articles.latimes.com/2010/sep/04/science/la-sci-dispersants-20100905>.

Kirkeleit, J., Riise, T., Bjorge, T., Moen, B., Bratveit, M., & Christiani, D. C. (2009). Increased risk of oesophageal adenocarcinoma among upstream petroleum workers. *Occupational and Environmental Medicine*, *67*, 335–340.

Paine, R. T., Ruesink, J. L., Sun, A., Soulanille, E. L., Wonham, M. J., Harley, C. D. G., Brumbaugh, D. R., & Secord, D. L. (1996). Trouble on oiled waters: Lessons from the Exxon Valdez oil spill. *Annual Review of Ecology and Systematics*, *27*, 197–235.

Palinkas, L., Petterson, J. S., Russell, J., & Downs, M. A. (1993). Community patterns of psychiatric disorders after the Exxon Valdez oil spill. *American Journal of Psychiatry*, *150*, 1517–1523.

Picou, J. (2000). The 'Talking Circle' as a sociological practice: Cultural transformation of chronic disaster impacts. *Sociological Practice: A Journal of Clinical and Applied Sociology*, *2*(2), 77–97.

Porter, A. (2005). Bank Says Saudi's Top Field in Decline. English Al-Jazeera, September 5, 2006.

Robertson, C. (2010). Gulf spill is the largest of its kind, scientists say. *NY Times*. Retrieved August 2 from <http://www.nytimes.com/2010/08/03/us/03spill.html?_r=2&fta=y/>.

Santa Barbara Wildlife Care Network (SBWCN) (2008). *Santa Barbara's 1969 oil spill*. Retrieved August 16 from <http://www.sbwcn.org/spill.shtml/>.

Sissell, K. (2001). Activists tour 'Cancer Alley'. *Chemical Week*, *163*(24), 39.

Sorahan, T. (2007). Mortality of UK oil refinery and petroleum distribution workers, 1951–2003. *Occupational Medicine*, *57*(3), 177–185.

State of Alaska (1990). *Details about the accident*. Alaska Oil Spill Commission. Retrieved July 14, 2010 from <http://www.evostc.state.ak.us/facts/details.cfm/>.

Sullivan, J., & Krieger, G. (2001). *Clinical environmental health and toxic exposures*. Lippincott Williams & Wilkins.

US Environmental Protection Agency (EPA) (2000). *2000 Toxics Release Inventory (TRI) public data release report*. Retrieved July 14, 2010 from <http://www.nepis.epa.gov/Exe/ZyPURL.cgi?Dockey=100049HZ.txt/>.

US Environmental Protection Agency (EPA) (2008). Methyl tertiary butyl ether (MTBE). Transportation & Air Quality. Retrieved July 7, 2010 from <http://www.epa.gov/mtbe/gas.htm/>.

US Environmental Protection Agency (EPA) (2009). *CITGO global refinery settlement*. Compliance and Enforcement. Retrieved July 7, 2010 from <http://www.epa.gov/compliance/resources/cases/civil/caa/citgo.html/>.

US Environmental Protection Agency (EPA) (2010a). *BP oil spill*. Retrieved September 13 from <http://www.epa.gov/BPSpill/>.

US Environmental Protection Agency (EPA) (2010b). *Dioxins and furans*. Retrieved July 19 from <http://www.epa.gov/pbt/pubs/dioxins.htm/>.

US Environmental Protection Agency (EPA) (2010c). *Listed Wastes. Hazardous Waste*. Retrieved August 30 from <http://www.epa.gov/osw/hazard/wastetypes/listed.htm/>.

Waxman report (1999, November). Oil refineries fail to report millions of pounds of harmful emissions. Report prepared for Rep. Henry A. Waxman, Minority Staff: Special Investigations Division, Committee on Government Reform, US House of Representatives.

Yale, S. (2008, February 26). Residents settle suits against oil companies. *Santa Maria Times*.

6 Coal-Fired Power Plants

Chapter Contents

6.1 Overview

Consisting mostly of carbon and hydrocarbons, coal is a combustible black or brownish-black sedimentary rock and is the most abundant fossil fuel in the US. According to official records, about 52% of the nation's electricity is produced using coal as a fuel source (EPA, 2010a). The four major types of coal are anthracite, bituminous, subbituminous, and lignite, but most power plants in the US burn bituminous and subbituminous coal to generate electricity. Although coal is abundant and relatively inexpensive, it also contains a large number of impurities, and upon combustion it releases harmful gases into the atmosphere. Sulfur dioxide, nitrogen oxides, and mercury are some of the emitted pollutants, which can lead to a variety of environmental issues ranging from acid deposition to ecosystem toxicity. Dangerous emissions as well as solid wastes from coal-fired power plants most strongly affect those living and working near the plant; however, due to the mobility of these particles (e.g. during storm events and natural transport phenomena), both humans and aquatic life are placed at risk for respiratory and neurological disorders. The purpose of this chapter is to discuss the processes involved in coal-fired operations, industry-related emissions and wastes, environmental impacts, lawsuits and regulations against coal power plants, and potential strategies to reduce pollution.

Risks of Hazardous Wastes. DOI: 10.1016/B978-1-4377-7842-7.00006-4

6.2 Power Production, Emissions, and Waste

6.2.1 Production

The energy stored in coal provides the electric power generated by many plants today, and the conversion process requires multiple steps. In order to increase surface area, the coal is first pulverized into a fine powder which allows it to burn more easily. The powder is then fired at high temperatures in the combustion chamber of a boiler, where hot gases and heat energy convert water into steam. Under high pressure and temperature, the steam can be used to drive propellers on a turbine. A generator may be mounted on the other end of the turbine shaft which produces the electricity. Afterward, the steam passing through the turbine can then be condensed or returned to the boiler to be re-heated. It is important to note that the thermal efficiency of these plants is low, ranging from only 35–47% (Al-Ali et al., 2008).

6.2.2 Emissions

During the combustion process, the main air pollutants coming from coal power plants are sulfur dioxide (SO_2), nitrogen oxides (NO_x), carbon dioxide (CO_2), hard metals, and VOCs. Elements may be volatilized in the stack, or condensed and enriched in fine particles that escape from particulate retention devices (Agrawal et al., 2010). Emissions can be direct (from exhaust gas streams), indirect (from industry-related activities like production and distribution), or fugitive (leaks). Lakes and catchment basins act as sinks for particles and compounds from these emission sources. As such, trace elements may enter water bodies through such means as surface runoff, groundwater, waste outlets, and atmospheric deposition (Sanei et al., 2010). Once they reach an aquatic ecosystem, even trace amounts of these pollutants can trigger serious consequences for the environment and the living organisms that depend on it.

Within the coal-fired power plant industry, one of the primary pollutants emitted from these activities is sulfur dioxide. Coal-fired power plants are responsible for approximately 59% of total US sulfur dioxide emissions (Sueyoshi et al., 2010). If released via exhaust gases during combustion, sulfur dioxide may exacerbate asthma, reduce lung function, and cause respiratory illnesses and premature death. Other significant air pollutants include carbon dioxide, nitrogen oxides, mercury, lead, chromium, and VOCs. The major volatile organic compounds emitted are 1,2-dichloroethane, benzene, carbon tetrachloride, chloroform, and trichloro-ethylene (Pudasainee et al., 2010).

The amount of air pollutants emitted is dependent upon removal technologies, as well as mining practices, coal beneficiation, and utility coal quality specifications. To achieve emission reduction, coal-fired power plants require several clean-up processes to remove carbon dioxide and sulfur compounds before combustion. Various technologies are available that are able to capture either a single pollutant or multiple pollutants present in the flue gas streams. Seasonal changes also influence the effect of emissions on environmental contamination. For instance, one study (Agrawal et al., 2010) discovered that samples of soil surrounding a coal-fired power plant contained higher concentrations of lead, cadmium, and nickel after rain events. This effect may be attributed to the settling of atmospheric particles to the ground during a storm. Thus these findings suggest that rainy winter seasons may exhibit greater soil contamination in nearby areas.

6.2.3 *Waste*

As these power plants seek to reduce their air emissions, more pollutants are left behind in solid waste residues in the form of coal ash, sludge from desulfurization systems, and particles trapped by other control systems. A typical 1,000 megawatt power plant consumes 12,000 tonnes of coal per day and produces an alarming amount of 1,000,000 tonnes of wastes per year (Agrawal et al., 2010). This coal combustion waste consists of radionuclides, lead, mercury, arsenic, and other hard metals, which can be extremely toxic to human health. These wastes may leach into downstream water supplies or soils where they can enter the food chain and potentially induce chronic toxic effects in humans and other organisms.

Each year, coal burning generates over 125 million tons of waste in the form of ash and sludge in the US alone (EPA, 1999a). Of this amount, King and Smith (2010) report that about 40% of wastes is re-used for new products and 60% is stored in waste ponds or pits. It is also alarming to note that no federal standard exists to regulate the lining of waste pits to prevent leaching of the substances into streams or groundwater supplies. So far, the EPA has located 63 known sites where heavy metals from coal waste ponds have contaminated groundwater in the US. In 2008, a surface impoundment near Kingston, Tennessee, spilled and flooded over 300 acres including the nearby Emory and Clinch rivers. Since then, a need for better national management of coal combustion wastes has become extremely apparent.

6.3 Environmental Health Impacts

6.3.1 *Acid Deposition*

As a consequence of polluted air emissions, coal-burning power plants are also responsible for a trend known as acid deposition or acid rain. Sulfur dioxide and nitrogen oxides from plant emissions interact in the air with water, oxygen, carbon dioxide, and sunlight to form nitric (HNO_3) and sulfuric acids (H_2SO_4) which are the primary responsible agents. This phenomenon is of major concern because it can cause significant stress to lakes, streams, and forests.

Acid rain, which may also manifest in the form of snow or sleet, can have severe ecological impacts on soil, forest, freshwater, and estuary ecosystems, including effects like base nutrient depletion, aluminum toxicity, nitrogen saturation, and eutrophication (ESA, 2000). In base nutrient depletion, soil acidification removes nutrients known as base cations to make trees highly vulnerable to stress and diseases. Acid deposition also causes the aluminum that is naturally present in soils to leach into lakes and streams. As a result, many freshwater species and other types of aquatic life that cannot withstand such high concentrations of aluminum face aluminum toxicity issues. Similarly, nitrogen saturation and eutrophication are other consequences of acid rain that lead to ecosystem degradation and loss of biodiversity (ESA, 2000).

Many factors affect the extent of acid deposition effects, including: watershed bedrock composition, land use history, vegetation type, soil depth, and base nutrient reserves. For example, certain types of rock (e.g. granite) do not produce neutralizing chemicals and are therefore at greater risk of acidification. Ecosystems that have experienced clear cutting land use practices, with shallow soils containing low amounts of calcium, potassium, and magnesium, are also less able to buffer against changes in acidity and are more likely to be negatively affected by acid rain (ESA, 2000).

As a result, many ecosystems today are too degraded by acidification to successfully maintain fish populations. Lower water quality, acidified waters, and aluminum toxicity are effects that threaten the livelihood of species such as salmon and trout (EPA, 2008). Thus coal-fired power plants that emit high levels of sulfur dioxide and nitrogen oxides make acidification a significant ecological problem. Long-term monitoring can aid in evaluating the effectiveness of regulations that aim to reduce air pollutants, as well as in testing models of ecosystem reaction to changes in acidity (ESA, 2000).

6.3.2 Mercury

Mercury has ignited global concern among environmentalists due to its toxic effects on human health, bioaccumulation in the environment, and long-range transport (Wang et al., 2010). The risk of mercury poisoning is dependent upon the probability of exposure, the form of mercury present, and the geochemical and ecological factors that affect its transport in the environment (USGS, 2000). Today, coal-fired power plants contribute greatly to mercury concentrations in the environment, comprising over 50% of all domestic human-caused mercury emissions (EPA, 2010c). At high temperatures during coal combustion, trace amounts of mercury are released into the exhaust gas. Following various oxidation and thermochemical processes, the oxidized mercury becomes soluble and has a tendency to stick to air particles (Wang et al., 2010).

Once volatilized in the air, mercury can settle into lakes and streams, or onto land where it may be washed into such water bodies. Once deposited, the substance can be transformed by microorganisms into a highly toxic form called methylmercury, which builds up in the bodies of fish, shellfish, and animals that eat fish. Increased acidity and increased dissolved organic carbon content of lakes or streams can augment the methylation process in fish. Humans may then be exposed to this substance via fish consumption, and due to its bioaccumulation tendencies, the organisms at the top of the food chain are at the greatest risk for mercury toxicity. Therefore, most states publish advisories against excessive fish consumption due to elevated mercury levels. In animals, methylmercury can cause death, reduced reproduction, slower growth and development, as well as abnormal behavior (EPA, 2010c). In humans, methylmercury can disrupt the immune system and nervous system, affecting genetic and enzyme systems as well as coordination and senses of the touch, taste, and sight (USGS, 2000).

Amendments made to the Clean Air Act in 1990 list mercury compounds as hazardous air pollutants (HAPs). About 50 tons of mercury are emitted annually by coal-fired power plants, but in 2005 the Clean Air Mercury Rule required a 70% reduction of mercury emissions from power plants to 15 tons per year (EPA, 2009a).

6.3.3 Climate Change

Due to coal production and combustion, rising concentrations of carbon dioxide and other greenhouse gases in the atmosphere are contributing to increased global temperatures and climate changes. Additionally the production, transmission, storage, and distribution of coal further emit carbon dioxide, which leads to the trapping of heat in the Earth's atmosphere. In the US, coal power plants account for a third of US carbon dioxide emissions (MIT, 2009). Higher carbon dioxide emissions can have a significant long-term effect on the environment and human populations including: increase of precipitation and flooding in storm-affected areas, more intense hurricanes in warm sea surfaces, increases in heat waves, and rise in sea level due to polar ice cap melting. Vast amounts of this greenhouse gas also threaten to acidify oceans, destroying plankton life that forms the bottom link of the food chain and affecting shell formation of aquatic organisms (Holzman, 2008).

One possible solution to the rise in carbon dioxide emissions is a new system whereby carbon dioxide produced by burning coal is subsequently captured and sequestered, in a process known as carbon capture and storage (CCS). Large volumes of the gas can then be stored in natural facilities such as depleted oil and gas fields (Drake, 2009). Other options to reduce carbon dioxide are also available in the form of alternative fuel sources and more efficient emission controls.

6.4 Lawsuits and Regulations

6.4.1 Violation of the Clean Air Act (CAA)

The coal-fired power plant industry is a national enforcement priority of the US EPA. On November 3, 1999, the Department of Justice and the Environmental Protection Agency announced the filing of civil complaints against seven electric utility companies operating coal-fired power plants in the Midwest and Southeast. The organizations charged that their plants illegally released significant amounts of air pollutants over a time span of several years and contributed to some of the most severe environmental problems facing the US today (EPA, 2010b). American

Electric Power Company, Cinergy, FirstEnergy, Illinois Power, Southern Indiana Gas & Electric Company, Southern Company, and Tampa Electric Company were the companies involved.

The seven separate lawsuits allege that the electric utility companies violated the Clean Air Act by completing major modifications to many of their plants without installing the required equipment to regulate smog, acid rain, and soot. The US aims to dramatically reduce levels of sulfur dioxide, nitrogen oxides, and particulate matter emitted by these utility plants through the implementation of appropriate air pollution control technology. Failure of the utilities to install the required equipment resulted in tens of millions of tons of sulfur dioxide, nitrogen oxides, and particulate matter illegally emitted into the air, leading to a number of adverse environmental and health impacts (EPA, 1999b). According to the EPA (1999b), electric utility plants collectively account for approximately 70% of sulfur dioxide emissions each year and 30% of nitrogen oxides emissions. While power plant emissions have been linked to detrimental health effects on asthma sufferers, the elderly, and children, such releases also contribute to forest degradation, waterway damage, reservoir contamination, and deterioration of stone and copper in buildings.

The Clean Air Act authorizes civil penalties of up to $25,000 for each day of violation at each plant prior to January 30, 1997, and $27,500 for each day thereafter. Each utility company was required to pay appropriate fees for civil penalties, new emission control technologies, and community advancement projects. For instance, under the settlement, Tampa Electric Company was required to pay a $3.5 million civil penalty, and between $10 and $11 million on environmentally beneficial projects in the region designed to lessen the impact of emissions from the company's plants. Meanwhile, Southern Indiana Gas & Electric Company was mandated to pay a civil penalty of $600,000, with an additional $30 million to install state-of-the-art pollution controls and $2.5 million to fund environmentally beneficial projects around the area to reduce sulfuric acid (EPA, 2009b).

6.4.2 Current Efforts

In July 2010, the Obama administration renewed EPA's attempts to reduce pollution in areas around coal plants and in states downwind where air quality is degraded by sulfur dioxide and nitrogen oxide emissions. In cooperation with other state and federal regulations, the rule aims to reduce sulfur dioxide emissions by 71% from 2005 levels by 2014 (Hawthorne, 2010). Nitrogen oxide emissions would also drop by 52%. The rule requires aging coal-fired power plants to be upgraded with modern pollution control technology, including scrubbers and selective catalytic reduction systems. These controls target both sulfur dioxide and nitrogen oxide pollution.

The rule is expected to cost $2.8 billion a year but promises greater health benefits among nearby residents (Hawthorne, 2010). As a result, the rule has caused many coal plants to evaluate whether it is more cost-effective to retrofit their worst polluting power plants or to shut them down completely. Because states will be allowed to engage in limited trading of pollution credits, each state will remain below a set emissions limit while still providing companies with greater financial incentives.

6.5 Control Technologies and Alternatives

A variety of new technologies are currently available to reduce toxic emissions from coal-fired power plants. For instance, fabric filters (also called bag filters) can be used to collect particulate matter, and a tightly woven fabric acts as a screen through which the gas is forced. Additionally a device known as a scrubber is able to reduce sulfur dioxide emissions by neutralizing the gas through a packed bed of filter material. To lower nitrogen oxides, selective catalytic reduction (SCR) is available, and this technology combines the exhaust gases with ammonia or urea and reduces nitrogen oxides into nitrogen and water. As previously mentioned, removing sulfur dioxide, nitrogen oxides, and particulates already eradicates a fraction of the mercury before it is released from the stack. To enhance mercury removal, particles of activated carbon can be introduced into the exit gas flow to attract mercury elements. The activated carbon (and attracted mercury) can then be removed in a traditional device. This process is known as activated carbon injection.

In addition to control technologies, cleaner types of coal-fired power plants are emerging as well. For instance, integrated gasification combined cycle (IGCC) power plant systems can transform coal into gas to generate electricity more cleanly and efficiently. In the gasifier stage, coal is directly converted into gas, or 'syngas' (a mix of carbon monoxide, carbon dioxide, and hydrogen). The IGCC process also removes impurities from the coal gas before combustion, and turns sulfur, ammonia, and other impurities into re-usable by-products. While the steam cycle is similar to conventional plants, IGCC plants exhibit lower emissions of sulfur dioxide, carbon dioxide, particulates, and mercury. As a result, these plants experience lower energy penalties and minimized emission reduction costs.

References

Agrawal, P., Mittal, A., Prakash, R., Kumar, M., Singh, T., & Tripathi, S. (2010). Assessment of contamination of soil due to heavy metals around coal fired thermal power plants at Singrauli region of india. *Bulletin of Environmental Contamination and Toxicology*, *85*(2), 219–223.

Al-Ali, H., Saif, Y., Elkamel, A., & Lohi, A. (2008). Optimal production planning of electricity from coal-fired power plant networks under environmental considerations. *American Journal of Engineering and Applied Sciences*, *1*(4), 358–367.

Drake, F. (2009). Black gold to green gold: Regional energy policy and the rehabilitation of coal in response to climate change. *Area*, *41*(1), 43–54.

Ecological Society of America (ESA) (2000). *Acid Deposition*. Retrieved September 13 from <http://www.esa.org/education_diversity/pdfDocs/aciddeposition.pdf>.

Hawthorne, M. (2010). EPA targets coal-fired power plants. *Chicago Tribune*. Retrieved July 7 from <http://articles.chicagotribune.com/2010-07-07/health/ct-met-epa-air-pollution-rules-20100706_1_coal-fired-power-plants-midwest-generation-plants-clean-air-interstate-rule>.

Holzman, D. (2008). The carbon footprint of biofuels: Can we shrink it down to size in time? *Environmental Health Perspectives*, *116*(6), A246–A252.

King, N., & Smith, R. (2010). White House, EPA at odds over coal-waste rules. *Wall Street Journal*. Retrieved January 9 from <http://online.wsj.com/article/SB126300256672322625.html>.

Massachusetts Institute of Technology (2009). *Retrofitting of coal-fired power plants for* CO_2 *emissions reductions. MIT Energy Initiative Symposium*. Retrieved September 13, 2010 from <http://web.mit.edu/mitei/docs/reports/coal-paper.pdf>.

Pudasainee, D., Kim, J., Lee, S., Park, J., Jang, H., Song, G., et al. (2010). Hazardous air pollutants emission from coal and oil-fired power plants. *Asia-Pacific Journal of Chemical Engineering*, *5*(2), 299–303.

Sanei, H., Goodarzi, F., & Outridge, P. (2010). Spatial distribution of mercury and other trace elements in recent lake sediments from central Alberta, Canada: An assessment of the regional impact of coal-fired power plants. *International Journal of Coal Geology*, *82*(1-2), 105–115.

Sueyoshi, T., Goto, M., & Ueno, T. (2010). Performance analysis of US coal-fired power plants by measuring three DEA efficiencies. *Energy Policy*, *38*(4), 1675–1688.

US Environmental Protection Agency (EPA) (1999a). *Technical background document for the report to congress on remaining wastes from fossil fuel combustion: Industry statistics and waste management practices*. Retrieved September 2, 2010 from <http://www.epa.gov/osw/nonhaz/industrial/special/fossil/ffc2_398.pdf>.

US Environmental Protection Agency (EPA) (1999b). *US sues electric utilities in unprecedented action to enforce the Clean Air Act*. Retrieved September 2, 2010 from <http://www.justice.gov/opa/pr/1999/November/524enr.htm>.

US Environmental Protection Agency (EPA) (2008). *Effects of acid rain – Surface waters and aquatic animals*. Retrieved September 2, 2010 from <http://www.epa.gov/acidrain/effects/surface_water.html>.

US Environmental Protection Agency (EPA) (2009a). *Controlling power plant emissions: Overview*. Retrieved September 13, 2010 from <http://www.epa.gov/mercury/control_emissions/index.htm>.

US Environmental Protection Agency (EPA) (2009b). *Southern Indiana Gas and Electric Company (SIGECO) F.B. Culley plant Clean Air Act settlement*. Retrieved September 2, 2010 from <http://www.epa.gov/compliance/resources/cases/civil/caa/sigecofb.html>.

US Environmental Protection Agency (EPA) (2010a). *Coal-fired power plant emissions*. Retrieved September 2 from <http://www.epa.gov/radtown/coal-plant.html>.

US Environmental Protection Agency (EPA) (2010b). *Coal-fired power plant enforcement initiative*. Retrieved September 2 from <http://www.epa.gov/compliance/resources/cases/civil/caa/coal/index.html>.

US Environmental Protection Agency (EPA) (2010c). *Basic information – Mercury*. Retrieved September 2 from <http://www.epa.gov/mercury/about.htm>.

US Geological Survey (USGS) (2000). *Mercury in the environment.* Retrieved September 13, 2010 from <http://www.usgs.gov/themes/factsheet/146-00/>.

Wang, S., Zhang, L., Li, G., Wu, Y., Hao, J., Pirrone, N., et al. (2010). Mercury emission and speciation of coal-fired power plants in China. *Atmospheric Chemistry and Physics, 10*(3), 1183–1192.

7 Iron, Steel, and Coke

Chapter Contents

7.1 Introduction

Steel is an alloy of iron that is utilized for a wide range of daily applications, ranging from use in food cans and household containers to automobiles and office buildings. Today the US has over 1,000 manufacturing facilities, with the majority of steel mills located in the Great Lakes region (EPA, 2009a). The iron and steel industry is accountable for the manufacture of goods such as bars, strips, and sheets, as well as formed products like wire, rods, and pipes.

Simultaneously, however, the industry has generated significant amounts of hazardous waste and additionally has emitted vast quantities of toxic pollutants into the atmosphere, posing severe health risks to both workers and surrounding communities. Metal dusts, slag, carbon monoxide, nitrogen oxides, and ozone are

Risks of Hazardous Wastes. DOI: 10.1016/B978-1-4377-7842-7.00007-6

examples of substances generated during the steelmaking process. Additionally coke oven emissions contain harmful substances like polycyclic aromatic hydrocarbons (PAHs), volatile organic compounds (VOCs), benzene, particulate matter, and dioxins – some of which are known human carcinogens. Studies have shown that residents who experience long periods of exposure to these industrial emissions are at higher risk for breast and prostate cancers, among other respiratory illnesses.

In response, environmental legislation has been passed to force the industry to develop cleaner and more efficient steelmaking technologies and processes (EPA, 1995). This chapter aims to provide insight on the steel industry, such as processes involved, pollution regulations, environmental impacts from emissions and wastes, health risks, and environmental case studies.

7.2 Steel Production

The general purpose of steelmaking operations is to refine the iron that contains large amounts of carbon and other impurities, and convert it into a highly elastic metal that can be forged and fabricated (International Labour Organization, 2005). As a series of consecutive refinements, usually the earliest stages exhibit the greatest environmental impact (ERI, 2004). The process starts by refining the fuel required to separate the iron from its naturally occurring compounds. Once an efficient fuel is produced, the iron itself may be removed from the ore in a blast furnace. Finally the crude material is purified and finished to give the desired properties (ERI, 2004). These three phases are referred to as cokemaking, ironmaking, and steelmaking.

7.2.1 Cokemaking

The cokemaking process basically involves baking coal in a heated oven. High heat frees the oils and tars in coal, and drives off compounds such as ammonia, nitrogen oxides, and sulfur dioxide. Left behind is a relatively pure carbon substance, pieces that are about 1–4 inches across that are strong, hard, and porous (ERI, 2004). This substance is a carbon-rich fuel known as coke that may be added to the blast furnace to help maintain the gas flow essential to iron reduction. However, this practice generates large amounts of harmful residues and discharges. As the first step in refinement, cokemaking is the largest source of emissions amongst all the processes within the steel industry.

To allow it to burn more easily, the coal is first milled into a fine powder, which creates one particulate emission source. The powder is then heated in an oven to temperatures around 2,000°F for a time span of 14–36 hours in an oxygen-deficient atmosphere (ERI, 2004). As such the gases emanating from the matter contain a multitude of criteria air pollutants (carbon monoxide, nitrogen dioxide, and sulfur dioxide) as well as hazardous air pollutants (toluene, naphthalene, phenol, cyanide compounds, and hydrogen sulfide) (DOE, 2000). Although some

Table 7.1 Typical Breakdown of Products and By-Products from One Ton of Coal (DOE, 2000)

Product/By-product	Amount
Blast furnace coke	1,200–1,600 lb
Coke breeze (fine particulate coke)	100–200 lb
Coke oven gas	9,500–11,500 ft^3
Tar	8–12 gal
Ammonia sulfate	20–28 lb
Ammonia liquor	15–35 gal
Light oil	2.5–4.0 gal

of these by-products can be recovered, losses are unavoidable during each processing stage and a number of hazardous sludges are produced during recovery processes. The finished substance is then removed from the oven and is often cooled with vast amounts of quenching water, which must be handled afterward as contaminated wastewater. About 270 gallons of cooling water are required for each ton of coke (DOE, 2000). Additionally, quench water from cokemaking also calls for high energy consumption, and produces suspected carcinogenic particulates and VOCs (EPA, 1995).

In addition to creating harmful air emissions and hazardous discharges, the use of coke is also environmentally detrimental because it is considered inefficient compared to other alternatives. For instance, pulverized coal injection – which substitutes coal for coke in the blast furnace – can replace about 25–40% of coke, thus reducing the amount of coke required and the associated air emissions (EPA, 1995). Due to these problems, cokemaking is often viewed by industry experts as the most environmentally harmful stage of the whole steelmaking process (EPA, 1995). Table 7.1 lists the typical breakdown of products and by-products from one ton of coal.

7.2.2 Ironmaking

Once coke is produced, metallic iron must be reduced from the oxide form from which it is naturally found. While oxygen is transferred to the carbon supplied by the coke, extra oxygen may be added to burn additional carbon which sustains the intense temperatures that are required to drive the iron reduction reaction (ERI, 2004). The product is known as pig iron, which contains a large percentage of carbon and many other impurities.

The reaction is carried out in a blast furnace, which results in environmental impacts that are difficult to evade. Ironmaking processes require that the furnace operate continuously, in order to avoid inefficiency from allowing the system to cool and then having to reheat it. Additionally the blast furnace relies on gravity and convection because mechanical mixing devices would not be able to withstand its extremely hot temperatures (ERI, 2004). As such, ore and coke are fed through the top of the vessel and slowly move downward as hot air and carbon monoxide move upward through the porous media (DOE, 2000). Reduced molten iron is then left to flow down toward the bottom of the furnace, where it may be collected.

The furnace is a major source of carbon monoxide emissions which can leak from the system at various openings (ERI, 2004). Additionally it contributes greatly to particulate emissions, where the most significant source takes place during molten iron collection. Although instruments such as hoods have been installed above the tapholes to capture emissions, some losses are unavoidable (ERI, 2004). Ironmaking processes also generate large amounts of slag, which consists of impurities that are found and removed from the metal. Slag can be utilized in the construction sector for fill purposes, but also contributes to particulate emissions in handling.

7.2.3 Steelmaking

Referring to the final stage of refinement, steelmaking removes carbon and other impurities, and is followed by various forming and finishing processes to give end products the desired properties. There are two primary methods for producing steel: a blast oxygen furnace (BOF) or an electric arc furnace (EAF).

The BOF procedure utilizes 25–35% recovered steel to create new steel by combining molten iron from blast furnaces with an injection of very pure oxygen (EPA, 2009a). The oxygen burns off most of the carbon, and the system must be cooled by the addition of more metal for the desired temperature range (ERI, 2004). One advantage of the BOF process is that it produces steel that is easily flattened into sheets, making it ideal for use in products like automotive fenders, soup cans, and industrial drums. However, there are large environmental impacts due to the carbon monoxide escaping from the melt and the particulates that are released during material transfers (DOE, 2000). Other negative effects include the use of large quantities of water for cleaning and cooling, and the significant amounts of slag that are sent to landfills due to unfavorable properties (ERI, 2004).

In an EAF facility, the primary raw material is scrap metal, which is melted and refined using electrical energy. Since scrap metal is used in place of molten iron, no cokemaking or ironmaking operations are required for EAF steel production. Although the use of electricity is cleaner than with coal, the processes used to generate the electricity emit a number of criteria air pollutants such as nitrogen and sulfur oxides, carbon monoxides, particulates, and VOCs (ERI, 2004). This method also produces metal dusts, slag, and gaseous products, and its emissions may be cleaned using a wet or dry system. The two primary hazardous constituents in emission control are lead and cadmium (EPA, 1995).

However, the EAF process holds various advantages over the BOF process. Electric arc furnace facilities use virtually 100% recovered steel to produce new steel. As such scrap metal is completely melted and refined using electric currents. Consequently the produced steel is very strong and is mainly utilized for application in structural beams, steel plates, and reinforcement bars. Steel recycling saves money, reduces energy consumption, and reduces the amount of greenhouse gases released (EPA, 1995). Additionally the EAF process does not require cokemaking or ironmaking processes, so in multiple respects the EAF process may be considered more efficient.

Whether the molten steel is produced using a BOF or an EAF facility, the steel must be solidified and finished by a number of different processes. Forming and finishing operations can include molding the steel, hot rolling to further shape the steel, cold forming to obtain better mechanical steel properties, and applying a protective coating to prevent erosion.

As seen, the complete steelmaking process is exceptionally lengthy and complicated, and it may include the production of coke and iron in addition to the actual steel itself. The extensive chain of various sub-operations introduces a multitude of potential areas that can generate hazardous wastes and emit harmful air pollutants. Furthermore, it is important to note that the entire integrated production process must also consider heat and electricity supply, and the handling and transport of intermediate and waste materials. These activities further increase environmental stresses of the steelmaking process.

7.3 Human Health Impacts

7.3.1 *Air Emissions*

Every stage of the steelmaking process contributes to air emissions. Although control technologies may be implemented, some losses are inevitable (ERI, 2004). In particular, cokemaking and blast furnace operations result in the greatest amount of particulate generation and release. Coke oven emissions contain high concentrations of VOCs and carbon monoxide, and have been shown to cause lung and stomach cancers (Redmond and Mazumdar, 1993; Bye et al., 1998). Other non-cancer health effects include respiratory, cardiovascular, and neurological. Specifically, coke oven emissions are considered carcinogenic to humans by a number of authoritative bodies including the EPA, the US Department of Labor's Occupational Safety and Health Administration (OSHA), and the World Health Organization's International Agency for Cancer Research (IARC).

Meanwhile, blast furnace ironmaking accounts for significant releases of sulfur and nitrogen oxides, as well as carbon monoxide. Forming and finishing operations like casting and rolling additionally contribute to the release of sulfur oxides since sulfur is released from the metal surface (ERI, 2004).

7.3.2 *Wastewater*

Quenching water from cokemaking represents a significant environmental problem and it must be handled and disposed of as contaminated wastewater. In addition to high loads of suspended solids and chemical oxygen demand, wastewater emanating from iron and steel mills usually contains high levels of ammonia, sulfide, and cyanide (Barton et al., 1978). Untreated effluent that reaches lakes and streams could have detrimental environmental effects. Cyanide, for example, is a toxic chemical compound that can disrupt the respiratory and central nervous systems (ICMI, 2010).

7.3.3 Solid Wastes

During iron and steel production, a substance known as slag (or coke breeze) is created when the acid part of the ores reacts with the limestone, and it contains unwanted impurities from the fuels such as sulfur. The slag itself poses various health risks such as aluminum leaching in environmental applications, which can disrupt a range of different water systems and types of wildlife. Furthermore steel industry slag tends to be more stable than other aggregates and provides a high degree of adhesion in asphalt and concrete (ERI, 2004). The material is commonly sold as a by-product or sent to a landfill, and is used in a wide range of environmental applications such as fill, roadbase, erosion control, and landscaping (Proctor et al., 2002). There are three main types of steel industry slag: blast furnace (which is a co-product of pig iron production), and electric arc furnace and basic oxygen furnace (which are both co-products of steel production). All three types of slags are generally utilized for the same applications. However, because it contains metal concentrations that are higher than typical concentrations in US soil, slag products pose an environmental risk to human and aquatic populations that come into contact with these substances (Proctor et al., 2002).

In one study (Proctor et al., 2002), samples of slag were collected from different kinds of furnaces to determine the human health risks associated with such environmental applications. The tests found that constituent levels were measured above screening levels, including compounds like antimony, beryllium, cadmium, trivalent and hexavalent chromium, manganese, thallium, and vanadium. In addition, the study found that all three types of slag posed site-specific ecological risks for applications in small water bodies, because high pH and aluminum were found to leach at levels harmful to aquatic life. Leaching effects were due to limited dilution volume of the small water bodies. High acidity and aluminum disrupt the salt and water balance in fish and are directly related to fish toxicity (EPA, 2008a). However, there has been a lack of research on inhalation and oral exposure risks to both humans and aquatic life (Proctor et al., 2002), which is needed in order to promote environmental health against slag applications in the environment.

7.4 Standards and Regulations

7.4.1 EPA NESHAP Standard

In 2008, the EPA issued the National Emission Standards for Hazardous Air Pollutants (NESHAP) rule for iron and steelmaking facilities. Specifically, the NESHAP established emission limitations and work practice requirements for the control of hazardous air pollutants from iron- and steelmaking operations

(EPA, 2008b). The rule also implemented maximum achievable control technology (MACT), or the maximum degree of emission reduction that was determined by the EPA to be achievable. The NESHAP aims to reduce air emissions of benzene, chromium, dioxin, lead, mercury, manganese, methanol, nickel, and triethylamine, and decreases air toxics from iron and steel foundries by about 580 tons per year (EPA, 2008b). Reductions of these substances could lead to reduced cases of chronic and acute disorders of the respiratory, reproductive, and central nervous systems.

7.4.2 OSHA Coke Oven Standard

The Occupational Safety and Health Administration (OSHA) developed a standard to control employee exposure to coke oven emissions. Coke oven emissions are considered carcinogenic to humans by a number of authoritative bodies including the EPA, the National Institute of Occupational Safety and Health (NIOSH), and the National Toxicology Program (NTP). The regulation requires employers to control employee exposure to coke oven emissions by the use of engineering controls and work practices (OSHA, 2010). Additionally employee exposures must remain at or below the permissible exposure limit. As an example, the OSHA standard for coke oven emissions is 0.15 mg/m^3 for PAH concentrations.

7.4.3 EPA Clean Water Act

EPA requires coke industries to comply with discharge regulations, asserting that effluents must not exceed the company's permit limits for certain constituents. In 2010, EPA ordered Tonawanda Coke Corporation (TCC) in New York to install pollution controls and improve monitoring of its effluent discharges after TCC violated its limits for cyanide. In 2009 several leaks of tar and process wastewater were discovered by the EPA. The discharge of improperly treated wastewater can contaminate surface waters and create public exposure risks for humans, fish, and wildlife (EPA, 2010b).

7.4.4 Resource Conservation and Recovery Act (RCRA)

Approved in 1976, the Resource Conservation and Recovery Act (RCRA) is the main federal law in the US regulating the disposal of solid and hazardous waste. The EPA has compiled a list of substances that qualify as hazardous wastes and has recognized that wastes from the iron and steel industry are hazardous to human health. These substances must then follow RCRA waste management regulations. Specifically EPA's so-called K-list details source-specific wastes from particular industries, such as petroleum refining, iron and steel, and coking. Sludges and wastewaters from treatment and production processes in these sectors are examples of source-specific hazardous wastes. Table 7.2 lists EPA-recognized solid hazardous wastes from the iron, steel, and coking industries.

Table 7.2 Solid Wastes Listed as Hazardous Wastes by the EPA from Specific Sources (EPA, 2010d)

Industry and EPA Hazardous Waste No.	Hazardous Waste	Hazard Code
Iron and steel		
K061	Emission control dust/sludge from the primary production of steel in electric furnaces	(T)
K062	Spent pickle liquor generated by steel finishing operations of facilities within the iron and steel industry (SIC Codes 331 and 332)	(C,T)
Coking		
K060	Ammonia still lime sludge from coking operations	(T)
K087	Decanter tank tar sludge from coking operations	(T)
K141	Process residues from the recovery of coal tar, including, but not limited to, collecting sump residues from the production of coke from coal or the recovery of coke by-products produced from coal. This listing does not include K087 (decanter tank tar sludges from coking operations)	(T)
K142	Tar storage tank residues from the production of coke from coal or from the recovery of coke by-products produced from coal	(T)
K143	Process residues from the recovery of light oil, including, but not limited to, those generated in stills, decanters, and wash oil recovery units from the recovery of coke by-products produced from coal	(T)
K144	Wastewater sump residues from light oil refining, including, but not limited to, intercepting or contamination sump sludges from the recovery of coke by-products produced from coal	(T)
K145	Residues from naphthalene collection and recovery operations from the recovery of coke by-products produced from coal	(T)
K147	Tar storage tank residues from coal tar refining	(T)
K148	Residues from coal tar distillation, including but not limited to, still bottoms	(T)

The hazard codes in the third column are employed by administrators to signify the type of hazard that is associated with each hazardous waste: (T) Toxic, (C) Corrosive, and (I) Ignitable. From the table, you can see that most hazardous wastes from the industry are considered Toxic Wastes. The significance of substances being present under the list above is that those materials are now subject to hazardous waste regulations based on federal requirements, such as on-site accumulation quantity limits, management requirements, and reporting, among others (EPA, 2010c).

7.5 Alabama By-Products Corp. (ABC) Coke Case Study

7.5.1 Background and Overview

The Drummond Alabama By-Products Corporation (ABC) Coke plant is the largest merchant coke producer in the United States. Situated in Tarrant, Alabama, just north of Birmingham, the plant has over 132 ovens, with annual capacity of approximately 730,000 tons of saleable coke (Drummond, 2010). Although the facility was constructed in the 1920s, it was not acquired by Drummond until the early 1980s. The community surrounding the ABC facility contains approximately 35,123 persons living within 3 miles of the ABC facility (EPA, 2010a). This 3-mile area is estimated to contain 13,387 households and 15,312 housing units. Today the ABC facility is one of the worst polluters in Jefferson County. Emissions include harmful coke oven gases, polycyclic organic matter, VOCs, particulate matter, and dioxins. Detrimental effects of such pollutants can range from respiratory illnesses, cardiovascular illnesses, and cancers.

7.5.2 Pollution and Emissions

The ABC facility is one of the single largest polluters in Jefferson County. The site produces large quantities of coke oven gases and particulate matter that pollute the surrounding community. Coke plants emit a large variety of toxic air pollutants including coke oven gases containing volatile organic compounds such as benzene, toluene, xylenes, and benzene-soluble organics including PAHs. Particulate matter produced in the coking ovens and released from storage and process units at the site contain PAHs and toxic metals such as arsenic, cadmium, chromium, lead, nickel, selenium, and mercury.

Other emissions at the site may include dioxins formed from the combustion process. As a result of coke manufacturing activities at the facility, vast quantities of atmospheric emissions of unhealthy and toxic substances have been and continue to be released into the surrounding neighborhoods. These emissions from the site occur through various operations like coal handling and blending, cokemaking and handling, and by-products processing. Emissions emanate as stack and fugitive emissions from equipment and operations, including boilers, coke ovens (including charging, coking, soaking, offtakes, flares, etc.), by-products handling, coal and coke handling systems, and other equipment and facility areas at the site (Cheremisinoff, 2010; JCDOH, 2008).

7.5.3 Reported Data

The ABC facility reports emissions data that are compiled in the EPA Toxic Release Inventory (TRI) and National Emissions Inventory (NEI) databases. Emissions reported by the facility to TRI and NEI indicate that the site emits a broad variety of unhealthy and toxic air pollutants. TRI-reported facility emissions include 1,2,4-trimethylbenzene, ammonia, ammonium sulfate, anthracene, benzene, cyanide compounds, dibenzofuran, ethylene, lead compounds, manganese and mercury compounds, naphthalene, phenanthrene, phenol, PAHs, propylene, styrene, sulfuric acid, toluene, and xylenes. Additionally, facility emissions reported to the NEI include Criteria Air Pollutants (CAPs) including, but not limited to, VOCs, sulfur dioxide (SO_2), ammonia (NH_3), carbon monoxide (CO), nitrogen oxides (NO_x), and particulate matter (PM). Hazardous Air Pollutants (HAPs) are also emitted from the factory and reported to the NEI, including compounds such as acetaldehyde, benzene, coke oven emissions, cyanide compounds, dibenzofuran, formaldehyde, lead compounds, naphthalene, phenol, polycyclic organic matter (POM), styrene, toluene, and xylenes. Additionally it has been noted by Cheremisinoff (2010) that the emissions data reported by the facility to the EPA TRI are inconsistent with other data they have reported and are inaccurate and misleading.

7.5.4 Health Cases

Communities surrounding the ABC facility experienced a multitude of various disorders related to long-term emissions inhalation. Exposures to emissions of particulate matter and other contaminants from the ABC facility have increased residents' risk of respiratory illnesses such as asthma. Also, exposures to emissions of PAHs and VOCs have contributed to residents' risk of cancers such as prostate cancer and breast cancer.

7.6 Alternatives to Conventional Technology

Proper control technologies for emissions can effectively reduce the amount of pollution released from the coke, iron, and steel industries. Since the cokemaking process is the most environmentally harmful in the steelmaking sector, it is important to concentrate efforts on reducing coke oven emissions and advancing cokeless ironmaking techniques. Further development of such operations can substantially lower air emissions and contaminated wastewater discharges. For example, one cokeless technology being studied today is the Japanese Direct Iron Ore Smelting (DIOS) process, which produces molten iron directly with coal and sinter feed ore (EPA, 1995). Reducing the environmental impact of coke production can also be achieved through recycling coke by-products and reducing wastewater volume. With regards to the entire production process, other prevention techniques include reducing material inputs, re-engineering processes to reuse by-products, improving management practices, and employing substitution of toxic chemicals (EPA, 1995).

Additional measures that can be employed are thorough risk assessment techniques, proper health surveillance and training, and safe work practices.

References

Barton, P., Hammer, C., & Kennedy, D. (1978). Analysis of cyanides in coke plant wastewater effluents. *Water Pollution Control Federation*, *50*(2), 234–239.

Bye, T., Romundstad, P., Ronneberg, A., & Hilt, B. (1998). Health survey of former workers in a Norwegian coke plant: Part 2. Cancer incidence and cause specific mortality. *Occupational and Environmental Medicine*, *55*(9), 622–626.

Cheremisinoff (2010). *Responsible care assessment of the ABC coke chemical plant – source term and emissions analysis*. N.P. Cheremisinoff, Ph.D., N&P Ltd.

Drummond (2010). *ABC coke coke operations*. Retrieved June 2010 from <http://www.drummondco.com/operations/coke/ABCCoke.aspx>.

Environmental Roadmapping Initiative (ERI) (2004). *Iron & steel*. Retrieved September 14, 2010 from <http://ecm.ncms.org/ERI/new/IRRironsteel.htm/>.

International Cyanide Management Institute (ICMI) (2006). *Environmental and health effects of cyanide*. Retrieved September 15, 2010 from <http://www.cyanidecode.org/cyanide_environmental.php>.

International Labour Organization (2005). *Safety and health in the iron and steel industry* (2nd ed.). Geneva: International Labour Office. Retrieved August 30, 2010 from <http://www.ilo.org/public/english/dialogue/sector/techmeet/meisi05/code.pdf>.

JCDOH (2008). *Major source operating permit – ABC coke, a division of Drummond Company, Inc., coke/utilities plant*. Jefferson County Department of Health, Air Pollution Program. Retrieved November 17, 2008.

Occupational Safety and Health Administration (2010). *Coke oven emissions*. Occupational Safety and Health Standards. Retrieved August 30, 2010 from <http://www.osha.gov/pls/oshaweb/owadisp.show_document?p_table=STANDARDS&p_id=10048>.

Proctor, D. M., Shay, E. C., Fehling, K. A., Finley, B. L. (2002). Assessment of human health and ecological risks posed by the uses of steel-industry slags in the environment. *Human and Ecological Risk Assessment*, *8*(4), 681–711.

Redmond, C., & Mazumdar, S. (1993). Long-term mortality studies of coke oven workers. *International Statistical Review*, *61*(2), 207–221.

US Department of Energy (DOE) (2000, August). *Energy and environmental profile of the US iron and steel industry, DOE/EE-0229*. Office of Industrial Technologies.

US Environmental Protection Agency (EPA) (1995). *Profile of the iron and steel industry*. Retrieved August 30, 2010 from <http://www.epa.gov/compliance/resources/publications/assistance/sectors/notebooks/iron.html>.

US Environmental Protection Agency (EPA) (2008a). *Effects of acid rain – surface waters and aquatic animals*. Retrieved August 31, 2010 from <http://www.epa.gov/acidrain/effects/surface_water.html>.

US Environmental Protection Agency (EPA) (2008b). *National emission standards for hazardous air pollutants for iron and steel foundries*. Retrieved August 30, 2010 from <http://www.epa.gov/ttn/atw/area/fr07fe08.pdf>.

US Environmental Protection Agency (EPA) (2009a). *Steel*. Resource Conservation. Retrieved August 4, 2010 from <http://www.epa.gov/osw/conserve/materials/steel.htm>.

US Environmental Protection Agency (EPA) (2010a). *Enforcement and compliance history online (ECHO). Detailed facility report.* Retrieved June, 2010 from <http://www .epaecho.gov/echo/compliance_report.html>.

US Environmental Protection Agency (EPA) (2010b). *EPA issues order to Tonawanda coke for Clean Water Act violations.* Retrieved September 14, 2010 from <http://yosemite. epa.gov/opa/admpress.nsf/0/348B56B5E62F2F4F8525778B004814C2/>.

US Environmental Protection Agency (EPA) (2010c). *Hazardous waste generator regulations.* Version 4: August 2010. Retrieved September 15, 2010 from <http://www.epa. gov/epawaste/hazard/downloads/tool.pdf>.

US Environmental Protection Agency (EPA) (2010d). *Listed wastes. Hazardous waste.* Retrieved August 30, 2010 from <http://www.epa.gov/osw/hazard/wastetypes/listed. htm>.

8 The Wood Treatment Industry

Chapter Contents

8.1 Overview

The wood treatment industry preserves wood through pressure or thermal impregnation of chemicals. These processes require the use of many toxic chemicals such as creosote coal tars, pentachlorophenol, and arsenicals to stave off attacks by fungi, bacteria, insects, and marine borers. Although these chemicals are extremely effective as pesticides, they are associated with many negative health effects for exposed workers. Historically, this industry has been one of the worst polluters in the United States, and it has been lagging in good housekeeping and source reduction practices that have been available for several decades.

Wood preservation involves preparing the wood for treatment, treating it with compounds, and then drying. The wood needs to be thoroughly dried before treatment begins, namely through steaming and vacuum; boiling under vacuum, or Boulton process; and kiln drying. Wood kiln air emissions can be significant, and there can be a large amount of wastewater generated from preparation processes. Post-seasoning, the wood is transported into treating cylinders, also known as retorts. Treatment can occur through pressure and non-pressure processes. Pressure processes rely on wood impregnation through force, while non-pressure processes rely on oil-borne preservatives applied with or without heat. Preservatives are repeatedly reused, resulting in accumulation of contaminated debris at the bottom of the retorts. Aerosols and vapors can also be released during chemical storage and mixing, solution storage, and opening of the retorts. After treatment, the wood is required by law to drip excess chemicals onto drip pads.

Risks of Hazardous Wastes. DOI: 10.1016/B978-1-4377-7842-7.00008-8

8.2 Chemicals Involved in the Wood Preserving Industry

8.2.1 Chemical Emission Sources in the Wood Preserving Industry

Solid waste:

- Timber preparation processes
- Sludge generated in the retorts (treating cylinders)
- Sludge generated from wastewater treatment
- Off-spec and damaged treated wood
- Sweepings and adsorbents from incidental spills and droppings of treating chemicals
- Miscellaneous sources such as kiln sticks.

Liquid waste:

- Wastewater from seasoning
- Post-treatment kickback drippings.

Gas waste:

- Fugitive vapor emissions from leaking piping aperture
- Fugitive dust from heavy machinery traffic
- Point source emissions from the stacks of boilers
- Air emissions from cylinder venting episodes
- Fugitive air emissions from the opening of treating cylinder doors
- Off-gassing from the surfaces of freshly treated wood
- Vaporization losses from treated wood stored in on-site stockyards.

8.2.2 Chemicals Emitted in the Wood Preserving Industry

- From incinerated toxic sludge:
    - Polycyclic aromatic hydrocarbons (PAHs)
    - Dioxins
    - Furans
- From kiln sticks:
    - Pentachlorophenol
    - Coal-tar creosote
- From seasoning:
    - Wastewater containing sap water and treating chemicals.

8.2.3 Overview of Select Chemicals

Coal-tar creosote, also referred to as creosote by the EPA, is formed by fractional distillation of crude coal tars. This compound is used as a wood preservative that acts as a fungicide and pesticide applied in a pressure treatment process, which involves pressurized treating cylinders that force pesticide penetration into the wood. It appears to be a thick amber to black oily liquid. The chemical composition of this compound varies depending on the origin of the coal and the nature of the distillation process. A typical mixture can consist of up to 10,000 different

Table 8.1 Six Major Classes of Compounds in Coal-Tar Creosote; Adapted from Cheremisinoff and Rosenfeld (2008)

Class of Compounds	Contains
Aromatic hydrocarbons	Polycyclic aromatic hydrocarbons (PAHs), alkylated PAHs, benzene, toluene, ethylbenzene, and xylene compounds (BTEX)
Tar acids/phenolics	Phenols, cresols, xylenols, and naphthols
Tar bases/nitrogen-containing heterocycles	Pyridines, quinolines, benzoquinolines, acridines, indolines, and carbazoles
Aromatic amines	Aniline, aminonaphthalenes, diphenylamines, aminofluorenes, and aminophenanthrenes, cyano-PAHs, benzacridine, and its methyl-substituted congeners
Sulfur-containing heterocycles	Benzothiophenes and their derivatives
Oxygen-containing heterocycles	Dibenzofurans

chemicals. The six major classes of compounds of coal-tar creosote include: aromatic hydrocarbons, tar acids/phenolics, tar bases/nitrogen-containing heterocycles, aromatic amines, sulfur-containing heterocycles, and oxygen-containing heterocycles (Table 8.1). In Canada and the European Union, there is restricted use of coal-tar creosote and there has been a movement toward a total ban of the product. The International Agency for Research on Cancer and the US EPA have determined that coal-tar creosote is a probable human carcinogen. Skin cancer and cancer of the scrotum have resulted in workers after long-term exposure to low levels of these chemicals, especially due to direct contact with skin during wood treatment or manufacture of coal-tar-creosote-treated products or in coke or natural gas factories (IARC, 1987). Eating contaminated food or drinking contaminated water results in burning in the mouth and throat and stomach pains. Additionally, brief exposure to high amounts of the product can cause severe irritation of the skin, chemical burns of eye surfaces, convulsions, mental confusion, kidney or liver problems, unconsciousness, or death. Longer exposure to low amounts of the product can result in sun sensitivity of the skin or irritation of the respiratory tract.

Some of the compounds in coal-tar creosote that have the most serious health effects are *polycyclic aromatic hydrocarbons* (PAHs). Creosote is typically composed of about 85% PAHs and 2–17% phenolic compounds (Bedient et al., 1984). Burchiel and Luster (2001) stated that PAHs are known to cause diseases in organ systems during organ and lymphoid cell development. One of the most harmful PAH chemicals is benzo[a]pyrene, which is a 'fingerprint' chemical for detecting the presence of PAHs. Benzo[a]pyrene exposure results in the most serious health effects to exposed workers; it has been found to be carcinogenic and immunotoxic (Burchiel and Luster, 2001). Other PAHs that are major components in coal-tar creosote include anthracene, phenanthrene, and pyrene (US EPA, 2008a).

Chromate copper arsenate (CCA) is a water-borne preservative, which leaves the wood surface clean and paintable. It is also applied through the pressure treatment process. It has been widely used to protect utility poles, building lumber, and foundations. This preservative is composed of chromium, copper, and arsenic, and its production was limited to industrial use and a few residential settings by the US EPA in 2003 (US EPA, 2008b). It is a restricted use product for use by only certain pesticide applicators. Typically if CCA is used in a dual treatment in conjunction with creosote, it is one of the most effective methods of protection against marine borers according to the AWPA. Although CCA is effective, there are serious associated health risks for organisms and workers who come into contact with the compound. Wood that is treated with CCA can leach the preservative into the surrounding environment, and the three elements that compose CCA are toxic and carcinogenic to many aquatic organisms (Weis and Weis, 1999). The presence of copper and chlorine in CCA-treated wood leads to the formation of dioxins or furans when combusted, furthermore generating volatile arsenic compounds and toxic organics in ash. In addition, CCA is an inorganic arsenic compound, which means that it is a known human carcinogen. The arsenic in CCA can cause skin, liver, intestinal, and bladder cancer.

Pentachlorophenol (PCP) is a chlorinated hydrocarbon that is a synthetic fungicide historically used as a pesticide, herbicide, and wood preserver. It is an effective oil-borne organic solution that is applied through the pressure treatment process. After using this compound, the life of preserved wood increases dramatically from approximately 7 to 35 years. Currently, PCP is only registered for restricted use as a heavy duty wood preservative by the EPA. The restricted use is related to the health problems associated with PCP products. Humans can be exposed to PCP through skin contact or inhalation, the most dangerous route of exposure. PCP has been classified as a probable human carcinogen associated with elevated risks for chloracne, soft tissue sarcoma, and non-Hodgkin's lymphoma (Jorens and Schepens, 1993).

PCP is a very highly toxic compound that is produced through a process that results in 84–90% purity. The impurities formed during manufacture, including *hexachlorobenzene* (HCB) and *chlorinated dibenzodioxins* and *chlorinated dibenzofurans* (CDDs and CDFs), are more toxic than PCP itself (US EPA, 2008c). HCB is a persistent, bio-accumulative, and toxic environmental pollutant that contaminates water and food-chain sources. Human exposure to HCB has been associated with skin lesions, nerve and liver damage, liver and kidney damage, reproductive effects, and cancer (US EPA, 2008c). CDDs and CDFs have been shown to be carcinogenic and cause endocrine, reproductive, developmental effects, immunotoxicity, hepatotoxicity, neurotoxicity, and a variety of cancers in animals and humans (Kogeninas, 2001). For example, the dioxin 2,3,7,8-TCDD has been classified by the EPA as a human carcinogen. Levels of dioxins in attic dust in communities surrounding wood treatment facilities have been shown to be many times higher than the U.S. EPA regulatory threshold for dioxins in residential soil. Furthermore, the PCP-associated dioxin congeners, octa-chlorinated dibenzo-p-dioxin (OCDD) and 1,2,3,4,6,7,8-hepta-chlorinated dibenzo-p-dioxin (1,2,3,4,6,7,8-HpCDD), have been

Table 8.2 Solid Wastes Listed as Hazardous Wastes by the EPA from Specific Sources (US EPA, 2010)

Industry and EPA Hazardous Waste no.	Hazardous Waste	Hazard Code
Wood preservation		
K001	Bottom sediment sludge from the treatment of wastewaters from wood preserving processes that use creosote and/or pentachlorophenol	(T)

found to be significantly higher in the blood of residents living near wood treatment facilities than in the general population (Feng et al., 2010).

8.3 Associated Hazardous Waste Laws and Regulations

EPA's so-called K-list details source-specific wastes from particular industries, such as petroleum refining, iron and steel, and wood preservation industries. Sludges and wastewaters from treatment and production processes in these sectors are examples of source-specific hazardous wastes. Table 8.2 lists EPA-recognized solid hazardous wastes specifically from the wood preservation industry.

The hazard codes in the third column are employed by administrators to signify the type of hazard that is associated with each hazardous waste: (T) Toxic, (C) Corrosive, (H) Acute, and (I) Ignitable.

8.4 Case Study – Koppers Tie Treating Facility, Somerville, TX

Koppers, Inc. owns a railroad tie treatment facility in Somerville, Texas. Wood was originally treated there for the Santa Fe Railroad under the Texas Tie and Lumber Preserving Company, and then it was absorbed by the Santa Fe Railway in 1950. The facility was then sold to Koppers, Inc. in 1995. Historically, the facility has produced crossties, switch ties, bridge timbers, and road crossing panels. Currently, wood treatment is limited to crossties and switch ties. The facility is located immediately adjacent to the Somerville community.

In Somerville, residents have been exposed to hazardous smoke, dust, and soot from the facility. Due to site-wide investigations of the facility since the 1980s, extensive contamination at the facility property has been documented. Site investigations have found a variety of hazardous substances, many of which are potent carcinogens, present in soils and groundwater. Extensive contamination has been documented to exist at the facility resulting from various Resource Conservation and Recovery Act (RCRA) investigations that have been performed at the site. In the past, hazardous waste releases resulted from operations, chemical and waste handling, process waste streams, landfilling, operation of the boilers, and other on-site activities. Environmental investigations attempting to characterize these releases appear to have started in the early 1980s.

On-site workers and members of the community in Somerville have been exposed to toxic chemicals from the facility over many years. Exposure resulted from direct contact with wood treatment chemicals; fugitive releases during wood treatment process operations; fugitive releases resulting from handling treated wood ties; emissions associated with burning untreated and treated wood shavings, scrap, and process sludge in the facility boilers; and exposure to contaminated soil, sludge and other contaminated wastes. In addition, the facility burned untreated and treated wood shavings and scrap, and process sludge containing treating chemicals in wood waste boilers. All of these forms of wastes, when burned, create dioxins under poor combustion conditions. Reportedly, ash from the boilers was mixed with sludge and creosote and then used to fill the pot holes in roadways, spread about in heavily trafficked areas to suppress fugitive dust, and spread in landfill areas. Process sludge was allowed to drip onto the bare ground over many decades of operations and burned and spread around the property. Besides enormous chemical emissions from the facility, there are undefined volumes of contaminated soil and groundwater at the facility resulting from the poor housekeeping practices that were allowed for nearly 100 years of operations.

8.5 Hazardous Waste Mitigation – Best Management Practices and Technologies in the Wood Preservation Industry

The drip pads used to collect excess chemicals after the treatment process should be up-to-date on regulations and inspections. Drip pads should have an overhead shed or secondary containment device to prevent storm water contamination. In addition, workers who work with drip pads should wear respirators to reduce inhalation or ingestion of aerosols containing treatment chemicals.

To minimize excess chemical drippage in unprotected areas of the plant, treated wood should be held inside cylinders or on drip pads until all drippage is eliminated. This time period depends on experience and should not be a set time. In addition, containment barriers should be set up at wood stockpiling areas so that in case of any drippage, it will be restricted to an area that is easily cleaned with minimal environmental implications.

Fugitive emissions should be controlled to reduce or eliminate the amount of worker exposure. Operators should have respiratory protection and chemical protective clothing to minimize associated health hazards. In addition, wood should be stockpiled for the minimum amount of time because volatile emissions can persist for more than 90 days. The stacking configuration of treated wood should change with seasonal conditions because high temperatures make PAHs more volatile and increase emissions. The stockpiled wood should be covered to reduce fugitive air emissions.

Door seals on the retorts should be inspected to make sure catastrophic spills and air emission releases do not occur. Also, overhead hoods can be installed over the retort doors to collect fugitive emissions and divert them to a wet scrubber or selective catalytic reduction burner.

Sludge and other solid wastes can be reduced by a variety of options. Retorts should be cleaned regularly, and wood that has been prepared should be cleaned of foreign matter such as sand, grit, and sawdust to reduce sludge generation. This will also allow for better quality of wood and reduce costs related to retort cleanings. In addition, kiln sticks or other wooden spacers should be replaced with metal chains or reusable spacers to reduce a solid waste stream. High-purity chemical feedstocks should also be used to reduce sludge generation while increasing cost savings by purchasing in bulk. Frequent inspection and cleaning of storage tanks should occur so that retorts are not contaminated.

Point source emissions can be reduced through various methods. Conical and tepee burner use should be eliminated. Wood-burning boilers should be replaced with natural gas-fired boilers so that they are cleaner burning. In addition, if wood-burning boilers are used, chlorine levels in the wood should be tested so that dioxin formation is minimized. Secondary pollution controls should be installed to ensure collection of both particulate matter and gaseous compounds.

References

Bedient, P. B., Rodgers, A. C., Bouvette, T. C., et al. (1984). Groundwater quality at a creosote waste site. *Groundwater*, *22*(3), 318–329.

Burchiel, S. W., & Luster, M. L. (2001). Signaling by environmental polycyclic aromatic hydrocarbons in human lymphocytes. *Clinical Immunology*, *98*(1), 2–8.

Cheremisinoff, N. P., & Rosenfeld, P. E. (2010). *Best practices in the wood and paper industries*. Burlington, MA: Elsevier Inc.

Feng, L., Wu, C., Tam, L., Sutherland, A. J., Clark, J. J., & Rosenfeld, P. E. (2010). Dioxin and furan blood lipid and attic dust concentrations in populations living near four wood treatment facilities in the United States. *Journal of Environmental Health*. In press.

International Agency for Research on Cancer (IARC) (1987). *Monographs on the evaluation of carcinogenic risk to humans. Supplement 7* (Vols. 1–47). IARC, World Health Organization.

Jorens, P. G., & Schepens, P. J. C. (1993). Human pentachlorophenol poisoning. *Human & Experimental Toxicology*, *12*(6), 479–495.

Kogevinas, M. (2001). Human health effects of dioxins: cancer, reproductive and endocrine system effects. *Human Reproduction Update*, *7*(3), 331–339.

US Environmental Protection Agency (EPA) (2008a, January). *Polycyclic Aromatic Hydrocarbons (PAHs) fact sheet* (pp. 1–3). Washington, D.C.: Office of Solid Waste.

US Environmental Protection Agency (EPA) (2008b). *Chromated Copper Arsenate (CCA)*. Office of Pesticide Programs. Retrieved November from <http://www.epa.gov/oppad001/reregistration/cca/>.

US Environmental Protection Agency (EPA) (2008c, September). *Reregistration eligibility decision for pentachlorophenol, EPA/739-R-08-008* (pp. 1−103). Washington, D.C.: Office of Pesticide Programs.

Weis, P., & Weis, J. S. (1999). Accumulation of metals in consumers associated with chromated copper arsenate-treated wood panels. *Marine Environmental Research*, *48*(1), 73−81.

9 The Paper and Pulp Industry

Chapter Contents

9.1 Overview

The pulp and paper industry manufactures pulp and paper from wood or recycled fiber. Pulp and paper mills use and generate materials that can pollute the air, water, and land, mainly in the pulping and bleaching stages. This industry is the largest industrial process water user in the United States according to the EPA Sector Notebook. Manufacturing activities at pulp and paper mills result in the release of many toxic substances into the surrounding communities.

The process of pulping consists of initial processing, washing, and bleaching. Pulping reduces the wood into a fibrous mat by separating the cellulose component from the lignin. It can be classified as chemical, mechanical, or semi-chemical; and of the chemical pulping methods, the kraft and sulfite processes are the most widely used. Wood is cooked in a digester, mechanically ground, or a combination of the two. High-temperature washing then removes dissolved lignin and chemicals, releasing exhaust gases containing hazardous air pollutants. Bleaching is then accomplished through elemental chlorine, elemental chlorine free, or totally chlorine free bleaching. The waste stream from bleaching can contain chlorine compounds and organics that can result in the formation of dioxins, furans, and chlorinated organics. This waste stream typically does not get treated at water treatment plants, accumulating in the discharge body of water. This process results in releases of toxic air and water pollutants.

Risks of Hazardous Wastes. DOI: 10.1016/B978-1-4377-7842-7.00009-X

9.2 Chemicals Involved in the Pulp and Paper Industry

9.2.1 Chemical Emission Sources in the Pulp and Paper Industry

- Chemical wood pulping:
  - Kraft
  - Sulfite
  - Neutral sulfite
  - Semi-chemical
  - Soda
- Bleaching
- Log processing equipment
- Digesters
- Stock washers
- Evaporators
- Lime kilns
- Power boilers
- Recovery furnaces
- Smelt tanks
- Paper machines
- Storage tanks and vessels
- Effluent treatment systems
- Coal and coke handling systems.

9.2.2 Chemicals Emitted in the Pulp and Paper Industry

- Criteria air pollutants:
  - Particulate matter (PM)
  - Ground-level ozone
  - Carbon monoxide
  - Sulfur oxides (SO_x)
  - Nitrogen oxides (NO_x)
  - Lead
- Other air pollutants:
  - Ammonia
  - Carbon dioxide
  - Carbonyl sulfide
  - Chlorine and chlorine dioxide
  - Chloroform
  - Dioxins and furans
  - Hydrogen chloride (as part of PM)
  - Methanol
  - Phenols
  - Total reduced sulfur compounds
  - Volatile organic compounds (VOCs)
- Liquid and solid pollutants:
  - Adsorbable organic halides (AOXs)
  - Pulping liquors
  - Bleaching effluents (chlorinated dioxins and furans, chloroform, etc.).

9.2.3 Overview of Select Chemicals

Particulate matter (PM) can be composed of material such as wood, lime, road dust, or chemical compounds created with carbon, metallic oxides and salts, acids, or oils. In the pulp and paper industry, PM is typically generated from the recovery boiler, lime kiln, smelt dissolving tank, power boilers, wood chip yard, and dust from landfills. The majority of the health effects from PM are from particles smaller than 10 microns (PM_{10}). $PM_{2.5}$ has a diameter so small that it is able to penetrate the lungs and deliver toxic loads to the body. Related health effects include chronic bronchitis, decreased lung function, asthma, irregular heartbeat, and premature death (US EPA, 2010a). Secondary particles are formed from chemical reactions involving precursors nitrogen oxides (NO_x), volatile organic compounds (VOCs), sulfur oxides (SO_x), and ammonia (NH_3).

Nitrous oxides (NO_x) are a category of compounds that includes NO_2, N_2O_4, N_2O_3, and N_2O_5. They are emitted from the lime kiln, recovery boiler, power boiler, gas turbines, and brown stock washers. NO_x is a harmful air contaminant that is a precursor to smog, ground-level ozone, fine particulates, and acid rain. It is known to be harmful to human health and vegetation growth. In humans, NO_x has been shown to cause asthma, emphysema, bronchitis, and premature death, and aggravate heart disease and reduce lung function (US EPA, 2010b). NO_x has also been shown to be irritating to the upper respiratory tract and lungs at low concentration; severe toxicity can result from one or two breaths of a high concentration (ATSDR, 2008). In addition, if children are chronically exposed to NO_x, bronchiolar damage can cause permanent restrictive and obstructive lung disease. NO_2 has many reproductive and developmental consequences, such as DNA damage and aberrations (ATSDR, 2008).

Sulfur oxides (SO_x) are a group of highly reactive gases that include SO_2, SO_3, and solid sulfates. These gases are produced whenever sulfur-containing compounds are burned, such as in the recovery boiler, lime kiln, power boilers, brown stock washers, and chip bins. SO_x are eye and lung irritants, and cause a wide range of adverse respiratory effects like bronchoconstriction and increased asthma (US EPA, 2009). In addition, SO_x is a precursor to fine PM formation and there is a connection between short-term exposure and increased hospital admissions for respiratory illness. SO_2, or sulfur dioxide, is the component of SO_x that is of the greatest concern and is used as an indicator for SO_x. SO_2 inhalation causes corrosive irritation to the respiratory tract and mucous membranes. Excess exposure to concentrations above exposure limits can result in chemical pneumonitis, pulmonary hemorrhage, and edema fluid buildup. Sulfuric acid causes bronchitis, emphysema, eye, nose, and stomach irritations, and potentially lung cancer.

Carbon monoxide (CO) produces adverse effects on the respiratory system. These gases can be emitted from the kraft process, including the power boiler and lime kilns (US EPA, 1995). CO is a toxic gas that has detrimental effects on the cardiovascular, central nervous, and pulmonary systems (US EPA, 1998). In addition, short-term variation in ambient CO levels has been shown to correlate with respiratory illness frequency, and ambient levels of carbon monoxide in air may contribute to respiratory illness and aggravation of existing respiratory diseases including asthma (US EPA, 2000; ATSDR, 2009). Human studies show that there are relationships between carbon monoxide and respiratory illness (Gordian et al., 1996). CO can also contribute to the formation of smog ground-level ozone that can trigger serious respiratory problems (US EPA, 2009b). CO is important since there is human visual impact at 50 ppm for 1 hour, death at more than 750 ppm, and vegetation impact at higher levels.

Volatile organic compounds (VOCs) are emitted as gases from certain solids or liquids, and this group of compounds includes chemicals such as terpenes, alcohols, phenols, methanol, acetone, and chloroform. In pulp and paper mills, VOCs are emitted from chip digesters, liquor evaporation, and pulp drying (in non-integrated mills) (US EPA, 2002). Irritation of the eye, nose, and throat can occur when exposed to VOCs. Some organics can cause cancer in animals and humans. Signs of exposure to VOCs include conjunctival irritation, nose and throat discomfort, headache, allergic skin reaction, dyspnea, declines in serum cholinesterase levels, nausea, emesis, epistaxis, fatigue, and dizziness (US EPA, 2010c). In addition, they play a significant role in chemical reactions that form ozone. When VOCs are combined with emissions of nitrogen in the presence of sunlight, ozone is formed. Ozone is harmful in the lower atmosphere causing many health problems such as lung tissue damage, lung function reduction, and adversely sensitized lungs.

Total reduced sulfur (TRS) is associated with foul odors from pulp and paper mills. It is composed of four reduced sulfur gases: hydrogen sulfide, methyl mercaptan, dimethyl sulfide, and dimethyl disulfide. According to the EPA, TRS compounds are responsible for the malodors associated with the production of pulp and paper (US EPA, 1998). These compounds are emitted from wood chip digestion, black liquor evaporation, and chemical recovery boiler process (US EPA, 2002). In addition, odor pollution caused by pulp mills has supported a link between odor and health symptoms such as headaches, watery eyes, nasal problems, and breathing difficulties (US EPA, 1998). Some TRS compounds in the air can be detected as a foul 'rotten egg' odor by humans at levels as low as one part per billion (US EPA, 2002). Furthermore, odorant stimulants of the nasal receptors associated with TRS emissions have been related to adverse reactions such as headaches, shortness of breath, nasal irritation, and even nausea and sinus congestion (US EPA, 1998). TRS compounds are toxic and heavier than air, thus traveling long distances to ground level.

Adsorbable organic halides (AOXs) are generated during the bleaching process through reactions between residual lignin from wood fibers and chlorine compounds. Compounds that end up in the wastewater stream of the pulp and paper mill are collectively referred to as AOXs. These compounds include chloroform, chlorate, chlorinated hydrocarbons, phenols, furans, dioxins, and vanillins (Savant et al., 2006). Many of these compounds have long half-life periods or are carcinogens and mutagens. AOXs with a low molecular weight are major contributors to mutagenicity and bioaccumulation through their ability to penetrate cell membranes. AOXs can exhibit toxicity and can bioaccumulate in fish tissue, and human consumption of these fish can result in elevated human health risk. In general, AOX have toxic effects that range from carcinogenicity to very acute and chronic toxicity (Savant et al., 2006).

Chloroform has been shown to be a potential human carcinogen. Long-term exposure can affect the liver or cause hepatitis and jaundice, while short-term exposure can negatively affect the central nervous system or cause dizziness and headaches. Chloroform is emitted from vents in the pulp bleaching process and bleach plant (US EPA, 1997). Chloroform is listed as one of the highest emitted Hazardous Air Pollutants from pulp and paper mills (US EPA, 1998). These Hazardous Air Pollutants are classified as possible, probable, or known human carcinogens (US EPA, 1998). In addition, chloroform is a suspected respiratory, cardiovascular, liver and kidney toxicant, and endocrine and neurological disruptor

Dioxins and furans have been shown to cause skin disorders, cancer, and reproductive effects, as well as affect the immune system. Unfortunately, dioxins and furans can be produced from the mixture of chemicals in the waste steam from the bleaching process in pulp and paper mills (US EPA, 1997). Another source of dioxins and furans is the recovery boiler and power boiler if burning 'salty' hog fuel. Exposure to dioxins and furans can cause skin disorders, cancer, and reproductive effects, as well as affecting the immune system (US EPA, 1997). Dioxins are a family of toxic chemicals that share a similar chemical structure and a common mechanism of toxic action, including seven polychlorinated dibenzo dioxins (PCDDs), ten polychlorinated dibenzo furans (PCDFs), and 12 polychlorinated biphenyls (PCBs). The most widely studied dioxin 2,3,7,8,TCDD is a well-known human carcinogen (US EPA, 2010d). Health effects associated with dioxins and the chemically similar polychlorinated biphenyls (PCBs) include reproductive effects, hyperactivity, allergies and immune and endocrine system malfunctions, diabetes, low birth weight, and poor motor skills for children.

9.3 Associated Hazardous Waste Laws and Regulations

The paper and pulp sector is subject to the same environmental regulations as other industry sectors, but the most significant recent rule-making activities affecting the pulp and paper sector are referred to collectively as the Cluster Rules (1998). This refers to a set of air and water rules that were issued simultaneously. The rules include:

- the Pulp and Paper NESHAP, specifying air emission standards for pulping and bleaching operations;
- the Effluent Limitations Guidelines and related water-quality standards for the pulp, paper, and paperboard category.

Estimates of the emissions reductions expected include:

- a 64% reduction in HAPs (by 153,000 tons per year, down from 240,000 tons emitted during 1996);
- a 450,000 ton per year reduction in overall VOC emissions (as a consequence of implementing the technology needed to meet the HAP rules);
- an 87,000 ton per year reduction in emissions of odor-causing reduced sulfur compounds (as a consequence of new source performance standards).

The water-quality standards specified under the Cluster Rules regulate concentrations of dioxins and furans (specifically TCDD and TCDF), as well as adsorbable organic halogens (such as chloroform) and chemical oxygen demand (COD). The standards represent a compromise that allows the substitution of chlorine dioxide for elemental chlorine in the bleaching process. A stronger set of standards, which would have required technologies that avoided the use of chlorine altogether, was not chosen.

The Regional NO_x Transport Rule (1998) requires 22 eastern states to institute measures to decrease their overall emissions of nitrogen oxides. The affected states were required to have controls on large industrial sources in place by 2003, and to meet overall NO_x limits by 2007. Since individual states have considerable flexibility in devising their own specific implementation plans, the effect on pulp and paper facilities has varied considerably, depending on location.

New standards for emissions of ozone precursors and fine particles affect some pulp and paper mills, particularly those impacting on ozone non-attainment areas.

The Regional Haze Rule was finalized in 1999. This calls for states to establish goals and develop long-term strategies for improving visibility, particularly in national parks and wilderness areas. Some pulp and paper mills located in certain regions (such as mills directly upwind of sensitive areas) have been called upon to meet more stringent emission limits for particulates and aerosol precursors.

The Total Maximum Daily Load (TMDL) program defines the maximum amount of pollutants a given body of water can receive and still meet water quality standards. Due to their high water use, pulp mills generally tend to be among the most significant impactors of the water bodies on which they are situated. State agencies are responsible for establishing effluent reduction levels for individual facilities.

The reader can obtain details of each of these rules from the EPA website at http://www.epa.gov/owow/tmdl/index.html.

9.4 Case Study – International Paper Facility, Prattville, AL

International Paper Company, which opened as Union Camp in 1967, is the largest industrial facility in the area of Prattville. Union Camp operated the facility until the company was purchased by International Paper in 1999. The manufacturing facility at the site is immediately surrounded by timberlands, water treatment-related surface water features, swampland, and other intervening land areas owned by IP and others. The mill facility is located at 100 Jensen Road and is situated approximately 3 miles south of the city center of Prattville. Dense residential communities are located within approximately 1 mile of the facility. Autauga Creek runs through the city of Prattville from the north and passes within 1,500 feet to the east of the mill. The city of Prattville lies primarily north of the mill facility.

As a result of pulp and paper manufacturing activities conducted at the Prattville Mill, vast quantities of atmospheric emissions of unhealthful, toxic, and malodorous substances have been and continue to be released into the surrounding community of Prattville. These emissions from the site occur through the mill's various operations of log processing equipment, digesters, stock washers, evaporators, lime kilns, power boilers, recovery furnaces, smelt tanks, paper machines, storage tanks and vessels, effluent treatment systems, coal and coke handling systems, and other equipment and facilities at the site.

Emissions from the Prattville mill have been documented in numerous complaints filed by members of communities surrounding the site as far as Montgomery and employees or contractors of the mill. Documents present in the IP documents indicate that foul odors have been reported many miles away from the mill facility. Various odor complaints describe respiratory issues, bleeding nose, burning chest, disturbed sleep, and damage to property. Subsequently, IP coordinated and managed the collection of 'odor panels' in the community surrounding the mill site to determine the extent of odorous complaints in the region surrounding the mill. These odor panels indicate that odors were detected frequently and also demonstrate that malodorous chemicals detected in the area of Prattville were likely associated with emissions from the mill.

Emissions from the mill have been reported by IP to state and federal environmental regulatory authorities. These emissions include a wide variety of Criteria Air Pollutants (CAPs) and Hazardous Air Pollutants (HAPs) including, but not limited to, particulate matter (PM), nitrogen oxides (NO_x), sulfur dioxide (SO_2), carbon monoxide, acetaldehyde, formaldehyde, hydrochloric acid (HCl), sulfuric acid (H_2SO_4), volatile organic compounds (VOCs), polycyclic aromatic hydrocarbons (PAHs), total reduced sulfur (TRS) compounds, polychlorinated dibenzo-p-dioxins (CDDs or dioxins), and dibenzofurans (CDFs or furans).

These emissions generally fall into four categories: VOCs, PM, dioxins/furans, and acid emissions and deposition. VOCs, specifically odorous compounds, are detectable in the Prattville community many miles from the facility and can cause or aggravate common respiratory health conditions in the exposed population. PM emissions from the mill can cause and contribute to property damage and also cause and aggravate respiratory health conditions. Dioxins/furans are potent human carcinogens and human system toxicants that bioaccumulate and promote cancer. Acidic compounds are associated with the emissions of PM, NO_x, SO_2, and HCl, and they are known to cause damage to property, adverse health effects, and acid rain.

9.5 Hazardous Waste Mitigation – General Guidance on Pollution Prevention (P2) and Cleaner Production in the Pulp and Paper Industry

In Chapter 8 of the *Handbook of Pollution Prevention Practices*, Cheremisinoff (2001) provides case studies and examples of successful P2 practices in this industry sector. A summary of some of these efforts are provided below. The majority of these practices have been aimed at solid waste and wastewater reductions.

The reader should note that pulp and paper are manufactured from raw materials containing cellulose fibers, generally wood, recycled paper, and agricultural residues. The main steps in pulp and paper manufacturing are: raw material preparation, such as wood debarking and chip making; pulp manufacturing; pulp bleaching; paper manufacturing; and fiber recycling.

Pulp mills and paper mills may exist separately or as integrated operations. Manufactured pulp is used as a source of cellulose for fiber manufacture and for conversion into paper or cardboard. Pulp manufacturing starts with raw material preparation; this includes debarking (when wood is used as raw material), chipping, and other processes such as depithing (for example, when bagasse is used as the raw material). Cellulosic pulp is manufactured from the raw materials, using chemical and mechanical means.

The manufacture of pulp for paper and cardboard employs mechanical and thermomechanical, chemimechanical, and chemical technologies. Mechanical pulping separates fibers by various techniques, which include disk abrasion and billeting. Chemimechanical processes involve mechanical abrasion and the use of chemicals. Thermomechanical pulps used for making newsprint and similar products are manufactured from raw materials by the application of heat or steam, in addition to mechanical operations. Chemimechanical pulping and chemithermomechanical pulping (CTMP) are similar methods but use less mechanical energy; these methods soften the pulp with sodium sulfite, carbonate, or hydroxide.

Chemical pulps are manufactured by the method of cooking known as digesting. The raw materials are digested using the kraft (sulfate) and sulfite processes. Kraft processes produce a variety of pulps used mainly for packaging and high-strength papers and board. In this method, wood chips are cooked with caustic soda to produce brownstock, which is then washed with water to remove the cooking liquor for the recovery of chemicals and energy. This recovered cooking liquor is referred to as 'black liquor' and in the past was sometimes called 'black gold' because of its value for energy recovery. Note that pulp is also manufactured from recycled paper.

Mechanical pulp can be used without bleaching to make printing papers for applications in which low brightness is acceptable – primarily, newsprint. For most printing, for copying, and for some packaging grades, the pulp has to be bleached. For mechanical pulps, most of the original lignin in the raw pulp is retained but is bleached with peroxides and hydrosulfites. In the case of chemical pulps (kraft and sulfite), the objective of bleaching is to remove the small fraction of lignin remaining after the digestion process. Oxygen, hydrogen peroxide, ozone, peracetic acid, sodium hypochlorite, chlorine dioxide, chlorine, and other chemicals are used to transform lignin into an alkali-soluble form. An alkali, such as sodium hydroxide, is necessary in the bleaching process to extract the alkali-soluble form of lignin.

In the bleaching process the pulp is washed with water. Oxygen is most often used in the first stage of bleaching. The trend is to avoid the use of any kind of chlorine chemicals and employ 'totally chlorine-free' (TCF) bleaching. TCF processes allow the bleaching effluents to be fed to the recovery boiler for steam generation; the steam is then used to generate electricity.

Elemental chlorine-free (ECF) processes use chlorine dioxide to bleach certain grades of pulp. The use of elemental chlorine for bleaching is not recommended.

Only ECF processes are acceptable and, from an environmental perspective, TCF processes are preferred. The soluble organic substances removed from the pulp in bleaching stages that use chlorine or chlorine compounds, as well as the substances removed in the subsequent alkaline stages, are chlorinated. The chlorinated organic substances are toxic and include dioxins, chlorinated phenols, and other chemicals. It is not practical to recover chlorinated organics in effluents, since the chloride content causes excessive corrosion.

Finished pulp is dried for shipment or may be used to manufacture paper on site (in an 'integrated' mill). Paper and cardboard are made from pulp by deposition of fibers and fillers from a fluid suspension on to a moving forming device that also removes water from the pulp. The water remaining in the wet web is removed by pressing and then by drying, on a series of hollow-heated cylinders (for example, calender rolls). Chemical additives are added to impart specific properties to paper, and pigments may be added for color.

The negative environmental impacts of the process result from the pulping and bleaching processes. In some processes, sulfur compounds and nitrogen oxides are emitted to the air, and chlorinated and organic compounds, nutrients, and metals are discharged to the wastewaters.

As noted in the previous chapter, in the kraft pulping process, highly malodorous emissions of reduced sulfur compounds, measured as total reduced sulfur (TRS) and including hydrogen sulfide, methyl mercaptan, dimethyl sulfide, and dimethyl disulfide, are emitted, typically at a rate of 0.3–3 kilograms per metric ton (kg/t) of air-dried pulp (ADP; note that ADP is defined by convention as 90% bone-dry fiber and 10% water). The World Bank Organization (WBO, 1998) reports the following emissions as typical:

- particulate matter, 75–150 kg/t;
- sulfur oxides, 0.5–30 kg/t;
- nitrogen oxides, 1–3 kg/t;
- volatile organic compounds (VOCs), 15 kg/t from black liquor oxidation.

In the sulfite pulping process, sulfur oxides are emitted at rates ranging from 15 to over 30 kg/t. Other pulping processes, such as mechanical and thermomechanical methods, generate significantly lower quantities of air emissions. Steam- and electricity-generating units using coal or fuel oil emit fly ash, sulfur oxides, and nitrogen oxides. Coal burning can emit fly ash at the rate of 100 kg/t of ADP.

The WBO also notes that wastewaters are discharged at a rate of 20–250 cubic meters per metric ton (m^3/t) of ADP. Wastewaters are high in: biochemical oxygen demand (BOD), at 10–40 kg/t of ADP; total suspended solids, 10–50 kg/t of ADP; chemical oxygen demand (COD), 20–200 kg/t of ADP; and chlorinated organic compounds, which may include dioxins, furans, and other adsorbable organic halides (AOX), at 0–4 kg/t of ADP. Wastewater from chemical pulping contains 12–20 kg BOD/t of ADP, with values of up to 350 kg/t. For mechanical pulping, wastewater discharges are 15–25 kg BOD/t of ADP. For chemimechanical pulping, BOD discharges are three to ten times higher than those for mechanical pulping.

Phosphorus and nitrogen are also released into wastewaters. The main source of nutrients, nitrogen, and phosphorus compounds is raw material such as wood. The use of peroxide, ozone, and other chemicals in bleaching makes it necessary to use a complexing agent for heavy metals such as manganese.

One of the greatest sources of solid wastes is wastewater treatment sludge (50–150 kg/t of ADP). Solid materials that can be reused include waste paper, which can be recycled, and bark, which can be used as fuel, whereas lime sludge and ash are typically disposed of in an appropriate landfill.

The discharge of chlorine-based organic compounds (from bleaching) and of other toxic organics are of great concern from pulp and paper mills. The unchlorinated material is mostly black liquor that has escaped the mill recovery process. Some mills have been reporting 100% recovery, which is a major accomplishment compared to a decade earlier. Industry developments demonstrate that totally chlorine-free bleaching is feasible for many pulp and paper products but may not be able to produce certain grades of paper.

There are a variety of P2 programs that the industry has focused on for reducing wastewater discharges and in some instances in minimizing air emissions. Some of the general areas where there has been limited success include:

- Use of energy-efficient pulping processes.
- Lowering aesthetic specifications by accepting less bright products.
- For less bright products such as newsprint, thermomechanical processes and recycled fiber may have been used.
- The generation of effluents has been reduced through process modifications and recycle of wastewaters.
- Reduced effluent volume and treatment requirements have been achieved by using dry instead of wet debarking methods, through recovering pulping chemicals by concentrating black liquor and burning the concentrate in a recovery furnace.
- Recovering cooking chemicals by recausticizing the smelt from the recovery furnace and using high-efficiency washing and bleaching equipment.
- Minimize unplanned or non-routine discharges of wastewater and black liquor, caused by equipment failures, human error, and faulty maintenance procedures, by training operators, establishing good operating practices, and providing sumps and other facilities to recover liquor losses from the process.
- Reducing bleaching requirements by process design and operation. The following measures have reduced emissions of chlorinated compounds to the environment: before bleaching, reducing the lignin content in the pulp (Kappa number of 10) for hardwood by extended cooking and by oxygen delignification under elevated pressure; optimizing pulp washing prior to bleaching; using TCF or, at a minimum, ECF bleaching systems; using oxygen, ozone, peroxides (hydrogen peroxide), peracetic acid, or enzymes (cellulose-free xylanase) as substitutes for chlorine-based bleaching chemicals; recovering and incinerating maximum material removed from pulp bleaching; where chlorine bleaching is used, reducing the chlorine charge on the lignin by controlling pH and by splitting the addition of chlorine.
- Minimization of sulfur emissions to the atmosphere has been achieved by using a low-odor design black liquor recovery furnace.
- Using energy-efficient processes for black liquor chemical recovery, preferably aiming for a high solid content (say, 70%) has also proven effective in reducing emissions and achieving greater energy efficiency.

References

Agency for Toxic Substances and Disease Registry (ATSDR) (2009). *Draft toxicological profile for carbon monoxide*. Agency for Toxic Substances and Disease Registry. Retrieved September from <http://www.atsdr.cdc.gov/toxprofiles/tp201.pdf>.

Agency for Toxic Substances and Disease Registry (ATSDR) (2008). Nitrogen Oxides (NO, NO_2, and others); CAS 10102-43-9; UN 1660 (NO); CAS 10102-44-0; UN 1067 (NO_2); UN 1975 (Mixture). *Medical Management Guidelines for Nitrogen Oxides*. Retrieved August from <http://www.atsdr.cdc.gov/mhmi/mmg175.pdf>.

Gordian, M. E., Ozkaynak, H., Xue, J., Morris, S. S., & Spengler, J. D. (1996). Particulate air pollution and respiratory disease in anchorage, Alaska. *Environmental Health Perspectives*, *104*(3), 290–297.

Savant, D. V., Abdul-Rahman, R., & Ranade, D. R. (2006). Anaerobic degradation of adsorbable organic halides (AOX) from pulp and paper industry wastewater. *Bioresource Technology*, *97*(9), 1092–1104.

US Environmental Protection Agency (EPA) (1995, January). AP 42, 5th ed., Vol. I: Chapter 10.2: Chemical wood pulping. *Technology transfer network – clearinghouse for inventories and emission factors*.

US Environmental Protection Agency (EPA) (1997, November). *EPA fact sheet – the pulp and paper industry, the pulping process, and pollutant releases to the environment, EPA-821-F-97-011* (pp. 1–2). Washington, D.C.: Office of Water.

US Environmental Protection Agency (EPA) (1998). National emission standards for hazardous air pollutants for source category: Pulp and paper production; effluent limitations guidelines, Pretreatment standards, and New source performance standards: Pulp, paper, and paperboard category; Environmental protection agency; Final rules. *Federal Register*, *63*(72), April.

US Environmental Protection Agency (EPA) (2000). *Air quality criteria for carbon monoxide*. United States Environmental Protection Agency. EPA 600/P-99-001F. June.

US Environmental Protection Agency (EPA) (2002, November). *EPA office of compliance sector notebook project – profile of the pulp and paper industry, EPA/310-R-02-002* (pp. 6–10). Washington, D.C.: Office of Enforcement and Compliance Assurance.

US Environmental Protection Agency (EPA) (2009). *Sulfur dioxide – health*. Retrieved November from <http://www.epa.gov/air/sulfurdioxide/health.html>.

US Environmental Protection Agency (EPA) (2009b). *Carbon monoxide – health and environmental impacts of CO*. Retrieved November from <http://www.epa.gov/air/urbanair/co/hlth1.html>.

US Environmental Protection Agency (EPA) (2010a). *Particulate matter – health and environment*. Retrieved August from <http://www.epa.gov/air/particlepollution/health.html>.

US Environmental Protection Agency (EPA) (2010b). *Nitrogen dioxide – health*. Retrieved January from <http://www.epa.gov/air/nitrogenoxides/health.html>.

US Environmental Protection Agency (EPA) (2010c). *Volatile organic compounds – an introduction to indoor air quality*. Retrieved April from <http://www.epa.gov/iaq/voc.html#Health%20Effects>.

US Environmental Protection Agency (EPA) (2010d). *Dioxins and furans – Persistent Bioaccumulative and Toxic (PBT) chemical program*. Retrieved February from <http://www.epa.gov/pbt/pubs/dioxins.htm>.

10 Nuclear Waste and Tritium Releases

Chapter Contents

10.1 Introduction

As of July 2010, 14.0% of the world's total power output was produced by nuclear power plants (International Atomic Energy Agency, 2010). There are numerous other sources of nuclear waste, including medical waste, industrial waste, and tailings from naturally occurring radioactive materials in metallic ores, coal, oil, and gas. Nuclear waste, as with most other solid waste, is defined as radioactive materials for which there is no further use (NRC, 2009).

While the acute dangers of exposure to radioactive substances were first noted in 1896 by Nikola Tesla, the mutagenic effects of long-term exposure to radioactive substances were not discovered until 1927 by American geneticist Hermann Joseph Muller (EPA, 2010). By 1945, the United States was forced to consider the issue of nuclear waste disposal as the first nuclear weapons were being developed. In 1946, the first piece of federal legislation regulating nuclear waste, the Atomic

Risks of Hazardous Wastes. DOI: 10.1016/B978-1-4377-7842-7.00010-6

Energy Act, was passed by Congress. Since then, many countries have established regulations for managing nuclear waste and developed increasingly advanced technologies for containing and disposing of this waste. Nonetheless, there are still *no permanent solutions* anywhere on earth for the storage of high-level nuclear waste, such as spent fuel rods or nuclear weapons waste. While some storage solutions for high-level waste will be operational well into the foreseeable future, all of the solutions currently employed are considered interim solutions until permanent ones such as a deep geological repository are found (IAEA, 2000).

The disposal of radioactive waste is regulated in the United States by the Nuclear Regulatory Commission (NRC), in the European Union by the European Atomic Energy Community (EAEC), and globally by the United Nations Atomic Energy Commission (UNAEC) and the International Atomic Energy Agency (IAEA). Radioactive materials decay over time by emitting ionizing particles or radiation until they are no longer radioactive. Thus the goal of radioactive waste management is to protect the public from harm from these wastes until they no longer pose a threat. Because some materials require significant lengths of time to decay, oftentimes radioactive waste management involves the formidable task of attempting to protect the public for hundreds of thousands to millions of years from the dangers of these wastes.

10.2 Types and Sources of Nuclear Waste

10.2.1 Radioactive Elements

Nuclear wastes usually contain a mixture of different radioactive elements (radioisotopes), each with its own physical properties and degradation characteristics. The half-life of a radioisotope is the time it takes for that particular isotope to lose half of its radioactivity. Some materials have a short radioactive half-life and remain radioactive for only a few days, while others have relatively long half-lives and will remain radioactive for hundreds of thousands (plutonium-239) to millions of years (iondine-129) (NRC, 2009). Radioactive wastes consist of naturally occurring radioisotopes that were present in the material before it was used, as well as man-made radioisotopes that were produced during the use of the nuclear material (i.e. fission). Uranium 238 (^{238}U) is the most abundant radioisotope found in nature, has a half-life of 4.468 billion years, and is also the most abundant radioisotope found in nuclear wastes from power plants. Adding to the danger of radioisotopes is the fact that most decay into multiple lesser-weight radioisotopes before reaching a stable one. Each isotope in these 'decay chains' can have different chemical and physical properties from the parent element, generating a waste whose chemical and physical properties are constantly changing over time (NRC, 2009).

Radioactive waste is classified (in the US and in most other countries) as either high-level waste or low-level waste. These classifications are important in mandating how the waste is regulated, transported, disposed of, and stored.

10.2.2 Low-Level Waste

Low-level radioactive waste is made up of radioisotopes with shorter half-lives and generally less radioactivity. Low-level waste can be liquid, solid, or gaseous, although the largest volume of low-level waste produced is liquid. Liquid low-level waste typically consists of the cooling water that flows through a nuclear power plant and can pick up radioactivity from the leakage of radioactive fuel. This cooling water can also become radioactive when high-energy neutrons from the energy production process strike the cooling water and form radioisotopes. These radioisotopes are easily transported in the coolant system and are the largest source of occupational radiation exposure. Of particular interest are the isotopes nitrogen-16 (N-16) and tritium (H-3). N-16 is formed from normal oxygen and tritium is formed from deuterium (or 'heavy hydrogen'), which is an isotope of hydrogen that is found in varying concentrations in water. Solid low-level waste usually consists of objects such as tools, clothing, filter cartridges, etc., that come in to contact with the cooling water or from solid parts of decommissioned nuclear reactors. Gaseous low-level waste typically comes from the evaporation of coolant water, and is usually compressed and stored on-site. The level of radioactivity and half-lives of radioactive isotopes in low-level waste are relatively small. Storing the waste for a period of 10–50 years will allow most of the radioactive isotopes in low-level waste to decay, at which point the waste can be disposed of as normal refuse (NRC, 2009). However, low-level waste still poses a threat to environmental and human health, because there are such large volumes of it and because (unlike high-level waste) low-level waste is eventually released into the environment like normal refuse.

10.2.3 High-Level Waste

High-level nuclear waste is either higher in radioactivity than low-level waste, longer lived than low-level waste, or both. The vast majority of high-level nuclear waste consists of spent fuel rods and materials used in the reprocessing of spent nuclear fuel and decommissioned nuclear weapons. Nuclear fuel reprocessing is a controversial process by which spent nuclear fuel is reprocessed to extract uranium. Reprocessing of high-level waste is controversial because it produces large quantities of liquid high-level waste from the solvents used, and also because extracted plutonium or uranium is easily amenable to illicit nuclear weapons. While nuclear fuel reprocessing was officially banned in the United States in 1976, the Department of Defense still practices reprocessing at some of its defense reactors (NRC, 2010). A number of private applications for the commercial reprocessing of fuel were recently filed with the NRC and there is considerable governmental dialogue as to whether or not the ban on commercial reprocessing in the US should be lifted (NRC, 2010).

High-level nuclear wastes present a more severe exposure because they are usually high temperature, highly radioactive, and extremely long-lived. High-level waste is extremely dangerous to human and animal life, particularly if subjected to direct exposure. The average spent fuel assembly 10 years after being removed from production still emits 10,000 rem/hour of surface dose radiation (the röntgen equivalent in man (or mammal) or rem is a traditional historical unit of radiation dose equivalent. It is the product of the absorbed dose in rads and a weighting factor which accounts for the effectiveness of the radiation to cause biological damage. 1 rem = 0.01 sievert (Sv)); and a lethal dose for humans is about 500 rem in a single exposure (NRC, 2009). High-level wastes also represent an indirect exposure hazard from the possibility of leaking into ground water or rivers where they could enter food chains, and also from the low levels of radiation that will irradiate the environmental media surrounding the storage site during normal operation. At present, a large quantity of high-level nuclear waste has already been released into the environment from nuclear power plant accidents (e.g. Chernobyl) and from the willful dumping of said waste into the environment.

10.2.4 Mill Tailings

A large volume of radioactive waste comes from mill tailings, which are the sand-like residues left behind from the mining of uranium ore. Only 1% of the mined material actually contains uranium and the rest is discarded on-site as large piles of radioactive mill tailings. These piles pose a health hazard to anybody in the vicinity of the mine because the radioactive materials in these mill tailings release high-energy gamma rays. In addition, wind and water distribute these tailings into the environment where they can cause contamination of ground or surface water that may eventually be used for drinking (EPA, 2007). Radioactive mill tailings were used in the past for construction material and for building foundations. Uranium mill tailings contain the element radium which decays into radon gas, a highly radioactive gas that significantly increases one's risk for lung cancer when exposed. Radon exposure is the second leading cause of lung cancer in the United States, many cases of which are attributed to exposure to uranium mill tailings (National Cancer Institute, 2004).

Twenty-five of the 26 uranium mills in the United States are located west of the Mississippi, mostly in dry climates. The combined weight of all of the licensed mill tailings piles in the US is approximately 200 million metric tons. Since these tailings piles represent a significant public exposure hazard, monitoring and mitigation of these piles by the EPA, the NRC, and the Department of Energy continues (EPA, 2007).

10.3 Management and Storage

10.3.1 Our Nuclear Burden

The quantity of nuclear waste accumulating on earth is staggering. As of 2008, there was an estimated 5,479,825 cubic meters of low- to intermediate-level nuclear waste and 363,574 cubic meters of high-level nuclear waste in storage across the earth (IAEA, 2010). These numbers will continue to rise as concerns over air quality and carbon emissions typically associated with other forms of energy production motivate countries to construct new nuclear power plants. In China, for example, 24 large nuclear power plants are currently under construction, 11 of which are scheduled to be completed between 2012 and 2015 (World Nuclear Association, 2010). The United States is also poised for an escalation of nuclear power plant construction. The Energy Policy Act of 2005 included incentives such as production tax credits and loan guarantees for the construction of new power plants. More recently, in February of 2010, the administration of President Obama approved an $8 billion loan guarantee for the construction of two new nuclear reactors in Georgia (Fahey, 2010). If these plans go through, these would be the first nuclear power plants built in the US since the 1970s. Each nuclear power plant produces approximately 20 metric tons of waste in the form of spent nuclear fuel every year (Nuclear Energy Institute, 2010).

10.3.2 How Nuclear Waste is Stored Today

In 2006 the International Atomic Energy Agency (IAEA) released a document entitled the Fundamental Safety Principles of Radioactive Waste Management. In it, they stipulated the general principles that should be kept in mind when managing or storing radioactive waste (IAEA, 2006).

Among these are:

- *Protection of present and future generations:* People and the environment, present and future, must be protected against radiation risks;
- *Prevention of accidents:* All practical efforts must be made to prevent nuclear or radiation accidents;
- *Emergency preparedness and response:* Arrangements must be made for emergency preparedness and response in case of nuclear or radiation incidents;
- *Protective actions to reduce existing or unregulated radiation risks:* These must be justified and optimized.

It is questionable whether any country on earth has adhered to these principles in their entirety while managing their nuclear wastes. The most glaring violation of these principles seems to occur most frequently in the category of 'protection of future generations.' As stated earlier, there are no permanent solutions for the disposal of high-level radioactive waste. In the United States, nuclear waste is stored on-site at the nuclear power plant from which it originates, either in large water-cooled pools or in dry storage casks. In the past, spent nuclear fuel was stored exclusively in large water-cooled pools. The water allowed for the highly radioactive materials to cool down and to lose radioactivity from decay. Since the 1970s many on-site fuel storage pools have reached capacity and the NRC has approved two new techniques for the storage of spent nuclear fuel. The first is dry storage casks, which have been used since 1986 to store high-level wastes and which are encumbered by problems like questionable structural integrity in the event of a natural disaster, difficulty to transport, and a relatively low storage life of 100 years

(NRC, 2007). Even more concerning is a method of storage approved by the NRC called 're-racking,' which has been done on nearly all fuel storage pools in the US (NRC, 2007). During re-racking, fuel storage pools are packed beyond their original capacity to densities approaching the inside of an active reactor. Precautions are taken to ensure that the stored fuel will not undergo an uncontrolled nuclear chain reaction; however, a loss of coolant accident remains a serious concern. A study done at the Sandia National Laboratory in 1979 showed that if the water in a re-racked storage pool was lost, a fire would start resulting in the catastrophic release of nuclear materials, exceeding the amount released at Chernobyl (Benjamin et al., 1979; Alvarez et al., 2003).

A small quantity of nuclear waste in the US is shipped off-site to one of three NRC-commissioned storage facilities across the US. Liquid high-level wastes are stored in underground tanks made of either stainless steel or carbon steel, depending on whether the wastes are acidic or basic, respectively. Some high-level liquid waste has been solidified into glass or other ceramics in a process known as vitrification (NRC, 2009).

10.3.3 The Yucca Mountain Question

In 1983, with the passage of the Nuclear Waste Policy Act, the US government was mandated with providing a deep geologic repository for spent fuel and high-level nuclear waste. This deep geologic repository is expected to take the place of on-site waste storage which remains a local environmental hazard for the more than 100 nuclear power plants across the United States. Yucca Mountain, a volcanically formed ridge located in south-central Nevada, was studied as a possible nuclear waste repository in 1978 by the Department of Energy. The storage of waste at Yucca Mountain has yet to be realized due both to political opposition and to lingering environmental and national security concerns. According to the State of Nevada Agency for Nuclear Projects, the volcanic material of Yucca Mountain is unsuitable for the containment of radioactive waste because it is over 80% saturated with water, has billions of liquid-conducting fractures, and also puts the proposed storage site above the local water table (State of Nevada, 2003). There are also numerous environmental and security concerns about transporting high-level nuclear waste across the US to the Yucca Mountain site and about centralizing all of the country's waste in one location.

10.4 The Hazards of Nuclear Waste

10.4.1 Human Health Hazard

As stated earlier, the acute dangers of exposure to radioactive substances were first noted in 1896 by Nikola Tesla, but the mutagenic effects of long-term exposure to radioactive substances were not noted until much later. Ionizing radiation released by nuclear materials causes DNA damage, which often results in mutations that lead to cancer (EPA, 2010). Because any amount of radiation can cause potentially carcinogenic DNA damage, the health risk from radiation exposure has no minimum threshold. This point is underscored by the NRC, which sets public radiation exposure limits, and which states that 'any exposure to radiation poses some health risk ... no matter how small' (NRC, 2010).

Certain radioactive materials can pose an increased human health risk if they have particularly long half-lifes, or if they are biologically important materials that can concentrate in the body.

For example, the radioactive element strontium-90 is chemically similar to calcium, and if it is ingested will seek bone and lodge itself there. This greatly increases the risk of bone malignancies (Howard et al., 1969). Another example is the radioactive isotope of iodine (iodine-131) which concentrates in the thyroid as it is ingested over a period of years and is implicated in causing thyroid cancer (Nagataki and Nystrom, 2002). This compound is of particular importance because it has been released by many nuclear facilities in the US and was a major component of the radioactive material released from Chernobyl.

Another radioactive compound of interest is tritium, or H-3, which is the radioactive isotope of hydrogen. Tritium is an important radioisotope because it joins together with oxygen to form tritiated water, which acts almost identically to normal water. Thus, tritiated water distributes itself through the environment as regular water does, into rivers and lakes, into vegetation, into wildlife, into drinking water supplies, and eventually into humans (Schell and Sauzay, 1973). Tritium is an important component in nuclear weapons production and is also a by-product of nuclear power production and has been released into the environment from both of these enterprises. Tritium has a radioactive half-life of 12.3 years and was thought to have a biological half-life of only 10–14 days (Rudy and Jordan, 1977). However, many studies have shown that once tritium becomes part of an organic molecule it can remain in the body for up to a year or more (Moghissi et al., 1971; Hatch and Mazrimas, 1972).

10.4.2 Case Studies

In order to fully understand the hazards associated with nuclear power production and nuclear waste storage, a number of case studies from nuclear projects in the United States are presented below. There are countless examples of nuclear waste releases in other countries that will not be discussed here, although for a particularly troubling nuclear history one should examine Russia, where millions of gallons of nuclear waste, 18 nuclear reactors, and two nuclear warheads were indiscriminately dumped into the Arctic Sea (Broad 1993).

10.4.3 Rocky Flats Plant

From 1952 to 1992 the United States government operated a plutonium production facility at Rocky Flats, Colorado, a desolate plateau only 15 miles from Denver. This facility produced plutonium triggers used to detonate atomic bombs and at one point was the second largest plutonium repository in the United States. Like many other nuclear weapons production facilities, it has a troubling history of nuclear materials releases, local contamination, and worker safety issues.

In 1959 employees of the plant discovered that radioactive waste was leaking from storage drums onto an open field on which they were kept. This information was not revealed to authorities until 1970 when radioactive soil particles were found throughout the metro area of Denver. These particles, whose radioactivity levels were 15 million times higher than normal soil particles, were traced back to the Rocky Flats facility where they had been picked up by the wind and distributed for tens of miles around Colorado. A similar incident occurred in 1967 when 3,500 barrels of plutonium waste were found to be leaking large quantities of waste onto a storage pad. This waste was also carried away from the Rocky Flats area by winds and dispersed throughout Colorado. The storage pad was covered with gravel and paved over with asphalt, yet elevated plutonium levels were still found in the surrounding topsoil up to 5 years later (Department of Energy, 2010).

In 1973 it was discovered that tritium was flowing from on-site creeks into a nearby river and into a reservoir that supplied drinking water to nearby residents. Plutonium was also found to be washing off-site from improperly built holding ponds that were storing plutonium-contaminated wastes. It is not known exactly how much tritium or plutonium was released into the environment, but the highest tritium concentrations measured in a nearby reservoir were 20 times above the normal background tritium levels. In addition, urine samples taken from 36 people living in an adjacent town were found to have tritium levels of 4,300 picocuries per liter, seven times higher than the normal urinary concentration of 600 picocuries per liter (Colorado Department of Public Health and Environment, 2010).

In an attempt to make waste removal easier, the operators of the Rockwell plant mixed concrete into contaminated waste pools to form a solid radioactive waste called 'pondcrete.' In 1988 it was discovered that this 'pondcrete,' which was stored on-site in large blocks, was improperly cured and was leaking radioactive sludge with an average half-life of 24,000 years into the ground (DOE, 2010). After a PCB leak was discovered and a large quantity of chromic acid was accidentally released into the plant's water supply, Rocky Flats was shut down following a joint raid by the FBI and the EPA. Overall, the operations at the Rocky Flats plant resulted in the release of large quantities of radioactive waste into the environment and in exposures to the public that have been largely unquantified and underinvestigated.

10.4.4 Vermont Yankee Nuclear Power Plant

The Vermont Yankee nuclear power plant provided the state of Vermont with 73% of its power needs in 2006. Since then, its power production has been cut by 50% and it is set to go offline in 2012 due to various lapses in safety and episodes of local contamination (US Energy Information Administration, 2009). First, in 2004, a federal inspection revealed that two fuel rods that were supposed to be in the spent fuel storage pool were missing and unaccounted for (NRC, 2007). In 2007, a three-story portion of one of the plant's cooling towers collapsed, releasing cooling water back into the Connecticut river from which it is drawn. The plant came under intense scrutiny in January of 2010 when it was discovered that tritium was leaking into the groundwater near the site. At this time, the levels of radiation were at 20,000 picocuries per liter, which is the federal limit for radiation in drinking water (Gram, 2010). By February of 2010, sampling of groundwater at the Vermont Yankee site revealed radiation levels to be 834,000 picocuries per liter (approximately 42 times the federal limit). The source of this radiation was determined to be contaminated underground pipes that were pumping high-level waste throughout the plant and which the executives of the power plant denied of existence (Wald, 2010).

10.4.5 The Hanford Site

The Hanford nuclear production facility located in Hanford, Washington, was the largest plutonium production facility in the United States for most of its operational period from 1943–1989 and shortly thereafter became the largest and most complex environmental cleanup project in the US (EPA, 2010). It is also the most contaminated nuclear facility in the US. While manufacturing the plutonium used in over 60,000 nuclear weapons, the Hanford site produced more than 43 million cubic yards of radioactive waste, over 130 million cubic yards of contaminated soil and debris, 475 billion gallons of contaminated water deposited on soil, and over 80 square miles of contaminated groundwater (EPA 2010).

During normal operation of the eight nuclear reactors at Hanford, cooling water was drawn in from the Columbia River, the largest river in the Pacific Northwest. This cooling water was consistently contaminated with 11 different radioisotopes with half-lives ranging from 15 hours to 245 days, and then released back into the Columbia River. The public was exposed to high doses of radiation from eating contaminated fish and shellfish, drinking contaminated water, swimming or boating in the Columbia River, and spending time along the river shoreline (Washington State Department of Health, 2004).

Over 200 radioisotopes, some in large quantities, were also released into the air from the Hanford site. During only 5 months of operation from May to December of 1945, the Hanford site released more than half of all of the radioactive iodine (iodine-131) released during the next 30-year period (a total of 740,000 curies). For comparison, the accident at the Three Mile Island nuclear power plant in 1979 released only about 20 curies of iodine-131. The largest single release of radioactive iodine, however, was actually done intentionally in a secret military experiment called the 'Green Run,' which sought to test the air monitoring equipment of the US Air Force. During this experiment the US Air Force deliberately released 7,000–12,000 curies of iodine-131 over the course of a single day. Nothing was done to mitigate this exposure and it was kept entirely secret until 1986 when the US Department of Energy released 19,000 pages of classified documents relating to the Hanford site. An estimated 200,000 curies of tritium were released from the Hanford site into the air between 1949 and 1954. Tritium continued to be released into the air during the rest of the site's lifetime although this amount is currently unquantified.

Soil and groundwater contamination remains the largest problem at the Hanford site to this day. According to the Washington State Department of Ecology, more than 1 million gallons of high-level radioactive waste have leaked into the soil from storage tanks at the Hanford site (Washington State Department of Ecology, 2007). Despite significant cleanup efforts, 177 tanks containing 53 million gallons of high-level radioactive waste still remain stored underground on-site. Most of these tanks (149) are single-shelled tanks that are highly prone to leakage. The contamination plume has reached groundwater 200 feet below the Hanford site and is expected to reach the Columbia River in 12–50 years, where it will pose an enormous public health hazard. The solid high-level radioactive waste buried in landfills at the Hanford site contains 5–6 million curies of radioactivity (Washington State Department of Health and Environment, 2004).

The Hanford site has been divided into four separate Superfund sites designated for cleanup by the EPA, the Washington Department of Ecology, and the Department of Energy. This cleanup began in 1989 and was expected to be completed by 2029; however, with continuing delays in the cleanup process the completion year has been pushed back to 2047 (Stiffler, 2008). Currently, a workforce of 11,000 and a budget of $1.8 billion dollars per year are devoted to cleanup activities at the Hanford site (Washington State Department of Ecology, 2007). Nonetheless, the Hanford site remains the most contaminated nuclear facility in the United States and will continue to pose a significant public health threat for many years to come.

References

Alvarez, R., Beyea, J., Janberg, K., Kang, J., Lyman, E., Macfarlane, A., et al. (2003). Reducing the hazards from stored spent power-reactor fuel in the United States. *Science & Global Security*, *11*(1), 1–51.

Benjamin, A. S. (1979). *Spent fuel heatup following loss of water during storage*. Sandia National Laboratory. NUREG/CR-0649, SAND77-1371.

Broad, W. J. (1993). Russians describe extensive dumping of nuclear waste. *New York Times*. Retrieved April 27 from <http://www.nytimes.com/1993/04/27/science/russians-describe-extensive-dumping-of-nuclear-waste.html?pagewanted=all>.

Colorado Department of Public Health and Environment (2010). *Rocky Flats historical public exposure studies*. <http://www.cdphe.state.co.us/rf/index.htm>.

Department of Energy (DOE) (2010). *Rocky Flats history*. Retrieved August 10 from <http://www.lm.doe.gov/WorkArea/linkit.aspx?LinkIdentifier=id&ItemID=3026>.

US Environmental Protection Agency (EPA) (2010). Health effects of radiation exposure. *Radiation protection program*. Retrieved February 23, 2010 from <http://www.epa.gov/rpdweb00/understand/health_effects.html>.

EPA (2007). Uranium mill tailings. *Radiation protection program*. Retrieved October 4, 2010 from <http://www.epa.gov/rpdweb00/docs/radwaste/402-k-94-001-umt.html>.

EPA (2010). *Region 10 superfund*. Hanford, Washington. Retrieved May 14 from <http://yosemite.epa.gov/r10/cleanup.nsf/1887fc8b0c8f2aee8825648f00528583/2f133ac95a7d2684882564ff0078b367!OpenDocument>.

Fahey, J. (2010). *Southern Co., Westinghouse win first nuke loan guarantee*. Forbes Online. Retrieved February 16 from <http://blogs.forbes.com/energysource/2010/02/16/southern-co-westinghouse-win-first-nuke-loan-guarantee/>.

Gram, D. (2010). Vt. Yankee well test shows radioactive isotope. *ABC News*. Retrieved January 7 from <http://abcnews.go.com/Business/wireStory?id=9507774>.

Hatch, F. T., & Mazrimas, J. A. (1972). Tritiation of animals from tritiated water. *Radiation Research*, *50*(2), 339–357.

Howard, E. B., Clarke, W. J., Karagianes, M. T., & Palmer, R. F. (1969). Strontium-90-induced bone tumors in miniature swine. *Radiation Research*, *39*(3), 594–607.

IAEA (2010). Consolidated radioactive waste inventory. *Net-enabled radioactive waste management database*. <http://newmdb.iaea.org/datacentre.aspx>.

IAEA (2000, September 19–20). Radioactive waste management – turning options into solutions. *3rd Scientific Forum*. Vienna, Austria.

IAEA (2006, November 15). *Fundamental safety principles*. IAEA STI/PUB/1273.

International Atomic Energy Agency (IAEA) (2010, July). *Nuclear power reactors in the world*. IAEA-RDS-2/30, Reference Data Series No. 2. Vienna, Austria.

Moghissi, A. A., Carter, M. W., & Lieberman, R. (1971). Long-term evaluation of the biological half-life of tritium. *Health Physics*, *21*(1), 57–60.

Nagataki, S., & Nystrom, E. (2002). Epidemiology and primary prevention of thyroid cancer. *Thyroid*, *12*(10), 889–896.

National Cancer Institute (2004). *Radon and cancer: Questions and answers*. US National Institutes of Health. Retrieved July 13, 2004 from <http://www.cancer.gov/cancertopics/factsheet/Risk/radon>.

United States Nuclear Regulatory Commission (NRC) (2009, August 18). *Backgrounder on radioactive waste*. Rockyville, Maryland.

NRC (2010) *Backgrounder on tritium, radiation protection limits, and drinking water standards*. Retrieved February 17, 2010 from <http://www.nrc.gov/reading-rm/doc-collections/fact-sheets/tritium-radiation-fs.html>.

NRC (2007). *Dry cask storage*. Retrieved February 13, 2007 from <http://www.nrc.gov/waste/spent-fuel-storage/dry-cask-storage.html>.

NRC (2007). *Location of spent fuel rod segments*. Retrieved February 13, 2007 from <http://www.nrc.gov/reactors/plant-specific-items/vermont-yankee-issues/location-spent-fuel-rod.html>.

NRC (2010). *Reprocessing*. Retrieved August 26, 2010 from <http://www.nrc.gov/materials/reprocessing.html>.

NRC (2007). *Spent fuel pools*. Retrieved February 13, 2007 from <http://www.nrc.gov/waste/spent-fuel-storage/pools.html>.

Nuclear Energy Institute (2010). *Nuclear waste: Amounts and on-site storage*. <http://www.nei.org/resourcesandstats/nuclear_statistics/nuclearwasteamountsandonsitestorage/>.

Rudy, C. R., & Jordan, K. C., Tritium Half-Life. The Smithsonian/NASA Astrophysics Data System. <http://adsabs.harvard.edu/abs/1977thl..rept.....R>.

Schell, W. R., & Sauzay, G. (1973, November 12). World distribution of environmental tritium. *Conference: Symposium on the Physical Behavior of Radioactive Contaminants in the Atmosphere*. Vienna, Austria.

State of Nevada (2003). *What's wrong with putting nuclear waste in Yucca Mountain?* The Agency for Nuclear Projects.

Stiffler, L. (2008). Troubled Hanford cleanup has state mulling lawsuit. *Seattle Pi*. Retrieved March 20 from <http://www.seattlepi.com/local/355924_hanford21.html>.

US Energy Information Administration (2009). *Vermont Yankee nuclear power plant.* Department of Energy. Retrieved September 10 from <http://www.eia.doe.gov/cneaf/nuclear/page/at_a_glance/reactors/vermontyankee.html>.

Wald, M. L. (2010). As clock ticks, nuclear power plant searches for leak. *New York Times.* Retrieved February 26 from <http://www.nytimes.com/2010/02/27/us/27nuke.html>.

Washington State Department of Ecology (2007). *Hanford quick facts.* Retrieved January from <http://web.archive.org/web/20080624232748/><http://www.ecy.wa.gov/features/hanford/hanfordfacts.html>.

Washington State Department of Ecology (2007). *Hanford cleanup.* Retrieved January from <http://www.ecy.wa.gov/programs/nwp/cleanup.htm>.

Washington State Department of Public Health (2004). *The release of radioactive materials from Hanford: 1944–1972.* Retrieved July 16 from <http://www.doh.wa.gov/hanford/publications/history/release.html>.

World Nuclear Association (2010). *Nuclear power in China.* Retrieved August 20, 2010 from <http://www.world-nuclear.org/info/inf63.html>.

11 Pesticides

Chapter Contents

The fertilizer, pesticide, and agricultural chemical industry sector is diverse. The justification for this industry is as the world population increases, crop lands are increasingly strained and unable to meet the growing demand for food without employing some method of crop enhancement. There are generally believed to be five common practices capable of meeting the growing demand: increasing tilled acreage, improving plant strains, introducing or expanding irrigation, initiating or increasing fertilizer usage, and controlling pests by chemical or biological methods. The last two methods have created a large agrichemical industry which produces a wide variety of products designed to increase crop production and protect crops

Risks of Hazardous Wastes. DOI: 10.1016/B978-1-4377-7842-7.00011-8

from disease and pests. Along with the benefits there have been increasingly negative impacts on the environment, various natural resources, and to human health.

11.1 Current Regulatory Framework

In the United States, the EPA regulates the use and distribution of pesticides under two major federal statutes: the Federal Insecticide, Fungicide and Rodenticide Act (FIFRA) and the Federal Food, Drug and Cosmetic Act (FFDCA). First passed in 1947, FIFRA gave the US Department of Agriculture responsibility for the regulation of pesticides; in 1972, the responsibility was transferred to the EPA. FIFRA authorizes the EPA to approve the use of a pesticide if it can be proven that 'it will not cause unreasonable adverse effects on the environment,' and to suspend or cancel the pesticide if subsequent information shows otherwise (US EPA, 2004).

The job of the EPA's Office of Pesticide Programs (OPP) is challenging and complex. OPP regulates the use of pesticides in the United States and establishes the maximum levels of pesticide residues in food, in the agriculture industry, greenhouses, lawns, swimming pools, hospitals, and food services establishments. According to the OPP, there are about 20,000 registered pesticide formulations containing approximately 675 active ingredients (US EPA, 2004).

Inside the organization of the OPP, the Science Based Risk Management Division, specifically the Biopesticides and Pollution Prevention Division (BPPD), investigates and reports the risk and benefit, risk management and use of pesticides in the environment. They work in collaboration with the Pesticide Environmental Stewardship Program (PESP), which is a voluntary partnership between EPA and the pesticide user community. Their main goal is to reduce the risk from pesticide use in agricultural and non-agricultural settings (US EPA, 2004).

The *Regulatory Process* for a new pesticide to enter the market involves a registration of the product. The registration involves (a) required test data; (b) information concerning the manufacturing process; (c) product chemistry; (d) human and environmental risk data packages; (e) tolerance information about pesticide residues on food; and (f) labeling information. The EPA also requires re-registration of older pesticides which were registered when standards for government approval were less stringent. Additionally, each State, Tribe and Territory can place further restriction on EPA-registered pesticides used or sold within their own jurisdiction (US EPA, 2004).

The human health risk assessment for pesticides follows the National Research Council's four-step process for human health risk assessment, which entails: (1) hazard identification; (2) dose–response assessment; (3) exposure assessment; and (4) risk characterization. This general process is discussed in Chapter 14.

The ecological risk assessment is conducted to determine the risks associated with the pesticide and whether changes to the use of that pesticide need to be made to ensure an acceptable level of risk to the environment. The EPA requires pesticide applicants to conduct and submit a variety of environmental laboratory and field studies. These studies (1) examine the ecological effects or toxicity of a pesticide and its breakdown products to various aquatic and land animals and plants, and (2) examine the chemical fate and transport of a pesticide in soil, water, and air resources. The EPA reviews all fate and transport information available for a pesticide, and integrates the toxicity information with the exposure data to calculate the ecological risk from that pesticide. Provided the EPA determines that a pesticide poses an acceptable risk to human health and the environment, the pesticide will become available for sale and use (EPA, 2008).

EPA may also authorize the use of experimental or unregistered pesticides. The Experimental Use Permit (EUP) allows the use of the experimental or unregistered pesticides in an area involving 10 acres or more of land or 1 acre or more of surface water. The EUP limits the sale and distribution of the experimental or testing pesticide. Pesticide companies request EUP for efficacy testing and/or crop-specific residue testing (US EPA, 2004).

The OPP also gives guidelines for the *Pesticide Use Characterization*, product labeling providing warnings, hazard and restrictions, and instruction on how to use the pesticide. Information contained in the labels is used in determining the input parameters for exposure models and the magnitude of exposure to non-target organisms. Label information includes: (a) type of formulation: bait, granule, wettable powder, emulsifiable concentrate; (b) product purity; (c) proposed and/or existing application rate; (d) treated crop(s) and, if specified, target pests; (e) geographic limitations; (f) application methods such as aerial, ground, foliar, soil surface, soil incorporated; (g) frequency of application; (h) hazard advisory statements such as protective measures for wildlife/aquatic habitats, groundwater, etc. (US EPA, 2004).

Individual states also have the authority to register or license pesticides and may even have more stringent regulations than the EPA. All pesticides must be registered by both the EPA and the state before they can be distributed in the state (EPA, 2007; EPA, 2009).

11.2 Case Studies of Select Pesticides

11.2.1 DDT

Dichlorodiphenyltrichloroethane (DDT) is one of the most widely used and well-studied pesticides ever synthesized. DDT was introduced for commercial use in 1945 and was used heavily in populated areas for vector control and in agriculture for pest control.

In 1946, the US Fish and Wildlife Service published a report warning that DDT can cause damage to fish and aquatic invertebrates (USFWS, 1946). Since then, numerous studies have shown that DDT causes eggshell thinning and reproductive damage in birds and toxicity to fish as well (Fry, 1995; Henderson et al., 1959). Because DDT accumulates in fatty tissue, it has a tendency to biomagnify through trophic levels and can be found in birds in concentrations of up to 10 million times higher than the concentration of DDT in water in which the birds feed

(USFWS, 2009). By 1972, DDT use was banned in the United States and worldwide production and use began to decrease as well.

Despite the injurious impacts on the environment and potential adverse health effects in humans, DDT is still produced; in fact global production actually appears to be increasing (UNEP, 2008). A large quantity of the DDT produced today is used as an intermediate in the synthesis of the acaricide dicofol (trade name: Kelthane). In 1986, the EPA temporarily cancelled registration of dicofol because large quantities of DDT were ending up in the final product. A recent study of dicofol formulations in China found that dicofol might be a significant contributor to DDT pollution in China and could also be responsible for the unchanging DDT levels found in China more than two decades since technical DDT was banned (Qiu et al., 2005).

In the US, DDT can still be found as a residue in food products. The food sources with the highest DDT concentrations are meat, fish, poultry, and dairy products. DDT residues in food have declined since it was banned but because of the extreme persistence of DDT in the environment, it is anticipated that low levels of residues will be present in food products for decades (ATSDR, 2002). In the 1999 FDA Total Diet Study, DDT was found in 255 out of 1,040 (22%) items analyzed (FDA, 1999).

11.2.2 Agent Orange

Agent Orange refers to a specific blend of herbicides used during the Vietnam War from 1961 to 1971. The US military sprayed millions of gallons of Agent Orange and other herbicides in Vietnam to remove leaves from trees that provided cover for enemy forces (Department of Veteran Affairs, 2010). The ingredients in Agent Orange were equivalent amounts of: 2,4-dichlorophenoxyacetic acid (2,4-D) and 2,4,5-trichlorophenoxyacetic acid (2,4,5-T), with trace amounts of the toxic dioxin, 2,3,7,8-tetrachlorodibenzo-p-dioxin (also known as TCDD). TCDD is thought to be responsible for most of the medical problems associated with exposure and has been shown to cause a variety of illnesses in laboratory animals and humans (US Department of Veterans Affairs, 2009).

Adverse effects of the herbicide were observed as early as 1949 when a Monsanto-owned plant in West Virginia experienced an explosion and 228 workers developed chloracne (Veterans of the Vietnam War, 2010). In 1964, over 70 Dow Chemical workers similarly experienced an outbreak of chloracne following exposure. After running its own studies, the director of Dow's Midland Division stated that severe contact with Agent Orange could lead to internal organ damage and nervous system disorders (Veterans of the Vietnam War, 2010). In humans, mild exposure symptoms could include: fatigue, loss of appetite, stomach and kidney pain, and skin and eye irritations. At high levels, the substance was shown to be toxic and fatal to animals. In 1969, both the National Institutes of Health and the American Association for the Advancement of Science reported that Agent Orange caused birth defects in rodents (Institute of Medicine, 1994). The program using Agent Orange as a herbicide terminated in 1971 when various questions were raised regarding the legality, morality, and possible long-term consequences of its use.

11.2.3 Chloropicrin

Chloropicrin, also known as trichloronitromethane and nitrochloroform, is a halogenated fumigant that reacts with the amino acids that form the protein structure of different pests (Albritton et al., 1998; Barry et al., 2010). It has been steadily used in California since 1991. Chloropicrin has also been used as a fungicide in a variety of crops including strawberries, tomatoes, sweet potatoes, grapes, citrus, and almonds (Trout, 2002).

Barry et al. (2010) reported that residuals of chloropicrin used in irrigation lines for post-application water treatment have contributed to considerable health problems in local communities. The use of common irritant fumigants such as chloropicrin close to residential areas poses a risk due to their airborne capacity and ability to disseminate for distances more than 2 miles away from the site of application. Barry reported that people living in adjacent areas to an irrigation site reported increased eye irritation, asthma attacks, headaches, nausea, and vomiting. Chloropicrin has been shown to be highly irritating even at very low concentrations (Oriel et al., 2009; Trout, 2002).

11.2.4 Endosulfan

Endosulfan is an organochlorine insecticide and acaricide that is commonly used in the US for agricultural pest control (ATSDR, 2000). It is one of the most commonly used pesticides in the California Central Valley, one of the most cultivated areas in the United States (Bradford et al., 2010). Endosulfan is used on food crops such as grains, tea, fruits, and vegetables, and on non-food crops such as tobacco and cotton. It is also used as a wood preservative (ATSDR, 2000).

Endosulfan poses problems due to its persistence in the environment. The pesticide may travel long distances in the air before it lands on crops, soil, or water. Studies have found residues of endosulfan approximately 121 miles from the agricultural site where the pesticide was sprayed (ATSDR, 2000; Bradford et al., 2010). The pesticide may stay in soil for several years before breaking down. Rainwater can also wash endosulfan in soil into surface water. This is especially problematic because endosulfan does not dissolve easily in water and remains in the water attached to soil (ATSDR, 2000). The concentration of endosulfan in the bodies of animals living in endosulfan-contaminated waters has been found to be several times greater than in the surrounding water (ATSDR, 2000; Bradford et al., 2010).

Exposure to endosulfan has been reported to cause mutations and alteration of DNA in cattle, mice, fish, birds, and even humans (Bajpayee et al., 2006). Researchers have also found that endosulfan causes genetic toxicity on white clover and earthworm, key elements in maintaining the balance of the ecosystem (Liu et al., 2009).

According to the EPA, endosulfan has been found in close to 164 of 1,577 sites on the National Priorities List (NPL) containing hazardous or toxic materials (ATSDR, 2000).

11.2.5 Propargite

Propargite was registered in the US in 1969 and is an organosulfur miticide/acaricide pesticide used on a variety of agricultural and non-agricultural sites. Propargite is used on crops including grapes, walnuts, almonds, nectarines, and mint (US EPA, 2005). It is one of the most commonly used pesticides in the California Central Valley, one of the most active agricultural areas in the United States (Bradford et al., 2010).

Propargite has been shown to produce adverse health effects in animals and humans. Propargite is classified by the EPA as a probable human carcinogen based on the appearance of intestinal tumors in test animals. In addition, the EPA has identified reproductive risk to birds, mammals, fish, and invertebrates exposed to propargite (Donald et al., 1992; US EPA, 2005). Propargite also causes severe dermal irritation in animals and humans (US EPA, 2005).

11.2.6 Malathion

Malathion is a common pesticide in the United States (Bradford et al., 2010; Bradman et al., 2009). Malathion is used to kill insects on farm crops and in gardens, to treat lice on humans, and to treat fleas on pets. Malathion is also used to kill mosquitoes and Mediterranean fruit flies in large outdoor areas (ATSDR, 2010). Hertzman et al. (1990) reported that malathion, among other pesticides such as paraquat, is potentially associated with central nervous diseases such as Parkinson's.

11.3 Worker Exposure to Pesticides

Worker exposure is a particular concern in the agriculture industry. It has been demonstrated that significant inhalation and dermal exposure can occur during the application of pesticides or during the handling of crops treated with pesticides. And various studies of workers in the pesticide industry have demonstrated a particular risk of adverse health effects.

Pesticide exposure assessment is difficult due to the heterogeneous chemical group and exposure scenario. One of the major difficulties in agricultural epidemiology is that exposure to pure substances rarely occurs and occupational studies often involve exposures to complex mixtures. Recently, the emergence of the analysis of pesticide use patterns has helped in interpreting the inherently complex occupational exposure of pesticide workers (Samanic et al., 2005; Schilmann et al., 2010).

11.3.1 DBCP Worker Case Studies

DBCP was used extensively worldwide during the 1970s on banana plantations. In 1977, employees at California's Occidental Chemical Plant who had handled DBCP were found to be sterile. Six years later, a $4.9 million judgment against Dow Chemical Corporation was reached on behalf of six of the workers at the

California plant. Two years after the judgment, the EPA banned all DBCP use (Bananas, 2009).

In 1992, a shift in regulations permitted cases with a foreign location as the place of incident to be heard in a US court of law. These new regulations allowed 1,000 Costa Rican workers to bring a lawsuit against Standard Fruit (now Dole Corporation). The ruling stipulated that the corporation had to pay $20 million to workers who were sterile as a result of DBCP exposure. After legal fees, each worker received approximately $1,500 to $15,000 (Bananas, 2009).

In 1993, a class-action lawsuit was filed by over 16,000 banana plantation workers from Costa Rica, Ecuador, El Salvador, Guatemala, Honduras, Nicaragua, and the Philippines. The lawsuit named several companies: Dole, Chiquita, Del Monte, Dow, and Shell. The companies collectively agreed to pay $41.5 million to those who were sterile due to DBCP exposure. After the payment of legal fees, workers received relatively small payments (Bananas, 2009).

Other health effects aside from sterility have been demonstrated. A study conducted between 1972 and 1979 examining the incidence of cancer among a group of Costa Rican banana plantation workers (29,565 men and 4,892 women), found an excess number of melanoma, leukemia, and cervical cancers among women, and penile cancer among men. Excesses in lung cancer, melanoma, penile cancer, and brain cancer were seen among men who worked at the plantation for three or more years; excesses in leukemia were seen among women who worked at the plantation for three or more years (IARC, 1987).

11.3.2 California Cyfluthrin Case Study

California has the most agricultural activity in the United States and therefore uses significant amounts of pesticides in order to meet the high demand of agricultural production. In 2005, the California Department of Public Health, Occupational Health Branch investigated a case involving 27 vineyard workers who became ill after being exposed to a drift of cyfluthrin. A mixture of cyfluthrin with spinosad, petroleum oil, and water was applied by air blast sprayers in an effort to control katylids (insects) in Kern County (Weinberg, 2009). Twenty-seven workers of a nearby vineyard reported feeling ill and were admitted to the emergency room. Some of the symptoms reported by the vineyard workers included parathesia, headaches, nausea, eye irritation, muscle weakness, anxiety, and shortness of breath (Weinberg, 2009).

11.3.3 Farm Worker Children

Exposure to children of farm workers has been the subject of concern. Children of farm workers are exposed to pesticides as a result of take-home exposure, which results when agricultural chemicals are transported by workers from the field to their homes, via vehicles, clothing, and skin. A study examining organophosphorus exposure among 218 farm workers and their children in Washington state demonstrated the exposure to children from the take-home pathway (Curl et al., 2002). Many of the workers in this study worked in apple, pear, or cherry crop fields where azinphosmethyl, a Toxicity 1 organophosphorus insecticide, is applied. Organophosphorus pesticides cause concern because they are highly toxic and readily used in agriculture (Jaga, 2003). In the study, the pesticide was found in house dust, vehicle dust, and in urine samples collected from both the farm workers and their children, demonstrating exposure.

There have also been reported correlations between childhood cancers and the parents' occupational exposure to pesticides prior to and during pregnancy. After reviewing a number of studies that analyzed the relationship between childhood cancers and pesticide exposure, Daniels et al. (1997) found that:

> *Collectively, these studies suggest an increase in risk of brain cancer, leukemia, Wilms' tumor, Ewing's sarcoma, and germ cell tumors associated with paternal occupational exposure to pesticides prior to and during pregnancy. Maternal occupational exposure during pregnancy was studied less frequently, but was also associated with leukemia, Wilms' tumor, and germ cell tumors. Most of these cancers were only evaluated in one or two studies, and the number of exposed cases was often small. Childhood brain cancer and leukemia were the most studied, with fairly consistent, moderate increases in risk (27–31, 39–42, 44–46). Farm residence was associated with brain cancers, neuroblastoma, retinoblastoma, non-Hodgkin's lymphoma, and Wilms' tumor to varying degrees.*
>
> *(Daniels et al., 1997)*

11.3.4 Protecting Worker Safety

Recently, studies have reported the importance of training and protective equipment to reduce workers' exposure to pesticides. A study focusing on pesticide exposure among Italian agricultural workers demonstrated that the inadequate use of personal protective equipment among workers results in unnecessary exposure to pesticides. In the study, ten treatments were monitored in the spring and summer. The pesticides applied by these workers included azinphos-methyl, dicamba, dimethoate, terbuthylazine, and alachlor (Vitali, 2009). Results showed that dermal exposure was significantly greater than exposure from inhalation, and that dermal exposure was increased dramatically if general work clothes were worn rather than specific protective clothing (Vitali, 2009).

The EPA Worker Protection Standard requires pesticide training for farm workers combined with re-entry intervals (Bradman et al., 2009). Use of gear such as protective gloves, disposable coveralls, and containers for storage for shoes and work clothes, as well as laundry services for work clothes have been emphasized as important measures to prevent exposure. In a study conducted in Central California, farm workers who were provided with safety gear showed reduced levels of malathion contamination compared to a control group (Bradman et al., 2009). These interventions reduced the transport of pesticides from work and further exposure to children and spouses at home.

In addition, better pest control practices, such as chemical substitution, engineering control, and better management policies, should be implemented to protect worker safety and reduce occupational exposure to pesticides.

11.4 Pesticides in Groundwater, Surface Water, and Drinking Water

Pesticide contamination in water has become a pervasive problem in the US. Surface water runoff carries pesticides from agricultural fields into rivers, lakes, and reservoirs. Rain or snow carries pesticides through the soil into groundwater that is the source of drinking water.

Under the Safe Drinking Water Act, originally passed in 1974 to regulate the nation's drinking water, the US EPA sets standards for allowable pesticide levels in drinking water and requires water utility companies to monitor these levels. Setting these standards is a two-part process. First, the EPA sets a non-enforceable Maximum Contaminant Level Goal (MCLG), which is a goal based solely on health considerations. These MCLGs are set at levels 'at which no known or anticipated adverse effects on the health of persons occur, and which allows an adequate margin of safety.' Second, the EPA sets enforceable Maximum Contaminant Levels (MCLs), which are based on MCLGs but adjusted to ensure technical and financial feasibility. Due to this adjustment, federally enforceable MCLs are not as stringent as the MCLGs, allowing pesticides to legally remain in public drinking water. Therefore, federal regulations for pesticides commonly found in drinking water fail to adequately protect the public's health (EWG, 2010).

The EPA has set enforceable MCLs for three pesticides called atrazine, alachlor, and simazine. However, it has not set enforceable MCLs for cyanazine, metolachlor, and acetochlor, which are three major pesticides used in the US. Instead, the EPA has issued non-enforceable Lifetime Health Advisories (LHAs) for these contaminants. Consequently, water utility companies are not required to test for these contaminants or inform their customers if these contaminants are found at levels that exceed federal health advisories (EWG, 2010).

This section will provide an overview of five pesticides commonly found in drinking water: atrazine, alachlor, metolachlor, 1,2,3-trichloropropane, and 1,2-dibromo-3-chloropropane (DBCP). Two of these pesticides, alachlor and DBCP, have been classified by the EPA as probable human carcinogens and have been associated with increased tumor rates and organ damage in animals (ATSDR, 1995; Extoxnet, 1996a).

11.4.1 Atrazine

Atrazine is a widely used herbicide utilized for the control of broadleaf and grassy weeds in corn, sorghum, rangeland, sugarcane, macadamia orchards, pineapple, turf grass sod, asparagus, forestry grasslands, grass crops, and roses. It is used most extensively on corn crop in Illinois, Indiana, Iowa, Kansas, Missouri, Nebraska, Ohio, Texas, and Wisconsin. Total estimated agricultural use in the US is 76.4 million pounds annually, with 86% of that amount applied to corn alone (EPA, 2008a). Due to health concerns and persistent contamination of groundwater, atrazine was banned in the European Union in 2004.

Many factors contribute to atrazine runoff and drinking water contamination. First, atrazine does not bind well to soil particles, so it is easily carried off of fields with storm water. Second, many areas of heavy application (namely northern Missouri and southern Iowa) have high clay content in the soil. Clay soils have relatively low infiltration rates and thus promote runoff during storm or flood events. Third, many Midwest farms practice no-till farming, which increases runoff potential and loss of atrazine due to the restrictive layer of the claypan that limits infiltration. Finally, atrazine application periods typically occur during the months that receive the most rain. Heavy rainfall directly following application can greatly increase atrazine losses. Because of these factors and the sheer quantity and density of atrazine application, hundreds of water systems have atrazine detections in their finished drinking water, impacting millions of people across the country.

In 2004, Holiday Shores Sanitary District in Holiday Shores, Illinois, filed class-action lawsuits against the manufacturers and primary distributors of atrazine. These lawsuits have expanded to include over 60 cities from across the Midwest looking to hold the corporations accountable for the contamination of their drinking water systems. A judgment in the case has not yet been issued.

Health Effects of Atrazine

The endocrine-disrupting properties of atrazine have been demonstrated in studies. Atrazine has been found to alter the brain's pituitary functions, resulting in the suppression of two hormones, lutenizing hormone and prolactin hormone. Changes in these hormones have concerning consequences. Research has shown that even brief atrazine exposure to a lactating mother alters the endocrine makeup of the mother's milk, raising concerns about the subsequent development of the child. Exposure to atrazine and atrazine metabolites has caused delayed puberty in both male and female rats (US EPA, 2009) Additional studies showed health effects such as increased risk of intrauterine growth retardation, reduced semen quality, and spontaneous abortions in humans, as well as demasculinization and hermaphrodism in frogs (Munger, 1997; Arbuckle, 2001; Hayes, 2002; Swan, 2003).

The effects of atrazine in frogs have received particular attention. Hayes et al. (2002) examined atrazine exposure on the development of the African clawed frog. During larval development, the larvae of the frogs were exposed via immersion to 0.01 to 200 ppb of atrazine. Study results demonstrated that greater than 0.01 ppb of atrazine exposure resulted in hermaphroditism and demasculinization in male frogs. Furthermore males had a ten-fold decrease in testosterone levels when exposed to 25 ppb atrazine. The study concluded that atrazine converts testosterone to estrogen.

A growing body of evidence also indicates that agrichemical exposures may contribute to birth defects. Winchester (2009) investigated whether babies conceived during the months when surface water agrichemicals are highest are at greater risk for birth defects. In the study, concentrations of nitrates, atrazine, and other pesticides were measured in water samples from 186 stream sites representing 51 hydrological systems, accounting for 50% of the US's drinking water, from 1991 to 2002. The highest concentrations of pesticides were found in May and June, with annual peaks from April to July. Results from the study of approximately 30 million babies showed that total birth defects, as well as 11 of the 22 birth defect subcategories, were more likely to occur in babies that were conceived between April and July.

Ochoa-Acuna (2009) investigated atrazine exposure from drinking water and the prevalence of small-for-gestational-age (SGA) and preterm delivery. The study found that atrazine, and perhaps other co-occurring herbicides in drinking water, was associated with an increased prevalence of SGA. Alarmingly, SGA resulted from exposure to atrazine in drinking water at levels just above 0.1 μg/L, well below the current MCL of 3.0 μg/L.

Cancer has also been associated with exposure to atrazine. MacLennan (2002) evaluated cancer incidence among approximately 2,000 workers at a Louisiana plant that manufactured atrazine and other triazine herbicides. Incidences of prostate cancer among active company employees were statistically increased.

Syngenta and the Atrazine Monitoring Program

After the use of atrazine was banned in Europe in 2004, the EPA expressed concern over the presence of atrazine in some water systems in the US. Subsequently, the Atrazine Monitoring Program (AMP) was created. With assistance from the EPA, Syngenta, the primary manufacturer of atrazine, tested 134 public water systems weekly or bi-weekly for atrazine and three chlorotriazine breakdown products: DIA, DEA, and DAC.

The AMP data revealed that levels of atrazine and its chlorotriazine breakdown products during some periods of the year were much higher than levels reported by water systems. Many water systems tested as part of the AMP showed levels exceeding the MCL at some point during the year. In other words, values reported by water systems and values shown by the AMP did not match up.

The AMP data showed that atrazine 'spikes' likely occurred during weeks of atrazine application or heavy rainfall. These spikes of atrazine levels in post-treatment water exceeded the 3 ppb MCL for short periods of time, but averaging and infrequent testing allowed these levels to be overlooked by water systems. Figure 11.1 presents AMP data of several water systems with an atrazine 'spike.' State data from the same period dc not indicate a spike.

Inconsistencies between data reported to the state by water systems and data from the AMP sparked much political and media interest. In August 2009, the *New York Times* published an article titled 'Debating How Much Weed Killer is Safe in Your Water Glass,' detailing inconsistencies between the two data sets and

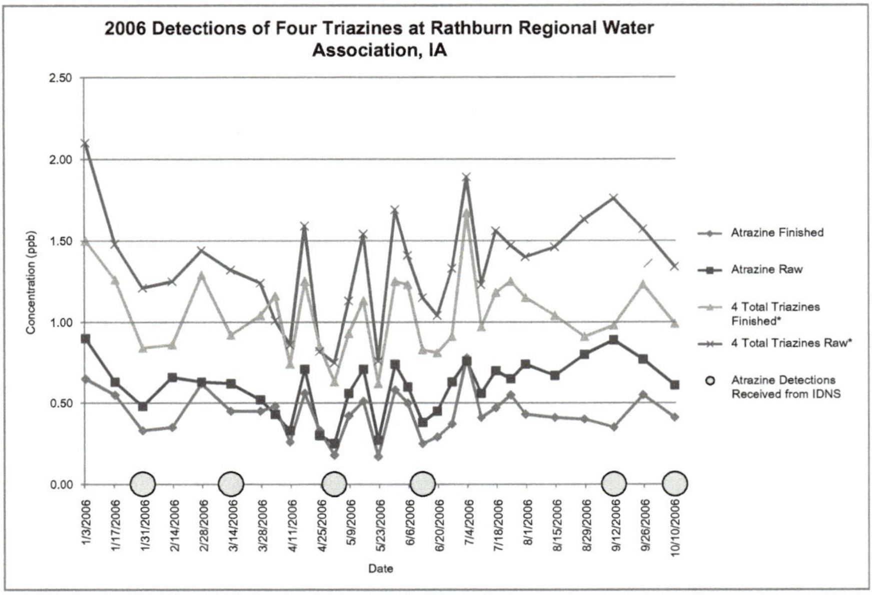

*Total triazines includes atrazine and 3 daughter products, although many other breakdown products may be present.
Dates of IDNS Atrazine Detections are approximate.

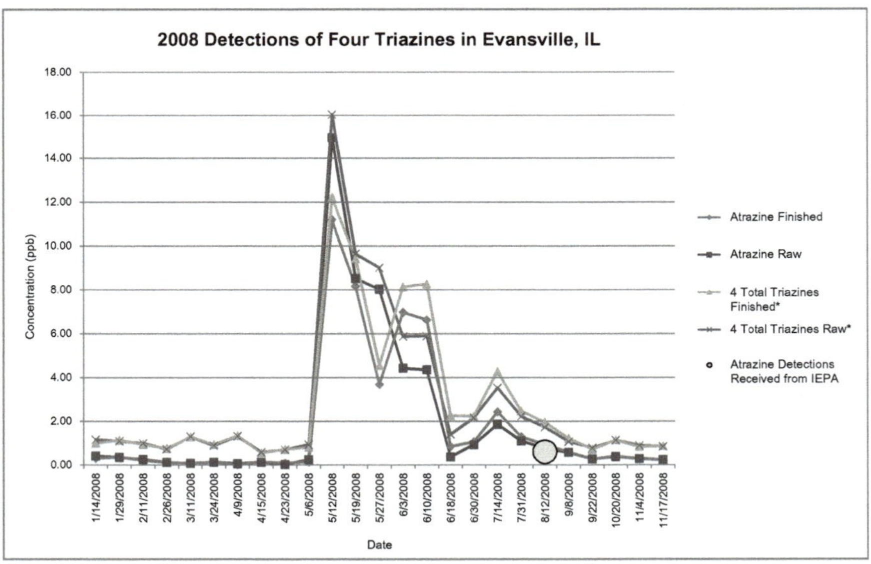

*Total triazines includes atrazine and 3 daughter products, although many other breakdown products may be present.
Dates of IEPA Atrazine Detections are approximate.

Figure 11.1 Comparison of Syngenta AMP data and state data (source: Syngenta AMP, Illinois EPA (IEPA) Bureau of Water, Iowa Department of Natural Resources Water Supply Program (IDNSWP)). Lines represent concentrations of atrazine or atrazine plus three degradate triazines detected under the Atrazine Monitoring Program. Dots represent atrazine concentrations reported to the state by water systems.

the overall danger of atrazine contamination in the US. Increased media coverage placed more pressure on Congress to address these issues, which in turn placed pressure on the EPA. The EPA responded with a press release stating it would begin a re-evaluation of atrazine. EPA's re-evaluation plan includes the review of atrazine effects to inform safety measures and plans for incorporating new epidemiologic and experimental studies into the atrazine risk assessment.

11.4.2 Alachlor

Alachlor, commonly known by its tradename Lasso, is a herbicide used to control grasses and weeds in corn, peanuts, soybeans, and other crops (DHSS, 2009). It is the second most widely used herbicide in the US. In 1990, over 50 million pounds of the herbicide were used (Extoxnet, 1996a). From 1993 to 1995, Alachlor was used on approximately 20–25% of the US corn cropland (CDC, 2010).

The use of Alachlor has resulted in significant groundwater contamination. Data on Midwest groundwater gathered by the USGS since 1991 indicated that alachlor ESA is present in over 40% of wells and that alachlor 2,6-diethylaniline is present in approximately 16% of wells. The EPA's Pesticides in Groundwater Database detected concentrations of alachlor in the ground water of 15 states, at concentrations higher than the MCL of 2 parts per billion (ppb) or 0.002 mg/L (EPA, 2010c).

Data suggest that groundwater contamination of alachlor usage occurs long after its use has stopped. In Florida, significant alachlor contamination is still present, even though alachlor use was halted in 1991. Additionally, prior to 1991, 16 wells sampled were found to have alachlor at concentrations greater that 2.0 ppb. These same wells have been sampled annually since 1991, yet the average concentration of alachlor has not changed (EPA, 1998).

An environmental assessment conducted by the EPA in 1998 concluded the following:

- Alachlor is highly mobile and moderately persistent.
- Alachlor presents a clear hazard to groundwater quality. Reliable monitoring studies have demonstrated that alachlor, even when used according to the label instructions, results in significant groundwater contamination.
- Alachlor use results in groundwater in the areas being contaminated with degradation products, which are also very mobile and persistent.
- Monitoring studies show that alachlor levels in surface water result in effects on aquatic plants and indirectly on aquatic animals (EPA, 1998).

In the US, concerns over possible health effects and groundwater contamination have resulted in some safety measures being taken. Alachlor is classified as a Restricted Use Pesticide (RUP), meaning that it can only be purchased and used by certified applicators (Extoxnet, 1996a). Other countries, however, have already banned the sale and use of alachlor. Due to public health and environmental concerns, the European Union (EU) in 2006 banned the sale of any product containing alachlor (Rotterdam, 2010). Similarly, alachlor use was banned in Canada in 1985 due to the product's carcinogenic potential and the existence of metolachlor, a lower-risk alternative (Rotterdam, 2010).

Health Effects of Alachlor

The EPA has classified alachlor as a probable human carcinogen. Studies in rats have demonstrated the development of stomach, thyroid, and nasal turbinate tumors in rats given high doses of alachlor. Another study on alachlor exposure in mice demonstrated an increase of lung tumors (Extoxnet, 1996a). In occupational studies, workers exposed to high levels of alachlor for several years or more demonstrated elevated rates of colorectal cancer (CEPA, 1997).

A study in the *American Journal of Epidemiology* assessed the incidence of cancer among pesticide applicators exposed to alachlor. The study observed a total of 49,980 pesticide applicators; of those, 26,510 used alachlor. A total of 1,466 malignant neoplasms were diagnosed during the study period from 1993–2000. An increasing trend in the incidence of all lymphohematopoietic cancers with lifetime and intensity exposure days among alachlor applicators was found. Leukemia and multiple myeloma risks were also increased among applicators with the highest exposures to the pesticide. These findings suggest an association between alachlor application and lymphohematopoietic cancer (Lee, 2004).

Non-cancer effects have also been reported. Animal studies on dogs have reported liver toxicity, and effects to the spleen and kidney, after the administration of alachlor. Rat studies have demonstrated irreversible iris and related eye structures degeneration (Extoxnet, 1996a).

11.4.3 Metolachlor

Metolachlor is a herbicide that was registered in 1976 to control weeds growing in non-crop areas. Today, its use has expanded. It is now registered for use on corn, cotton, peanuts, pod crops, potatoes, safflowers, soybeans, tree nuts, non-bearing citrus and grapes, cabbage, peppers, and in non-agricultural locations such as nurseries, hedgerows/fencerows, and landscape plantings (EPA, 1995a). Products containing metolachlor are sold under the tradenames Bicep, CGA-24705, Dual, Pennant, and Pimagram. Metolachlor is also found in other herbicides such as atrazine, cyanazine, and Fluometuron (DHSS, 2010). Metolachlor usage in the US is estimated to be between 60–65 million pounds per year. In 2000, over 125,000 pounds of herbicide were applied in California (CDPR, 2003).

Concerns over health effects have resulted in some safety measures. The EPA has classified metolachlor as a possible human carcinogen (DHSS, 2010). Metolachlor use is prohibited in various areas: in greenhouses, on muck or peat soils, on vegetables such as sweet potatoes or yams, on trees or vines that have been planted within the last 30 days, on sand or loamy sand soils, and on trees or vines that will bear fruit in the next year. Metolachlor use is also prohibited in areas where livestock graze (EPA, 1995b). Nevertheless, there is no established MCL, Occupational Safety and Health Administration (OSHA) standard or National Institute for Occupational Safety and Health (NIOSH) standard for metolachlor.

Contamination of groundwater and surface water has become a problem due to metolachlor's mobility and persistence in soil (CDPR, 2003), and because significant amounts of the herbicide present for several months after application can run off in surface water (EPA, 1995a). A study conducted in 1988 of both surface and groundwater in the US detected the herbicide in 50% of surface water (2,091 of 4,161 samples) and 13 of 596 samples of groundwater. Another survey found metolachlor residues in groundwater at varying levels from 0.1 to 0.4 μg/L (WHO, 2003b). Another study conducted in 1997 that examined surface water found the herbicide in 1,644 samples (312 locations in 14 states) at a concentration of up to 138 ppb, due to runoff (CDPR, 2003). Metolachlor residues have been found in wells in 20 states. Although the lifetime Health Advisory Level (HAL) for metolachlor is 100 ppb, levels exceeding this were found in three wells located in Wisconsin, New York, and Montana (EPA, 1995a).

In the Midwestern Corn Belt, metolachlor is among the top five most frequently detected pesticides in raw and finished surface water. A high percentage of surface water samples from numerous locations for several months post-application contain metolachlor. Detection percentages pre-application (early spring) and post-application (late fall and winter) are lower than during application months, but are still high due to its persistence (EPA, 1995a).

Health Effects of Metolachlor

Metolachlor has been classified as a possible human carcinogen by the EPA. This classification was based on a study that detected increased liver tumors in female rats and a study that replicated these findings (EPA, 1995a).

An Agricultural Health Study conducted from 1993 to 2001 in Iowa and North Carolina observed 57,311 pesticide applicators and 32,347 spouses. An elevated lung cancer risk was observed among applicators with the most exposure (AHS, 2006). Those who had used metolachlor over a certain length of time (over 457 lifetime days) had a four-fold risk of lung cancer compared to those who had never used the pesticide. Such a risk was observed even after factoring characteristics such as age, gender, smoking history, and total days of any pesticide application. This study was the first to suggest a link between metolachlor and lung cancer (AHS, 2006).

Metolachlor exposure results in a myriad of non-cancer health effects that include eye and skin irritation, stomach cramps, shortness of breath, weakness, sweating, diarrhea, dizziness, and nausea. Exposure can also result in anemia, convulsions, and jaundice (DHSS, 2010). Dermal exposure to metolachlor causes skin sensitization among workers exposed to the herbicide (Extoxnet, 1993b). In an animal study, adminstration of metolachlor led to decreased body weight gain and minor changes in the liver structure of rats (Extoxnet, 1993b).

A study published in *Environmental Health Perspectives* hypothesized that pesticides used in the Midwest for agricultural purposes resulted in semen quality differences in men. The study results showed elevated pesticide metabolite levels in cases from Missouri for alachlor and atrazine compared to the controls. Metolachlor was also linked to poor semen quality in men. These associations suggest that these pesticides and herbicides may contribute to a reduction in semen quality for men in the mid-Missouri region (Swan, 2003).

11.4.4 1,2,3-Trichloropropane (TCP)

1,2,3-Trichloropropane (1,2,3-TCP or TCP) is a synthetic chemical used primarily in the production of other chemicals. It is a chlorinated hydrocarbon that was used historically as an industrial solvent and degreasing agent. Currently, TCP is used as an intermediate in the production of polymer cross-linking agents, glycerol, and is produced in large quantities as a by-product of epichlorohydrin production. In the agrochemical industry, TCP is used as an intermediate in the production of pesticides. In 2002, total US production was estimated to be between 1–10 million pounds.

TCP is also formed as a by-product of the manufacture of nematicidal soil fumigants. Dichloropropene (DCP) soil fumigants have been marketed for use on citrus fruits, pineapples, soy beans, cotton, tomatoes, and potatoes, and are in use today on potatoes, tobacco, carrots, peanuts, cotton, and other fruits and vegetables. Before 1978, approximately 55 million pounds of 1,3-DCP were produced annually in the US and approximately 20 million pounds of 1,2-DCP and TCP were produced annually as by-products in the production of 1,3-DCP (EPA, 2005).

The use of fumigants containing TCP impurities has resulted in local ground water contamination (Oki & Giambelluca, 1987; Tesoriero et al., 2001; WHO, 2003c; Zebarth, 1998). When the fumigants are injected into soil, TCP is injected inadvertently. TCP in soil is subject to leaching and groundwater infiltration. Once in water, TCP will often concentrate at the bottom of an aquifer due to its insolubility and density, forming a dense, non-aqueous phase liquid (DNAPL). DNAPLs present difficulties to remediation because they travel deep into the ground.

TCP has been detected throughout the US in surface, drinking, and groundwater (ATSDR, 1992a). A sampling of US groundwater revealed TCP concentrations ranging from 2 μg/L in Hawaii to 100 μg/L in New York State (WHO, 2003c). Drinking water concentrations of TCP range from 0.1 μg/L to 0.24 μg/L (City of Shafter, 2000; WHO, 2003c). Despite groundwater contamination issues, there are currently no federal MCLs in place for TCP in drinking water.

TCP groundwater contamination associated with the use of DCP fumigants has been prevalent in California. In the agricultural Central Valley of California, TCP was detected in five of six active water supply wells in 1999 at concentrations ranging from 0.02 to 0.11 μg/L (City of Shafter, 2000; EPA, 2005). TCP was found in groundwater from all wells where 1,2-DCP was detected. A study of water wells in the heavily agricultural area of Merced found some wells to contain up to 150 μg/L of TCP (SWRCB, 2009), dramatically higher than the California Department of Health Services' advisory action level of 0.005 μg/L.

Health Effects of TCP

Although human carcinogenicity data are unavailable, the International Agency for Research on Cancer (IARC) classifies TCP as probably carcinogenic to humans (Group 2A) because it causes tumors in mice and rats, it is metabolized similarly in humans and in rodents, and it is mutagenic to bacterial and mammalian cells and forms deoxyribonucleic acid (DNA) adducts (IARC, 1995). TCP is classified by the American Conference of Governmental Industrial Hygenists (ACGIH) as a 'confirmed animal carcinogen with unknown relevance to humans' (ACGIH, 2009).

TCP is suspected to cause kidney, liver, cardiac, and reproductive organ damage. One study in which rats were fed TCP found abnormally increased liver and kidney weights. Hematological changes such as decreased erythrocyte levels were also observed (ACGIH, 2009). Another study revealed extensive cellular damage to the liver, kidneys, and nasal passages with TCP exposure (ACGIH, 2009). Marked necrosis of the heart was observed in rats fed TCP (Merrick et al., 1991). Studies have also found that there is a significant decrease in litter-size when rats are chronically exposed to TCP, indicative of reproductive harm (Chapin et al., 1997).

11.4.5 1,2-Dibromo-3-chloropropane (DBCP)

1,2-Dibromo-3-chloropropane (DBCP), once one of the most heavily used pesticides in the US, was used for over 20 years to control nematodes, worms that damage crops and plants. An estimated 9.8 million pounds of DBCP was applied to crops in 1974 (IARC, 1979), with greatest use in Arizona, Hawaii, Maryland, North Carolina, South Carolina, and California's Central Valley (Cornell University, 2005). DBCP was used on over 40 different crops including grapes, tomatoes, citrus fruits, cotton, and peaches, with its most extensive use on soybeans. DBCP was also used extensively on pineapples in Hawaii. In addition, the pesticide was used on plants in nurseries and greenhouses, and on lawns and golf courses (Cornell University, 2005). The chemical was sold under multiple tradenames, such as Fumagon, Fumazone, Nemabrom, Nemafum, Nemagon, Nemanax, Nemapaz, Nemaset, Nemazon, and Gro-Tone Nematode. Besides its use in agriculture, DBCP is also used as an intermediate in the synthesis of organic chemicals.

After discovery of deleterious health effects in humans, the US EPA banned the use of DBCP in 1977 (except for use on pineapples, which was subsequently banned in 1985). However, due to its extensive manufacture and use on a variety of crops in the past, environmental contamination by DBCP persists today. DBCP has been detected in various environments: ambient and urban air, groundwater, drinking-water, and soil (IARC, 1987).

DBCP use has resulted in widespread contamination of groundwater and surface water. DBCP's release into groundwater has occurred primarily through leaching of agricultural soil treated with DBCP, and leaching of DBCP-treated soil at hazardous waste and spill sites. DBCP's release into surface water has occurred primarily from runoff at farmland and waste sites, and its direct release during production and use as an intermediate for organic synthesis (ATSDR, 2010).

A survey conducted by the EPA in 1990 detected DBCP in approximately 370 community water system wells and 38,000 rural domestic wells. A survey conducted 2 years later found DBCP in 10% of the 20,545 groundwater wells tested (Cornell University, 2005). Concentrations of up to 95 μg/L in California and 137 μg/L in Arizona have been detected in drinking water (ATSDR, 1992), many times higher than the EPA's MCL for DBCP in drinking water of 0.2 parts per billion (ppb).

DBCP contamination has also been found in areas where usage of the pesticide is not known to have occurred. A water sampling study conducted between 1979 and 1980 from one of three municipal water supplies in South Carolina detected DBCP in concentrations ranging from 0.008 μg/L (detection limit) to 0.05 μg/L in an area where DBCP had never been used (ATSDR, 1992).

Health Effects of DBCP

The EPA has classified DBCP as a probable human carcinogen (EPA, 2000). Animal studies have found nose cancer after the inhalation of DBCP, stomach cancer after ingestion of DBCP, and skin and stomach cancer after direct dermal contact with DBCP. One study conducted on rodents exposed to DBCP via inhalation resulted in tumors of the nasal tract, tongue, adrenal cortex, and lungs (ATSDR, 1995).

Non-cancer effects are many. Short-term human exposure to DBCP can lead to non-cancer effects such as depression of the central nervous system, pulmonary congestion, gastrointestinal distress, and pulmonary edema. It can also cause headaches, nausea, lightheadedness, and weakness in workers exposed to DBCP, as well as skin and eye damage from direct contact (ATSDR, 1995). Long-term human exposure to DBCP negatively impacts the male reproductive system, leading to decreased sperm count (EPA, 2000). Workers exposed to DBCP may produce less sperm, produce sperm that results in more female babies, and even lose the ability to father children (ATSDR, 1995).

11.4.6 Removal Technologies

Investment in removal technologies has obvious benefits. Many pesticides can be removed from water using a technology called granular activated carbon (GAC). GAC has a random porous structure, containing a broad range of pore sizes ranging from visible cracks and crevices down to molecular dimensions. GAC uses this porous structure to remove dissolved contaminants from water in a process known as adsorption. This porous structure results in a large adsorption surface area (USBR, 2009).

GAC is found in many applications ranging from personal in-home use to industrial, commercial, and municipal treatment systems. GAC treatment technologies include:

- Pour-through devices for treating small volumes, such as a hand-held Brita filter.
- Faucet-mounted (with or without by-pass) for treating water at a single faucet.
- In-line filter (with or without by-pass) for treating large volumes for several faucets.
- High-volume commercial units for treating community water supply systems. Typically they are gravity-fed (larger volumes) or pressure-driven (smaller volumes) contactors. These high-volume units can be sequenced in parallel or in series. GAC filters can be used alone or can also be combined with media filters (USBR, 2009).

Installation and operation of granular activated carbon systems, however, are expensive. Capital costs to install GAC systems are in the order of millions of dollars. Purchase of land (if necessary), and operation and maintenance costs, including reactivation or the purchasing of new carbon columns can add to the overall cost.

Many public water providers cannot afford to install this type of advanced treatment system. In order to recover costs for the installation of new treatment systems, water systems have filed lawsuits against the manufacturers of chemicals found in drinking water. For example, Holiday Shores Sanitary District (in Holiday Shores, Illinois) has filed class-action lawsuits against the manufacturers and primary distributors of atrazine, to recover costs of treatment.

11.5 Conclusion

Pesticides will continue to be part of human life and the environment in order to increase crop production. It is imperative for public health authorities to educate the public, farmers, and farm workers on the use of and risks from pesticides. Improvement of human quality of life by means of more efficient and environmentally friendly food production will clearly be a challenge for years to come. Reduction in the annoyance produced by pests is also part of the equation and poses major challenges to balance the well-being of the ecosystem. Rigorous testing and more stringent rules need to be adopted to address the harms posed by pesticides. Scientists, legislators, public health officials, and other stakeholders should familiarize themselves with the different pesticides that are used in their environment and invest in research and development for safer alternatives.

Acknowledgments

Jasmin Abou-Diwan, Alvaro Camacho, and Marlene Gomez contributed to this chapter.

Jasmin Abou-Diwan is a Masters in Public Health (MPH) degree candidate at the University of California, Los Angeles. She works as a Quality Control Specialist in Valencia, California, at Boston Scientific Neuromodulation. Prior to Boston Scientific, she worked for 5 years in the Quality Assurance department at Amgen. She received a Bachelor of Science degree in Business Administration from California Lutheran University in Thousand Oaks, California.

Alvaro Camacho, MD, is an Assistant Clinical Professor of Psychiatry for the University of California, San Diego and the Medical Director for Imperial County Behavioral Health. He has an interest in the longitudinal evolution of psychiatric disorders and comorbidities in Hispanics living in rural underserved areas.

Marlene Gomez is a Masters in Public Health (MPH) degree candidate at the University of California, Los Angeles.

References

Ackerman, F. (2007). The economics of atrazine. *International Journal of Occupational and Environmental Health*, *13*(4), 441–449.

Agency for Toxic Substances and Disease Registry (ATSDR) (1992, September). *Toxicological profile for 1,2,3-Trichloropropane*. Atlanta, GA: Agency for Toxic Substances and Disease Registry, US Department of Health and Human Services.

Agency for Toxic Substances and Disease Registry (ATSDR) (1992). *Toxicological profile for 1,2-dibromo-3-chloro-propane (final report).* Atlanta, GA: Agency for Toxic Substances and Disease Registry. NTIS Accession No. PB93-110906.

Agency for Toxic Substances and Disease Registry (ATSDR) (1995). *ToxFAQs™ for 1,2-dibromo-3-chloropropane (1,2-dibromo-3-cloropropano).* Retrieved September 1995 from <http://www.atsdr.cdc.gov/tfacts36.html>.

Agency for Toxic Substances and Disease Registry (ATSDR) (2003). *Toxicological profile for atrazine.* Retrieved September 2003 from <http://www.atsdr.cdc.gov/toxprofiles/tp153.pdf>.

Agency for Toxic Substances and Disease Registry (ATSDR) (2006). *Interaction profile for: Atrazine, deethylatrazine, diazinon, nitrate and simazine.* Retrieved August 2006 from <http://www.atsdr.cdc.gov/interactionprofiles/IP-10/ip10-p.pdf>.

Agency for Toxic Substances and Disease Registry (ATSDR) (2010). *3. Chemical and physical information (1,2-dibromo-3-chloropropane).* Retrieved March 1, 2010 from <http://www.atsdr.cdc.gov/toxprofiles/tp36-c3.pdf>.

Agricultural Health Study (AHS) (2006). *Lung cancer in the agricultural health study.* National Institutes of Health. Retrieved September 8, 2006 from <aghealth.nci.nih.gov/pdfs/IALungCancer2006.pdf>.

Albritton, D. P., Watson, R. T., & Aucamp, P. J. (1998). *WMO/UNEP. Scientific assessment of ozone depletion: Executive summary, World Meteorological Organization global ozone research and monitoring project.* Report No. 37. Available at: <http://www.AL.NOAA.GOV/wwwhd/PUBDOC/ASSESSMENT98/eXECsUM98.PDF>.

American Conference of Governmental Industrial Hygienists (ACGIH) (2009). *Threshold Limit Values of Chemical Substances and Biological Exposure Indices* (p. 58). Cincinnati, OH: ACGIH.

American Conference of Governmental Industrial Hygienists, Inc. (ACGIH) (1991). *Documentation of the threshold limit values and biological exposure indices* (6th ed.) (Vols. I–III) (pp. 1626). Cincinnati, OH.

Arbuckle, T., Lin, Z., & Mery, L. S. (2001). An exploratory analysis of the effect of pesticide exposure on the risk of spontaneous abortion in an Ontario farm population. *Environmental Health Perspectives, 109*(8), 851–857.

ATSDR (2000). *Toxicological profile for endosulfan.* US Department of Health and Human Services. Publich Health Service. Available at: <http://www.atsdr.cdc.gov/ToxProfiles/tp41.pdf>.

ATSDR (2003). *Malathion. ToxFaqs.* Available at: <http://www.atsdr.cdc.gov/toxfaqs/TF.asp?id=521&tid=92>.

ATSDR (2006). *Atrazine, deethylatrazine, diazinon, nitrate, and simazine.* Available at: <http://www.atsdr.cdc.gov/interactionprofiles/IP-10/ip10-c2.pdf>.

ATSDR (2010). *Agency for toxic substances and disease registry.* Malathion. April. Available at: <http://www.atsdr.cdc.gov/substances/toxsubstance.asp?toxid=92>.

Bajpayee, M., Pandey, A. K., Zaidi, S., Musarrat, J., Parmar, D., Mathur, N., et al. (2006). DNA damage and mutagenicity induced by endosulfan and its metabolites. *Environmental and Molecular Mutagenesis, 47*(9), 682–692.

Bananas (2009). *Pesticide lawsuits – a DBCP overview.* Retrieved May 5, 2009 from <http://www.bananasthemovie.com/pesticide-lawsuits-%E2%80%93-a-dbcp-overview>.

Barchan'ska, H., & Baranowska, I. (2009). Procedures for analysis of atrazine and simazine in environmental matrices. In D. M. Whitacre (Ed.), *Reviews of environmental contamination and toxicology volume 200.* Springer Science & Business Media.

Barry, T., Oriel, M., Verder-Carlos, M., Mehler, L., Edmiston, S., & O'Malley, M. (2010). Community exposure following a drip-application of chloropicrin. *Journal of Agromedicine*, *15*(1), 24–37.

Beaulieu, A., & Fessele, K. (2003). Agent orange: Management of patients exposed in Vietnam. *Clinical Journal of Oncology Nursing*, *7*(3), 320–323.

Beltran, F., Acedo, B., & Rivas, J. (1999). Use of ozone to remove alachlor from surface water. *Bulletin of Environmental Contamination and Toxicology*, *62*, 324–329.

Beyond Pesticides (2010). *Organic agriculture. National coalition against the misuse of pesticides.* Retrieved 2010 from <http://www.beyondpesticides.org/organicfood/index.htm>.

Boesten, J. J. T. I., & Van der Linden, A. M. A. (2001, November). Effect of long term sorption kinetics on leaching as calculated with PEARL model for FOCUS scenarios. In: *Pesticide behaviour in soils and water.* Alton, UK: Proc of a Symp organized by the British Crop Protection Council.

Bonner, M. R., Williams, B. A., Rusiecki, J. A., Blair, A., Beane Freeman, L. E., Hoppin, J. A., et al. (2010). Occupatinal exposure to terbufos and the incidence of cancer in the agricultural health study. *Cancer Causes Control*, *21*(6), 871–877.

Bosma, T. (2002). *Engineering bacteria for the degradation of halopropanes*. Dissertation, University of Groningen, Netherlands. <http://dissertations.ub.rug.nl/faculties/science/2002/t.bosma/>.

Boul, H. L. (1995). DDT residues in the environment – A review with a New Zealand perspective. *New Zealand Journal of Agricultural Research*, *38*(2), 257–277.

Bradford, D. F., Heithmar, E. M., Tallent-Halsell, N. G., Momplaisir, G.-M., Rosal, C. G., Varner, K. E., et al. (2010). Temporal patterns and sources of atmospherically deposited pesticides in alpine lakes of the Sierra Nevada, California, U.S.A. *Environmental Science & Technology*, *44*(12), 4609–4614.

Bradman, A., Salvatore, A. L., Boeniger, M., Castorina, R., Snyder, J., Barr, D. B., et al. (2009). Community-based intervention to reduce pesticide exposure to farmworkers and potential take-home exposure to their families. *Journal of Exposure Science & Environmental Epidemiology*, *1*, 79–89.

California Department of Pesticide Regulation (CDPR) (2003). *Environmental fate of metolachlor.* Environmental Monitoring Branch, Department of Pesticide Regulation. Retrieved April 2003 from <http://www.cdpr.ca.gov/docs/emon/pubs/fatememo/metolachlor.pdf>.

California Department of Public Health (CDPH) (2008, February). *DBCP consumer fact sheet.* Berkeley, CA: Hazard Evaluation Section and Public Water Supply Branch of the California Department of Health Services.

California EPA (CEPA) (1997). *Public health goal for ALACHLOR in drinking water.* Office of Environmental Health Hazard Assessment. Retrieved December 1997 from <http://oehha.ca.gov/water/phg/pdf/alach_c.pdf>.

Centers for Disease Control and Prevention (CDC) (2010). *Chemical information: Alachlor, CAS No. 15972-60-8.* National Report on Human Exposure to Environmental Chemicals. Retrieved April 13, 2010 from <http://www.cdc.gov/ExposureReport/data_tables/Alachlor_ChemicalInformation.html>.

Chapin, R., Gulati, D., Mounce, R., et al. (1997). Trichloropropane. *Environmental Health Perspectives*, *105*(S1), 1–2.

Chemical Safety Information from Intergovernmental Organizations (INCHEM) (1996). *WHO/FAO data sheets on pesticides: No. 86 alachlor.* International Programme on Chemical Safety. Retrieved July 1996 from <http://www.inchem.org/documents/pds/pds/pest86_e.htm#2.2>.

Chen, C., Yang, S., Guo, Y., et al. (2009). Photolytic destruction of endocrine disruptor atrazine in aqueous solution under UV irradiation: Products and pathways. *Journal of Hazardous Materials*, *172*(2-3), 675–684.

Cirillo, T., Viscardi, V., Fasano, E., Farina, A., & Amodio-Cocchieri, R. (2009). Polychlorinated biphenyls, organochlorine pesticides, and polycyclic aromatic hydrocarbons in wild, farmed, and frozen marine seafood marketed in Campania, Italy. *Journal of Food Protection*, *72*(8), 1677–1685.

City of Shafter (2007). *Consumer confidence report for water quality*. Shafter, California.

Collivignarelli, C., & Sorlini, S. (2004). AOPs with ozone and UV radiation in drinking water: Contaminants removal and effects on disinfection by-products formation. *Water Science and Technology*, *49*(4), 51–56.

Cornell University (2005). *Pesticides and breast cancer risk: Dibromochloropropane (DBCP) fact sheet #50*. Program on Breast Cancer and Environmental Risk Factors. Retrieved January 26, 2005 from <http://envirocancer.cornell.edu/FactSheet/pesticide/fs50.dbcp.cfm>.

Curl, C. L., Fenske, R. A., Kissel, J. C., Shirai, J. H., Moate, T. F., Griffith, W., et al. (2002). Evaluation of take home organophosphorus pesticide exposure among agricultural workers and their children. *Environmental Health Perspectives*, *110*(12), A787–A792.

Daniels, J. L., Olshan, A. F., & Savitz, D. A. (1997). Pesticides and childhood cancers. *Environmental Health Perspectives*, *105*(10).

Delaware Health and Social Services (DHSS) (2009). *Alachlor frequently asked questions*. Division of Public Health. Retrieved October 2009 from <http://www.dhss.delaware.gov/dhss/dph/files/alachlorfaq.pdf>.

Delaware Health and Social Services (DHSS) (2010). *Metolachlor frequently asked questions*. Division of Public Health. Retrieved January 2010 from <www.dhss.delaware.gov/dhss/dph/files/metofaq.pdf>.

DoItYourself (2010). *How does ozone purification work?* Retrieved 2010 from <http://www.doityourself.com/stry/how-does-ozone-water-purification-work#ixzz0wd9bIPQ9>.

Donald, J. M., Monserrat, L. E., Hooper, K., Book, S. A., & Chernoff, G. F. (1992). Prioritizing candidate reproductive/developmental toxicants for evaluation. *Reproductive Toxicology*, *6*(1), 99–108.

Engel, L. S., Hill, D. A., Hoppin, J. A., Lubin, J. H., Lynch, C. F., Pierce, J., et al. (2005). Pesticide use and breast cancer risk among farmers' wives in the agricultural health study. *American Journal of Epidemiology*, *161*(2), 121–135.

EPA (1998). R. E. D. F. D. P., Pesticides, and Toxic Substances. EPA-738-F-98-016, November. R.E.D. Facts: Dicofol. Prevention, Pesticides, and Toxic Substances. EPA-738-F-98-016, November.

EPA (2005). *Reregistration eligibility decision for propargite*. Available at: <http://www.epa.gov/oppsrrd1/REDs/propargite_amend_red.pdf>.

EPA (2007). *Pesticides and food: Health problems pesticides may cause*. Available at: <http://www.epa.gov/pesticides/food/risks.htm>.

EPA (2008). *Ecological risk assessments*. Available at: <http://www.epa.gov/pesticides/ecosystem/ecorisk.htm>.

EPA (2009). *Assessing health risks from pesticides*. Available at: <http://www.epa.gov/pesticides/factsheets/riskassess.htm>.

Environmental Working Group (EWG) (2010). *Weed killers by the glass: Chapter 1. Health effects of herbicides*. Retrieved August 2, 2010 from <http://www.ewg.org/node/7635>.

Extension Toxicology Network (Extoxnet) (1993). *Metolachlor*. Retrieved September 1993 from <http://pmep.cce.cornell.edu/profiles/extoxnet/metiram-propoxur/metolachlor-ext.html>.

Extension Toxicology Network (Extoxnet) (1996). *Alachlor*. Retrieved June 1996 from <http://extoxnet.orst.edu/pips/alachlor.htm>.

Extension Toxicology Network (Extoxnet) (1996). *Metolachlor*. Retrieved 1996 from <http://extoxnet.orst.edu/pips/metolach.htm>.

FDA (1999). *Food and drug administration pesticide program: Residue monitoring*. Available at: <http://www.cfsan.fda.gov/~dms/pesrpts.html>.

FDA (2001). *Food and drug administration total diet study*. College Park, MD: Center for Food Safety. Available at: <http://www.cfsan.fda.gov/~comm/lds_toc.html>.

Frueh, F. W. (1999). *Dioxin*. Stanford University. Available at: <http://www.stanford.edu/group/whitlock/dioxin.html>.

Fry, D. M. (1995). Reproductive effects in birds exposed to pesticides and industrial chemicals. *Environmental Health Perspectives*, *103*, 165–171.

Global Development and Environment Institute (GDEI) (2003). *Free trade, corn, and the environment: Environmental impacts of US–Mexico corn trade under NAFTA*. Tufts University. Retrieved June 2003 from <http://ase.tufts.edu/gdae/pubs/wp/03-06-nafta-corn.pdf>.

Griffith, M., Tajik, M., & Wing, S. (2007). Patterns of agricultural pesticide use in relation to socioeconomic characteristics of the population in the rural U.S. South. *International Journal of Health Services*, *37*(2), 259–277.

Hayes, T., Collins, A., Lee, M., et al. (2002). Hermaphroditic, demasculinized frogs after exposure to the herbicide atrazine at low ecology relevant doses. *Proceedings of the National Academy of Sciences of the United States of America*, *99*(8), 5476–5480.

Hayes, W. J. (1982). *Pesticide studies in man*. Baltimore, MD: Williams & Wilkins.

Health Canada (1990). Metolachlor. Environmental and Workplace Health. Retrieved November 1990 from <http://www.hc-sc.gc.ca/ewh-semt/alt_formats/hecs-sesc/pdf/pubs/water-eau/metolachlor-metolachlore/metolachlor-metolachlore-eng.pdf>.

Henderson, C., Pickering, Q. H., & Tarzwell, C. M. (1959). Relative toxicity of ten chlorinated hydrocarbon insecticides to four species of fish. *Transactions of the American Fisheries Society*, *88*, 23–32.

Hertzman, C., Wiens, M., Bowering, D., Snow, B., & Calne, D. (1990). Parkinson's disease: A case-control study of occupational and environmental risk factors. *American Journal of Industrial Medicine*, *17*(3), 349–355.

Institute of Medicine (1994). *Veterans and agent orange*. Washington DC: National Academy Press.

International Agency for Research on Cancer (IARC) (1979). *Some halogenated hydrocarbons. IARC monographs on the evaluation of carcinogenic risk of chemicals to humans, vol. 20*. Lyon, France: International Agency for Research on Cancer, p. 609.

International Agency for Research on Cancer (IARC) (1987). *1,2-DIBROMO-3-CHLOROPROPANE. IARC monographs on the evaluation of the carcinogenic risk of chemicals to humans*. Lyon, France: International Agency for Research on Cancer, pp. 479–500.

International Agency for Research on Cancer (IARC) (1995). *1,2,3-Trichloropropane. IARC monographs on the evaluation of the carcinogenic risk of chemicals to humans, vol. 63*. Lyon, France: International Agency for Research on Cancer, pp. 223–244.

Jaga, K., & Dharmani, C. (2003). Sources of exposure to and public health implications of organophosphate pesticides. *Revista Panamericana De Salud Publica/Pan American Journal Public Health*, *14*, 3.

Kannan, K., Holcombe, R., Jain, S., Alvarez-Hernandez, X., Chervenak, R., Wolf, R., et al. (2000). Evidence for the induction of apoptosis by endosulfan in a human T-cell leukemic line. *Molecular and Cellular Biochemistry*, *205*(1), 53–66.

Keller, A. E., & Ruessler, D. S. (1997). The toxicity of malathion to unionid mussels: Relationship to expected environmental concentrations. *Environmental Toxicology and Chemistry*, *16*(5), 1028–1033.

Lee, J. W., Hoppin, J. A., Blair, A., et al. (2004). Cancer incidence among pesticide applicators exposed to alachlor in the agricultural health study. *American Journal of Epidemiology*, *159*(4), 373–380.

Lenntech (2009). *Ozone applications drinking water*. Retrieved 2009 from <http://www.lenntech.com/library/ozone/drinking/ozone-applications-drinking-water.htm>.

Liu, W., Zhu, L.-S., Wang, J., Wang, J.-H., Xie, H., & Song, Y. (2009). Assessment of the genotoxicity of endosulfan in earthworm and white clover plants using the comet assay. *Archives of Environmental Contamination and Toxicology*, *56*(4), 742–746.

Lu, Y., Zhou, S.-B., & Li, B.-X. (2010). Exposure to environmental hexachlorocyclohexane (HCH) and dichlorodiphenyltrichloroethane (DDT) among rural children in north eastern China. *Biomedical and Environmental Sciences*, *23*(3), 230–233.

Luo, Y., & Zhang, M. (2009). A geo-referenced modeling environment for ecosystem risk assessment: Organophosphate pesticides in an agriculturally dominated watershed. *Journal of Environmental Quality*, *38*(2), 664–674.

Luo, Y., & Zhang, M. (2010). Spatially distributed pesticide exposure assessment in the Central Valley. *Environmental Pollution*, *158*(5), 1629–1637.

Mackay, D., Shiu, W., & Ma, K. (1993). *Illustrated handbook of physical-chemical properties and environmental fate for organic chemicals. Vol. 3. Volatile organic chemicals*. Boca Raton, FL: Lewis Publishers.

MacLennan, P., Delzell, E., Sathiakumar, N., et al. (2002). Cancer incidence among triazine herbicide manufacturing workers. *Journal of Environmental and Occupational Medicine*, *44*(11), 1048–1058.

Munger, R., Isacson, P., Hu, S., et al. (1997). Intrauterine growth retardation in Iowa communities with herbicide-contaminated drinking water supplies. *Environmental Health Perspectives*, *105*(3), 308–314.

National Institute of Occupational Safety and Health (NIOSH) (2004). *1,2,3-Trichloropropane. Pocket guide to chemical hazards*. Cincinnati, Ohio: National Institute of Occupational Safety and Health (NIOSH), Centers for Disease Control and Prevention, US Department of Health and Human Services. <http://www.cdc.gov/niosh/npg/npg.html>.

Ochoa-Acuna, H., Frankenberger, J., Leighanne, H., et al. (2009). Drinking-water herbicide exposure in Indiana and prevalence of small-for-gestational-age and preterm delivery. *Environmental Health Perspectives*, *117*(10), 1619–1924.

Oki, D. S., & Giambelluca, T. W. (1987, November–December). DBCP, EDBB, and TCP contamination of ground water in Hawaii. *Ground Water*, *25*(6), 693–702.

Oriel, M., Edmiston, S., Beauvais, S., Barry, T., & O'Malley, M. (2009). Illnesses associated with chloropicrin use in California agriculture, 1992–2003. In D. M. Whitacre (Ed.), *Reviews of environmental contamination and toxicology Vol. 200* (Vol. 200, pp. 1–31). US: Springer.

Qiu, X. H., Zhu, T., Yao, B., Hu, J. X., & Hu, S. W. (2005). Contribution of dicofol to the current DDT pollution in China. *Environmental Science & Technology*, *39*, 4385–4390.

Rasenberg M. H. C., & Van de Plassche E. J. (2003, January). Dicofol Dossier prepared for the meeting March 17–19 in Norway of the UNECE Ad-hoc Expert Group on POPs, 4L0002.A1/R0011/EVDP/Nijm.

Rotterdam (2010). *Draft decision guidance document: ALACHLOR*. Retrieved 2010 from <http://www.pic.int/home.php?type = t&id = 335>.

Samanic, C., Hoppin, J. A., Lubin, J. H., Blair, A., & Alavanja, M. C. (2005). Factor analysis of pesticide use patterns among pesticide applicators in the agricultural health study. *Journal of Exposure Analysis and Environmental Epidemiology*, *15*(3), 225–233.

Schilmann, A., Lacasaña, M., Blanco-Muñoz, J., Aguilar-Garduño, C., Salinas-Rodríguez, A., Flores-Aldana, M., et al. (2010). Identifying pesticide use patterns among flower growers to assess occupational exposure to mixtures. *Occupational and Environmental Medicine*, *67*(5), 323–329.

Simoniello, M. F., Kleinsorge, E. C., Scagnetti, J. A., Grogolato, R. A., Poletta, G. L., & Carballo, M. A. (2008). DNA damage in workers occupationally exposed to pesticide mixtures. *Journal of Applied Toxicology*, *28*, 957–965.

State Water Resources Control Board (SWRCB) (2009, November). *1,2,3-Trichloropropane (TCP)*. California: Division of Water Quality, GAMA Program.

State Water Resources Control Board (SWRCB) (2010, June). *Dibromochloropropane (DBCP)*. California: Division of Water Quality, GAMA Program.

Swan, S., Kruse, R. L., Liu, F., et al. (2003). Semen quality in relation to biomarkers of pesticide exposure. *Environmental Health Perspectives*, *111*(12), 1478–1484.

Tesoriero, A. J., Loffler, F. E., & Liebscher, H. (2001). Fate and origin of 1,2-dichloropropane in an unconfined shallow aquifer. *Environmental Science and Technology*, *35*, 455–461.

Trout T. (2002). Fumigant use in California. USDA. Available at: <http://mbao.org/2002proc/094troutt%20mb-conf02-fumuse.pdf>.

UNEP (2008, October 23). United Nations Environmental Programme. Global Status of DDT and its alternatives for use in vector control to prevent disease, UNEP/POPS/DDT.1/2. Stockholm Convention on Persistent Organic Pollutants, Geneva, Switzerland.

United States Department of the Interior, Bureau of Reclamation (USBR) (2009). *Granular activated carbon (GAC)*. Retrieved 2009 from <http://www.usbr.gov/pmts/water/publications/reportpdfs/Primer%20Files/07%20-%20Granular%20Activated%20Carbon.pdf>.

United States EPA (EPA) (1987). Metolachlor Health Advisory Draft. Office of Drinking Water. Retrieved August 1987 from <http://nepis.epa.gov/EPA/html/DLwait.htm?url=/Exe/ZyNETexe/2000SP8V.PDF?ZyActionP=PDF&Client=EPA&Index=1986%20Thru%201990&File=D%3A/ZYFILES/INDEX%20DATA/86THRU90/TXT/00000014/2000SP8V.txt&Query=820K88116&SearchMethod=1&FuzzyDegree=0&User=ANONYMOUS&Password=anonymous&QField=pubnumber^%22820K88116%22&UseQField=pubnumber&IntQFieldOp=1&ExtQFieldOp=1&Docs=>.

United States EPA (EPA) (1995). R.E.D FACTS: Metolachlor. Prevention, Pesticides and Toxic Substances. Retrieved April 1995 from <http://epa.gov/oppsrrd1/REDs/factsheets/0001fact.pdf>.

United States EPA (EPA) (1995). Reregistration Eligibility Decision (RED): Metolachlor. Prevention, Pesticides and Toxic Substances. Retrieved April 1995 from <http://www.epa.gov/oppsrrd1/REDs/0001.pdf>.

United States EPA (EPA) (1997, June). EnviroMetal Technologies, Inc., Metal Enhanced Dechlorination of Volatile Organic Compounds Using an Above-Ground Reactor, Innovative Technology Evaluation Report, EPA/540/R-96/503. National Risk Management Research Laboratory, Office of Research and Development, Cincinnati, Ohio.

United States EPA (EPA) (1998). Reregistration Eligibility Decision (RED): Alachlor. Prevention, Pesticides and Toxic Substances. Retrieved December 1998 from <http://www.epa.gov/oppsrrd1/REDs/0063.pdf>.

United States EPA (EPA) (2000). 1,2-Dibromo-3-Chloropropane (DBCP). Technology Transfer Network. Retrieved January 2000 from <http://www.epa.gov/ttnatw01/hlthef/dibromo-.html>.

United States EPA (EPA) (2001). The Incorporation of Water Treatment Effects on Pesticide Removal and Transformations in Food Quality Protection Act (FQPA) Drinking Water Assessments. Office of Pesticide Programs. Retrieved October 25, 2001 from <http://www.epa.gov/oppfead1/trac/science/water_treatment.pdf>.

United States EPA (EPA) (2005, July 27). Interim Guidance for Investigating Potential 1,2,3-Trichloropropane Sources in San Gabriel Valley, E072005013SCO. San Francisco, California.

United States EPA (EPA) (2007). What are Endocrine Disruptors? Endocrine Disruptor Screening Program. Retrieved 2007 from <http://www.epa.gov/endo/pubs/edspoverview/whatare/htm>.

United States EPA (EPA) (2008). *Atrazine background*. Retrieved 2008 from <http://www.epa.gov/opp00001/factsheets/atrazine_background.htm>.

United States EPA (EPA) (2008). Regulatory Determinations Support Document for Selected Contaminants from the Second Drinking Water Contaminant Candidate List (CCL 2). The Office of Groundwater and Drinking Water. Retrieved June 2008 from <http://www.epa.gov/safewater/ccl/pdfs/reg_determine2/report_ccl2-reg2_supportdocument_ch12_metolachlor.pdf>.

United States EPA (EPA) (2009). *Research addresses potential risks of atrazine*. Retrieved 2009 from <http://www.epa.gov/hhrp/factsheets/potential-risks-of-atrazine-and-related-pesticides.html>.

United States EPA (EPA) (2010). *Atrazine updates*. Retrieved June 2010 from <http://www.epa.gov/oppsrrd1/reregistration/atrazine/atrazine_update.htm#atrazine>.

United States EPA (EPA) (2010). *Basic information about atrazine in drinking water*. Retrieved March 5, 2010 from <http://www.epa.gov/ogwdw000/contaminants/basicinformation/atrazine.html>.

United States EPA (EPA) (2010). *Basic information about 1,2-dibromo-3-chloropropane (DBCP) in drinking water*. Retrieved 2010 from <http://www.epa.gov/ogwdw000/contaminants/basicinformation/1-2-dibromo-3-chloropropane.html>.

United States EPA (EPA) (2010). *Technical factsheet on: ALACHLOR*. Retrieved 2010 from <http://www.epa.gov/ogwdw000/pdfs/factsheets/soc/tech/alachlor.pdf>.

United States Geological Survey (USGS) (2006). Pesticides in the Nation's Streams and Groundwater, 1992–2001. Chapter 5, 67–86.

United States Geological Survey (USGS) (2007, February). *Pesticides in the nation's streams and groundwater, 1992–2001*. Reston, VA: National Water-Quality Assessment Program. p. 1.

United States Geological Survey (USGS) (2010). 2002 Pesticide Use Maps – Atrazine. Pesticide National Synthesis Project. Retrieved 2010 from <http://water.usgs.gov/nawqa/pnsp/usage/maps/show_map.php?year=02&map=m1980>.

United States Geological Survey (USGS) (2010). 2002 Pesticide Use Maps – Alachlor. Pesticide National Synthesis Project. Retrieved August 2010 from <http://water.usgs.gov/nawqa/pnsp/usage/maps/show_map.php?year=02&map=m1863>.

United States Geological Survey (USGS) (2010, August) 2002 Pesticide Use Maps – Metolachlor. Pesticide National Synthesis Project. Retrieved August 2010 from <http://water.usgs.gov/nawqa/pnsp/usage/maps/show_map.php?year=02&map=m9124>.

US Department of Veterans Affairs (2009). Agent Orange: Basics. Office of Public Health and Environmental Hazards. Available at: <http://www.publichealth.va.gov/exposures/agentorange/basics.asp>.

US Department of Veterans Affairs (2010). Agent Orange. Office of Public Health and Environmental Hazards. Available at: <http://www.publichealth.va.gov/exposures/agentorange>.

US EPA (1990). National Pesticide Survey. Simazine. Available at: <http://nepis.epa.gov/Exe/ZyPURL.cgi?Dockey=P100134T.txt>.

US EPA (2004). *Overview of the ecological risk assessment process in the office of pesticide programs*. Washington, D.C.: Office of Prevention, Pesticides and Toxic Substances Office of Pesticide Programs.

US EPA (2005). *Reregistration elegibility decision for propargite*. Available at: <http://www.epa.gov/pesticides/reregistration/REDs/propargite_amend_red.pdf>.

US EPA (2009). *Types of pesticides*. Available at <http://www.epa.gov/pesticides/about/types.htm>.

US EPA *Ecological risk assessment*. Available at: <http://www.epa.gov/pesticides/ecosystem/ecorisk.htm>. (Accessed July 2010).

US National Library of Medicine (2010). Pesticides. Medline Plus. Available at: <http://www.nlm.nih.gov/medlineplus/pesticides.html>.

USFWS (1946). United States Fish and Wildlife Service. New Publication Announcement – DDT: Its Effects on Fish and Wildlife. Available at: <http://www.fws.gov/contaminants/pdf/historic/19460518.pdf>.

USFWS (2009). United States Fish and Wildlife Service. Chlorinated Hydrocarbons (Organochlorines) – DDT. Available at: <http://www.fws.gov/contaminants/Info/ddt.html>.

Van Maele-Fabry, G., & Willems, J. L. (2003). Occupation related pesticide exposure and cancer of the prostate: A meta-analysis. *Occupational and Environmental Medicine*, *60*(9), 634–642.

Veterans of the Vietnam War (2010). Agent Orange Fact Sheet: A Historical Perspective. Available at: <http://www.vvnw.org/educational_material/agent_orange.htm#Service>.

Vitali, M., Protano, C., Del Monte, A., Ensabella, F., & Guidotti, M. (2009). Operative modalities and exposure to pesticides during open field treatments among a group of agricultural subcontractors. *Archives of Environmental Contamination and Toxicology*, *57*, 193–202.

Volp, R. F., Sipes, I. G., Falcoz, C., et al. (1984). Disposition of 1,2,3-trichloropropane in the Fischer 344 rat: Conventional and physiological pharmacokinetics. *Toxicology and Applied Pharmacology*, *75*(1), 8–17.

Baz, W., El-Soueidi, R., Nakhl, F., Aoun, N., Chin, N., & Dhar, M. (2010). Desmoplastic small round-cell tumor: An adult with previous exposure to Agent Orange. *Japanese Journal of Clinical Oncology*, *40*(6), 593–595.

Weinberg, J. L., Bunn, L. J., & Das, R. (2009). Application of the industrial hygiene hierarchy of controls to prioritize and promote safer methods of pest control: A case study. *Public Health Reports*, *124*(suppl 1), 53–62.

Winchester, P., Huskins, J., & Ying, J. (2009). Agrichemicals in surface water and birth defects in the United States. *Acta Paediatrica*, *98*(4), 664–669.

World Health Organization (WHO) (2003). 1,2,3-Trichloropropane. Concise International Chemical Assessment Document 56, Geneva Switzerland. <http://www.who.int/ipcs/publications/cicad/en/cicad56.pdf>.

World Health Organization (WHO) (2003). *Metolachlor in drinking-water.* Retrieved 2003 from <http://www.who.int/water_sanitation_health/dwq/chemicals/metolachlor.pdf>.

World Health Organization (WHO) (2003). *Terbuthylazine (TBA) in drinking-water.* Retrieved 2003 from <http://www.who.int/water_sanitation_health/dwq/chemicals/terbuthylazine.pdf>.

World Health Organization (2004). The WHO Classification of Pesticides by Hazard. Guideliness to Classification. International Program on Chemical Safety. Available at: <http://www.inchem.org/documents/pds/pdsother/class.pdf>.

World Health Organization (2010). *Dioxins and their effects on human health.* Available at: <http://www.who.int/mediacentre/factsheets/fs225/en>.

Zebarth, B. J., Szeto, S. Y., Hii, B., et al. (1998). Groundwater contamination by chlorinated hydrocarbon impurities present in soil fumigant formulations. *Water Quality Research Journal of Canada, 33*(1), 31–50.

12 Current Practices in Hazardous Waste Treatment and Disposal

Chapter Contents

12.1 Introduction

The final stages of the hazardous waste management sequence, namely the treatment and disposal steps, are perhaps the most complex and technologically demanding. A myriad of treatment and disposal options for hazardous waste exist; however, selecting the best choice for a particular hazardous waste is both difficult and crucial to the proper management of that waste. Many of the disposal strategies currently in use are outdated and are slowly being replaced by newer and cleaner technologies. These advancements are discussed later in this book.

The purpose of hazardous waste treatment and disposal is to mitigate the characteristics that make this waste hazardous, and to permanently contain the wastes, respectively. Often times, hazardous waste is disposed of without prior treatment with the hope that the waste will eventually transmute into less hazardous wastes over time. In addition, these treatment and disposal strategies usually seek to reduce the volume of the waste and to concentrate the waste for more efficient storage.

Deepwell or underground injection was the most common hazardous waste management technique in 2007, representing 42.6% of the techniques used, followed by other disposal (24.5%), aqueous organic treatment (6.2%), incineration (6%), landfill and surface impoundments (3.8%), and other techniques representing the remaining 17% (Table 12.1).

Risks of Hazardous Wastes. DOI: 10.1016/B978-1-4377-7842-7.00012-X

Table 12.1 Top 15 Hazardous Waste Management Strategies in 2007 (EPA, 2008a)

Rank	Management Method	Tons Managed	Percentage of Total Managed	Number of Facilities
1	Deepwell or underground injection	21,505,921	42.6	42
2	Other disposal	12,363,634	24.5	91
3	Aqueous organic treatment	3,106,828	6.2	67
4	Incineration	3,047,982	6	140
5	Landfill and surface impoundment	1,939,712	3.8	67
6	Aqueous inorganic treatment	1,879,946	3.7	170
7	Energy recovery	1,764,693	3.5	91
8	Other treatment	1,298,339	2.6	353
9	Metals recovery	1,116,357	2.2	137
10	Fuel blending	737,397	1.5	112
11	Stabilization	658,249	1.3	107
12	Sludge treatment	397,863	0.8	59
13	Other recovery	335,093	0.7	65
14	Solvents recovery	328,931	0.7	456
15	Land treatment/application/farming	1,981	0	16
		50,482,926	100	1,973

12.2 Underground Injection

Underground injection (or 'deepwell injection') is currently the most commonly practiced method of hazardous waste disposal in the United States. In 2007, over 21 million tons of hazardous waste was disposed of using deepwell injection (EPA, 2008b). Deepwell injection began in Texas in the 1930s to dispose of oil field brine generated from the petroleum industry. Today, the oil industry is still the biggest user of deepwell injection systems for regular waste disposal, with a total of 143,951 wells registered for the oil and gas industry as of the writing of this text. Wastes from oil and gas production are excluded wastes (non-hazardous) under RCRA, and thus chemical manufacturers (such as DOW, Solutia, and Occidental) are the largest users of deepwell injection for hazardous waste disposal. The first recorded use of underground injection wells for hazardous waste disposal by the chemical industry was by the DuPont Chemical Company, in Texas, in 1951 (Knape, 2005).

In a deepwell injection system, hazardous wastes in liquid form are injected several thousand feet below the surface through a reinforced well shaft into porous injection zones that are confined by impermeable rock layers. In 1984, with the enactment of the Hazardous Waste and Solid Waste Amendments (HSWA), the disposal of untreated hazardous waste on land was banned. Thus, the HSWA was poised to halt all disposal of untreated hazardous waste through underground injection throughout the United States. However, the EPA, which has primary regulatory authority for injection wells, changed its Underground Injection Control regulations to exempt deep-well injection systems from the HSWA ban. Consequently, since

1988, the EPA has allowed hazardous waste to get pumped straight from its source of creation into an underground location without any pretreatment. Well operators must demonstrate that the hazardous liquids will not migrate from the disposal site for 10,000 years, although the modeling process for hazardous waste migration is subject to great uncertainties (EPA, 2010a).

12.2.1 Problems with Underground Injection

Underground injection is beset with a number of limitations in its use. For example, underground injection cannot be used in any area that is prone to seismic activity. In addition, underground injection can only be used with highly liquefied wastes since the presence of suspended solids at a concentration greater than 2 ppm leads to clogging of the injection well. Hazardous wastes with high organic carbon content must be monitored carefully because the added carbon content can lead to a flourishing of bacterial populations in the injection zone or well shaft resulting in fouling of the injection well (EPA, 2001).

While EPA regulations governing the construction and operation of underground injection wells for hazardous waste disposal have become more stringent over the past few decades, a review of the most severe leakages from past use of these systems is warranted. From 1964 to 1972 the Hammermill Paper Company pumped 1.1 billion gallons of spent pulping liquor into three underground injection wells located in or around Lake Erie in Presque Island State Park, Pennsylvania. In March of 1979 it was discovered that these pulping liquors were leaking from an abandoned injection well located on a small island in the lake. These wastes contaminated both Lake Erie (a primary source of drinking water for the surrounding population) as well as other underground sources of drinking water. At this time it was also discovered that a large vertical fracture system existed that could potentially bring wastes up to the surface and that wastes had already migrated horizontally from the injection zone, a distance more than eight times over the government mandated limit. In 1982 this site was added to the Superfund National Priority List (Gordon and Bloom, 1985).

Incompatibilities between the liquid hazardous waste and the materials of the well shaft are a serious concern of well operators and regulatory organizations. In 1976, in Beaumont, Texas, the Velsicol Chemical Corporation was permitted to inject highly acidic herbicide wastes into an underground rock formation. The wastes had a pH of less than 4 and contained a range of toxic substances, including dioxins. Soon after pumping began, it was discovered that the wastes had corroded through both the inner and outer well casings, as well as the surrounding concrete barrier. As a result, an estimated 5 million gallons of these herbicide wastes contaminated a nearby freshwater aquifer (Gordon and Bloom, 1985).

In a more recent study from 1988 to 1991, the EPA found that there were 130 cases of mechanical failures leading to internal leakage of underground injection wells. The next time period that was studied (1993–1998) saw a decrease in the number of mechanical failures of injection wells in most of the United States. Although in Texas, one of the largest users of underground injection wells, the number of mechanical well failures had nearly doubled to 65% of all operating wells during this period (EPA, 2001).

Injection wells have also been shown to trigger earthquakes and other seismic activity in hundreds of recorded incidents in the US and around the world (Gordon and Bloom, 1985; Nicholson and Wesson, 1992; Lei et al., 2008; Assumpcao et al., 2010; Frohlich et al., 2010). The improper abandonment of injection wells is another serious concern because it leads to lapses in monitoring and a significantly increased risk of hazardous waste leakage over time.

12.3 Aqueous Organic Treatment

Aqueous organic treatment refers to treatments done to liquid hazardous wastes to reduce their toxicity, ignitability, corrosivity, or reactivity. Aqueous organic treatment can be done prior to underground injection if a waste is deemed too toxic to inject by the EPA, or prior to landfilling. There are a number of aqueous organic treatment options which can be used in conjunction or alone; however, particular care must be taken to ensure that the treatment methods used are applicable to the hazardous contaminant to be mitigated. Biological treatment, a technique used often on wastewater streams, refers to the use of bacteria or algae to metabolize hazardous compounds and concentrate these compounds in the biomass of the organisms for more efficient disposal as solid waste. This treatment technology has proven to be widely applicable and can be used in the removal of alcohols, aliphatics, amines, aromatics, halocarbons, metals, phenols, phthalates, polycyclic hydrocarbons, and various other compounds from aqueous hazardous waste. Reverse osmosis is also available for hazardous waste streams, although this is most effective for removing inorganic salts and less effective for the removal of organic compounds. Air stripping can be used for the removal of volatile organic compounds (VOCs) from hazardous wastes. The most heavily used aqueous organic treatment method is adsorption via granular activated carbon (GAC). GAC is effective for the removal of many organic compounds, although the efficacy of GAC decreases with decreasing molecular weight of the compound. Numerous other treatment methods, including centrifugation, crystallization, dialysis, distillation, evaporation, ion exchange, and solvent extraction are also available for the treatment of hazardous waste streams (EPA, 1985).

12.4 Incineration

In 2007, over 3 million tons of hazardous wastes were incinerated at 140 facilities across the United States (EPA, 2008b). Incineration is used for hazardous wastes which cannot be reused or recycled and cannot be disposed of safely in a landfill because of excessive toxicity or risk of infectious transmission. This technique allows for significant volume reduction and varying magnitudes of toxicity reduction in the hazardous waste being treated. During incineration, the wastes are combusted and converted into gases containing trace amounts of organic contaminants and a highly toxic solid residue that is typically disposed of in a hazardous waste landfill. Incineration is often used for biological wastes to reduce the risk of infectious transmission can also be used to recover energy from combustible hazardous wastes.

While some studies have shown that hazardous waste incineration is a reliable hazardous waste treatment strategy (ATSDR, 1992), there are concerns about its effect on public health. The two primary risks associated with hazardous waste incinerators are hazardous organic compounds and toxic heavy metals in either stack emissions or the leftover solid residue. Organic contaminants of incinerator emissions or residues are attributed to uncombusted or partially combusted wastes known as products of incomplete combustion (PICs) or to new compounds that are formed during the combustion process. PICs encompass a range of highly toxic compounds many of which are known carcinogens, mutagens, and endocrine effectors. Dioxins, including the particularly deleterious TCDD, polychlorinated dibenzofurans (PCDFs), polychlorinated biphenyls (PCBs), and polycyclic aromatic hydrocarbons (PAHs) are among the most hazardous PICs released from hazardous waste incinerators (Sedman and Esparza, 1991). While most organic compounds are destroyed during the incineration process, it is important to note that waste streams entering hazardous waste incinerators are heterogeneous and their chemical makeup is never completely determined. Therefore, combustion conditions can change rapidly in any incinerator, leading to the creation of unwanted toxicants in the gaseous emissions or leftover residue.

The release of neurotoxic metals such as lead, mercury, and arsenic, and of carcinogenic metals such as beryllium, cadmium, and chromium, is another public health risk associated with hazardous waste incineration. Because metallic compounds are not destroyed in incinerators, they are simply oxidized and transformed and can be found in nearly the same concentrations in incinerator residues as in the incoming waste stream (Jakob et al., 1995). Sedman and Esparza (1991) evaluated 20 incinerators across the United States for emissions of either organic compounds or metals. They found that two incinerators were releasing organic compounds at an amount leading to 'significant risk' for the surrounding population and that five incinerators were releasing heavy metals at an amount constituting a significant risk (Sedman and Esparza, 1991). Hazardous waste incinerator emissions are regulated under the RCRA (40 CFR Part 264/265), the Clean Air Act, and the EPA's National Emission Standards for Hazardous Air Pollutants (NESHAPs). None of these regulations currently requires continuous emissions monitoring systems (CEMS) for any contaminant, meaning that harmful emissions can occur between the periodic stack testing that takes place (EPA, 2009a).

12.5 Land Disposal

The hazardous waste history of the United States is marked by a series of large accidental contaminations followed by public outcry and a governmental response. Perhaps the most infamous of such incidents is the discovery of the Love Canal hazardous waste landfill in 1978 which led to public outcry about the lurking health risks of hazardous waste disposal sites. Subsequent regulatory changes in 1980 were implemented in order to improve the safe handling and disposal of hazardous wastes in the United States. The design specifications of hazardous waste landfills were modified in 1985 and again later in 1992 in order to introduce additional safeguards meant to reduce the risk of leaks or other escape of toxins from the landfills from causing harm to the environment or nearby population. Unfortunately, a large number of hazardous waste landfills are expected to eventually leak because they were either built prior to the introduction of modern designs or because of errors in the construction or management of modern-day sites. Exposure to toxins contained within or leaking from hazardous waste landfills appears to pose a significant risk to human health and numerous studies have suggested an increase in the occurrence of congenital anomalies, decreased birth weight, and increased risk of respiratory diseases among individuals living near hazardous waste landfills. It is not surprising that hazardous waste landfills are often located in socioeconomically disadvantaged minority neighborhoods which increases the health risks for this population. With the multitude of environmental and health risks associated with this form of hazardous waste disposal, future waste management strategies must seek to limit the use of hazardous waste landfilling.

12.5.1 Love Canal

The Love Canal incident brought the danger of hazardous waste landfills to the public's attention in 1978 and was a large motivating factor for political action leading to the passage of CERCLA in 1980 and the establishment of the Superfund program for hazardous waste cleanup. In the early 1900s, William T. Love intended to create a model city near Niagara Falls, New York. To power the small three-block tract of land, he proposed building a short canal between the upper and lower Niagara Rivers. Construction on the canal was begun but only a partial ditch was completed before the project was abandoned due to financial reasons. In the 1920s, the ditch was turned into a municipal and industrial chemical dumpsite. In 1953, the Hooker Chemical Company covered the canal with soil and sold the land to the city for one dollar. This land then became the site of about 100 homes and a school in the late 1950s. Record rainfall in 1978 caused the chemicals to leach into the community with disastrous effects. Eckardt C. Beck, then an administrator for the EPA, described his visit to the site:

> *visited the canal area at that time. Corroding waste-disposal drums could be seen breaking up through the grounds of backyards. Trees and gardens were turning black and dying. One entire swimming pool had popped up from its foundation, afloat now on a small sea of chemicals. Puddles of noxious substances were pointed out to me by the residents. Some of these puddles were in their yards, some were in their basements, others yet were on the school grounds. Everywhere the air had a faint, choking smell. Children returned from play with burns on their hands and faces.*
>
> *(Beck, 1979)*

In addition to these immediate injuries, an unusually high rate of miscarriages and birth defects were also seen in the small community. Eighty-two different compounds, including known carcinogens such as benzene were detected in the ground and homes of the community. Residents eventually were all evacuated and their homes purchased by the state government. In all, 221 families were initially moved. Congressional funding subsequently allowed an additional larger relocation area, termed the Emergency Declaration Area, totaling 814 single-family homes and an undetermined number of additional public housing apartments (EPA, 2009b).

Researchers have found evidence suggesting an increase in stillbirths, birth defects, and other adverse reproductive outcomes related to Love Canal (Goldman et al., 1985). However, an increase in long-term cancer risks from hazardous waste exposure among Love Canal residents has been more difficult to quantify. Gensburg and colleagues identified 5,052 former Canal residents in 1996 and compared their incidence of cancer as compared to a control cohort (Gensburg et al., 2009). A total of 304 new cancer cases were found in this cohort. While this study was limited by a small sample size, there were an elevated number of bladder and kidney cancer cases as compared to a control population.

12.5.2 Methods of Land Disposal

The term 'land disposal units' is used to refer to land-based disposal facilities and includes landfills, surface impoundments, waste piles, land treatment units, injection wells, salt dome formations, salt bed formations, underground mines, and underground caves. The first four types of land disposal units have established technical standards and are regulated by the EPA under the treatment, storage, and disposal facility (TSDF) requirements of the RCRA. In an RCRA training module, the EPA provided the following definitions (EPA 2005).

Surface Impoundments

Surface impoundments consist of a natural topographic depression, man-made excavation, or diked area formed from earthen materials into which hazardous wastes are placed. They are required to have a double liner, a leachate collection and removal system, and a leak detection system. An important distinction between surface impoundments and landfills is the temporary nature of impoundments. These units are used for the temporary storage and treatment of wastes and therefore a cap or cover is not required as is the case in landfills. However, if the operator chooses to permanently store wastes in an impoundment, landfill requirements will then be applicable, including post-closure care.

Waste Piles

Waste piles are temporary non-containerized piles of solid non-flowing hazardous wastes. They must be located under or in a structure and must be protected from water run-on. Similar to an impoundment, they are required to have a double liner and a leachate collection and removal system. Unlike an impoundment, they do not require a leak detection system but require a second leachate collection and removal system above the top liner.

Land Treatment Units

Land treatment units differ in that the waste is directly applied to the soil surface or upper layers of soil in order to degrade, transform, or immobilize the hazardous constituents of the wastes with soil microbes or sunlight acting to degrade the waste. Liners are not generally used in land treatment units. Prior to use, operators are required to perform a treatment demonstration to insure that the hazardous components of the wastes can be effectively degraded or immobilized using the treatment unit.

Hazardous Waste Landfills

Hazardous waste landfills, also known as 'secure' landfills, are intended to be the final disposal site for hazardous wastes. They are subjected to additional requirements for permanent monitoring in the closure and post-closure period. They consist of a natural or man-made depression with a soil foundation. They must have a double liner and a leak detection system. Like waste piles, they are required to have two leachate collection and removal systems. Landfills additionally must have stormwater run-on and run-off controls to withstand at least a 25-year storm and a cover to prevent wind dispersal. The final cover must minimize water migration through the landfill, promote drainage, and accommodate settling. Post-closure, the owner of the landfill is required to maintain the final cover, leak detection system, groundwater monitoring system, and water run-on/run-off protection systems. Prior to placement in a landfill, wastes often undergo treatment, a process that changes the physical, chemical, or biological nature of a waste to make it less of an environmental threat. Bulk or non-containerized liquid hazardous wastes are not allowed in landfills because of the risk of these liquids leaching into the ground. Free liquids must be removed or treated with absorbents or otherwise solidified prior to placement in a landfill. Certain liquid containers such as batteries or small containers including laboratory ampules are allowed.

According to the EPA, landfill and surface impoundments represented only 3.8% of the techniques used for hazardous waste management in 2007, with a total of 67 registered facilities. A total of 21 active commercial hazardous waste landfills operating in the United States were identified by the US Army Corps of Engineers in 2006. Laidlaw Waste Systems and Chemical Waste Management, Inc. were the two largest operators, managing seven and six of the operating landfills, respectively. In addition to these regulated hazardous waste landfills, there also exist a significant number of additional hazardous waste sites that pose a significant risk to the population and environment. The EPA uses a National Priorities List (NPL) to identify and rank these hazardous wastes sites. As of August 2010, the NPL listed 1,277 identified active sites, 343 deleted (remedied) sites, and 61 proposed new sites (EPA 2010b).

Liners and Leaks

Two primary environmental risks exist with hazardous waste landfills according to the US Army Corps of Engineers. The first risk is that of the mismanagement of reactive, ignitable, or incompatible wastes stored within landfills. This can lead to fires, explosions, or the release of toxic fumes. The second environmental risk is that of contamination of subsoil, groundwater, or surface water from leachate, run-off, or wind erosion resulting from the landfill. This second type of environmental risk is of greater concern, as the effects may be more insidious and less likely to be detected or rectified (US Army Corps of Engineers, 2010). Additionally, the risks of contamination are long-lived and generally amplify over time.

Many hazardous waste landfills, both current and historical, were unregulated or illegal and were never designed to limit the escape of wastes into the environment via direct means such as airborne spread and leaking of liquid wastes, or through indirect means, such as leaching. Leaching occurs when water enters a landfill through rainfall or flooding. This water, if not adequately drained or captured, then migrates through the landfill layers and contents, picking up dissolved toxins and carrying them out of the landfill into surrounding soil and groundwater. In the absence of measures designed to prevent either direct or indirect escape of wastes, contamination of the environment and population from these non-lined or improperly lined sites is a persistent risk. Examples of landfills that had no protective measures in place include the Love Canal in New York and the Lipari landfill in New Jersey. In both instances wastes were dumped into ditches without a liner or even a compacted clay soil base to contain them.

Regulated hazardous waste landfills, on the other hand, are designed to prevent the wastes contained within them from causing harm to the environment or population. However, these too may fail. Landfills constructed prior to 1985 generally used outdated or inadequate designs that are prone to leakage. Historical designs relied on a compacted clay soil barrier alone to prevent leakage of contents. Suter and colleagues raised several concerns with the use of only a compacted clay soil barrier in landfill design (Suter et al., 1993). They listed the following as possible mechanisms of long-term failure of compacted clay soil barriers:

- Initial flaws in barrier construction
- Shrink–swell cycles
- Freeze–thaw cycles
- Erosion
- Subsidence
- Root intrusion
- Animal intrusion.

The authors estimated that compacted clay soil barrier designs may function effectively for 30–50 years but that ‘natural physical and biological processes can be expected to cause barriers to fail in the long term.' This observation highlights the troubling legacy that the nation will likely face as landfills using this type of design begin failing, allowing their contents to leak out and contaminate the surrounding environment.

In 1985, the EPA introduced standards requiring hazardous waste landfills to use double liner systems and leachate collection systems in order to overcome the recognized inadequacies of a clay barrier alone. In these regulations, the inner liner consisted of a synthetic membrane while the outer liner was made from 3 feet of compacted clay. This arrangement is now typically referred to as a composite single liner (Hughes et al., 2010).

In 1992, the EPA introduced additional regulations for hazardous waste land disposal units. These regulations amended prior guidelines by augmenting them with the currently used design standards. This included enhanced double liners, double leachate collection systems, leak detection systems, and requirements for the use of monitoring requirements during active service and post-closure (Office of Environmental Guidance, 1992). While both guidelines use the term 'double liners,' the 1985 regulations used this term to describe a composite liner consisting of a synthetic membrane and a compacted clay soil base while the 1992 guidelines specified both a composite liner and an additional separate synthetic liner to further reduce the risk of leakage.

Modern hazardous waste landfill designs, while an important improvement over previous approaches, are not infallible. Failures have occurred when they are incorrectly constructed, managed, or due to unforeseen circumstances. The Kettleman Hills Hazardous Waste Facility in Kettleman City, California, has been the site of multiple incidents. The operators, Waste Management Inc., were fined 2.5 million dollars in 1984 and 4 million dollars in 1985 by the EPA for the mishandling of hazardous wastes including allowing toxins to contaminate groundwater (Miller, 1992). In 1988, the facility was the site of one of the largest failures of a hazardous waste liner system to date. A landslide occurred on one of the site's slopes and tore out part of the liner system. This resulted in a displacement of over a million cubic yards of hazardous waste. A huge swath of land had to be excavated before the landfill could resume operations. Subsequent analysis suggests that the landslide resulted from design and construction issues (Mitchell et al., 1990). Failures of other lining systems related to errors during the construction process have also been reported (Blight, 2007).

12.5.3 Human Health Effects of Land Disposal

A review of 50 reports on the adverse health effects of living near hazardous waste sites by Vrijheid and associates concluded that the evidence is suggestive of a true association but that issues of confounding and biases remain problematic (Vrijheid, 2000). Direct exposure information has been difficult to obtain. The authors identified additional research needs including more research on individual chemicals and chemical mixtures, well-designed prospective studies, development of biomarkers, and better research on risk perception and sociologic determinants of ill health.

Birth Weight Effects

Berry and Bove used birth certificate data over a 25-year period to determine if a mother's residence near the Lipari Landfill in New Jersey affected her child's birth weight (Berry and Bove, 1997). The landfill they investigated had previously been an excavation site for sand and gravel. In 1958 it became an unlined hazardous waste landfill as municipal, household, chemical, and industrial wastes were dumped into the pit. Liquid wastes were emptied directly into the landfill from 1958 to 1969 and solid wastes were placed into the landfill until 1971. The EPA found that leachate migrated from the landfill into two nearby streams and a lake in the community. Exposure to the community occurred through inhalation of volatilized chemicals from the landfill and contaminated waters and through direct contact with soil and waters. The authors found that births to mothers living closest to the landfill were significantly more likely to be of lower weight during the years of highest exposure (OR 5.1, 95% confidence interval 2.1–12.3). This relationship remained significant after controlling for other maternal variables including demographics and use of prenatal care. In addition, birth weights among mothers living in closest proximity recovered in later time periods when exposure had been reduced through mitigation, supporting a causal relationship between birth weight and residence near a hazardous waste landfill.

Congenital Defects

In a multicenter study covering 21 hazardous waste landfill sites in Europe, Dolk and colleagues compared pregnancies within a 3 km proximate zone against those occurring within 7 km but outside the proximate zone (Dolk et al., 1998). A total of 295 cases of congenital anomalies and 511 normal pregnancies occurred within the proximate zone while 794 cases and 1,855 normal pregnancies occurred within the 3–7 km zone, yielding an adjusted odds ratio of 1.33 (95% CI 1.11–1.59) for having any type of anomaly. A dose relationship was suggested, with a trend towards increased cases with increasing proximity to the landfills.

Respiratory Diseases

Rates of hospitalization for respiratory diseases in communities close to hazardous waste disposal sites in New York along the Hudson River were increased as compared to control communities in a report by Kudyakov and associates (Kudyakov et al., 2004). Exposure was related to the presence of persistent organic pollutants, including PCBs and persistent pesticides at the hazardous waste sites. The authors hypothesized that the increased risk of bronchitis and chronic airway obstruction they found might have been due to suppression of the immune system in affected individuals.

Social Inequity

Given the negative public opinion associated with hazardous waste landfills, they are often located in economically disadvantaged minority communities, raising concerns of social inequity. A review by Martuzzi et al. found that the literature supports an inequitable distribution of waste facilities, including hazardous waste landfills, towards economically disadvantaged individuals and ethnic minorities (Martuzzi et al., 2010). On a related topic, Anderton and colleagues examined the location of Superfund sites and analyzed their distribution in relationship to economic status and race (Anderton et al., 1997). While they found that the location of Superfund sites was not significantly different for minorities or economically disadvantaged persons, the prioritization of these sites on the National Priority List was not equitable. The National Priority List is a register of Superfund sites that are considered the most dangerous and also indicates which sites are scheduled for cleanup. The researchers found that listed sites were less likely to include black families and poor families, suggesting that cleanup efforts are socially disproportionate. This effect was small but significant.

Acknowledgment

The section on land disposal was contributed by Mike M. Nguyen, M.D. Mike is an Assistant Professor of Surgery/Urology at the University of Arizona College of Medicine, and is an MPH candidate at the University of Los Angeles School of Public Health.

References

Anderton, D. L., Oakes, J. M., & Egan, K. L. (1997). Environmental equity in superfund-Demographics of the discovery and prioritization of abandoned toxic sites. *Evaluation Review*, *21*(1), 3–26.

Agency for Toxic Substances and Disease Registry (ATSDR) (1992, February). *The public health overview of incineration as a means to destroy hazardous wastes*. Atlanta, GA: Division of Health Assessment and Consultation.

Assumpcao, M., Yamabe, H., Barbosa, R., Hamza, V., Lopes, A., Balancin, L., et al. (2010). Seismic activity triggered by water wells in the Parana Basin, Brazil. *Water Resources Research*, 46.

Beck, E. C. (1979). The Love Canal tragedy. *EPA Journal*. <http://www.epa.gov/history/topics/lovecanal/01.htm>

Berry, M., & Bove, F. (1997). Birth weight reduction associated with residence near a hazardous waste landfill. *Environmental Health Perspectives*, *105*(8), 856–861.

Blight, G. E. (2007). Failures during construction of a landfill lining: a case analysis. *Waste Management & Research*, *25*(4), 327–333.

Dolk, H., Vrijheid, M., Armstrong, B., Abramsky, L., Bianchi, F., Garne, E., et al. (1998). Risk of congenital anomalies near hazardous-waste landfill sites in Europe: the EUROHAZCON study. *Lancet*, *352*(9126), 423–427.

EPA (2001). *Class I Underground injection control program: Study of the risks associated with class I Underground injection wells*. Washington DC: Office of Water. EPA 816-R-01-007.

EPA (2005). *Introduction to land disposal units*. (40 CFR Parts 264/265, Subparts K, L, M, N).

EPA (2008a, November). *The national biennial RCRA Hazardous Waste Report (based on 2007 data)*. EPA530-R-08-012, Solid Waste and Emergency Response.

EPA (2008b). *National Priorities Chemicals Trends Report (2004–2006) executive summary*.

EPA (2009a). *Combustion*. Retrieved February 10 from <http://www.epa.gov/osw/hazard/tsd/td/combustion.htm>.

EPA (2009b). *Love Canal*. Retrieved August 12 from <http://www.epa.gov/history/topics/lovecanal/>.

EPA (2010a). *EPA national priorities list website*. Retrieved August 24, 2010 from <http://www.epa.gov/superfund/sites/npl/index.htm>.

EPA (2010b). *Basic Information about injection wells*. Underground Injection Control. Retrieved June 24 from <http://water.epa.gov/type/groundwater/uic/basicinformation.cfm>.

Frohlich, C., Potter, E., Hayward, C., & Stump, B. (2010). Dallas-Fort Worth earthquakes coincident with activity associated with natural gas production. *The Leading Edge*, *29*(3), 270–275.

Gensburg, L. J., Pantea, C., Kielb, C., Fitzgerald, E., Stark, A., & Kim, N. (2009). Cancer incidence among former Love Canal residents. *Environ Health Perspect*, *117*(8), 1265–1271.

Goldman, L. R., Paigen, B., Magnant, M. M., & Highland, J. H. (1985). Low birth-weight, prematurity and birth-defects in children living near the hazardous-waste site, Love-Canal. *Hazardous Waste & Hazardous Materials*, *2*(2), 209–223.

Gordon, W., & Bloom, J. (1985). Limits to underground injection as a hazardous waste disposal method. *Hazardous Waste & Hazardous Materials*, *2*(4), 411–451.

Hughes, K. L., Christy, A. D., & Heimlich, J. E. (2010). *Ohio State University fact sheet: landfill types and liner systems*.

Knape, B. (2005). Applications of deep-well injection of industrial and municipal wastewater in Texas. *Developments in Water Science*, *52*(3), 21–28.

Kudyakov, R., Baibergenova, A., Zdeb, M., & Carpenter, D. O. (2004). Respiratory disease in relation to patient residence near to hazardous waste sites. *Environmental Toxicology and Pharmacology*, *18*(3), 249–257.

Lei, X., Yu, G., Shengli, M., Xueze, W., & Qiang, W. (2008). Earthquakes induced by water injection at ~3 km depth within the Rongchang gas field, Chongqing, China. *Journal of Geophysical Research*, 113.

Martuzzi, M., Mitis, F., & Forastiere, F. (2010). Inequalities, inequities, environmental justice in waste management and health. *European Journal of Public Health*, *20*(1), 21–26.

Miller, E. L. (1992). *Final report: Waste Management, Inc*. San Diego: District Attorney office.

Mitchell, J. K., Seed, R. B., & Seed, H. B. (1990). Kettleman-Hills Waste Landfill Slope Failure. 1. Liner-System Properties. *Journal of Geotechnical Engineering-Asce*, *116*(4), 647–668.

Nicholson, C., & Wesson, R. L. (1992). Triggered earthquakes and deep well activities. *Pageoph*, *139*(3), 561–578.

Office of Environmental Guidance (1992). *Liners and leak detection systems for hazardous waste land disposal units*. Department of Energy.

Suter, G. W., Luxmoore, R. J., & Smith, E. D. (1993). Compacted soil barriers at abandoned landfill sites are likely to fail in the long-term. *Journal of Environmental Quality*, *22*(2), 217–226.

US Army Corps of Engineers (2010). *Report on treatment, storage & disposal facilities (TDSF) for hazardous, toxic, and radioactive waste*. 2006 Update.

United States Environmental Protection Agency (EPA) (1985). *Concentration technologies for hazardous aqueous waste treatment*. Cincinnati, Ohio: Solid and Hazardous Waste Research Division. 600281019.

Vrijheid, M. (2000). Health effects of residence near hazardous waste landfill sites: A review of epidemiologic literature. *Environmental Health Perspectives*, *108*, 101–112.

13 The Export of Hazardous Waste

Chapter Contents

13.1 Overview, Main Drivers, and Types of Exported Waste

13.1.1 Overview

With increasing amounts and toxicity of waste, predominantly produced by industrialized nations, it has become more common to remove or export a society's toxic by-products to keep it out of sight and mind. Kikuchi and Gerardo (2009) among others have referred to the growing local resistance to processing hazardous waste as the 'NIMBY syndrome' or 'not in my backyard.' The export or trade of hazardous waste involves the transboundary removal of hazards to a recipient country for disposal or management.

Environmental protection against uncontrolled, transboundary, waste disposal began to gain momentum in the 1970s (Wilson, 2007); however, by then the practice of dumping hazardous waste was already impacting developing countries in almost every region of the world (Clapp, 2001). In recent decades, with the benefits of globalization weighted heavily toward developed countries, there has been a persistent trend toward both dumping hazardous waste and exporting 'hazardous investment capital' or polluting industries into less developed nations (Faber, 2008). This logic follows the idea that developing countries are desperate for

Risks of Hazardous Wastes. DOI: 10.1016/B978-1-4377-7842-7.00013-1

materials and capital, thus it is considered an opportunity to manage second-hand waste; essentially the poor make a living from discarded toxic exports from wealthier nations (Wilson, 2007). Unfortunately, these destinations often do not have proper facilities or technology for processing or disposing of the waste (Robinson, 2009). In all, such arrangements cause harm to poor populations around the world, reports Basel Action Network (BAN), a non-governmental organization monitoring international agreements on waste trade (Puckett, 2009).

Due to the sharp distinction of economic power representing the ability to produce and externalize hazardous waste, the vast majority of this waste – estimated at 90% – is generated from the group of over 30 highly developed or industrialized countries called the Organisation for Economic Co-operation and Development or OECD (BAN, 2010a). Apart from globalized trade as a driver, in many OECD countries the increasing amount of hazardous waste generated, the ever-tightening domestic regulations and the rising processing or disposal costs in-country have led to the ongoing and dangerous export of these pollutants (Henrik, 2006). The ability of international law to protect and regulate this unfair trade along with the responsibility of polluting industries to ensure the safety of all products through the lifecycle to disposal or recycling continue to be the major questions for the future of hazardous waste management.

This chapter will give an overview of the history of toxic trade, the main types of exported hazards, the pertinent international agreements attempting to provide regulation, the challenge of electronic recycling and risks for recipient communities and, finally, recommendations for the future of this industry.

13.1.2 Main Drivers Toward Awareness

Key events in the 1980s served to awaken the international community to the issue of illicit and toxic transboundary dumping, the associated risks for health, and the need to control pollutants released into the environment. While the following are two of the major events connected to the United States, many other headlines at the time reported multiple incidents of toxic barrels dumped on coastlines and 'ships of death' embarking to find any open port that could be convinced to accept a hazardous load (Puckett, 1997).

Union Carbide in Bhopal, India

The deadly chemical gas release at the Union Carbide India Limited plant in Bhopal in 1984 was an international toxic disaster. The plant was a subsidiary of the US company Union Carbide Corporation, which is now owned by Dow Chemical, and produced pesticides with the dangerous chemical methyl isocyanate. The gas release immediately killed thousands of nearby residents and permanently injured or killed hundreds of thousands in the decades following the accident. Unfortunately, the repercussions of this particular waste management disaster continue for the local community as the health effects, diffusion of responsibility, and environmental remediation have still not been fully addressed more than 25 years later (Henrik, 2006; Unk, 2010).

It took a similar event in the United States to compound the urgency of preventing more chemical disasters. Shortly after Bhopal, there was a serious chemical release at a sister plant in West Virginia, leading to public demand for information on releases of toxic chemicals near these facilities. In 1986, the United States Emergency Planning and Community Right-to-Know Act was enacted, which – among other mandates – created the Toxics Release Inventory (TRI) that collects data and informs the public of releases or transfers of potentially hazardous material (Henrik, 2006). The events around Bhopal awoke the international community as well. According to Henrik of the OECD (2006), the incident led to the OECD implementation of the Pollutant Release and Transfer Registers (PRTR) in 1994, which built upon the database of the TRI, and became more standardized in 2003 with the Aarhus Convention of the OECD and 30 other parties. The PRTR is an environmental database or centralized listing of any release or transfer of a potentially harmful pollutant, naming the sources and location, information that must be available to the public (Henrik, 2006).

Khian Sea, *Toxic Cargo*

Another infamous event revealed the complacency of waste generators and the need to regulate the exportation of hazardous materials. In August 1986, the *Khian Sea* barge left Philadelphia with nearly 14,000 tons of municipal incinerator ash full of toxic chemicals, dioxins and heavy metals, such as arsenic, mercury, and cadmium. Under the guise of delivering topsoil fertilizer, the vessel began to dump more than 3,000 tons on a Haitian beach during the night, until local government officials intervened and ordered the ash to be reloaded; but the ship fled instead. The ship then attempted to dock in five continents over 27 months, searching for a place to dump the remaining waste. It is suspected to have off-loaded somewhere in the Indian Ocean before arriving to the next port, empty and with a changed name (Schmidt, 1999).

The toxic ash and contaminated soil remained on the beach in Haiti for more than a decade until a deal was brokered with the US for its removal. It was finally removed from the beach in April 2000 in the barge called *Santa Lucia*, headed to Florida where the state EPA agreed to handle what remained. But Florida revoked the offer several days later and the *Santa Lucia* waited offshore for two years until finding a resting place in a landfill in Franklin County, Pennsylvania (Athans, 2002; Clapp, 2002). This incident is a prime example of the protracted struggle for regulation, responsibility for health risk and protection around hazardous waste management – particularly when transferred from OECD to non-OECD countries.

13.1.3 Types of Waste and Toxic Trade

There are many examples of both the legal and ongoing illicit practice of hazards disposal for each type of waste and the risks they pose to people and the environment. While illicit incidents of dumping municipal or biohazardous (medical) wastes certainly exist, they are not among the common hazardous exports of OECD countries. Also, the exportation of toxic industries will not be discussed in depth, but should be mentioned as a form of hazards dumping. An example of this includes the highly polluting American-owned factories lining the US–Mexico border, operating with disregard for community health and prior bilateral agreements to prevent pollution (Clapp, 2002; Garcia and Simpson, 2006). Other hazards commonly exported from OECD countries include agricultural waste and shipbreaking. A brief explanation follows of these types of waste with a more detailed description of the risks of exporting e-waste addressed subsequently in Section 13.3.

Ag-waste and POPs

> *There is an established link between the condition of poverty and increased risks of exposure to toxic and hazardous chemicals, as they affect predominantly the poor who routinely face unacceptable high risk of poisoning because of their occupation, living location and lack of knowledge of proper chemicals management.*
>
> *United Nations Environment Programme (UNEP)*

Exported agricultural and chemical waste, including pesticides and their byproducts that remain in an environment for long periods of time, comprise what are called persistent organic pollutants (POPs). Through the 1980s, it was very common for European countries and the United States to export large loads of pesticides, with Africa being the main destination as well as Central and Eastern Europe (Bernstorff and Stairs, 2000). Even as many chemicals have now become banned in developed countries, the history of dumping remains present in Obsolete Pesticide Stockpiles of POPs, with little ability of the recipient to properly dispose of or handle the chemicals.

According to the Food and Agriculture Organization (FAO) of the United Nations (1999), pesticides become obsolete, unwanted or waste when they can no longer be used, practically or for safety reasons, and require disposal. The total amount of stockpiles calculated in the mid-1990s in non-OECD countries greatly exceeded 100,000 tons and by 2001 the estimate was between 400,000–500,000 tons, not taking into account the huge amounts of heavily contaminated soil and water that should also be regarded as hazardous waste and highly poisonous (FAO, 1999, 2001). Greenpeace International attributes the majority of the stockpiles to have been produced and exported by nearly a dozen multinational OECD-based companies, including American Cyanamid, Bayer, DowElanco, Dupont, and Monsanto among others (Bernstorff and Stairs, 2000). Common chemicals in the stockpiles include (sources: Bernstorff and Stairs, 2000; FAO, 1999):

- Common chemicals in stockpiles: aldrin, carbaryl, chlordane, DDT, dieldrin, dimethoate, endrin, fenitrothion, heptachlor, malathion, mirex, toxaphene, hexachlorobenzene, polychlorinated biphenyls (PCBs), propoxur;
- Unintended byproducts: dioxins, furans.

In 2006, the Strategic Approach to International Chemicals Management was created through the UNEP as a policy framework for global chemical safety with the goal of minimizing 'significant adverse impacts on the environment and human health' by 2020 (SAICM). The Stockholm Convention, which addresses the production and trade of POPs is described in Section 13.2.

Shipbreaking

The export of retired ships to be destroyed on 'breaking beaches' is not commonly known as hazardous waste yet it is certainly part of the toxic trade. The average lifespan of a ship is 25–30 years and the structure of the vessel is primarily steel, a valuable resource. Shipbreaking includes the dismantling of old ships that contain asbestos, PCBs, lead paint, fuel residue and other chemicals, and it entails frequent lethal accidents and explosions (Schulling, 2005).

While the Clinton Administration declared a moratorium on the practice in 1998 after Pulitzer Prize-winning media attention covering the issue, there have been attempts both under the Bush (2003) and Obama (2009) administrations to continue exporting obsolete navy vessels to locations such as Bangladesh, India, China, and Pakistan under the jurisdiction of the US Maritime Administration and the EPA (BAN, 2003; Shipbreaking Platform, 2009; Unk, 2010). It is estimated that up to 60% of global shipbreaking occurs in Bangladesh, indirectly employing up to 200,000 workers, and child labor comprises one-quarter of the workforce (Dao, 2008).

The NGO Platform on Shipbreaking is a global coalition created in 2005 to promote the standard that toxic wastes be removed from ships by the exporting party prior to transport and recycling (Puckett, 2009), and to promote compliance with the Stockholm and Basel Conventions as well as the US Toxic Substances Control Act that forbids PCB transfers (see Section 13.2 for listing of international law). International environmental and human rights organizations recommend an enforceable regime for clean steel-scrapping to ensure the export of pre-cleaned ships and the safety, health and rights of workers in the breaking yards and the protection of the coastal environment of recipient countries (Schulling, 2005). In 2009, the International Maritime Organization adopted the 'Convention on the Safe and Environmentally Sound Recycling of Ships' and the process of ratification is currently in progress (IMO, 2010). Policy efforts for these types of waste are a step in the right direction for global regulation and protection.

13.2 International Law and the Loophole

13.2.1 International Law, Regional Agreements, and Compliance

For every major toxic waste there is a legal framework that exists for managing the hazard. The most substantial international agreement is the Basel Convention on the Control of Transboundary Movements of Hazardous Wastes and their Disposal

(1989), organized by the United Nations Environment Programme and entered into force in 1992. The convention prohibits export of waste to recipients who do not have the technical or legal processes to manage hazards in an environmentally sound manner, and there are currently 170 parties to the agreement (UNEP). The treaty also establishes procedures for notifying the importing countries about the elements of the hazardous waste and the risks involved. While the convention was a landmark beginning, it was not a complete safeguard. A critique is that the convention involved weak monitoring of hazardous waste trade to developing countries and only required prior informed consent to be obtained with the signature of just one government official. Jim Puckett (1997), Executive Director of BAN, explains:

> *The use of the word 'control' in the Convention's title – rather than prevention or prohibition – was telling. During the negotiations leading up to the Basel Convention, the vast majority of nations made it clear that they wanted to ban waste trafficking entirely, particularly from developed to developing countries. Certain heavily industrialized countries, however, most notably the United States, fought to reject any such prohibition.*

Basel Ban Amendment

During the second meeting of Basel in 1994, the parties moved forward to address those elements missing from the original convention. The Basel Ban Decision II/12 was passed through consensus of 66 parties, which banned all hazardous waste exports from OECD to non-OECD countries including an immediate ban for final disposal exports with recycling wastes to be banned within 3 years. But the US and others in opposition claimed that such a decision must be an amendment to the convention, which would instead require three-quarters adoption of the 62 countries party to the convention and a more laborious process for full implementation. Accordingly, in the Third Conference of Parties of the Basel Convention in 1995 the Ban Decision III/1 was passed by consensus of 82 present parties to establish the amendment to the convention as Article 4A. Similar to the decision just a year prior, this article prohibits all hazardous waste exports by parties listed in Annex VII (OECD countries, EU, and Liechtenstein) to all countries outside that list, addressing final disposal and recycling destinations. The amendment would enter into force with ratification by 62 countries; however, debate has ensued around the legitimate counting of ratifications, which has now effectively delayed the legal mandate of the Basel Ban Amendment for 15 years (BAN, 2010a).

Three primary international agreements since Basel include (BAN, 2010c):

- The Protocol to the London Convention on the Prevention of Marine Pollution (1996), in force in 2006 – bans dumping or incineration at sea of most industrial and radioactive waste.
- The Rotterdam Convention on the Prior Informed Consent for Certain Hazardous Chemicals (1998), in force in 2004 – mandates that consent is given by recipient countries before the export of dangerous chemicals, instead of the prior use of voluntary consent. This is the first international treaty on chemical products, covering some 40 chemicals and pesticides.

- The Stockholm Convention on Persistent Organic Pollutants (2001), in force in 2004 – a landmark law seeking to end both commercial use and spread of up to 21 of the most toxic pollutants known to earth, which includes by-products of chlorine, dioxins, and furans.

Compliance with International Agreements

Currently, the United States is the only developed or OECD country that has not ratified the Basel Convention. For the US to be an official party to the convention, congressional legislation and a presidential letter of ratification delivered to the UN are required (Schmidt, 1999). However, the US argues that the Basel Convention mischaracterizes wastes, not allowing for agreement between what is considered non-hazardous recyclables by US national legislation (RCRA) and hazards under Basel (Schmidt, 1999).

US industry and political influences have indeed obstructed ratification of Basel, not only because of increased regulation but for economic pursuits – especially in regard to recyclables, or secondary materials. These 'recyclable shipments,' according to Schmidt (1999), such as used lead-acid batteries, include a large and profitable trade market with Mexico, Canada, Malaysia, Costa Rica, and OECD countries. As a non-party, the US must make bilateral or multilateral agreements outside of international accountability, exploiting the loopholes of legality.

For the Basel Amendments, apart from the US the other OECD countries that have not ratified the Basel Ban Decision III/1 include New Zealand, Mexico, South Korea, Japan, Israel, Canada, and Australia (BAN, 2010a). Further, BAN has given the following countries a 'failing scorecard' for their broad non-compliance, and failure to ratify the entirety of Basel and the other three primary international treaties: Russia, Israel, Malta, and the US (BAN, 2010c).

Regional Agreements

There are broad efforts by developing nations to prohibit the import of hazardous wastes, but there are still developing regions that remain vulnerable to illicit trade and open to toxic exports. In part as a response to the failed global consensus through the Basel negotiations, many developing countries and non-OECD regions passed legislation to safeguard against future toxic trade arrangements. A partial listing of regional agreements can be found in Table 13.1 (BAN, 2010a; Henrik, 2006). National legislation to ban hazardous imports and to adopt regional conventions has been enacted by over 100 countries (BAN, 2010a).

13.2.2 The Loophole: Recycling, Repair, and Reassembly

'International regulation may put a damper on certain forms of hazard transfer, but the practice finds new outlets in response,' explains Clapp (2001). 'Plugging one hole in the dike tends to create a new one elsewhere.' Hazardous waste exports definitively declined in the 1990s, attributed to the increased media attention, public awareness, and the numerous international regulations that were enacted

Table 13.1 Regional Agreements

Agreement	Year	Region	Significance
Barcelona Convention and Izmir Protocol	(1989) Adopted 1996	Mediterranean	Prohibits export of hazardous and radioactive wastes to non-OECD and prohibits import by non-EU members
Lomé IV Convention	(1989)	70 countries in Africa, Caribbean, and Pacific	Prohibits EU export of nuclear or hazardous wastes to ACP parties, who also banned waste imports from any country
Bamako Convention	(1991) In force 1996	Members of Organization of African Unity	Bans all imports of hazardous and nuclear waste to continent of Africa including toxic products banned in country of manufacture
Rio Declaration on Environment and Development	(1992)	UNEP	States must attempt to end transfer of substances that cause damage to the environment and human health
Central American Agreement on Haz Waste	(1992) In force	6 countries of Central America	Similar to Bamako
Association of South East Asian Nations (ASEAN)	(1993) not yet adopted	Members of ASEAN	Voted for regional convention to prohibit import of hazardous wastes, but not adopted
Waigani Convention	(1995) In force	Members of South Pacific Forum	Prohibits the islands from import of hazards and Australia and New Zealand banned from export to all other Forum countries

(Clapp, 2001). However, the amount of toxic waste generated surely did not disappear. While waste exported for final disposal declined, there was a sharp increase in the waste shipped to be 'recycled' (Clapp, 2001). For most of the 1980s, the waste destined for 'further use' accounted for approximately 37% of exports, which rose to 88% by the year the Basel Convention went into force, and by the late 1990s, the percentage was estimated to be over 95% (Puckett, 1997).

The Basel Convention covers recycling; however, a 'loophole' exists that makes shipments labeled as recyclables instead of waste more difficult to track and regulate (Clapp, 2001). And while exporting hazardous wastes for recycling from

OECD to non-OECD countries would be banned under the Basel Amendment, it still has not entered into force after more than a decade.

Tension between environmental law and international trade are certainly involved as well. Ibitayo (2008) explains that the effectiveness of the Basel Convention is obstructed by laws such as the General Agreement on Tariffs and Trade (GATT was replaced by the World Trade Organization in 1995), 'If hazardous waste is viewed as a product with materials that can be recycled and reused, and if waste disposal is considered as a service, limiting or restricting the export of hazardous waste to developing countries may be contrary to the requirements of GATT [and] precludes industrialized nations from using nontariff barriers to limit the access of developing countries to any new kind of commodity or service.' Clapp (2001) argues that this controversy 'does not analyze the dynamic nature of the global liberal trade order, now enforced by the WTO, that gave rise to the hazard transfer problem in the first place.'

Meanwhile, the market for recycling hazardous exports has boomed, particularly for electronics waste, which will be discussed for the remainder of the chapter. Yet, reprocessing almost never occurs and non-recyclables are mislabeled (Schmidt, 1999). Types of recycling, according to BAN include 'sham' and 'dirty' methods: Sham recycling does not actually involve recycling, it means dumping and burning the waste; while dirty recycling denotes the dangerous and polluting process in which workers of poor nations, often minors, attempt to recover any material from the waste that could be of value, often without the required technology and equipment (BAN, 2010b). For hazardous waste exports and management, new strategies are rapidly needed because the biggest challenge of toxic waste management our world has seen is on the horizon with the growth of e-waste.

13.3 E-Waste – The New Export Challenge

13.3.1 Background of E-Waste Exports

Electric or electronic equipment that is disposed of or no longer valued by the owner is referred to as e-waste or WEEE (Widmer et al., 2005). Worldwide, the generation of e-waste is increasing by 40 million tons per year, and Schluep et al. of the UNEP (2009) estimate that the United States produces 3 million tons of e-waste annually, the highest of any single country. In 2005, the US produced more e-waste than Switzerland, Germany, the UK, and Canada combined (Osibanjo and Nnorom, 2007). And China is not far behind, already producing approximately 2.3 million tons domestically while continuing to be the dumping ground for e-waste exports (Schluep et al., 2009).

The US e-waste industry has reported that the majority of the waste collected was exported to Asia with 90% going to China, but these exports are often received in violation of international law (Osibanjo and Nnorom, 2007). Most of the waste is disposed in landfills but not before valuable metals have been extracted or incinerated by informal recyclers and after harm is done to workers and the environment

(Robinson, 2009; Schluep et al., 2009). According to LaDou and Lovegrove (2008), 'The United States and other developed countries export e-waste primarily to Asia, knowing it carries a real harm to the poor communities where it will be discarded.' Although the export of e-waste is illegal under the Basel Convention, Robinson (2009) explains, developed countries continue the practice because of the expense of ensuring legal compliance, effective and safe reprocessing for the millions of tons that must be managed each year, and as mentioned previously, there is economic gain in the new recycling industry.

The report by Schluep (2009) reveals the urgency of establishing regulations for e-waste management, especially in China, and provides a framework for efficiently recycling the imports. Other countries already facing serious environmental degradation and health problems from informal sector imports and improper e-waste recycling are India, Brazil, and Mexico, among others. In recent years, European Union directives have banned the sale of electronics containing lead, mercury, cadmium, and other dangerous elements in order to encourage development of clean and safe electronics products, but these initiatives have not been broadly implemented both by governments and electronics manufacturers (LaDou and Lovegrove, 2008).

13.3.2 Hazards of E-Waste Exports

The vast majority of electronics destined for reuse, recycle, or reassembly are not processed in a manner that is safe for workers, the environment and the surrounding community, thus these exports are above all hazardous waste – despite the guise of new labels. According to LaDou and Lovegrove (2008) less than 10% of e-waste is recycled. Some of the harmful elements of e-waste include (sources: Brigen et al., 2005; Robinson, 2009; Schluep et al., 2009):

- *Primary – direct exposure by workers:* Lead, beryllium, mercury, arsenic, antimony trioxide, cadmium, nickel, chromium, cobalt, zinc, barium, silver, polybrominated diphenyl ethers (PBDEs), polychlorinated biphenyls (PCBs), fluorinated cooling fluids, etc.
- *Secondary – by-products or reactions:* Dioxins and furans (including the most toxic dioxin congener TCDD or 2,3,7,8-tetrachlorodibenzo-p-dioxin), polycyclic aromatic hydrocarbons (PAHs), polyhalogenated aromatic hydrocarbons (PHAHs), hydrogen chloride, etc.
- *Tertiary – products used for recycling and toxic release from the process:* Cyanide or other leaching agents and mercury, which are used in the process of acid baths and 'cooking circuit boards' that releases pollutants into the air.

Due to the informal nature and location of e-waste processing and disposal, exposure by the workers to these dangerous elements spreads to contamination of local water and food sources. 'E-waste workers suffer negative health effects through skin contact and inhalation, while the wider community are exposed to the contaminants through smoke, dust, drinking water and food' (Robinson, 2009). Two e-waste destinations will be briefly highlighted in the following: China and India.

China's Poverty and Poison

Ash piles from electronics processing sampled in Guiyu, Guangdong Province of southern China, contain concentrations of lead up to 200 times the US regulatory limits, and 82% of small children tested in Guiyu were found to have clinical lead poisoning (Puckett, 2009). Dust samples collected by Greenpeace International (2005) from floors of solder recovery workshops in the same region revealed high metal levels in comparison to background levels, particularly lead (up to 7.6% by weight) and tin (29.3%), as well as copper, antimony and, in some cases, cadmium and mercury; and for all dust samples, the lead concentrations were hundreds of times higher than typical and safe indoor dusts. The same study found high levels of the same contaminants in the homes of the workers and in the discharge channels and waterways, along with dozens of congeners of PBDEs and other endocrine-disrupting or carcinogenic agents – nonylphenol (NP), dibutyl phthalate (DBP), di(2-ethylhexyl) phthalate (DEHP), triphenyl phosphate (TPP), and mixed chlorinated/brominated benzenes (Brigen et al., 2005). Evidence of severe local contamination as a result of e-waste streams in China begets the need for more research to determine the extent of the impact on human and environmental health (see Figures 13.1 and 13.2).

India's 'Digital Dump'

'Recyclers are the new faces of urban poverty,' states Satish Sinha of Toxics Link, a Delhi-based organization (Mercier, 2007). In an article in *Quebec Science* that won the 2008 Reuters Media Award, journalist Noemi Mercier explains in great detail the contamination in poor communities of India from processing old computers and e-waste. The workers are often very young, and the article describes children using their bare hands to strip, heat or treat with acid the mountains of wire, keyboards and hard drives in order to recover every possible material of value, such as small amounts of copper or gold. The workers here are also exposed to lead, cadmium, PCBs, and brominated flame retardants. She reports that local

Figure 13.1 Ash piles from burning operations covered with sand and dumped beside the Langjiang River. Guiyu, China. May 2008 ©2008 Basel Action Network (BAN).

Figure 13.2 Women processing wires torn out of computers. The wires are sorted by day and burned by night in this village in Guiyu, China. The families live near the burn-yards. Cancer-causing polycyclic aromatic hydrocarbons and dioxins will result from burning wires made from PVC and brominated flame retardants. Photo taken December 2001. ©2006 Basel Action Network (BAN).

physicians are not prepared to diagnose and treat the physical conditions resulting from this industry, such as respiratory and bronchial problems, irritated eyes and skin and increased miscarriages, and the groundwater is no longer safe to drink.

> *For these workers, earning a living is equivalent to poisoning yourself slowly. They do not wear gloves or masks to protect themselves from the dust that accumulates on the ground and on their hands. Each computer contains a terrifying amount of toxic substances. There are metals such as beryllium, which can cause skin lesions, as well as berylliosis, a serious lung disease; cadmium, which, if breathed in, can damage the kidneys and the bones; mercury, which attacks the brain and the kidneys; and antimony, similar to arsenic in its composition and toxicity. There are also chromium derivates, which, like beryllium and cadmium, are associated with lung cancer. And then there is lead, lots of lead, whose effects on the nervous system, particularly serious in children, are irreversible.*
>
> *(Mercier, 2007)*

According to Mercier, in 2007 there were only three authorized recyclers in India with a great need for national and international support for mainstreaming the backyard recycling industry into the formal sector.

13.4 Recommendations and Conclusion

The elements to address e-waste recycling and the export of other hazards from OECD to non-OECD countries include adequate technology, increased transparency and responsibility on behalf of the industry, remediation of known contamination, enforceable international policy and sustainable development. According to Osibanjo (2007) and Babu et al. (2007) there is first a need for more transparency in labeling waste exports and a standardized system for correctly characterizing second-hand appliances as well as the urgent need for safe recycling technology and value added products.

LaDou and Lovegrove (2008) summarize some of the necessary actions that must be taken by the manufacturers of hazardous products as: taking responsibility for the products through their lifecycle, mandates for compliance with regulations on transboundary movement and toxic trade – including the process of recycling, and the commitment to develop clean, durable, and repairable products. Alternatives to replace the heavy metals that require destructive mining and other contaminants found in electronics products must be introduced as well as product development that increases the ability to reuse and upgrade existing equipment (BAN, 2010b).

In all, effective policy and new social standards must be a priority. The national and international policies that will govern the e-waste system must include extended producer responsibility (EPR) or end-of-life (EOL) takeback with definitive roles and expectations for all participants in the waste management process (Osibanjo and Nnorom, 2007). This legal responsibility should also address the fate of environments and communities in developing countries already consumed by contamination (Yu et al., 2010). In a new campaign by Basel Action Network, recycling agencies can be positively recognized with E-Stewards Certification for their compliance with standards (2010). Further, in October 2010 the OECD Global Forum is discussing the necessary shift from end-of-life hazardous waste transfers to a lifecycle approach of Sustainable Materials Management, and a manual has been issued to guide the practice (OECD, 2007). These are all small steps toward a major shift in responsibility and compliance with transboundary e-waste management.

There is not a quick and easy solution, as the underlying issues of poverty and local governance for recycling destinations and the role of globalized trade must be recognized for all waste exports beyond electronics – from retired ships to obsolete chemicals as well. According to Clapp (2001), globalization is responsible for the context where toxic waste is transferred despite regulations, using economic disparity to its advantage. The complexity of global economics and poverty is described in a BAN briefing:

> *This is not simply a matter of a lack of adequate technology but involves many additional factors that might be taken for granted in developed countries. Social, financial and infrastructural factors are at least as important to protecting the populace and environment as technical criteria. These factors include adequate legislation, resources, manpower, and political will, to enforce such legislation, including monitoring and inspecting operations. It involves infrastructure to provide emergency response, adequate roads and services to ensure safe transport, and adequate medical facilities to monitor worker and community health. It involves the public and workforce having democratic capability to redress environmental and occupational concerns and to be able, if necessary, to protest hazardous working or living conditions.*
>
> *(BAN, 2010b)*

In conclusion, there is much work to be done in order to safely and effectively manage the global load of hazardous waste that already exists, especially in developing countries. This includes promoting legal compliance with international agreements, particularly the full ratification of the Basel Convention and Amendment, creating new standards for product responsibility and innovation that embraces non-toxic design. This approach for modern hazardous waste management should be multisectoral, involving international non-governmental groups and health professionals as well as the government and private sectors.

Most of all, there must be the recognition that the solution for a healthier world, without the risks of hazardous waste, requires the commitment by every member of society to reduce the consumption and eventually eliminate generation of the products. As the predominant society that uses these products prior to disposal, it is equally our responsibility to end the assumption that our harmful by-products can be externalized to developing nations. Public awareness and participation is crucial for waste management accountability and to prevent more shameful international incidents. A strong and informed society is needed to promote safe and responsible management of hazardous waste around the world.

Acknowledgment

This work was contributed by Maryada Vallet, who is currently completing a Masters in Public Health at the University of California Los Angeles and holds Bachelors of Arts in Global Studies and International Relations. She has worked in emergency medicine in Southern Arizona, as a volunteer aid worker on the US–Mexico border and with humanitarian affairs of World Vision International. She was first introduced to the toxic trade by supporting efforts to stop a hazardous waste landfill in Quitovac, Sonora, Mexico, near sacred grounds of the indigenous group the Tohono O'odham.

References

Athans, M. (2002, June 28). Pa ash from '80s finds final dump site. *The Baltimore Sun*.

Babu, B., Parande, A., & Basha, C. (2007). Electrical and electronic waste: A global environmental problem. *Waste Management & Research*, *25*, 307–318.

BAN (2003). *Marada alert: Stop the US from exporting toxic waste ships to Asia*. Basel Action Network Media Release. Retrieved November from <http://www.ban.org/COP6/MARADALERT.pdf>.

BAN (2010a). *Country status/waste trade ban agreements*. Retrieved August 15, 2010 from <http://www.ban.org/country_status/country_status_chart.html>.

BAN (2010b). *Hazardous waste recycling: No justification for toxic trade*. Basel Action Network. Briefing Paper 7.

BAN (2010c). *Package deal: Four treaties to fight pollution*. Basel Action Network. Briefing Paper 6.

Bernstorff, A., & Stairs, K. (2000). *POPs in Africa: Hazardous waste trade 1980–2000 obsolete pesticide stockpiles*. Greenpeace International.

Brigen, K., Labunska, I., Santillo, D., & Allsopp, M. (2005). *Recycling of electronic wastes in China & India: Workplace & environmental contamination*. Greenpeace International.

Clapp, J. (2001). *Toxic exports: The transfer of hazardous wastes from rich to poor countries*. Ithaca, NY: Cornell Press.

Clapp, J. (2002). *Distancing of waste: Overconsumption in a global economy*. Trent International Political Economy Centre.

Dao, R. (2008). *Childbreaking yards: Child labor in the ship recycling industry in Bangladesh*. The International Federation for Human Rights (FIDH) with YPSA.

Faber, D. (2008). *Capitalizing on environmental injustice: The polluter-industrial complex in the age of globalization*. Plymouth, UK: Rowman & Littlefield Publishers Inc.

FAO (1999). *Prevention of accumulation of obsolete pesticide stocks – provisional guidelines*. Plant Production and Protection Division United Nations.

FAO (2001). *Baseline study on the problem of obsolete pesticide stocks*. Retrieved August 15, 2010 from <http://www.fao.org/docrep/003/x8639e/x8639e02.htm>.

Garcia, C., & Simpson, A. (2006). Community-based organizing for labor rights, health, and the environment: Television manufacturing on the Mexico-U.S. Border. In T. Smith (Ed.), *Challenging the chip* (pp. 150–160). Philadelphia: Temple University Press.

Henrik, H. (2006). Hazardous waste. *Annals of the New York Academy of Sciences*, *1076* (Living in a Chemical World: Framing the Future in Light of the Past), 462–477.

Ibitayo, O. (2008). *Transboundary dumping of hazardous waste*. International Environmental Issues and Waste Management. Retrieved August 2010 from <http://www.eoearth.org/article/Transboundary_dumping_of_hazardous_waste>.

IMO (2010). *Information resources on recycling of ships*. Retrieved August 2010 from <www.imo.org>.

LaDou, J., & Lovegrove, S. (2008). Export of electronics equipment waste. *International Journal of Occupational and Environmental Health*, *14*(1), 1–10.

Mercier, N. (2007). Our computers are poisoning the planet. *Quebec Science*, 20–27.

OECD (2007). *Guidance manual for the implementation of the OECD recommendation on environmentally sound management of waste*. Organisation for Economic Co-operation and Development.

Osibanjo, O., & Nnorom, I. (2007). The challenge of electronic waste (e-waste) management in developing countries. *Waste Management & Research*, *25*(6), 489–501.

Puckett, J. (1997). *The Basel Ban: A triumph over business-as-usual. About Basel Ban*. Retrieved August 9, 2010 from <http://www.ban.org/about_basel_ban/jims_article.html>.

Puckett, J. (2009). *Annual report: Preventing toxic trade & promoting a toxic-free future*. Seattle, WA: Basel Action Network.

Robinson, B. H. (2009). E-waste: An assessment of global production and environmental impacts. *Science of the Total Environment*, *408*(2), 183–191.

Schluep, M., Hagelueken, C., Kuehr, R., Magalini, F., Maurer, C., Meskers, C., et al. *Recycling – from e-waste to resources*. United Nations Environment Programme.

Schmidt (1999). Trading trash: Why the U.S. won't sign on to the Basel convention. *Environmental Health Perspectives*, *107*(8), A410–A412.

Schulling, J. (2005). *End of life ships: The human cost of breaking ships*. Greenpeace International and FIDH.

Shipbreaking Platform (2009). *Obama's EPA allows toxic navy ships to be illegally dumped in Bangladesh*. Retrieved August 2010 from <http://www.shipbreakingplatform.com/>.

Unk (2010, January 19). *Bhopal recalls gas leak disaster. Bhopal: 25 years on*. BBC. Retrieved August 2010 from <http://news.bbc.co.uk/2/hi/in_depth/business/2009/bhopal/default.stm>.

Widmer, R., Oswald-Krapf, H., Sinha-Khetriwal, A., Scnellmann, M., & Boni, H. (2005). Global perspectives on the e-waste. *Environmental Impact Assessment Review*, *25*, 436–458.

Wilson, D. (2007). Development drivers for waste management. *Waste Management & Research*, *25*, 198–207.

Yu, J., Williams, E., Ju, M., & Yang, Y. (2010). Forecasting global generation of obsolete personal computers. *Environmental Science & Technology*, *44*(9), 3232–3237.

14 Introduction to Human Exposure, Toxicology, and Risk Assessment

Chapter Contents

The consequences of human exposure to the myriad of chemicals released into the environment may be impossible to predict, due to the sheer number of chemicals that humans may be exposed to, the many pathways by which they can be exposed, the relative lack of toxicity information available, and the complex interactions between chemicals in the environment and in the human body. Moreover, wide variations in the human population present additional difficulties in predicting potential dangers. Nevertheless, exposure and risk assessment have been developed to attempt to characterize the human health risks associated with exposure to chemicals. This chapter introduces the basic concepts of toxicology, human exposure and risk assessment, finishing with a discussion of the limitations of these efforts.

14.1 Exposure Pathways

Humans can be exposed to chemicals and physical agents through various exposure pathways. An exposure pathway describes the course that a substance takes from

Risks of Hazardous Wastes. DOI: 10.1016/B978-1-4377-7842-7.00014-3

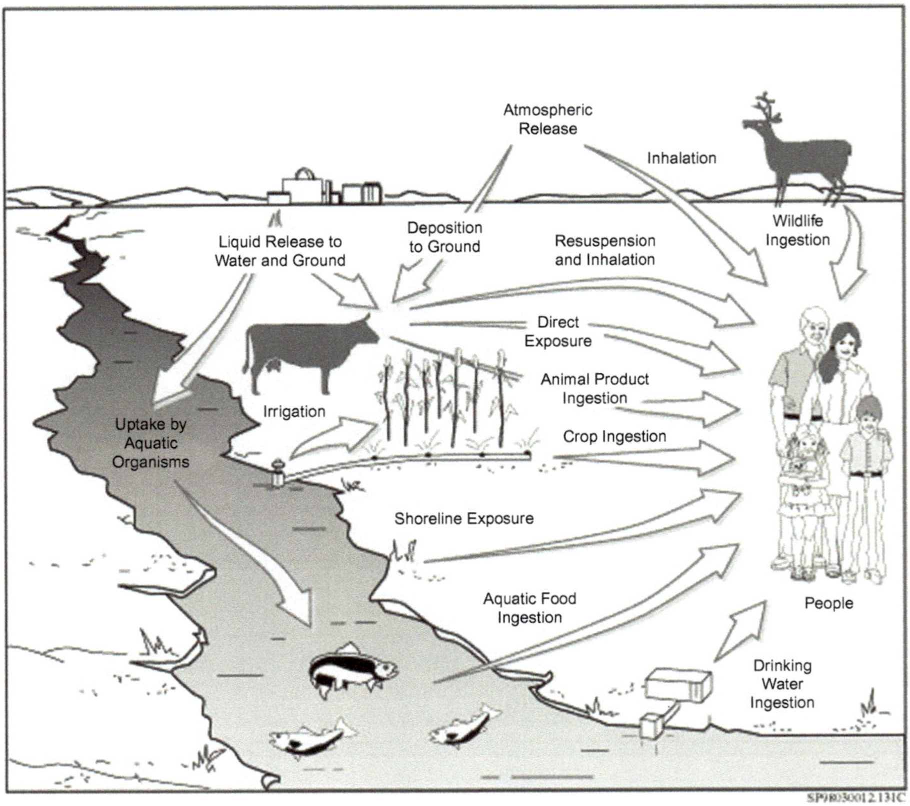

Figure 14.1 Illustration of exposure pathways (United States Department of Energy).

the source of the chemical to the exposed individual. An exposure pathway (see Figure 14.1) generally consists of the following: (1) the source of the chemical; (2) a medium (i.e. soil, water, or air); (3) an exposure point, the potential point of contact between the human and the contaminated medium; and (4) a route of exposure (inhalation, ingestion, dermal absorption), the way in which a substance enters the human body (US EPA, 1989).

14.1.1 Soil, Water, Air, and Biota – Media

Humans are exposed to toxic chemicals in a number of ways. Sometimes, exposure occurs directly at the source, as in the case of workers. Generally, however, people are exposed to harmful chemicals by contact with a medium. The primary media, soil, water, air, and biota, become contaminated by various mechanisms. Table 14.1 presents some typical release sources, release mechanisms, and receiving media.

Table 14.1 Common Mechanisms by Which Contaminants are Released to Media (US EPA, 1997, 1989)

Release Source	Release Mechanism	Receiving Medium
Smoke stacks, furnaces	Direct	Air
Vehicles, aircraft, marine vessels		
Surface wastes – lagoons, ponds, pits, spills	Volatilization	
Contaminated surface water		
Contaminated surface soil		
Contaminated wetlands		
Leaking drums		
Contaminated surface soil	Fugitive dust generation	
Landfills, waste piles		
Contaminated surface soil	Surface runoff	Surface water
Lagoon overflow	Episodic overland flow	
Spills, leaking containers		
Contaminated groundwater	Groundwater seepage	
Surface or buried wastes	Leaching	Groundwater
Contaminated soil		
Addition of pesticides, fertilizers	Direct	Soil
Contaminated air	Deposition, rain	
Surface or buried wastes	Leaching	
Contaminated surface soil	Surface runoff	
Lagoon overflow	Episodic overland flow	
Spills, leaking containers		
Contaminated surface soil	Fugitive dust generation/ deposition	
Waste		
Contaminated surface soil	Tracking	
Surface wastes – lagoons, ponds, pits, spills	Surface runoff, episodic overland flow	Sediment
Contaminated surface soil		
Contaminated groundwater	Groundwater seepage	
Surface or buried wastes	Leaching	
Contaminated soil		
Contaminated soil, surface water, sediment, groundwater or air	Uptake	Fruits, vegetables, grains
Contaminated water and sediments	Uptake	Fish and shellfish
Contaminated soil, water, or food	Uptake	Meat and dairy products

14.1.2 Inhalation, Ingestion, and Dermal Absorption

After soil, water, and air become contaminated, human exposure to the contaminants can occur through contact with the medium. These potential points of contact are called exposure points. Any sources or contaminated media at a source site can be considered a point of exposure. If contaminants have traveled offsite, significant exposure points would typically be locations close to or downgradient or downwind of the site; however, chemicals can sometimes be transported from a site and deposited in a distant water body. At the point of exposure, chemicals can enter the human body typically through the following exposure routes: inhalation, ingestion, and dermal absorption.

Inhalation

Human exposure to toxic substances can occur via inhalation of polluted air. Inhalation of chemicals in air essentially occurs continuously because ambient air contains a variety of gaseous and particulate pollutants. During inhalation, both gaseous and particulate pollutants in the air come into contact with the surface of the lung and can deposit in various regions of the lung. Water-soluble and reactive gases tend to deposit in the upper respiratory tract, while lipid-soluble or non-reactive gases tend to deposit in the farther regions. Gases can be absorbed into the blood stream and/or react with lung tissue. Particulate pollutants can become deposited throughout the respiratory tract, with very small particles having the ability to enter the blood stream (US EPA, 1997).

Ingestion

Human exposure to toxic substances can occur via ingestion, the process of taking a substance into the body through the mouth. Ingestion of mediums such as water, food, and soil, can result in the inadvertent consumption of toxic substances. Ingestion of contaminated drinking water, as well as ingestion of contaminated groundwater and surface water (such as during swimming) can result in exposure to toxic chemicals. Ingestion of biota including fruits, vegetables, grains, meats and dairy, fish and shellfish, can all result in chemical exposure. Exposure to chemicals through the ingestion of contaminated soil, although typically inadvertent, occurs frequently. Soil particles can adhere to food and hands and subsequently come into contact with the mouth, with children more likely to ingest more soil than adults as a result of childhood behaviors (US EPA, 1997).

Dermal Absorption

Dermal exposure to toxic substances can occur during various activities. Dermal exposure to contaminated water can occur during bathing or swimming; dermal exposure to soil can occur during gardening, construction, or recreation; dermal exposure to sediment can occur during wading or fishing; dermal exposure to liquids can occur during the use of commercial products; dermal exposure to vapors can occur during the use of commercial products; and dermal exposure to toxic substances can occur during contact with indoor surfaces such as carpets, floors, and countertops (US EPA, 1997).

Exposure to toxic chemicals in one medium, such as soil, water, or air, can occur through more than one route of exposure at once. For example, PCBs in soil can reach an individual through the direct ingestion of the soil, through the

Table 14.2 Typical Routes of Exposure to Contaminated Media (US EPA, 1989, 1991)

Medium	Routes of Exposure
Groundwater	Ingestion from drinking
	Inhalation of volatiles
	Dermal absorption, such as during bathing
Surface water	Ingestion from drinking
	Inhalation of volatiles
	Dermal absorption, such as during bathing
	Ingestion during swimming
	Ingestion of contaminated fish
Soil	Direct ingestion
	Inhalation of particulates
	Inhalation of volatiles
	Exposure to groundwater contaminated by soil leachate
	Ingestion via plant uptake
	Dermal absorption, such as during gardening
Sediment	Ingestion
	Dermal absorption, such as during bathing or swimming
Air	Inhalation
	Dermal absorption
Food	Ingestion of fish, shellfish, meat, dairy, eggs, fruits, vegetables, and grain

inhalation of particulates originating from the soil, through the inhalation of volatile gases from the soil, through exposure to groundwater contaminated by the soil, through ingestion of contaminated fish from a stream receiving surface runoff containing the contaminated soil, through the ingestion of plants grown in the soil, or through dermal contact with the soil (US EPA, 1992).

14.2 Quantifying Exposure

Determining the potential risks of exposure to toxic substances requires the quantification of the exposure. There are generally two approaches to quantifying exposure: direct quantification and indirect quantification. Direct methods of exposure quantification involve directly measuring the concentrations of chemicals at or within an exposed individual. Direct methods include point-of-contact measurement and biological monitoring. Indirect methods of exposure quantification involve the predicting of exposure based on determination of chemical concentrations in a medium (not necessarily at the point of contact) and assumptions about the time spent in contact with the medium.

14.2.1 Point-of-Contact and Biological Monitoring

The point-of-contact approach involves the measurement of exposure as it occurs, meaning chemical concentrations at the exposure point are recorded along with the length of contact time. Examples include radiation badges to measure the amount of radiation to which someone has been exposed, and personal continuous-sampling devices which can analyze airborne vapors or dust concentrations in the individual's breathing zone. Point-of-contact exposure assessments are usually conducted in an occupational setting to determine if exposures are within regulatory levels (US EPA, 1992).

Biological monitoring involves the measurement of chemicals of concern or biomarkers in the breath, tissue, or fluids of the exposed individual, or the measurement of a biological effect resulting from exposure. Biological monitoring is sometimes referred to as 'estimating exposure from reconstructed dose,' because data on body burden levels or biomarkers can provide information about past exposures. Examples include the testing of breath for volatile organic compounds from relatively recent exposures, testing of blood for metals that accumulate over time or for lipophilic compounds (i.e. some pesticides) from recent exposures, testing of adipose tissue for bioaccumulative, lipophilic compounds such as dioxins and PCBs, testing of nails and hair for heavy metals, and the testing of urine for various chemicals, typically from recent exposures (US EPA, 1992).

14.2.2 Exposure Scenarios

Indirect methods of exposure quantification involve measuring or modeling the chemical (i.e. its locations, concentrations, and times) in the contaminant media and subsequently estimating an individual's exposure based on information or assumptions about the individual's physical characteristics and the individual's contact with the medium. This approach is sometimes referred to as exposure scenario evaluation (US EPA, 1992).

Assumptions about an individual's contact with a medium and assumptions about the individual's physical characteristics are referred to as exposure factors. Table 14.3 provides a list of common default exposure factors. For more information, the US EPA's Exposure Factors Handbook (US EPA, 1997) is recommended as a resource for detailed statistical data on many factors such as body weight, soil ingestion rates, inhalation rates, drinking water consumption rates, consumption rates of fruits, vegetables, fish, meat, dairy, and breast milk.

Exposures can be presented as an average exposure over a period of time, or as a function of body weight. Often, the exposure is normalized for both time and weight, and presented in units of the mass of a chemical per mass of body weight per unit time, typically mg chemical/kg body weight/day. The generic equation for calculating chemical intakes is presented in Figure 14.2.

For inhalation exposure, the chemical concentration in the air time-weighted over the duration of exposure may be sufficient to characterize exposure. The generic equation is presented in Figure 14.3.

For dermal exposure, the equation in Figure 14.4 is used to evaluate the absorbed dose from chemicals in water or soil. Simply, the estimated dose of a chemical that is absorbed through the skin depends on various chemical-specific

Table 14.3 Common Default Exposure Factors (US EPA, 1991, 2004, 2010)

Exposure Factor	Default Value
Body weight – Adult	70 kg
Body weight – Child	15 kg
Drinking water ingestion rate – Child	1 L/day
Drinking water ingestion rate – Adult	2 L/day
Inhalation rate – Adult	20 m^3/day
Inhalation rate – Child	10 m^3/day
Soil ingestion rate – Child resident	200 mg/day
Soil ingestion rate – Adult resident	100 mg/day
Soil ingestion rate – Adult worker	50 mg/day
Skin surface area for soil/dust exposure – Adult resident	5,700 cm^2
Skin surface area for soil/dust exposure – Adult worker	3,300 cm^2
Skin surface area for soil/dust exposure – Child resident	2,800 cm^2
Skin surface area for water exposure – Adult	18,000 cm^2
Skin surface area for water exposure – Child	6,600 cm^2
Consumption of homegrown fruit	42 g/day
Consumption of homegrown vegetable	80 g/day
Consumption of locally caught fish	54 g/day
Exposure frequency – Residential	350 days/year
Exposure frequency – Worker	250 days/year
Exposure duration – Residential	30 years
Exposure duration – Child resident	6 years
Exposure duration – Adult resident	24 years
Exposure duration – Worker	25 years
Exposure time – Resident	1 hr/hr

$$I = \frac{C \times CR \times EF \times ED}{BW \times AT}$$

Where:

I = Intake of the chemical that is available for absorption (mg/kg body weight-day)
C = Chemical concentration in the medium (i.e. mg/L water)
CR = Contact rate, the amount of contaminated medium contacted (i.e. L water/day)
EF = Exposure frequency (days/year)
ED = Exposure duration (years)
BW = Body weight (kg)
AT = Averaging time, the period over which the exposure is averaged (days)

Figure 14.2 Generic equation for calculating chemical intakes (US EPA, 1989, 1991).

$$EC = \frac{CA \times ET \times EF \times ED}{AT}$$

Where:

EC = Exposure concentration (μg/m^3)
CA = Chemical concentration in air (μg/m^3)
ET = Exposure time (hours/day)
EF = Exposure frequency (days/year)
ED = Exposure duration (years)
AT = Averaging time, the period over which the exposure is averaged (total hrs in lifetime)

Figure 14.3 Equation for calculating inhalation exposure concentration (US EPA, 2009).

$$DAD = \frac{DA \times EV \times EF \times ED \times SA}{BW \times AT}$$

Where:

DAD = Dermally absorbed dose (mg/kg-day)
DA = Absorbed dose per event (mg/cm^2-event)
EV = Event frequency (events/day)
EF = Exposure frequency (days/year)
ED = Exposure duration (years)
SA = Skin surface area available for contact (cm^2)
BW = Body weight (kg)
AT = Averaging time, the period over which the exposure is averaged (days)

Figure 14.4 Equation for calculating the dermally absorbed dose of a chemical in water (US EPA, 2004).

factors, as well as exposure factors relating to exposure time and the individual's physical characteristics.

14.3 Toxicity Assessment

In order to determine the risks posed to exposed individuals from chemicals or physical agents, it is necessary to understand the potential for the chemicals or physical agents to cause adverse effects, a process known as a toxicity assessment. Toxicity assessment consists of hazard identification and a dose–response evaluation.

Hazard identification is the determination of whether exposure to a chemical or physical agent can cause an increased incidence of a particular adverse health effect in humans, such as cancer or birth defects. Different types of studies are performed to collect these data, including epidemiological studies and clinical studies, but the bulk of our information comes from laboratory studies on animals.

The dose–response evaluation determines the quantitative relationship between the level of exposure to a chemical and the incidence of adverse health effects in the exposed population. Critical details such as the specific exposure route (inhalation or oral), as well as the length of exposure (chronic, subchronic, or single event), are noted. Key terms in the dose–response evaluation include the following:

- Lowest-observed-adverse-effect-level (LOAEL) – The LOAEL is the exposure level at which the frequency or severity of adverse effects in an exposed population is statistically elevated compared to a control group.
- No-observed-adverse-effect-level (NOAEL) – The NOAEL is the highest exposure level at which no statistically or biologically significant adverse effects are increased among the exposed population.
- No-observed-effect-level (NOEL) – The NOEL is the exposure level at which no statistically or biologically significant effects of *any* kind are increased among the exposed population (US EPA, 1989).

14.4 Estimating Risks

Once the exposure to a chemical or agent has been characterized and toxicological information about the chemical has been gathered, the risk of adverse effects occurring from the exposure can be estimated by integrating the exposure information with the toxicological information. From the dose–response data, useful toxicity values, such as RfDs and slope factors, are derived that can be used to estimate the risk of adverse health effects from a certain level of exposure. (A discussion of dermal exposure risk is omitted here due to its complexity, but a detailed guideline for calculating risks from dermal exposure is available in the US EPA's 2004 *Supplemental Guidance for Dermal Risk Assessment*.)

Carcinogenic effects and non-carcinogenic effects are evaluated separately in risk assessment. For non-carcinogenic effects, it is assumed that a threshold can be estimated below which the risk is zero, based on the idea that protective mechanisms must be overcome before the adverse effect can be expressed. However, for carcinogenic effects, it is assumed that a small number of molecular events can cause changes in a single cell leading to cellular proliferation; thus, any exposure is considered to contribute to the risk of developing cancer. A threshold does not exist which relates to a zero risk, except for zero exposure. As such, distinct toxicity values exist for carcinogenic and non-carcinogenic effects, even for the same chemical (US EPA, 1989).

14.4.1 Non-Carcinogenic Effects

RfD_o

An oral reference dose (RfD_o) is an upper-bound estimate of a daily oral exposure level, typically expressed as mg/kg/day, below which non-carcinogenic adverse effects are considered unlikely during a lifetime. The RfD_o is derived from the NOAEL (or LOAEL if the NOAEL is not available) by (1) making adjustments to the dose to account for differences in the exposure time and (2) dividing by uncertainty factors, multiples of 10, to account for uncertainties such as variations in the exposed population, interspecies variability between humans and the studied animals, and uncertainty when extrapolating from LOAELs to NOAELs. Various types of RfD_o are available depending on the effect (developmental or other) and the length of exposure (chronic or subchronic) (US EPA, 1989).

RfC_i

An inhalation reference concentration (RfC_i) is an upper-bound estimate of the concentration of a chemical in air that is unlikely to cause adverse effects during a lifetime of continuous exposure. It is derived from the NOAEL or LOAEL, after adjustments are made using uncertainty factors. Various types of RfCs are available depending on effect (developmental or other) and the length of exposure (chronic or subchronic) (US EPA, 2009, 2010).

14.4.2 Carcinogenic Effects

SFO

A slope factor, derived for potential carcinogens, is a toxicity value that quantifies the relationship between the dose of a carcinogen and the response. The oral slope factor (SFO), the most common slope factor, is an upper-bound estimate of the probability of an individual developing cancer as a result of a lifetime of oral exposure to one unit intake (mg/kg/day) of the potential carcinogen. Deriving the slope factor from animal or human studies involves extrapolating from actual doses in the studies to lower doses expected in the environment and determining the physiologically equivalent human dose from animal data. When a slope factor is multiplied by an individual's projected intake of the chemical, the result is an estimate of the probability of the individual developing cancer over a lifetime, as a result of the exposure. The slope factor is always accompanied by the chemical's weight-of-evidence designation, indicating the strength of evidence of carcinogencity (US EPA, 1989).

IUR

An inhalation unit risk (IUR), sometimes referred to as an inhalation cancer slope factor, represents the upper-bound cancer risk associated with continuous exposure to 1 ug/m^3 of a carcinogen in the air. When an IUR is multiplied by the concentration that the individual is exposed to (i.e the EC), the result is an estimate of the probability of the individual developing cancer over a lifetime as a result of the exposure (US EPA, 2009, 2010).

14.5 Risk-Based Regulatory Levels

Using available toxicity information and exposure assumptions, the US EPA sets forth screening levels for chemical contaminants in soil, drinking water, and air. US EPA Regional Screening Levels (RSLs), developed for CERCLA and RCRA Corrective Action sites, are screening concentrations for specific chemicals in residential and industrial soils, residential and industrial air, and drinking water. For carcinogens, the RSLs correspond to a fixed level of cancer risk of one-in-one-million (10^{-6}), that is, a lifetime of exposure to the RSL level of a contaminant corresponds to the probability of one additional person developing cancer in a population of one million equally exposed people (on top of the cancer incidences that would normally occur in an unexposed population). The 10^{-6} incremental risk is called the 'Target Risk' by the US EPA, considering it to be an acceptable level of risk. For non-carcinogens, the RSLs are derived from the RfD_o or RfC_i values, and therefore exposure to this concentration of the chemical in the medium is considered unlikely to produce adverse effects in a population (US EPA, 2010).

14.6 Resources for Toxicity Information

Human health toxicity values are available from the following sources, presented in their order in the US EPA's hierarchy (US EPA, 2010).

1. EPA's Integrated Risk Information System (IRIS)
2. The Provisional Peer Reviewed Toxicity Values (PPRTVs) derived by EPA's Superfund Health Risk Technical Support Center (STSC) for the EPA Superfund program. (The PPRTV website is not open to users outside of EPA, but can be obtained upon request.)
3. The Agency for Toxic Substances and Disease Registry (ATSDR)
4. The California Environmental Protection Agency (OEHHA) Office of Environmental Health Hazard Assessment's Chronic Reference Exposure Levels (RELS) and Cancer Potency Values
5. Screening toxicity values in an appendix to certain PPRTV assessments
6. The Health Effects Assessment Summary Tables (HEAST). (Not open to users outside of EPA, but can be obtained upon request.)

14.7 Uncertainties in Risk Assessment

Despite the efforts that have been made to assess the toxicity of chemicals, and the large amount of information that has been gathered so far, there is a long way to go before we can fully understand the risks of human exposure to the myriad chemicals in the environment. Numerous chemicals have yet to be evaluated. Moreover, many gaps in data exist for the chemicals for which toxicity values are currently available. A study conducted by the European Chemical Bureau on high production volume chemicals (HPVCs), referring to chemicals which are produced in quantities greater than 1,000 tons per year, revealed that only 28% of these more than 100,000 chemicals had 'some data relevant for the evaluation of risk to man and the environment' (Allanou et al., 2003). A summary of data availability for these HPVCs shown in Table 14.4 illustrates the vast data inadequacies that promote an incomplete understanding of many contaminants.

Table 14.4 Data Availability for all HPVCs in the EU (Allanou et al., 2003)

Chapter	Chapter Description	Availability
2	**Physico-chemical Data**	
2.1	Melting Point	75.46%
2.2	Boiling Point	68.76%
2.3	Density	84.54%
2.4	Vapour Pressure	61.14%
2.5	Partition Coefficient	58.38%
2.6	Water Solubility	76.23%
2.7	Flash Point	65.56%
2.8	Auto Flammability	41.38%
2.9	Flammability	40.37%
2.10	Explosive Properties	44.83%
2.11	Oxidizing Properties	27.42%
2.12	Additional Remarks	51.03%
3	**Environmental Fate and Pathways**	
3.1.1	Photodegradation	47.59%
3.1.2	Stability in Water	40.81%
3.1.3	Stability in Soil	23.16%
3.2	Monitoring Data (Environment)	22.88%
3.3.1	Transport between Environ. Compart.	25.48%
3.3.2	Distribution	31.24%
3.4	Mode of Degradation in Actual Use	25.52%
3.5	Biodegradation	60.57%
3.6	BOD5, COD or BOD5/COD Ratio	26.29%
3.7	Bioaccumulation	29.94%
3.8	Additional Remarks	25.23%
4	**Ecotoxicity**	
4.1	Acute/Prolonged Toxicity to Fish	67.95%
4.2	Acute Tox. to Aquatic Invertebrates	54.65%
4.3	Toxicity to Aquatic Plants e.g. Algae	45.56%
4.4	Tox. to Microorganisms e.g. Bacteria	56.92%
4.5.1	Chronic Toxicity to Fish	13.71%
4.5.2	Chronic Tox. to Aquatic Invertebrates	17.77%
4.6.1	Toxicity to Soil Dwelling Organisms	30.30%
4.6.2	Toxicity to Terrestrial Plants	31.76%
4.6.3	Tox. to Other Non-mamm. Terr. Species	32.70%
4.7	Biological Effects Monitoring	25.80%

(Continued)

Table 14.4 (Continued)

Chapter	Chapter Description	Availability
4.8	Biotransformation and Kinetics	26.98%
4.9	Additional Remarks	35.82%
5	**Toxicity**	
5.1.1	Acute Oral Toxicity	76.96%
5.1.2	Acute Inhalation Toxicity	50.75%
5.1.3	Acute Dermal Toxicity	52.94%
5.1.4	Acute Toxicity, Other Routes	35.01%
5.2.1	Skin Irritation	73.27%
5.2.2	Eye Irritation	72.90%
5.3	Sensitisation	48.32%
5.4	Repeated Dose Toxicity	58.17%
5.5	Genetic Toxicity in Vitro	66.94%
5.6	Genetic Toxicity in Vivo	37.89%
5.7	Carcinogencity	43.89%
5.8	Toxicity to Reproduction	26.00%
5.9	Developmental Toxicity/Teratogenicity	32.01%
5.10	Other Relevant Information	51.93%
5.11	Experience with Human Exposure	55.94%

Our current toxicity database has limited data on the toxic effects of chemicals in humans (as opposed to animals). The EPA explains:

> *Well-conducted epidemiologic studies that show a positive association between an agent and a disease are accepted as the most convincing evidence about human risk. At present, however, human data adequate to serve as the sole basis of a dose-response assessment are available for only a few chemicals. Humans are generally exposed in the workplace or by accident, and because these types of exposures are not intentional, the circumstances of the exposures (concentration and time) may not be well known. Often the incidence of effects is low, the number of exposed individuals is small, the latent period between exposure and disease is long, and exposures are to mixed and multiple substances. Exposed populations may be heterogeneous, varying in age, sex, genetic constitution, diet, occupational and home environment, activity patterns, and other cultural factors affecting susceptibility.*
>
> *(US EPA, 1989)*

The correlation between animal data and human health effects is one obstacle to understanding the risks of chemical exposure. While toxicology relies on the assumption that mammals are similar in their susceptibility to toxic chemicals, there is an inherent uncertainty in the use of animal data for the evaluation of human health risks (even after making adjustments for known differences in physiology).

Additionally, other uncertainties arise from the extrapolations that are made from available data to derive toxicity values. Using studies with high doses to predict the adverse health effects that may occur from actual low-level exposure in the environment presents uncertainty, as does using a study at one exposure frequency and duration to estimate the effect at another exposure frequency and duration.

Another obstacle to understanding the risks to human health is the difficulty in characterizing all exposure pathways by which a population may be exposed. As previously discussed, individuals are exposed to pollutants through various mediums and exposure pathways, and only considering a select few of these in a risk assessment can result in an underestimate of the projected dangers.

Moreover, chemicals within a medium are rarely found independently of other chemicals. Many uncertainties exist when it comes to chemical mixtures, with respect to the potential interactions within a medium and within the body. The main types of interactions are (Khan, 2007):

- Additivity – Exposure to two or more chemicals produces the same effect in the body as the sum of the expected individual effects. Example: A formulation containing chlorinated insecticides and halogenated solvents produces liver toxicity equal to the sum of the individual toxicities.
- Antagonism – Exposure to one chemical results in a reduction in the effect of the other chemical. Example: Chlorinated insecticides stimulate the central nervous system, while halogenated solvents cause central nervous system depression.
- Potentiation – Exposure to one chemical results in another chemical producing a greater effect than expected. Example: The liver toxicity of carbon tetrachloride is enhanced by simultaneous exposure to isopropanol.
- Synergism – Exposure to one chemical results in another chemical producing a dramatically greater effect than expected. Example: The liver toxicity of a combination of carbon tetrachloride and ethanol is much greater than the sum of the individual effects.

The potential interactions that may occur in the environment and in the human body are vast and largely unknown. In risk assessment practice, the risks or hazards from exposures to different substances and different exposure pathways are typically treated as additive, and so it is believed that only considering one chemical or exposure pathway at a time can result in a significant underestimate of the potential dangers (US EPA, 1989, 1992).

Another problem arises in the use of dose–response information from homogeneous animal populations or healthy human populations to predict the effects in a heterogeneous population that consists of sensitive subpopulations such as children, the elderly, and people with pre-existing conditions. While uncertainty factors are incorporated into toxicity values to account for sensitive groups, and exposure factors can be adjusted according to general considerations such as age and sex, it is extremely difficult to predict the risks posed to all members of a variegated population.

The limitations in exposure and risk assessment are many. Unfortunately, the ability of regulatory agencies to monitor and protect human health depends on the understanding of chemicals risks. Therefore, any limitation in the knowledge base impairs the ability of regulatory agencies to adequately protect human populations. The subsequent chapters on Emerging Contaminants, Childhood Exposures, and Failures of Regulatory Agencies contain further discussion of these issues.

References

Allanou, R., Hansen, B. G., & Van Der Bilt, Y. (2003, June). Public availability of data on EU high production volume chemicals – Part 1. *Chemistry Today.*

Khan, M. (2007, July 27). *Interaction of chemicals in the body.* National Library of Medicine.

United States Department of Energy. *Image from the habitable planet. Unit 6: Risk, exposure, and health // Section 3: Measuring exposure to environmental hazards.* Available at: <http://www.learner.org/courses/envsci/unit/text.php?unit=6&secNum=3#exposure_pathways>.

US EPA (1989, December). *Risk assessment guidance for superfund. Volume I: Human health evaluation manual (Part A). Interim final.* United States Environmental Protection Agency. EPA/540/1-89/002.

US EPA (1991, March 25). *Risk assessment guidance for superfund. Volume I: Human health evaluation manual. Supplemental guidance. 'Standard default exposure factors.' Interim Final.* United States Environmental Protection Agency. OSWER Directive 9285.6-03.

US EPA (1992, May). *Guidelines for exposure assessment.* United States Environmental Protection Agency. EPA/600/Z-92/001.

US EPA (1997, August 1997). *Exposure factors handbook.* National Center for Environmental Assessment. United States Environmental Protection Agency.

US EPA (2004, July). *Risk assessment guidance for superfund. Volume I: Human health evaluation manual. Part E, Supplemental guidance for dermal risk assessment. Final.* United States Environmental Protection Agency. EPA/540/R/99/005. OSWER 9285.7-02EP. PB99-963312.

US EPA (2009, January). *Risk Assessment Guidance for Superfund. Volume I: Human Health Evaluation Manual. Part F, Supplemental Guidance for Inhalation Risk Assessment. Final.* United States Environmental Protection Agency. EPA-540-R-070-002. OSWER 9285.7-82.

US EPA (2010). *Regional screening table – User's guide. Mid-Atlantic risk assessment.* United States Environmental Protection Agency. Retrieved May 2010 from <http://www.epa.gov/reg3hscd/risk/human/rb-concentration_table/usersguide.htm>.

15 Bioaccumulation of Dioxins, PCBs, and PAHs

Chapter Contents

A particular class of pollutants has garnered attention not only due to their toxic properties, but also because of their tendency to remain in the environment for long periods of time and accumulate in the bodies of animals and humans. These chemicals, referred to as persistent, bioaccumulative, and toxic (PBT) chemicals, are often by-products of incineration, combustion, or industrial processes, or sometimes intentionally produced for commercial use. Their ability to bioaccumulate and to biomagnify in organisms upward through the food chain means that organisms located at the top of the food chain, such as humans, develop the highest concentrations of these chemicals within their bodies. This class of substances includes

Risks of Hazardous Wastes. DOI: 10.1016/B978-1-4377-7842-7.00015-5

dioxins and dioxin-like compounds, polychlorinated biphenyls (PCBs), and polycyclic aromatic hydrocarbons (PAHs), chemicals which have been linked to various types of cancers, respiratory problems, cardiovascular diseases, neurological diseases, and reproductive disorders.

15.1 Overview of Persistent, Bioaccumulative, and Toxic Chemicals

15.1.1 Bioaccumulation, Bioconcentration, and Biomagnification

Bioaccumulative chemicals have specific properties that enable them to be stored in organisms. These chemicals are hydrophobic and have low water solubility, allowing them to be stored in adipose tissue where metabolism is slower (Mackay et al., 2000; Kelly et al., 2007).

Bioaccumulation is the process by which chemicals in the environment are taken up by an organism either directly from exposure to a contaminated medium (e.g. water, air) or by consumption of food containing that chemical (EPA, 2010a). The *bioaccumulation factor (BAF)* is the ratio of the contaminant in an organism to the concentration in the surrounding environment from which the organism can take in the contaminant either though ingestion or through direct contact (EPA, 2010a). Bioaccumulation takes into account both bioconcentration and biomagnification. Total accumulation of the chemical is dependent on the rate of intake versus the rate of elimination or metabolic degradation.

Bioconcentration refers to the process by which organisms absorb chemicals in the environment via direct exposure to a contaminated medium, to levels exceeding those in the surrounding environment. The concept is similar to bioaccumulation; however, it does not include dietary uptake and is a term generally applied to aquatic organisms. The *bioconcentration factor (BCF)* is the ratio of the contaminant in an organism to the concentration in the surrounding environment from which the organism can take in the contaminant through direct contact. Bioconcentration is dependent on factors such as toxic effects, bioavailability, concentration of the chemical in the water, pH of the water, and the lipid content of the organism (Geyer et al., 2000).

Biomagnification is the process by which the concentration of a chemical in an organism increases at successively higher trophic levels of the food chain. The producers at the bottom of the food chain absorb nutrients and chemical pollutants that may be present from the surrounding environment. A consumer then eats the producer and the concentration of the chemical is transferred from the producer to the consumer and stored. If another organism eats this consumer, this concentration of the chemical is now transferred to the new consumer and stored. This process repeats itself thus lending higher concentrations of chemical pollutants to those at higher trophic levels. Those at the top of the food chain, such as marine mammals and humans, are at the greatest risk of being exposed to toxic levels of chemicals.

15.1.2 Persistent, Bioaccumulative, and Toxic Chemicals

Persistent, bioaccumulative, and toxic (PBT) chemicals pose significant health concerns to the environment and to human health. As their name suggests, this class of chemicals persists in the environment, with long half-lives spanning years to decades depending on the reservoir. Additionally they have the potential to bioaccumulate in the food chain, and they have been shown to have toxic effects on wildlife and human health, including effects on the nervous system, reproductive and developmental problems, cancer, and genetic impacts. The populations considered most at risk to PBTs are children and the developing fetus (US EPA, 2010b).

Programs have been established in an effort to reduce the effects of these environmental pollutants. The US EPA's National Waste Minimization Program has targeted 31 'Priority Chemicals' (PCs) for reduction from industrial products and wastestreams, based on their tendency to enter and persist in the environment, and the hazards they pose to human health. Of the 31 Priority Chemicals, 27 are PBT chemicals. PCBs, PAHs, and dioxins are among those listed. Chapter 22 contains the entire list of Priority Chemicals.

15.1.3 Persistent Organic Pollutants (POPs)

Persistent organic pollutants (POPs) are a subset of PBT chemicals that have the potential for long-range transport involving soil, water, and air (Gouin, 2003; Muir et al., 2006). The Stockholm Convention, adopted in 2001 and administered by the United Nations Environment Programme, is a global agreement to protect human health and the environment from specific POPs (Stockholm Convention on POPs, 2010). The treaty requires parties to take measures to (1) eliminate the production and use, (2) restrict the production and use, and/or (3) reduce the unintentional releases of the chemicals as described by the treaty. The 12 POPs initially targeted, referred to as the 'dirty dozen,' are: aldrin, chlordane, DDT, dieldrin, endrin, heptachlor, hexachlorobenzene, mirex, toxaphene, PCB, polychlorinated dibenzo-p-dioxins, and polychlorinated dibenzofurans. At its fourth meeting in May 2009, nine additional POPs were added to the Stockholm Convention.

15.2 Dioxins

15.2.1 Overview

Dioxins and dioxin-like compounds (referred to collectively as DLCs) are some of the most toxic chemicals ever identified and have been the subject of a great deal of scientific research and environmental and health policies. 2,3,7,8-Tetrachlorodibenzo-p-dioxin, also known as 2,3,7,8-TCDD or just TCDD, is considered the most toxic dioxin, and is classified by the EPA as 'carcinogenic to humans,' while other DLCs are classified as 'likely to be carcinogenic to humans' (NRC, 2006). DLCs are created from activities like combustion, metal processing, herbicide use, and chemical manufacturing and processing. Although DLC levels in the environment have been declining over the past three decades, DLCs do not readily degrade and continue to persist in the environment (FDA, 2010).

15.2.2 Sources and Exposure

Food is the main source of human exposure to dioxins and dioxin-like compounds (NRC, 2003). DLCs enter the food chain through direct and indirect pathways, including transfer from air to plants and soil, transfer from plants and soil to animals, and transfer from water and sediment to fish (EPA, 2000). Both land and aquatic food animals may be exposed to dioxin compounds primarily through soil-based ecological pathways (McLachan et al., 1990; NRC, 2003). For example, animal agriculture practices in the United States can contribute to DLC exposure via contamination of plant and animal by-products used to produce animal feeds (NRC, 2003).

A major route by which DLCs can enter the food supply is through aquatic animals that can accumulate elevated DLC levels from the environment. In wild populations of fish, DLCs accumulate up the food chain from smaller to larger predatory fish via bioaccumulation. In general, older, larger, and oily fish have higher DLC levels (NRC, 2003). Fish and seafood populations living near coastal areas have higher DLC levels than deep-water populations (NRC, 2003). Fish and seafood aquaculture practices may also cause low levels of DLC exposure due to the use of animal products in fish feed (NRC, 2003).

DLCs may be found in all foods, with a range of content within each food group (NRC, 2003). The foods with the highest concentration of DLCs are animal products, with red meat, fish, and dairy products having concentrations over 0.10 ppt (NRC, 2003). Foods with higher fat contents such as animal products tend to have higher DLC concentrations (NRC, 2003). Foods with low-fat content including lean meats, fruits, vegetables, and grain products have significantly lower DLCs concentrations; however, because larger volumes may be consumed, these foods can contribute to total DLC exposure (NAS, 2000). Currently, there are no standards for allowable DLC levels in foods (NRC, 2003).

15.2.3 Health Effects

Toxicity among the congeners is variable with the most toxic congener being TCDD. Because exposure is typically due to variable mixtures of dioxins and dioxin-like compounds, US EPA and others use toxicity equivalency factors (TEFs) to compare the toxicity of each of the individual dioxins against the toxicity of TCDD. TCDD has a TEF value of one, while all other congeners have TEF values equal to or less than one. The toxicity equivalents (TEQs) value is the amount of TCDD it would take to match the combined toxic effect of all the dioxin-like compounds present in the mixture. TEFs are used to calculate the toxicity of a mixture of dioxins and other dioxin-like compounds in terms of its TEQs relative to TCDD (Van den Berg et al., 2006). The EPA and World Health Organization (WHO) have implemented the use of TEFs for quantifying the toxicity of multiple dioxins/furans as a TCDD TEQ.

The International Agency for Research on Cancer (IARC) has classified dioxins as carcinogenic to humans based on data in humans and extensive evidence in experimental animals (IARC, 1997). In animals, TCDD has been shown to be a multiple site carcinogen when administered at low doses and by various routes in mice, rats, and monkeys (IARC, 1997). Carcinomas and adenomas that affect the liver, thyroid, thymus, lungs, lymph nodes, and soft tissue such as skin have been found in these experiments (IARC, 1997). In addition, endocrine-related tumors have also been seen affecting the pituitary gland, adrenal gland, and pancreas. TCDD has also been shown to have reproductive and developmental effects such as immune suppression and genital malformations (IARC, 1997). Occupational studies have also shown that TCDD also appears to affect multiple sites, and that exposure has been associated with breast cancer, endometrial cancer, testicular cancer, liver cancer, lung cancer, multiple myeloma, and cancers affecting soft tissue and the lymphopoietic system (Kogevinas, 2001).

Epidemiological data on the effect of dioxins include studies conducted following the 1976 Seveso disaster in Italy. The Seveso disaster was an industrial accident that occurred in a small Italian chemical manufacturing plant, resulting in the highest known exposure to TCDD in residential populations (Eskenazi et al., 2004). In follow-up studies, excess cancer risks were seen in Seveso residents. Among men, mortality from total cancers, lung cancer, and rectal cancer was elevated (Bertazzi, 1997; Bertazzi, 2001). Additionally an increased risk of breast cancer and liver cancer was found in women. Mortality from lymphopoietic neoplasms was increased in both men and women (Bertazzi, 1997; Bertazzi, 2001; Warner et al., 2002).

In addition to potential carcinogenic effects, there are other non-cancer health effects. Acute exposure to TCDD causes alterations in liver enzymes and dermatological skin conditions such as hyperpigmentation and chloracne (WHO, 2010). Chloracne resembles acne with its eruption of blackheads, pustules, and cysts. Other effects that have been noted include changes in thyroid function, the reproductive function, and neurodevelopment (Kogevinas, 2001). TCDD has been shown to interfere with the endocrine system causing adverse developmental, reproductive, neurological, and immune effects (NIEHS, 2010). Examples of these effects include decreased testosterone and increased gonadotropin production, changes in glucose metabolism, alterations in thyroid hormones, spontaneous abortions, birth and weight defects, and developmental dental effects (IARC, 1997; Kogevinas, 2001).

15.3 Polychlorinated Biphenyls (PCBs)

15.3.1 Overview

Polychlorinated biphenyls belong to a group of man-made organic chemicals known as chlorinated hydrocarbons. Between the 1930s and 1970s, PCB mixtures were used in capacitors and transformers, heat transfer fluids, hydraulic fluids, lubricating oils, and as additives in pesticides, paints, paper, adhesives, and plastics (ATSDR, 2001; Henry and DeVito, 2003). Like all PBTs, PCBs are pervasive in the environment. PCBs occur in air, water, soil, and plants, and travel up the food chain through fish, mammals, birds, and ultimately to humans (ASDTR, 2001).

15.3.2 Sources and Exposure

Although PCB manufacturing and use has been banned since the 1970s, PCBs continue to be released into the environment from poorly maintained hazardous waste sites, improper dumping of PCB wastes, leaks from electrical transformers containing PCBs, and disposal of PCB-containing consumer products into landfills not designed to handle hazardous waste (EPA, 2008). PCBs may also be released into the environment by the incineration of wastes (EPA, 2008).

Similar to mercury and dioxins, higher concentrations of PCBs are found in foods of animal origin (particularly predatory mammals and fish) compared to fruits and vegetables (EFSA, 2005). Foods of animal origin that contain high amounts of PCBs include oily fish such as farmed salmon, butter, pork, hamburgers, bacon, chicken/pork/beef fat, eggs, and milk (EFSA, 2005). PCBs in the context of food contamination are divided into two groups based on their structure and toxicological effects: dioxin-like PCBs (DL-PCBs) and the non-dioxin-like PCBs (NDL-PCBs). DL-PCBs are more toxic than NDL-PCBs and are therefore considered to be of more concern in food safety (EFSA, 2005).

15.3.3 Health Effects

Due in part to compelling evidence of carcinogenicity among animals, as well as various epidemiological studies, the US EPA and the IARC have classified PCBs as probable carcinogens to humans (EPA, 2007a; IARC, 1987). Cancers that have been reported in animals include liver cancer, renal cancer, and thyroid cancer (EPA, 2007a). In humans, PCB health effects have been evaluated in occupational populations and those who have been accidently exposed to large doses of the substance. Cancers that have been reported include melanomas, liver cancer, gastrointestinal, and hematopoietic cancers (IARC, 1987). Male workers were found to have excess gastrointestinal cancer including stomach, pancreatic, and liver cancer, while female workers were found to have excess hematopoietic neoplasms (Bertazzi, 1987). Although more research is required in this area, suggestive evidence has also been found associating PCBs with breast cancer as PCBs have estrogenic activity (Aronson, 2000).

Non-carcinogenic effects include effects to the immune, reproductive, nervous, and endocrine systems. Adults exposed to PCBs showed decreased concentrations of lymphocytes, which can result in immune deficiency (Chang, 1981). Children born to women exposed to PCBs have decreased birth weight and a decrease in gestational weight, while poorer cognitive functioning has also been seen in preschool children with prenatal exposure to PCBs (Fein et al., 1984; Patandin et al., 1999). PCBs have been shown to affect the neurological development and cognitive function in the developing fetus, infants, and children (Shantz et al., 2003). PCBs from prenatal exposure were found to persist in childhood and were associated with an altered immunity and a greater susceptibility to certain infectious diseases (Weisglas-Kuperus et al., 2000). PCBs are endocrine disrupting chemicals and altered thyroid hormones have been found in children (Osius et al., 1999). Other effects include respiratory tract symptoms, gastrointestinal effects including anorexia, weight loss, nausea, vomiting, and abdominal pain, liver effects, anemia, behavioral alterations, impaired reproduction, and effects on the skin and eyes such as chloracne, skin rashes, and eye irritation (ATSDR, 2001; EPA, 2007a).

15.3.4 Regulation of PCBs

Some regulations are in place to limit the amount of PCBs in the food supply. The FDA regulates infant foods, eggs, milk, and mandates tolerance levels in fish of 2 ppm (ATSDR, 2001). The FDA limits PCBs in plastic food-packaging materials to 10 ppm. Additionally, the EPA requires the level of PCBs in waters contaminated with PCBs to be no greater than 0.17 parts of PCBs per trillion parts (ppt) of water (EPA, 2000).

In March 2007, the European Union (EU) developed new regulations on contaminant levels in foods such as dioxins and DL-PCBs and required tougher safety controls and improved testing in food manufacturing plants. In 2010, the European Food Safety Authority (EFSA) published additional research on DL-PCBs and dioxins in food and found that 8% of over 7,000 food samples exceeded different maximum DL-PCB levels and 4% exceeded some action levels (EFSA, 2010). The EFSA concluded 'continuous random testing of a sufficient number of samples in each food and feed group is needed to ensure accurate assessments of the presence of dioxins and dioxin-like PCBs' (EFSA, 2010).

15.4 Polycyclic Aromatic Hydrocarbons (PAHs)

15.4.1 Overview

Polycyclic aromatic hydrocarbons (PAHs) are a group of semi-volatile organic compounds present in oil, coal, and tar, or formed during incomplete burning of coal, oil and gas, garbage, or other organic substances (ATSDR, 1995). PAHs are persistent, toxic, bioaccumulative, and have the potential for long-range transport, commonly traveling in the atmosphere in particulate form. There are over one hundred different PAH congeners, but benzo[a]pyrene is the most commonly monitored congener due to its toxicity. Along with dioxins and PCBs, PAHs have been listed by the EPA as a pollutant of concern in the Great Waters Program, which targets pollution in the Great Lakes, Lake Champlain, Chesapeake Bay, and many of the coastal estuaries (EPA, 2007c).

15.4.2 Sources and Exposure

PAHs can be found throughout the environment in soil, water, and air. Sources that contribute to PAHs in the environment include vehicle exhausts, wildfires, agricultural burning, municipal and industrial waste incineration, and hazardous waste sites. The main pathways of exposure to PAHs are inhalation of the compounds in tobacco smoke, wood smoke, and ambient air, and consumption of PAHs in foods (ATSDR, 1995). Those living near hazardous waste or industrial sites may be exposed to PAHs through contaminated air, water, and soil. Exposure to PAHs can occur from eating foods that have grown in contaminated soil or air (ATSDR, 1995; Phillips, 1999), or by cooking meat or other foods over an open flame, such as during grilling or charring (Phillips, 1999). In addition, workers at facilities such as coal tar production plants, coking plants, asphalt production plants, coal-gasification sites, smoke houses, aluminum production plants, and municipal trash incinerators are exposed to PAHs (ATSDR, 1995).

15.4.3 Health Effects

The US EPA and the IARC have evaluated a number of PAHs for carcinogenicity. The US EPA has classified benzo[a]pyrene, benz[a]anthracene, chrysene, benzo[b]fluoranthene, benzo[k]fluoranthene, dibenz[a,h]anthracene, and indeno[1,2,3-cd]pyrene, as probable human carcinogens (EPA, 2007b). The IARC has classified benzo[a]pyrene as a carcinogen to humans; cyclopenta[cd]pyrene, dibenz[a,h]anthracene, and dibenzo[a,l]pyrene have been classified as probable carcinogens to humans (IARC, 2010).

Animal studies have shown an array of carcinogenic and non-carcinogenic health effects with exposure to PAHs. Animal studies have reported excess lung cancer, stomach cancer, and leukemia (EPA, 2007b). Immunologic, renal, neurological, and reproductive toxicity have also been noted (ATSDR, 1995; EPA, 2007b). Decreased fertility, reduced birth weight, and increased birth defects have been shown to occur with exposure to PAHs (ATSDR, 1995; EPA, 2007b).

Epidemiologic studies with humans have demonstrated the carcinogenic effects of PAHs. Lung, bladder, and skin cancer have been reported following high exposures in occupational settings (Boffetta, 1997). An increased risk of lung cancer has been reported in industries with PAH exposure, possibly because PAHs often bind to particulates, which can then get deep into the lungs (Boffetta, 1997; Mastrangelo et al., 1996; Armstrong, 2004; IARC, 2010). Skin cancer has been reported with dermal exposure to PAHs, and bladder cancer has been reported with exposure to PAHs from coal tars and pitches from aluminum production, coal gasification, tar distillation, and from driving trucks (Boffetta, 1997; IARC, 2010). Other cancers that have been reported include stomach, colorectal, pharyngeal, esophageal, laryngeal, scrotal, pancreatic, and hematopoietic cancers (IARC, 2010). The IARC states that there is sufficient evidence of carcinogenicity from occupational exposures to PAHs such as in coal gasification, coke production, coal-tar distillation, aluminum production, paving and roofing with coal-tar pitch, and chimney sweeping (IARC, 2010).

Non-cancer health effects have also been reported with exposure to PAHs, but require further studies. PAH is an endocrine-disrupting chemical and can have reproductive effects. PAHs in air pollution have been shown to retard fetal growth (Dejmek, 2000) and result in reduced birth weights (Choi, 2006). Respiratory disorders have also been associated with exposures to PAHs in air pollutant mixtures (ATSDR, 1995; Delfino, 2002).

The US EPA evaluates exposure to PAHs using the benzo[a]pyrene equivalent (BaP TEQ) concentration. The US EPA has assigned TEFs to seven carcinogenic PAHs. The concentrations of these seven PAHs detected in a sample are multiplied by their respective TEFs and the products are summed to determine the BaP TEQ for the mixture of PAHs in the sample.

15.5 Case Studies

15.5.1 Bioaccumulation of PCBs

From 1940 to 1977, General Electric (GE) discharged approximately 1.3 million pounds of PCBs into the Hudson River from two capacitor manufacturing plants located at Fort Edward and Hudson Falls, New York (EPA, 2010b). Fish in the Hudson River subsequently accumulated dangerous levels of PCBs in their bodies. Between 1970 and 1974, the level of PCBs found in Hudson River fish exceeded the FDA's 5.0 ppm tolerance level (Brown, 1985). Due to the levels of PCBs found in fish, the New York State Department of Environmental Conservation (NYSDEC) banned all fishing in the upper Hudson River in 1976 (EPA, 2010b) In 1983, 90% of fish still exceeded the stricter FDA tolerance level of 2.0 ppm (Brown, 1985). In 1995, the ban was lifted and replaced with a catch and release fishing program as fish from the Hudson River still could not be safely consumed (EPA, 2010c). Currently, phased-in efforts are being taken to remove PCBs from contaminated sediments by environmental dredging.

15.5.2 Bioaccumulation of POPs

Adverse health effects from persistent organic pollutants have been noted in the Inuit population of Nunavik in northern Quebec, where POPs have turned up via long-range atmospheric transport, waterways, and ocean currents. The lipophilic properties of POPs have resulted in the bioaccumulation of POPs in the fatty tissues of predator species at the top of the food web, including polar bears and beluga whales (Dewailly, 1993).

Because the Inuit population relies heavily on local seafood, they are at an increased risk of exposure to POPs. The mean consumption of marine mammals, freshwater fish, and sea fish is ten, eight, and nine times a month, respectively, among the Inuit population. Marine mammals include the ringed seal, bearded seal, polar bear, and walrus (Dewailly, 1989)

A study measuring PCB levels in breast milk found PCB levels in Inuit women nearly five times higher than levels in Caucasian women (Dewailly, 1989). Levels of PCB in breast milk were found to be similar to levels in the fatty tissue of the beluga and seven times greater than levels in the breast milk of women in southern Quebec (Dewailly, 1993). Inuit children were found to have lower T-helper cells and a 10–15-fold increase in infections (Dewailly, 1989). An increased risk of the ear infection otitis media was found in those children with organochlorine exposure (Dewailly, 2000).

Acknowledgments

This chapter was contributed by Kristine Huang, OD. Kristine Huang is an Assistant Professor at the Southern California College of Optometry. She is currently pursuing her Masters in Public Health at the University of California, Los Angeles.

References

Agency for Toxic Substances and Disease Registry (ATSDR) (1995). *Polycyclic aromatic hydrocarbons (PAHs): Public health statement*. Retrieved August 21, 2010 from <http://www.atsdr.cdc.gov/PHS/PHS.asp?id=120&tid=25>.

Agency for Toxic Substances and Disease Registry (ATSDR) (2001). *Polychlorinated biphenyls*. Retrieved August 21, 2010 from <http://www.atsdr.cdc.gov/tfacts17.pdf>.

Armstrong, B., Hutchinson, E., Unwin, J., Fletcher, T. (2004). Lung cancer risk after exposure to polycyclic aromatic hydrocarbons: A review and meta-analysis. *112*(9), 970–978.

Arnot, J. A., & Gobas, F. (2006). A review of bioconcentration factor (BCF) and bioaccumulation factor (BAF) assessments for organic chemicals in aquatic organisms. *Environmental Review*, *14*, 257–297.

Aronson, K. J., Miller, A. B., & Woolcott, C. G. (2000). Breast adipose tissue concentrations of polychlorinated biphenyls and other organochlorines and breast cancer risk. *Cancer Epidemiology, Biomarkers & Prevention*, *9*, 55–63.

Bertazzi, P. A., Riboldi, L., Pesatori, A., Radice, L., & Zocchetti, C. (1987). Cancer mortality of capacitor manufacturing workers. *American Journal of Industrial Medicine*, *11*, 165–176.

Bertazzi, P. A., Zocchetti, C., Guercilena, S., Consonni, D., Tironi, A., Landi, M. T., & Pesatori, A. C. (1997). Dioxin exposure and cancer risk: A 15-year mortality study after the 'seveso accident'. *Epidemiology*, *8*(6), 646–652.

Bertazzi, P. A., Consonni, D., Bachetti, S., Rubagotti, M., Baccarelli, A, Zocchetti, C., & Pesatori, A. C. (2001). Health effects of dioxin exposure: A 20-year mortality study. *American Journal of Epidemiology*, *153*(11), 1031–1044.

Boffetta, P., Jourenkova, N., & Gustavsson, P. (1997). Cancer risk from occupational and environmental exposure to polycyclic aromatic hydrocarbons. *Cancer Causes and Control*, *8*, 444–472.

Bostrom, C., Gerde, P., Hanberg, A., Jernstrom, B., Johansson, C., Kyrklund, T., et al. (2002). Cancer risk assessment, indicators, and guidelines for polycyclic aromatic hydrocarbons in the ambient air. *Environmental Health Perspectives*, *110*(3), 451–488.

Brown, M. P., Werner, M. B., Sloan, R. J., & Simpson, K. W. (1985). Polychlorinated biphenyls in the Hudson River. *Environmental Science & Technology*, *19*(8), 656–661.
Chang, K., Hsieh, K., Lee, T., Tang, S., & Tung, T. (1981). Immunologic evaluation of patients with polychlorinated biphenyl poisoning: Determination of lymphocyte subpopulations. *Toxicology and Applied Pharmacology*, *61*, 58–63.
Choi, H., Jedrychowski, W., Spengler, J., Camann, D. E., Whyatt, R. M., Rauh, V., et al. (2006). International studies of prenatal exposure to polycyclic aromatic hydrocarbons and fetal growth. *Environmental Health Perspectives*, *114*(11), 1744–1750.
Delfino, R. J. (2002). Epidemiologic evidence for asthma and exposure to air toxics; linkages between occupational, indoor, and community air pollution research. *Environmental Health Perspectives*, *110*(Suppl. 4), 573–589.
Dejmek, J., Solansky, I., Benes, I., Lenicek, J., & Sram, R. J. (2000). The impact of polycyclic aromatic hydrocarbons and fine particles on pregnancy outcome. *Environmental Health Perspectives*, *108*(12), 1159–1164.
Dewailly, E., Nantel, A., Weber, J., & Meyer, F. (1989). High levels of PCBs in breast milk of Inuit women from arctic quebec. *Bulletin of Environmental Contaminated Toxicology*, *43*, 641–646.
Dewailly, E., Ayotte, P., Bruneau, S., Laliberte, C., Muir, D., & Norstrom, R. (1993). Inuit exposure to organochlorines through the aquatic food chain in arctic quebec. *Environmental Health Perspectives*, *101*(7), 618–620.
Dewailly, E., Ayotte, P., Bruneau, S., Gingras, S., Belles-Isles, M., & Roy, R. (2000). Susceptibility to infections and immune status in Inuit infants exposed to organochlorines. *Environmental Health Perspectives*, *108*(3), 205–211.
Dewailly, E., Blanchet, C., Lemieux, S., Sauve, L., Gingras, S., Ayotte, P., & Holub, B. (2001). n-3 fatty acids and cardiovascular disease risk factors among Inuit of Nunavik. *American Journal of Clinical Nutrition*, *74*, 464–473.
Eskenazi, B., Mocarelli, P., Warner, M., Needham, L., Patterson, D., Samuels, S., et al. (2004). Relationship of serum TCDD concentrations and age at exposure of female residents of seveso, Italy. *Environmental Health Perspectives*, *112*(1), 22–27.
European Food Safety Authority (EFSA) (2005). *Panel on contaminants in the food chain. Non dioxin like PCB contaminant levels decreasing in food and animal feed – continuing effort needed to further reduce possible risks to human health.* Parma, Italy: European Food Safety Authority. Retrieved August 2010 from <http://www.efsa.europa.eu/en/scdocs/scdoc/284.htm>.
European Food Safety Authority (EFSA) (2010, July). *Results of the monitoring of dioxin levels in food and feed*. Parma, Italy: European Food Safety Authority. Retrieved August 15, 2010 from <http://www.efsa.europa.eu/en/scdocs/doc/1385.pdf>.
Fein, G. G., Jacobson, J. L., Jacobson, S. W., Schwartz, P. M., & Dowler, J. K. (1984). Prenatal exposure to polychlorinated biphenyls: Effects on birth size and gestational age. *The Journal of Pediatrics*, *105*(2), 315–320.
Geyer, H. J., Rimkus, G. G., Scheunert, I., Kaune, A., Schramm, K-W., Kettrup, A., et al. (2000). Bioaccumulation and occurrence of endocrine-disrupting chemicals (EDCs), persistent organic pollutants (POPs), and other organic compounds in fish and other organisms including humans. *The Handbook of Environmental Chemistry*, *2*(J), 1–166
Henry, T., & DeVito, M. (2003, June). *Non-dioxin like PCBs: Effects and consideration in ecological risk assessment.* United States Environmental Protection Agency. Experimental Toxicology Division National Health and Environmental Effects Research Laboratory. Office of Research and Development. NCEA-C-1340 ERASC-003. Retrieved August 14, 2010 from <http://www.epa.gov/cswer/riskassessment/pdf/1340-erasc-003.pdf>.

International Agency for Research on Cancer (IARC) (1987). *Polychlorinated Biphenyls (Group 2A)* (Suppl. 7), 322. Retrieved August 21, 2010 from <http://www.inchem.org/documents/iarc/suppl7/polychlorinatedbiphenyls.html>.

International Agency for Research on Cancer (IARC) (1997). *Volume 69 polychlorinated dibenzo-para-dioxins and polychlorinated dibenzofurans: summary of data reported and evaluation.* Retrieved August 21, 2010 from <http://monographs.iarc.fr/ENG/Monographs/vol69/volume69.pdf>.

International Agency for Research on Cancer (IARC) (2010). *Summary of data reported and evaluation. Volume 92.* Retrieved August 21, 2010 from <http://monographs.iarc.fr/ENG/Monographs/vol92/mono92-10.pdf>.

Kelly, B. C., Ikonomou, M. G., Blair, J. D., Morin, A. E., & Gobas, F. (2007). Food web-specific biomagnification of persistent organic pollutants. *Science*, *317*, 236–239.

Kogevinas, M. (2001). Human health effects of dioxins: cancer, reproductive, and endocrine system effects. *Human Reproduction Update*, *7*(3), 331–339.

Mackay, D., & Fraser, A. (2000). Bioaccumulation of persistent organic chemicals: Mechanisms and models. *Environmental Pollution*, *110*, 375–391.

Mastrangelo, G., Fadda, E., & Marzia, V. (1996). Polycyclic aromatic hydrocarbons and cancer in man. *Environmental Health Perspectives*, *104*(11), 1166–1170.

Muir, D. C. G., & Howard, P. H. (2006). Are there other persistent organic pollutants? A challenge for environmental chemists. *Environmental Science and Technology*, *40*, 7157–7166.

Mumford, J. L., He, X. Z., Chapman, R. S., Cao, S. R., Harris, D. B., Li, X. M., et al. (1987). Lung cancer and indoor air pollution in Xuan Wei, China. *Science*, *235*, 217–220.

National Academy of Sciences (NAS) (2000). *Toxicologic effects of methylmercury.* Washington DC: Committee on the Toxicological Effects of Methylmercury. Commission on Life Sciences. National Research Council. Retrieved August 14, 2010 from <http://www.nap.edu/openbook.php?isbn=0309071402>.

National Institute of Environmental Health Sciences. *Dioxins.* Retrieved August 13, 2010 from <http://www.niehs.nih.gov/health/topics/agents/dioxins/>.

National Research Council (NRC) (2003). Reassessment of TCDD and related compounds. *Board on Environmental Studies and Toxicology-Division on Earth and Life Studies.* 1–240. Retrieved August 12, 2010 from <http://www.nap.edu/catalog.php?record_id = 11688>.

National Research Council (2006). *Health risks from dioxin and related compounds: Evaluation of the EPA reassessment.* Committee on EPA's Exposure and Human Health. Retrieved August 20, 2010 from <http://www.sph.umich.edu/dioxin/PDF/NRC_HealthRisksfromDioxins.pdf>.

Osius, N., Karmaus, W., Kruse, H., Witten, J. Exposure to polychlorinated biphenyls and levels of thyroid hormones in children. *Environmental Health Perspectives*, *107*(10), 843–849.

Patandin, S., Lanting, C. I., Mulder, P. G., Boersma, E. R., Sauer, P. J., & Weisglas-Kuperus, N. (1999). Effects of environmental exposure to polychlorinated biphenyls and dioxins on cognitive abilities in Dutch children at 42 months of age. *The Journal of Pediatrics*, *134*, 33–41.

Phillips, D. H. (1999). Polycyclic aromatic hydrocarbons in the diet. *Mutation Research*, *443*, 139–147.

Rehwagen, M., Muller, A., Massolo, L., Herbarth, O., & Ronco, A. (2005). Polycyclic aromatic hydrocarbons associated with particles in ambient air from urban and industrial areas. *Science of the Total Environment*, *248*, 199–210.

Skea, J. C., Simonin, H. A., Dean, H. J., Colquhoun, J. R., Spagnoli, J. J., & Veith, G. D. (1979). Bioaccumulation of aroclor 1016 in hudson river fish. *Bulletin of Environmental Contamination and Toxicology, 22*(1), 332–336.

Stockholm Convention on Persistent Organic Pollutants (POPs) (2010). *What are POPs?* Retrieved August 13, 2010 from <http://chm.pops.int/Convention/ThePOPs/tabid/673/language/en-US/Default.aspx>.

US Environmental Protection Agency (EPA) (2000). *Exposure and human health reassessment of 2,3,7,8-tetrachlorodibenzo-p-dioxin (TCDD) and relation compounds.* Washington DC: Draft Final Report. Retrieved August 12, 2010 from <http://www.epa.gov/ncea/pdfs/dioxin/nas-review/>.

US Environmental Protection Agency (EPA) (2007a). *Polychlorinated biphenyls (PCBs) (arochlors).* Retrieved August 21, 2010 from <http://www.epa.gov/ttnatw01/hlthef/polychlo.html>.

US Environmental Protection Agency (EPA) (2007b). *Polycyclic organic matter (POM).* Retrieved August 21, 2010 from <http://www.epa.gov/ttnatw01/hlthef/polycycl.html>.

US Environmental Protection Agency (EPA) (2007c). *Survey of new findings in scientific literature related to atmospheric deposition to the great waters: Polycyclic aromatic hydrocarbons (PAH).* Retrieved August 21, 2010 from <http://www.epa.gov/oaqps001/gr8water/pdfs/PAHSurveyReport.pdf>.

US Environmental Protection Agency (EPA) (2008). *Health effects of PCBs.* Retrieved August 20, 2010 from <http://www.epa.gov/epawaste/hazard/tsd/pcbs/pubs/effects.htm>.

US Environmental Protection Agency (EPA) (2009). *Priority chemicals.* Retrieved September 20, 2010 from <http://www.epa.gov/wastes/hazard/wastemin/priority.htm>.

US Food and Drug Administration (FDA) (2010). *Questions and answers about dioxins.* Retrieved August 21, 2010 from <http://www.fda.gov/Food/FoodSafety/FoodContaminantsAdulteration/ChemicalContaminants/DioxinsPCBs/ucm077524.htm#g11>.

US Environmental Protection Agency (EPA) (2010a). *Waste and cleanup risk assessment glossary.* Retrieved August 13, 2010 from <http://epa.gov/oswer/riskassessment/glossary.htm>.

US Environmental Protection Agency (EPA) (2010b). *Priority PBT profiles.* Retrieved August 27, 2010 from <http://www.epa.gov/pbt/pubs/cheminfo.htm>.

US Environmental Protection Agency (EPA) (2010c). *Hudson River PCBs.* Retrieved August 27, 2010 from <http://www.epa.gov/hudson/faqs.htm>.

Van den Berg, M., Birnbaum, L. S., Denison, M., De Vito, M., Farland, W., Feeley, M., et al. (2006). The 2005 World Health Organization reevaluation F hyman and mammalian toxic equivalency factors for dioxins and dioxin-like compounds. *Toxicological Sciences, 93*(2), 223–241.

Warner, M, et al. (2002). Serum dioxin concentrations and breast cancer risk in the Seveso Women's Health study. *Environmental Health Perspectives, 110*(7), 625–628.

Weisglas-Kuperus, N., et al. (2000). Immunological effects of background exposure to polychlorinated biphenyls and dioxins in Dutch preschool children. *Environmental Health Perspectives, 108*(12), 1203–1207.

16 Emerging Contaminants

Chapter Contents

16.1 Overview of Chemicals of Emerging Concern

Before the widespread use of pharmaceutical drugs and the mass production of a myriad of industrial chemicals, the primary concern of drinking and waste water treatment was the elimination of viral or bacterial pathogens. Today, a new class of chemical compounds consisting of pharmaceuticals, pesticides, industrial chemicals, surfactants, and personal care products are consistently being found in groundwater, surface water, municipal waste water, drinking water, and food sources. These 'emerging contaminants' include endocrine-disrupting compounds, analgesics, antibiotics, hormones, and a whole range of other pharmaceutical compounds including anti-inflammatory, antidiabetic, and antiepileptic drugs. The threat of emerging contaminants lies in the fact that the environmental and human toxicology of most of these compounds have not yet been studied and in the fact that many of these compounds are not or cannot be tested for in municipal water systems. In addition, when these contaminants pass through drinking water treatment systems, by-products are generated whose chemical properties are as yet undetermined. The danger presented by our lack of knowledge with regards to these compounds is underscored by the *Handbook of Environmental Chemistry* (2008) which states:

> *every day new potential emerging contaminants are discovered and new disinfection by-products are also generated during treatment, with a total ignorance of their potential toxicity or effect on human health.*

Risks of Hazardous Wastes. DOI: 10.1016/B978-1-4377-7842-7.00016-7

The ecotoxicological perspectives typically used to characterize environmental contaminants have fallen short when attempting to classify the complex network of emerging contaminants. The typical approach taken by toxicologists of characterizing a compound by its persistence, lypophilicity (preference for fat-tissue), and toxicity (both acute and chronic) has proven to be inadequate with regards to emerging contaminants. An emerging contaminant can exhibit what is known as 'pseudo-persistence' whereby a contaminant is continually present in the environment because it is continually being released by sewage treatment plants. Lypophilicity, typically used to determine how easily a contaminant will cross cell membranes or enter tissues, is an incomplete characterization tool because many emerging contaminants are pharmaceutically engineered to be actively transported into cells and tissues.

There are even more characteristics of emerging contaminants that limit our reliable characterization of their risk to humans and the environment. Our power to detect and measure these compounds currently surpasses our knowledge of their toxic effects on humans or wildlife. In addition, to establish a cause–effect relationship, toxicologists cannot simply cite the co-occurrence of contaminant and adverse effects. Rather, the uptake, mode of action, and biological endpoints of each emerging contaminant must be researched and documented to establish a correlation between contaminant and consequence.

Emerging contaminants are defined strictly as 'any synthetic or naturally occurring chemical or any microorganism that is not commonly monitored in the environment, but has the potential to enter the environment and cause known or suspected adverse ecological and/or human health effects' (Smital, 2008). It is important to note that the majority of emerging contaminants are *not* pollutants that are totally new or have just gained entry into the environment. Rather, most emerging contaminants are well-established pollutants with a newly demonstrated toxic effect or mode of action. Thus, the word 'emerging' refers not only to the contaminant itself but also to an emerging *concern* about the contaminant. As such, emerging contaminants are often referred to as 'chemicals of emerging concern' or 'contaminants of emerging concern.'

Since there is little toxicological information for the vast majority of the chemicals we use, particularly with regards to long-term, low-level exposure, it is possible that many chemicals with a long history of use (such as MTBE) could contribute to the future burden of emerging contaminants. A study of all high production volume chemicals (HPVC's) (>1,000 tons produced per year) conducted by the European Chemical Bureau revealed that only 28% of these more than 100,000 chemicals had 'some data relevant for the evaluation of risk to man and the environment' (Allanou et al., 2003). A review of the data availability for these HPVC's highlights the vast data inadequacies that promote an incomplete understanding of many emerging contaminants.

Emerging contaminants are typically divided into the following categories: pharmaceuticals and personal care products (PPCPs), surfactants, plasticizers, pesticides, and flame retardants (Yan et al., 2010). Biological emerging contaminants are also an issue of significant concern and will be discussed below. The issue of pesticides in drinking water is addressed in previous chapters.

16.2 Pharmaceuticals and Personal Care Products

Health products (e.g. supplements, over-the-counter drugs, and prescription pharmaceuticals) and cosmetic products comprise the diverse range of chemicals that are PPCPs. These chemicals enter the environment and drinking water supplies after being discharged from sewage treatment plants, or after being leached from manure applied to farmland (Derksen et al., 2004). They are often found in relatively low concentrations of $1-10$ g/L however, one should not overlook their potential to cause harm. PPCPs may have more deleterious effects on non-target organisms present in the environment than they have on humans. In addition, the effects on wildlife of continuous low-dose exposure to these compounds have not been studied and cannot simply be extrapolated from acute exposure data. Finally, certain PPCPs such as synthetic estrogens have been shown to be toxic even at the low concentrations at which they are found. Analgesics, anti-inflammatories, antimicrobials, β-blockers, and lipid regulators were the most commonly found pharmaceuticals in a large wastewater sampling conducted in the Western Balkan Region (Terzić et al., 2008).

Triclosan is the primary antibacterial/antiseptic agent used in liquid hand soaps and is now being incorporated into children's toys, cutting boards, utensils, and various other plastic products under the product name Microban®. While triclosan itself is not indicated as a major health concern for humans, it has been shown to be acutely toxic to some aquatic species, and it has numerous degradation products that could result in increased health risks to humans or wildlife. In tap water, triclosan forms chlorinated by-products such as chloroform (a suspected human carcinogen) and in the environment it can be photochemically or microbially degraded into certain dioxins (particularly 2,8-dichlorodibenzo-p-dioxin [2,8-DCDD] and 2,4-dichlorophenol [2,4-DCP]) and into a potentially bioaccumulative form called methyl triclosan (Smital, 2008). Triclosan was detected in 58% of 139 US streams in a recent study and it is likely that triclosan is not removed by standard water treatment methods (Kolpin et al., 2002). In addition, safety standards for triclosan in drinking water have not yet been set. Another set of personal care products that are rousing the concern of the scientific community are fragrances (also referred to as musks) used in soaps, shampoos, perfumes, detergents, air fresheners, and cleaning agents. Nitromusk and polycyclic musk fragrances are synthetic fragrances commonly found in steams, lakes, and fish in the US and Europe as well as in human breast milk and blood (Smital, 2008). Some nitromusks have been associated with cancer as well as reproductive and fertility problems in humans. Studies have also demonstrated neurotoxic effects of nitromusks in animals and endocrine effects of polycyclic musks. Many nitromusks have been banned from cosmetic and personal care use in the EU, but polycyclic musks remain largely unregulated.

In the US, nearly all musk chemicals are unregulated and exposure limits have not been created.

16.3 Surfactants

Surfactants encompass a wide range of organic chemicals added to soaps, dishwashing liquids, laundry detergents, and shampoos. They are also used in industrial applications as constituents in lubricants, hydraulic fracturing fluids, and fluids used in the petroleum industry (e.g. cleaning oil spills). Surfactants enter aquatic ecosystems when they are washed down the drain and not entirely removed by wastewater treatment systems. Alkylphenols are a widely used class of surfactants of particular concern and are being found consistently in urban runoff and wastewater streams. They are highly toxic to aquatic organisms and have been shown to possess estrogenic properties in human cell-line experiments.

16.4 Plasticizers

Plasticizers are low-molecular-weight organic compounds used ubiquitously in the production of plastics. The most studied, most abundant, and perhaps most well-known plasticizers that are also emerging contaminants are bisphenol A (BPA) and its derivative bisphenol A diglycidyl ether (BADGE). According to the *Handbook of Environmental Chemistry* (2008), BPA and BADGE are found in 'metal food and drink cans, plastic baby bottles, pacifiers, baby toys, dental sealants, computers, cell phones, hard plastic water bottles (such as Nalgene), paints, adhesives, enamels, varnishes, CDs and DVDs, and certain microwaveable or reusable food and drink containers.' BPA and BADGE enter the environment when they are leached from food and drink containers or when they are released from aging plastic. BPA has endocrine-disrupting effects such as altered prostate and uterus development, reduced sperm count, and early puberty. Other health effects include prostate tumor proliferation, reduced immune response, and alterations in brain chemistry and behavior (Smital, 2008). BADGE enters the environment in a similar way to BPA, although the toxicology of BADGE is less studied. BADGE has been shown to be a potential carcinogen in rodents and an effector of pregnancy ability and sex ratio in rabbits. Human data on these categories are not yet available (Smital, 2008). Phthalates are another group of plasticizers found in clear food wrap, detergents, soaps, and pesticides that are an emerging cause of concern. They have been shown to cause developmental toxicity in animals, particularly in males (Smital, 2008).

16.5 Fire Retardants

Brominated fire retardants added to various types of furniture, vehicle upholstery, and electronic products constitute an emerging contaminant. Brominated fire retardants, particularly polybrominated diphenyl ethers (PBDEs), are classified as an emerging contaminant because they are extremely stable in the environment and because of increasing health concerns. PBDEs are known to bioaccumulate in fat tissue due to their lipophilicity and are also linked to adverse health effects in animals such as decreased sperm count, fetal malformations, and hormone disruption (Smital, 2008).

16.6 Biological Emerging Contaminants

Both healthy and unhealthy humans have high viral loads in their feces and in their urine, contributing to high viral loads in municipal sewage. These viruses are only partially eliminated by wastewater treatment systems and are now being considered an emerging contaminant. Some of the most prominent emergent viruses being found in water supplies are human polyomaviruses, hepatitis E virus, and human adenoviruses (Smital, 2008).

16.7 Odor as a Potential Health Issue

Consistent with the idea that emerging contaminants may be established pollutants with newly demonstrated modes of action, a discussion is warranted regarding the role of chemical odors in the occurrence of adverse health effects. Environmental odor sources include fertilizer factories, pesticide operations, industrial and hazardous waste sites, sanitary landfills, paper mills, petroleum refineries, foundries, chemical (plastics, adhesives, solvents) manufacturing factories, livestock feed lots, sewage treatment plants, composting and biomass operations (Shiffman, 2005). While odors have typically been treated as nuisances or indicators of potential risk, rather than hazards in and of themselves, scientific studies demonstrate that odors themselves may indeed cause health effects.

Relatively consistent patterns of negative health effects have been found among individuals who live near environmental odor sources; common symptoms reported among individuals living near agricultural and industrial odor sources include irritation of the eye, nose, and throat, dizziness, nausea, loss of appetite, headache, increased blood pressure, wheezing, development of asthmatic conditions, and exacerbation of existing asthma. Odors have also been reported to cause emotional responses, such as depression, anxiety, and aggression. Even small concentrations of odor can trigger physical and emotional symptoms in people with special sensitivities (Liu, 2007; Nimmermark, 2004; Schiffman, 2005; Shusterman, 1992, 2001; Stein, 1957).

Odor, along with noise, crowding, and heat, is considered an environmental stressor, and it has been shown that odors trigger negative health effects even in the absence of toxicological irritation. For instance, highly odorous compounds like hydrogen sulfide (which can be detected at concentrations as low as 3 ppb, or

0.004 mg/m^3), found in industrial and hazardous waste materials, have been reported to cause symptoms at levels many orders of magnitude lower than those that are known to cause symptoms by toxicologic and irritative mechanisms. In these cases, the acute odor-triggered symptoms can be explained by a variety of pathophysiologic mechanisms, including exacerbation of underlying medical conditions, innate odor aversions, aversive conditioning, pheromonal reactions, and stress-induced illness (Shusterman, 1992, 2001; Smeets, 2005; Stein 1958). Long-term exposure to stressors can ultimately lead to heart and blood vessel diseases together with other disorders on account of a weakened immune system (Nimmermark, 2004).

Multiple authoritative bodies recognize the health harms associated with odor. In a Federal Register promulgating national emission standards for hazardous air pollutants in the pulp and paper production source category, the US EPA stated, 'Surveys of odor pollution caused by pulp mills have supported a link between odor and health symptoms such as headaches, watery eyes, nasal problems, and breathing difficulties.' Furthermore, 'Odorant stimulants of the nasal receptors that are associated with TRS emissions have been associated with marked respiratory and cardiovascular responses .... There are numerous anecdotal reports of adverse reactions related to odors associated with TRS, including headaches, shortness of breath, nasal irritation, and in some cases, nausea and sinus congestion' (US EPA, 1998).

The Agency for Toxic Substances and Disease Registry (ATSDR) has also noted that odors can cause negative health effects, stating, 'In public health practice, odors are not only warning signs of potential health effects, but may in themselves be the direct cause of some symptoms in humans' (ATSDR, 2009). In a document describing the environmental triggers of asthma, ATSDR explains that noxious odors associated with hazardous air pollution can cause exacerbation of asthma symptoms (ATSDR, 2007). Furthermore, ATSDR states that people with the following conditions have increased sensitivity to odors: asthma, chronic obstructive pulmonary disease (COPD), depression, and hypersensitivity. Ultimately, ATSDR points out that odor can cause negative effects to numerous sites of the body, including mucous membranes, upper and lower respiratory airways, heart and blood vessels, stomach and intestines, brain, as well as psychological and general well-being (ATSDR, 2009).

16.8 Future Research

A successful approach to solving the problem of emerging contaminants will have to be highly interdisciplinary. There exist huge information gaps in our understanding of emerging contaminants that must be filled in by chemical, biological, and environmental researchers. Toxicologists will play an extremely important role in determining how these compounds affect our health and the health of the environment. Additional research must be done into how production of and exposure to these compounds can be prevented. Finally, it is important that the public become aware of the rising problem of emerging contaminants so that pressure can be placed on regulatory organizations such as the EPA to manage this problem.

References

Allanou, R., Hansen, B. G., & Van Der Bilt, Y. (2003, June). Public availability of data on EU high production volume chemicals – Part 1. *Chemistry Today*.

ATSDR (2007). *Environmental triggers of asthma*. Agency for Toxic Substances and Disease Registry. Retrieved October 17, 2007 from <http://www.atsdr.cdc.gov/csem/asthma/>.

ATSDR (2009). *The Health Effects of Environmental Odors*. Agency for Toxic Substances and Disease Registry. Available at: <http://www.atsdr.cdc.gov/sites/intel/docs/Intel%20Package%20-%20Odors%20Health%20Effects%20Generic%20for%20Communities.pdf>.

Derksen, J. G. M., Rijs, G. B. J., & Jongbloed, R. H. (2004). Diffuse pollution of surface water by pharmaceutical products. *Water Science and Technology*, *49*(3), 213–221.

Kolpin, D., Furlong, E. T., Meyer, M. T., Thurman, E. M., Zaugg, S. D., Barber, L. B., et al. (2002). Pharmaceuticals, hormones, and other organic wastewater contaminants in U.S. Streams, 1999–2000: A national reconnaissance. *Environmental Science and Technology*, *36*(2), 1202–1211.

Liu, H. L., Lam, L. T., Wang, T., Hou, C. C., Zeng, Q., Feng, L. H., et al. (2007). The health effects of odor emission from domestic renovation on household residents in Tianjin, China. *Journal of Environmental Health*, *69*(7), 46–50.

Nimmermark, S. (2004). Odour influence on well-being and health with specific focus on animal production emissions. *Annals of Agricultural and Environmental Medicine*, *11*, 163–173.

Schiffman, S. S., & Williams, C. M. (2005). Science of odor as a potential health issue. *Journal of Environmental Quality*, *34*, 129–138.

Shusterman, D. (1992). Critical review: The health significance of environmental odor pollution. *Archives of Environmental Health*, *47*(1), 76–87.

Shusterman, D. (2001). Odor-associated health complaints: Competing explanatory models. *Chemical Senses*, *26*, 339–343.

Smeets, M. A. M., & Dalton, P. H. (2005). Evaluating the human response to chemicals: Odor, irritation and non-sensory factors. *Environmental Toxicology and Pharmacology*, *19*, 581–588.

Smital, T. (2008). Acute and chronic effects of emerging contaminants. *Handbook of Environmental Chemistry*, *5*(S/1), 105–142.

Stein, M, & Ottenberg, P. (1958). Role of odors in asthma. *Psychosomatic Medicine*, 61–65.

Terzić, S., Senta, I., Ahel, M., Gros, M., Petrović, M., Barcelo, D., et al. (2008). Occurrence and fate of emerging wastewater contaminants in Western Balkan Region. *The Science of the Total Environment*, *399*(1–3), 66–77.

US EPA (1998, April 15). National emission standards for hazardous air pollutants for source category: Pulp and paper production; Effluent limitations guidelines, Pretreatment standards, and New source performance standards: Pulp, paper, and paperboard category; Environmental protection agency; final rules. *Federal Register*, *63*(72).

Yan, S., Subramanian, S. B., Tyagi, R. D., Surampalli, R. Y., & Zhang, T. C. (2010). Emerging contaminants of environmental concern: Source, transport, fate, and treatment. *Pract. Periodical of Haz., Toxic, and Radioactive Waste Mgmt. 14*(1), 2–20.

17 Mercury, BPA, and Pesticides in Food

Chapter Contents

Today, food safety is an issue of global health concern, and contamination may occur due to pollution of the air, water, and soil where food may be cultivated, as well as through the use of pesticides (WHO, 2010). As an EPA requirement, food products from the agricultural industry must meet governmental food safety standards before entering the marketplace (EPA, 2008). Governmental agencies have the responsibility to protect residents from the health risks of eating contaminated fish and wildlife, to establish maximum pesticide residue levels, and to ensure that food manufacturers maintain accurate records (EPA, 2008). However, toxins in food remain a large issue today as seen by the prevalence of fish food advisories and illnesses due to food contamination. Toxic chemicals found in food that are currently of the utmost concern include mercury, bisphenol A, pesticide residues, dioxins, PCBs, and PAHs. Human exposure to these chemicals at toxic levels has been linked to numerous health effects including cancers, cardiovascular diseases, kidney and liver dysfunction, hormonal imbalances, reproductive disorders, and birth defects.

Risks of Hazardous Wastes. DOI: 10.1016/B978-1-4377-7842-7.00017-9

For many chemicals, food consumption is the main pathway of human exposure. This chapter provides an overview of the most prevalent issues concerned.

17.1 Mercury

17.1.1 Mercury in the Environment

Mercury is an element that is found naturally in the environment, existing in several forms: elemental or metallic mercury, inorganic mercury compounds, and organic mercury compounds (EPA, 2010c). Mercury enters the environment as the result of the normal degradation of minerals in rocks and soil from exposure to wind, water, and volcanic activity (EPA, 2010c). Human activities have also resulted in significant releases of mercury to the environment. Monosson (2007) reports that 80% of the mercury released from human activities is released to the air from fossil fuel combustion, mining and smelting, and from solid waste burning. Another 15% is released to the soil from fertilizers, fungicides, and municipal solid waste while the remaining 5% is released from industrial wastewater to water in the environment. Other means by which mercury is released into the environment include burning hazardous wastes, producing chlorine, breaking mercury products, spilling mercury, and the improper treatment and disposal of products or wastes (EPA, 2010c). Currently the largest single source of mercury emissions to the air is coal-fired power plants (EPA, 2010c). The main concern with the volatilization of mercury is that the substance is then able to settle into water or onto land where it can be washed into water and thereby contaminate food sources.

17.1.2 Occurrence in Food

In aquatic ecosystems, environmental bacteria transform mercury from its inorganic form into a highly toxic form of organic mercury called methylmercury (USGS, 2009). The methylmercury-containing bacteria are then consumed by the next higher-level species in the food chain. Animals generally accrue methylmercury faster than they remove it, so organisms consume higher and higher concentrations of the toxic substance at each consecutive level of the food chain, which results in an accumulation effect (USGS, 2009). Small environmental concentrations of methylmercury can therefore easily build up to toxic concentrations in fish, fish-eating wildlife, and people (USGS, 2009). Additionally studies indicate that consumption of ocean fish and shellfish accounts for over 90% of human methylmercury exposure (USGS, 2009).

Methylmercury builds up more in some types of fish and shellfish than others depending on factors such as diet, lifespan, and place in the food chain (USGS, 2009). Accordingly large predatory fish such as shark, swordfish, king mackerel, and tilefish contain higher amounts of methylmercury than do other species (Sunderland, 2009). Tables 17.1 and 17.2 document mercury levels in various commercial fish and shellfish species, as published by the US Food and Drug Administration (2009). As

Table 17.1 Fish and Shellfish with Lower Levels of Mercury (FDA, 2009)

Species	Mercury Concentration (ppm)					Number of Samples
	Mean	Median	Standard Deviation	Min	Max	
Anchovies	0.043	N/A	N/A	ND	0.340	40
Clam	ND	ND	ND	ND	ND	6
Cod	0.095	0.087	0.080	ND	0.420	39
Crab	0.060	0.030	0.112	ND	0.610	63
Crawfish	0.033	0.035	0.012	ND	0.051	44
Herring	0.044	N/A	N/A	ND	0.135	38
Lobster	0.09	0.14	N/A	ND	0.27	9
Oyster	0.013	ND	0.042	ND	0.250	38
Salmon (fresh/frozen)	0.014	ND	0.041	ND	0.190	34
Shrimp	ND	ND	ND	ND	0.050	24
Tilapia	0.010	ND	0.023	ND	0.070	9
Tuna (canned light)	0.118	0.075	0.119	ND	0.852	347

Table 17.2 Fish and Shellfish with the Highest Levels of Mercury (FDA, 2009)

Species	Mercury Concentration (ppm)					Number of Samples
	Mean	Median	Standard Deviation	Min	Max	
Mackerel King	0.730	N/A	N/A	0.230	1.670	213
Shark	0.988	0.830	0.631	ND	4.540	351
Swordfish	0.976	0.860	0.510	ND	3.220	618
Tilefish	1.450	N/A	N/A	0.650	3.730	60

the table indicates, tilefish has the highest mean mercury concentration of 1.450 ppm, while a smaller species like tilapia has a value of only 0.010 ppm.

17.1.3 Toxicity and Health Effects

Most methylmercury is absorbed by the gastrointestinal tract, after which it enters the bloodstream and disperses throughout the body (Counter et al., 2004). Methylmercury in blood and tissues can be transported across the blood–brain barrier and accumulate in the brain, or cross the placenta, accumulating in fetal blood, tissues, and brain (Counter et al., 2004). In the fetal brain, methylmercury affects neuronal migration and cell division (Castolidi et al., 2001; Clarkson, 1993). High levels of exposure can result in developmental effects such as hyperactive reflexes, deafness, blindness, cerebral palsy, mental retardation, and general paralysis (Amin-Zaki et al., 1974; NRC, 2000). At low levels of exposure, neuro developmental defects include language, learning, and visual-spatial organizational deficits as well as deficits in fine motor skills (NRC, 2000).

Methylmercury concentration levels in humans can vary according to amount and type of fish consumption as well as geography. Since fish consumption has gained popularity as a health food, pregnant women and women of childbearing age have increased their consumption of fish, resulting in an increase in blood mercury concentrations (Kim et al., 2006). Methylmercury concentration levels are higher for those in island areas where fish consumption is higher than for residents in the mainland US, despite equal levels of mercury in fish of both areas (Myers et al., 2003). Women living in coastal areas in the northeastern United States had three to four times the risk of exceeding acceptable blood mercury concentration levels than women who lived in inland areas (US News and World Report, 2009). Furthermore, women of Asian, Native American, Alaskan, Pacific Island, and Caribbean descent were also more likely to have elevated mercury levels (US News and World Report, 2009).

Methylmercury toxicity was first discovered in the 1950s in Minamata, Japan, where consumption of fish containing high levels of methylmercury by pregnant women resulted in at least 30 cases of cerebral palsy in children (Harada, 1968). Similar reports of methylmercury toxicity were found in Niigata, Japan, in 1965 and Iraq in 1972 (Bakir et al., 1973; Myers et al., 2003). Although deaths and significant defects resulting from long-term high-exposure occurrence as in these events are now uncommon, chronic low-level methylmercury exposure from fish consumption is a concern because of the evidence that low-level exposure produces subtle neurodevelopmental disabilities (Counter et al., 2003).

In 2000, the National Academy of Sciences (NAS) and National Research Council (NRC) reviewed health studies on methylmercury. Using NRC quantitative analyses, the Environmental Protection Agency (EPA) established the methylmercury reference dose (RfD), an 'estimate of a daily exposure of the humans that is likely to not increase risk of harmful effects during a life span' (EPA, 2002). The reference dose is based on measures of mercury in cord blood and is the level 'assumed to be without considerable harm, most specifically neurological effects' (EPA, 2002). Methylmercury exposure closer to or less than the reference dose could establish a low-level risk that the EPA deems not significant (EPA, 2005). Some environmental groups claim that there is no guaranteed safe level of exposure to toxins like mercury as the risks may not be completely removed until exposure to the toxin itself is eliminated (Clearwater, 1997).

17.1.4 Regulating Mercury Exposure

Regulations were first issued in the 1990s to control mercury emissions from waste incineration with Congress limiting the use of mercury in batteries and the EPA placing regulatory limits on the use of mercury in paint (EPA, 2010b). The passage of the Mercury Export Ban Act in 2008 included strict regulations on the storage, importing, and exporting of mercury (EPA, 2010b). Today, efforts have been applied towards controlling the largest sources of mercury emissions such as waste incinerators and coal-fired utilities in order to reduce methylmercury exposures.

Since the passage of the 1990 Clean Air Act (CAA) Amendments in 2005, air mercury emissions have declined more than 58% (EPA, 2010b). In 2005, the Clear Air Act Section 112(n) Revision Rule and the Clean Air Mercury Rule (CAMR) were enacted to regulate mercury emissions from coal-fired power plants. However, in 2008, the DC Circuit Court of Appeals repealed both pieces of legislation. In 2010, the EPA issued regulations requiring reductions of mercury emissions from cement plants, and proposed that gold mine ore processing and production should be subject to regulation under the hazardous air pollutant section of the Clean Air Act (CAA) due to their mercury emissions (EPA, 2010b). Currently the EPA is developing air toxics emissions standards for power plants and cement plants plans to finalize air toxic standards for coal- and oil-fired electric generating units by the end of 2011 (EPA, 2010b).

Methylmercury regulations also have taken the form of advisories regarding consumer consumption of high methylmercury fish species. According to the EPA, the increasing number of advisories is due to the increased monitoring and testing of previously untested waters rather than increased levels of environmental mercury contamination (EPA, 2009). Currently, 50 states, one territory, and three tribes have fish consumption advisories for mercury spanning 16.8 million lake acres and 1.3 million river miles (EPA, 2009). Twenty-seven states have issued advisories for mercury in all freshwater lakes and rivers, 13 states have statewide advisories for mercury in their coastal waters, and one state has a deep sea advisory (EPA, 2009). While varying among states, the action levels of mercury in fish that prompt an advisory generally range from 0.1 to 0.3 parts per million (ppm) (EPA, 2009).

EPA action levels are based on 'protecting the health of humans and wildlife' while the FDA action level 'targets the average adult consumer and balances risks to human health with economic and other considerations' (Oceana, 2008). In 1979, the FDA set the action level for total mercury content in fish at 1 ppm, which is currently considered the safe maximum limit (PBS, 2005). In 1984, the FDA shifted its focus from total mercury to methylmercury concentrations (PBS, 2005).

17.1.5 Limiting Mercury Exposure in the Diet

Although fish represents a relatively common food source, mercury contamination remains a serious threat and can result in a multitude of health disorders if consumption is not properly moderated. National fish consumption guidelines as well as local advisories on safety of fish caught in local waters are available and easily accessible by the public. For example, the US Food and Drug Administration (FDA) and the EPA (2004) recommend that children and pregnant women refrain from consuming certain types of fish (particularly shark, swordfish, king mackerel, and tilefish) containing high levels of mercury. Instead, it is advised to eat shrimp, canned light tuna, salmon, and catfish which tend to contain lower mercury content. Per week, the FDA and EPA suggest eating only up to 6 ounces of fish from local waters.

The Institute of Medicine (IOM) of the National Academies developed expert committee report recommendations to direct consumers in making seafood selections.

The guidelines for fish consumption are most stringent for pregnant women and children under age 12, which advise avoiding predatory fish and eating only two 3-ounce (cooked) servings of fish per week. Additionally the IOM advises that individuals who consume more than two servings of fish per week choose a variety of types of seafood to reduce risk of exposure from a single source (IOM, 2007). At the end of the report the IOM notes that the risk of methylmercury depends on various factors such as type, size, and amount of species consumed, where the species came from, and age of species that may not be accounted for (IOM, 2007).

17.2 Bisphenol A

17.2.1 Overview

Bisphenol A (BPA) is a chemical used in large volumes for the production of polycarbonate plastics and the epoxy resin lining of metal-based food and beverage cans. Used to add strength and resilience, BPA is also utilized in products such as toys, water supply pipes, and medical tubing (Braun et al., 2009). At high temperature, acidic, and basic conditions, polymers made from BPA can be hydrolyzed which causes BPA to leach into nearby materials. As a result, commonly used items containing BPA, like plastic water bottles and baby bottles, may face leaching and contamination issues. In 2003, over 6 billion pounds of BPA were manufactured worldwide, and the EPA estimates that each year over 1 million pounds of BPA are released into the environment (Braun et al., 2009; EPA, 2010a).

Due to the pervasive use of BPA-containing plastics in developed countries, over 90% of the US population has exhibited detectable levels of urinary BPA (Braun et al., 2009). Additionally it has been discovered in the urine of pregnant women, as well as newborn infants. BPA is a contaminant of utmost concern because it is an endocrine disruptor that may also affect neurodevelopment in children (Braun et al., 2009). The primary source of BPA exposure is through the diet, which can occur through direct contact or by exposure to food or drink that has been in contact with a material containing BPA (NTP, 2010).

17.2.2 Toxicity and Health Effects

Concerns about BPA and its health risks have heightened, with consumers increasingly demanding BPA-free plastic products. BPA behaves like estrogen in the human body and binds to the same receptors as do natural female hormones, which can 'promote human breast cancer cell growth as well as decrease sperm count in rats' (Biello, 2008). In various animal studies, BPA affected reproductive and developmental systems, which raises questions about the potential impact on children's health and the environment (EPA, 2010a). Low-dose studies have suggested that sensitive aquatic organisms could be at risk for adverse effects (EPA, 2010a). In addition to affecting the growth of reproductive organs during development, studies also suggest that BPA can increase the risk of infertility, cardiovascular problems, diabetes, and breast and prostate type cancers (Layton and Lee, 2008; Adams, 2010).

In 2007, the National Institute of Health's National Toxicology Program (NIH NTP) Center for Evaluation of Risks to Human Reproduction completed a review of BPA and reported 'concern for effects on the brain, behavior, and prostate gland in fetuses, infants, and children at current human exposures' (NTP, 2007). In addition, in 2008, Canada deemed BPA a hazardous substance and became the first country to ban its use in baby bottles. In January 2010, the FDA agreed with the NTP review findings and raised concerns about the potential effects of BPA on the developing neurological system in infants and children (FDA, 2010). The FDA supports the reduction of BPA exposure, recommending changes in industry and changes in consumer food preparation methods, such as the stopping of production of infant bottles made with BPA or developing alternatives to linings of canned infant formula and other foods (FDA, 2010).

Risk of BPA toxicity depends on factors such as daily intake and the individual's dose response. Scientists are currently still debating whether or not a rapid metabolism affects BPA risk (Ginsberg and Rice, 2009). Nonetheless, serious risks posed by this chemical can be avoided by reducing contact with BPA. The FDA and NTP recommend limiting exposure by *avoiding* the following: microwaving polycarbonate plastic containers, using plastic containers with #7 on the bottom, washing polycarbonate plastic containers in the dishwasher, and consumption of canned foods. Additionally it is advised to choose glass, porcelain, or stainless steel containers, and to use infant bottles and toys that are BPA-free (NPT, 2010).

17.3 Pesticide Residues

17.3.1 Overview

Beginning in 1962 with Rachel Carson's groundbreaking book *Silent Spring*, the American public has been aware of, and increasingly concerned about the possibility and hazards of pesticide exposure. While 80% of US citizens live in urban or suburban areas (UN, 2007) removed from agriculture and direct pesticide exposure, the specter of pesticides traveling in foods from fields to the dinner tables has concerned the public. Since 1984, several fundamental changes have occurred in both government regulation and public purchasing. Pesticide residue monitoring and regulation are primarily handled by three governmental organizations: the US Environmental Protection Agency (US EPA), the US Department of Agriculture (USDA) and the Food and Drug Administration (FDA). The enhancement of government surveillance and monitoring and the passing of the Food Quality Protection Act in 1996 have all contributed to increased understanding of pesticide deposition, transport, and fate in food.

17.3.2 Monitoring and Surveillance

The US EPA registers pesticides, sets pesticide maximum residue limits (MRLs) for raw, unprocessed commodities, and researches the environmental fate of residues (NRC, 1993). The maximum residue limits, or tolerances, are established for consumer foods and beverages, which are then used by the USDA to enforce pesticide levels in meat, poultry, and some egg products and the FDA for enforcement in other commodities. New tolerances and changes to old tolerances (e.g. for pesticide re-registration) are published in the Federal Register as required in the Food Quality Protection Act (FQPA). The FQPA (1996) required the EPA by 2006 to review the 9,700 tolerances approved before the Act was passed, and determine potential risks posed by pesticides to humans and the environment.

Additionally, the US Department of Agriculture monitors the incidence of pesticide contamination of uncooked food in the United States within its Pesticide Data Program (PDP). The PDP data are primarily used by the EPA to establish tolerances for pesticide residue levels found in food, and the PDP is required by law to focus on products frequently consumed by infants and children (USDA, 2008). This database supports pesticide residue data on 85 different commodities tested for 440 different pesticides in ten different states (USDA, 2008). In their most recent report (2008), the PDP tested 13,381 samples of food and water including (in descending order) fruit and vegetables, corn grain, drinking water, honey, catfish, potable groundwater, almonds, and rice. Of the food samples analyzed in the 2008 report (excluding catfish), 70% contained either a pesticide or metabolite. Out of that 70%, 46% contained two or more pesticides. Of the groundwater sampled, 44% contained trace levels of pesticide. Of all the food tested, 0.5% of the samples violated tolerance levels set by the EPA (USDA, 2008).

The FDA enforces EPA pesticide tolerance levels in foods (except in meat and poultry), and monitors pesticide residues in national and international foods in domestic markets. Unlike the PDP, which washes commodities to be sampled (to emulate grocery store conditions), the FDA's Pesticide Monitoring Program primarily examines raw, unwashed, unpeeled agricultural product. The foods are tested and if pesticide levels on the commodity surpass either the EPA's tolerance levels, or the FDA's enforcement levels, the FDA will act to remove the product from the market (FDA, 2007). In the FDA's most recently available Pesticide Monitoring Summary (2007), they reported that 42% of the tested foods contained pesticide residue. In the same report, 2.3% of domestic foods tested had residue levels that violated federal regulations.

The Total Diet Study, also known as the market basket study, has also been conducted by the FDA annually since 1961, and it presents data on food that is 'table-ready.' The testing of prepared food can produce more realistic results in that it can account for secondary contaminants which can occur during cooking. Pesticides are among the many chemicals tested for in the report. In the most recent available market basket study of pesticides (FDA, 2003), 4,747 of 8,402 food samples, or 56%, tested positive for pesticide residue.

Congress unanimously passed the Food Quality Protection Act (FQPA) in 1996 to address pesticide contamination in food directly. The FQPA grants the EPA the power to (1) set pesticide residue limits in food and (2) control the registration and re-evaluation of new and existing pesticides. The law required the EPA to set tolerance levels that are based on health effects. Other highlights of the FQPA include: special considerations for infants and children when setting tolerances, required testing for endocrine disruptors and implementing pesticide re-registrations with the EPA.

17.3.3 The Dirty Dozen

Between 2000 and 2008, the Food and Drug Administration and US Department of Agriculture conducted 89,000 pesticide-contamination tests on the most consumed 50 fruits and vegetables in the United States. From these studies, a Washington think-tank, the Environmental Working Group, synthesized the data and published their findings as the 'Shopper's Guide to Pesticides.' The group found that certain produce absorbed and retained pesticides more effectively than others, even after peeling and/or washing. Of the 50 fruits and vegetables tested, 12 appeared to retain pesticides effectively enough to be dubbed 'the Dirty Dozen.'

The *Dozen* include seven fruits (peaches, strawberries, apples, domestic blueberries, nectarines, cherries and imported grapes) and five vegetables (celery, sweet bell peppers, spinach, kale/collard greens, and potatoes). Celery, the worst of the *Dozen*, was shown to have pesticides in 95% of 2,953 tests, with an average of 3.79 different pesticides found per single sample. These 12 fruits and vegetables have softer, more absorbent skin than other produce resulting in more pesticide pickup (Dellorto, 2010). By purchasing the *Dozen* from organic sources rather than ones that use pesticides, the group claims that an individual's pesticide intake can be reduced by four-fifths (from an average exposure of ten pesticides per day, to two pesticides per day) (EWG, 2010).

Conversely, the EWG also established a *Clean Fifteen* grouping of fruits and vegetables – those that can be treated with pesticides, but which are safe to eat after washing and peeling. These include: onions, avocado, sweet corn, pineapple, mangos, sweet peas, asparagus, kiwi, cabbage, eggplant, cantaloupe, watermelon, grapefruit, sweet potato, honeydew, melon. The tough outer skin fortifies these 15 from pesticide absorption.

17.3.4 Exposure and Health Risks

Exposures of consumers to residual pesticides can occur in both long-term (chronic) exposures and short-term (acute) exposures. Though high-level exposures are more common as occupational hazards for agricultural workers, acute exposures often occur at the residual, consumer level. There are four main ways that pesticide outbreaks can start: contamination of foodstuff during transport or storage; ingestion of seed dressed for sowing; use of pesticides in food preparation; and presence of pesticides due to misuse during harvest, misuse of containers, or excessive application (Ferrer and Cabral, 1995). Contamination occurs when the commodity is placed in an area that once contained pesticides, or near an area housing pesticides.

All produce varies in the pesticide load it carries to market. Even produce grown in the same field, or transported in the same container, can contain more or fewer residuals than its cohorts. In one review examining variability factors (Hamilton et al., 2004), the authors found that deposition of pesticides during crop application was the most significant contributor to pesticide level variability in grocery store produce. For example, pesticide levels on the bottom two-thirds of apple tree foliage were found to be three times higher than levels on the top of the foliage after treatment with pesticides. In another example described by Hamilton et al. (2004), apples in transit to grocery stores were treated with a post-harvest pesticide. It was found that the apples at the bottom of the crate contained more of the pesticide than those at the top of the crate.

A textbook example of acute exposure to consumers can be found in the pesticide Aldicarb, a substance used to protect cucumbers, watermelons, bananas, potatoes, and other crops from nematodes and insects. From 1978 to 1988, there were three food poisoning outbreaks, where all instances reported were due to produce purchased at the consumer level (e.g. at a grocery store or warehouse). Residual levels of the pesticide in the contaminated produce ranged from 0.3 ppm to 10.7 ppm, but were high enough to induce vomiting, diarrhea, dizziness, disorientation, and other symptoms within roughly one hour of consumption.

Food processing generally results in the substantial decrease of residual pesticide levels. Washing, blanching, and canning are three typical methods used by processors to prepare a food for market. In a report presented by the National Resource Council, these three processes reduced malation levels by 99%, and carbaryl levels by 99% in treated tomatoes. In green beans, levels of malathion and carbaryl were reduced by 94% and 73%, respectively.

17.4 Dioxins, PCBs, and PAHs

Dioxins, polychlorinated biphenyls (PCBs), and polyaromatic hydrocarbons (PAHs) are other toxic chemicals that may reach human populations through food exposure. A detailed discussion of these constituents can be found in Chapter 15.

Acknowledgment

Katherine Phillips, who conttributed to this chapter, is a Registered Dietitian in Long Beach, California. She is currently pursuing a Masters in Public Health at the University of California Los Angeles.

References

Adams, J. (2010). Bisphenol A and its potential health risks. *Los Angeles Times*. Retrieved August 24 from <http://articles.latimes.com/2010/aug/09/health/la-he-closer-look-receipts-20100809>.

Amin-Zaki, L., Elhassani, S., & Majeed, M. A. (1974). Intra-uterine methylmercury poisoning in Iraq. *Pediatrics*, *54*(5), 587–595.

Bakir, F., Damluji, S. F., Amin-Zaki, A., Murtadha, A. K., Al-Rawi, S., Tikriti, S., et al. (1973). Methylmercury poisoning in Iraq – an interuniversity report. *Science*, *181*, 230–241.

Biello, D. (2008). Plastic (Not) Fantastic: Food Containers Leach a Potentially Harmful Chemical. *Scientific American*. Retrieved February 19 from <http://www.scientificamerican.com/article.cfm?id=plastic-not-fantastic-with-bisphenol-a>.

Braun, J., Yolton, K., Dietrich, K., Hornung, R., Ye, X., Calafat, A., et al. (2009). Prenatal bisphenol a exposure and early childhood behavior. *Environmental Health Perspectives*, *117*(12), 1945–1952.

Clarkson, T. W. (1993). Mercury: Major issues in environmental health. *Environmental Health Perspectives*, *100*, 31–38.

Clearwater (Husdon River Sloop Clearwater) (1997). *Fact sheet 6: PCB contamination of the Hudson – A health hazard*. Retrieved August 20, 2010 from <http://www.clearwater.org/news/hazard.html>.

Committee on Pesticides in the Diets of Infants and Children, National Research Council (NRC) (1993). *Pesticides in the diets of infants and children*. Washington D.C.: National Academy Press.

Counter, S., & Buchanan, L. (2004). Mercury exposure in children: A review. *Toxicology and Pharmacology*, *198*, 209–230.

Environmental Working Group (EWG's) (2010). *Shopper's guide to pesticides*. Retrieved July 2010 from <http://www.foodnews.org/>.

Ferrer, A., & Cabral, R. (1995). Recent epidemics of poisoning by pesticides. *Toxicology Letters*, 55–63.

Ginsberg, G., & Rice, D. (2009). Does rapid metabolism ensure negligible risk from bisphenol A? *Environmental Health Perspectives*, *117*(11), 1639–1643.

Hamilton, D., Ambrus, A., Dieterle, R., Felsot, A., Harris, C., Petersen, B., et al. (2004). Pesticide residues in food – Acute dietary exposure. *Pest Management Science*, *60*, 311–338.

Harada, Y. (1968). *Infantile Minamata disease. Minamata Disease* (*Study Group of Minamata Disease, Kumamoto University ed.*) pp. 73–91. Kumamoto: Kumamoto University.

Institute of Medicine (IOM), National Academy of Science (NAS) (2007). Seafood Choices: Balancing Benefits and Risks. Executive Summary. Retrieved August 20, 2010 from <http://books.nap.edu/catalog/11762.html>.

Kim, E., Kim, I., & Kwon, J. (2006). The effect of fish consumption on blood mercury levels of pregnant women. *Yonsei Medical Journal*, *47*(5), 626–633.

Layton, L., & Lee, C. (2008). *Canada bans BPA from baby bottles. Washington Post*. Retrieved April 19 from <http://www.washingtonpost.com/wp-dyn/content/article/2008/04/18/AR2008041803036.html>.

Monosson, E. (2007). Mercury. *Encyclopedia of earth*. Retrieved August 14, 2010 from <http://www.eoearth.org/article/Mercury>.

Myers, G., Davidson, P., Cox, C., et al. (2003). Prenatal methylmercury exposure from ocean fish consumption in the Seychelles child development study. *The Lancet*, *361*(9370), 1686–1692.

National Research Council (NRC) (2000). Reassessment of TCDD and Related Compounds. Board on Environmental Studies and Toxicology – Division on Earth and Life Studies. 1–240. Retrieved August 12, 2010 from <http://www.nap.edu/catalog.php?record_id=11688>.

National Toxicology Program (NTP) (2007). Center for the Evaluation of Risks to Human Reproduction (CERHR). US Department of Health and Human Services. Expert Panel

Report on the Reproductive and Developmental Toxicity of Bisphenol A. Retrieved August 22, 2010 from <http://cerhr.niehs.nih.gov/evals/bisphenol/BPAFinalEPVF112607.pdf>.

National Toxicology Program (NTP) (2010). National Institute of Health (NIH). Bisphenol A (BPA). Retrieved August 22, 2010 from <http://www.niehs.nih.gov/health/docs/bpa-factsheet.pdf>.

Public Broadcasting System (2005). *Mercury in fish*. Retrieved August 27, 2010 from <http://www.pbs.org/now/science/mercuryinfish.html>.

United Nations (UN) (2007). World Urbanization Prospects: The 2007 Revision. Department of Economic and Social Affairs. Retrieved July from <http://www.un.org/esa/population/publications/wup2007/2007WUP_Highlights_web.pdf>.

US Department of Agriculture (USDA) (2008). Quick Facts about the Pesticide Data Program (PDP). USDA – AMS Pesticide Data Program. Retrieved July from <http://www.ams.usda.gov/AMSv1.0/getfile?dDocName = STELDEV3003252>.

US Environmental Protection Agency (EPA) (2002). Integrated Risk Information System (IRIS). Methylmercury. Oral RfD and inhalation RfC assessments last revised 7/27/2001. Carcinogenicity assessment last revised 5/1/1995. US Environmental Protection Agency, Office of Research and Development, National Center for Environmental Assessment. Retrieved August 14, 2010 from <http://www.epa.gov/iris/subst/0073.htm>.

US Environmental Protection Agency (EPA) and the Food and Drug Administration (FDA) (2004). *What you need to know about mercury in fish and shellfish: Advice for women who might become pregnant, women who are pregnant, nursing mothers, and young children*. Retrieved August 14 from <http://www.fda.gov/Food/FoodSafety/ProductSpecificInformation/Seafood/FoodbornePathogensContaminants/Methylmercury/ucm115662.htm>.

US Environmental Protection Agency (EPA) (2005). Regulatory Impact Analysis of the Clean Air Mercury Rule. Final report. EPA-452/R-05-003. Research Triangle Park, NC: US Environmental Protection Agency, Office of Air Quality Planning and Standards, Air Quality Strategies and Standards Division. Retrieved August 14, 2010 from <http://www.epa.gov/ttn/atw/utility/ria_final.pdf>.

US Environmental Protection Agency (EPA) (2008). *Food safety*. Retrieved September 21, 2010 from <http://www.epa.gov/agriculture/tfsy.html>.

US Environmental Protection Agency (EPA) (2009). 2008 Biennial National Listing of Fish Advisories. EPA 823-F-09-007. Washington, DC: US Environmental Protection Agency, Office of Water. Retrieved August 14, 2010 from <http://www.epa.gov/waterscience/fish/advisories/tech2008.pdf>.

US Environmental Protection Agency (EPA) (2010a). *Bisphenol A (BPA) action plan summary*. Retrieved September 21, 2010 from <http://www.epa.gov/oppt/existingchemicals/pubs/actionplans/bpa.html>.

US Environmental Protection Agency (EPA) (2010b, April). *Guidance for implementing the January 2001 methylmercury water quality criterion*. Retrieved August 11, 2010 from <http://water.epa.gov/scitech/swguidance/waterquality/standards/criteria/aqlife/pollutants/methylmercury/index.cfm>.

US Environmental Protection Agency (EPA) (2010c). *Mercury*. Retrieved August 14, 2010 from <http://www.epa.gov/mercury/about.htm>.

US Food and Drug Administration (FDA) (2007). Pesticide Monitoring Program FY 2007. Food Safety. Retrieved July 2010 from <http://www.fda.gov/Food/FoodSafety/

FoodContaminantsAdulteration/Pesticides/ResidueMonitoringReports/ucm169577.htm#tdsM>.

US Food and Drug Administration (FDA) (2009). *Mercury levels in commercial fish and shellfish.* Retrieved September 22, 2010 from <http://www.fda.gov/Food/FoodSafety/Product-SpecificInformation/Seafood/FoodbornePathogensContaminants/Methylmercury/ucm115644.htm>.

US Food and Drug Administration (FDA) (2010, January). *Update on bisphenol A for use in food contact applications.* Retrieved August 22, 2010 from <http://www.fda.gov/NewsEvents/PublicHealthFocus/ucm197739.htm>.

United States Geological Survey (USGS) (2009). *Factsheet 146-00: Mercury in the environment.* Retrieved August 14, 2010 from <http://www.usgs.gov/themes/factsheet/146-00/>.

US News and World Report (2009). *Women in coastal areas have higher mercury levels.* Retrieved August 27, 2010 from <http://health.usnews.com/health-news/family-health/womens-health/articles/2009/01/19/women-in-coastal-areas-have-higher-mercury-levels.html>.

World Health Organization (WHO) (2010). *Chemical risks in food.* Retrieved August 14, 2010 from <http://www.who.int/foodsafety/chem/en/>.

18 Childhood Exposure to Environmental Toxins

Chapter Contents

Alan Greene, M.D. states in his book *Raising Baby Green* that 'if a chemical is "out there," it may also be "in there,"' meaning that any harmful chemical found in our environment may also enter our children's bodies (Greene, 2007). He cites the preliminary results of a 2005 study conducted by the Environmental Health Group in which umbilical cord blood of ten babies born between August and September 2004 was examined for toxins. The findings were dramatic: each baby's cord blood showed an average of 200 chemicals, which included fire retardants, mercury, and pesticides. A total of 287 different chemicals was detected, 180 of which are known to cause cancer, and 217 of which are known to be neurotoxins (EWG, 2005). The factors involved in fetal and childhood chemical exposures are discussed in this chapter, along with discussions of the implications of early-life exposure for health.

18.1 What Makes Children Vulnerable

Environmental toxins can affect children even before birth as low-molecular-weight toxins, including carbon monoxide and lead, readily cross the placenta. Carbon monoxide can bind with high affinity to fetal hemoglobin resulting in reduced oxygen delivered to tissues, leading to organ damage (Visnjevac, 1986). The developing brain is particularly sensitive to environmental contaminants. Relatively low blood lead levels and prenatal exposures have been shown to lead to reductions in intellectual function and behavior disorders. Exposure to high levels of

Risks of Hazardous Wastes. DOI: 10.1016/B978-1-4377-7842-7.00018-0

methylmercury, a toxic form of organic mercury found in food, can also result in developmental disabilities (Myers, 2000).

After birth, the particular physiology and behaviors of children increase their exposures to environmental toxins. Children consume more food and water per unit of body weight, they have higher inhalation rates per unit of body weight, and they have higher surface area to volume ratios than adults, all factors that result in higher doses of environmental pollutants in the body. Children's propensity for oral exploration also exposes children to more toxins. Young children mouth objects or explore with their fingers as they navigate their environment, resulting in the ingestion of toxins in their environment at a higher rate than adults. Additionally, young children spend a significant amount of time on the floor and ground, where dust and tracked-in soil accumulates (Gurunathan, 1998).

After exposure occurs, the rapid development in children makes them more susceptible to damage from absorbed pollutants. A child's metabolism allows ready absorption of nutrients, which also implies that uptake of toxins is accelerated in children. For instance, children need greater amounts of calcium than adults due to growing bones, translating to increased absorption of calcium, and other chemicals, from food sources (US EPA, 1997a). Overall children's cellular immaturity and ongoing growth and development can lead to elevated risk from toxins (AAP, 1997a).

Children are especially susceptible to the effects of toxic exposure due to their more permeable blood–brain barrier. To illustrate, after exposure to lead, the chemical persists in blood and bone tissue, and is only slowly eliminated from the body. In adults, lead is not likely to penetrate the blood–brain barrier; however, this barrier is less developed in children, and consequently they are more susceptible to subsequent brain damage. Lead poisoning in children manifests as behavioral disturbances, learning, and concentration difficulties (WHO, 1995).

A further problem is the difficulty in assessing exposures and risks to chemicals in children. The potential for ingestion of contaminants from objects, hands, and surfaces in the environment is assessed by characterizing and quantifying children's mouthing behaviors. However, data on children's mouthing activities are extremely limited. Also, current toxicological testing of chemicals that use endpoints such as death or organ dysfunction may lead risk assessors to overlook low-level concentrations that can lead to health effects such as intellectual impairment, mood changes, behavioral and attention problems, conditions that are becoming associated with exposure to environmental toxins (Cohen-Hubal, 2000).

18.2 Breastfeeding and Transfer of Organochlorine Compounds

There is growing awareness of infants' exposure to toxins during breastfeeding. While the benefits of breastfeeding are many, and include nutritional benefits, protection against infections, promotion of brain development, decreased risk of diseases such as allergies and diabetes, there is also unambiguous data that breast milk accumulates chemicals such as organochlorine compounds, heavy metals, and volatile solvents (AAP, 1997b; Jensen, 1991).

Organochlorine compounds like polychlorinated biphenyls (PCBs), hexachlorobenzene (HCB), and 1,1,1-trichloro-2,2-bis(p-chlorophenyl) ethane (DDT) are ubiquitous in the environment and persist because of their chemical stability and the absence of metabolizing enzymes that can lead to their breakdown in mammals (Butte, 1984). Organochlorine compounds have a great affinity to fat, as well as a long half-life. Exposure to organochlorine compounds leads to the accumulation of these chemicals in adipose, contributing to the individual's body burden. Since milk is high in lipid content, these toxins readily accumulate in human breast milk. Breastfeeding can then lead to the transfer of toxins to the child. Breastfeeding has been shown to lead to a dose- and time-dependent increase in PCBs, HCB, and DTT in children's serum during the first 6 months of life (Lackmann, 2006). While pregnancy also exposes the fetus to minor amounts of organochlorine compounds in the mother's blood, a more significant exposure occurs through breastfeeding (Jensen, 1991; Jacobson, 1990).

The implications of these exposures are often understated, because concentrations of chemicals in breast milk are unlikely to cause acute toxic symptoms. It remains to be seen whether the exposures to organochlorine compounds in nursing infants today will have serious effects on our children in the future. For the time being, reducing exposure to organochlorine compounds among expectant mothers should be encouraged. The available data suggest that the banning of organohalogen production and use has resulted in reductions in their levels in breast milk over time (Smith, 1999).

18.3 Children in Agricultural Areas

Children living in agricultural areas are especially vulnerable to pesticide exposure. Members of the family who work on farms may track pesticides into their homes. Young children may then be exposed to pesticides as a result of hand-to-mouth contact in contaminated areas. Studies have shown that young children can be exposed to pesticides by touching floors, surfaces, and other contaminated objects (Eskenazi, 1999; Gurunathan, 1998; Nishioka, 1999). Children living in agricultural areas may also play in nearby fields with pesticide contamination. In addition, it has also been shown that babies can be exposed to pesticides via the consumption of contaminated breast milk from their farm worker mother (Visnjevac, 1986).

18.4 The Effects of Air Pollution on Children's Respiratory Health

Children's exposure to air pollution is a special concern because their immune systems and lungs are not fully developed at birth, and full functionality of their lungs is not apparent until at least 6 years of age. To illustrate, the number of alveoli in the human lung increases from 24 million at birth to 257 million by

4 years of age (Dunnill, 1962). Growth of the lung during early childhood means a greater permeability of the lung's epithelial layer. Children also have a larger lung surface area per body weight compared to adults and breathe more air per body weight relative to adults. The child's immune system, which is immature at birth, also develops at the same time as the child's lungs. All of these factors contribute to children's susceptibility to respiratory illnesses.

The increased incidence of respiratory illnesses, such as asthma, allergies, bronchitis, and respiratory infections, among children has garnered attention in recent years. Asthma prevalence approximately doubled between 1980 and 1995 in the United States (Woodruff, 2000, 2003). The tendency to develop asthma can be inherited, but genetic factors are unlikely to explain the significant increases that have occurred in the past 20 years (NAS, 2000). The cause of these increases may be due at least in part to air pollution. There is some evidence that indoor air pollutants such as nitrogen dioxide, pesticides, plasticizers, and volatile organic pollutants may play a role in asthma (NAS, 2000). A number of recent studies have suggested that chronic exposure to ozone may be associated with the development of asthma in children, and chronic exposure to particulate matter may affect lung development, growth, and function (Avol, 2001; McConnell, 2002; Gauderman, 2002). One study found that exposure to hazardous air pollutants is linked to increases in chronic respiratory symptoms characteristic of asthma (Ware, 1993). There are also several studies that link short-term changes in air pollution to changes in pulmonary health of children (Berry, 1991; Kinney, 1989; Spektor, 1991) and others that link long-term exposure of children to air pollution with significant decreases in lung function (ACS, 2007; Schwartz, 2004).

18.5 Childhood Cancers and the Link to Environmental Toxins

Although childhood cancers are relatively rare, there has been some increase in the incidence of children diagnosed with all forms of invasive cancer, from 11.5 cases per 100,000 children in 1975 to 14.8 per 100,000 children in 2004 (NCI, 2008).

Childhood cancer is not only less frequent than adult cancer, it is also mechanistically different. Adult cancers are thought to occur because of years of cumulative damage to the cells. In children, this kind of long-term damage has not had a chance to take place. Consequently, the most common adult cancers – prostate, breast, lung, and colon cancers – are not found in children. Instead, children tend to get leukemias, brain tumors, and cancers of the blood and connective tissues. Environmental causes of childhood cancer have long been suspected but have been difficult to pin down because childhood cancer is generally uncommon, and exposure levels during important periods of developmental activity are difficult to measure retrospectively. Moreover, the different types of childhood cancers develop in a variety of ways, with a potentially wide diversity of causes. Genetically predisposed children may form a subpopulation that is sensitive to environmental toxins.

There are some well-established environmental causes of childhood cancer, including pharmaceuticals such as diethylstilbestrol (DES), an estrogen prescribed to prevent miscarriage, chemotherapeutic agents, and ionizing radiation (Gouveia-Vigeant, 2003). Increasing evidence indicates that parental and childhood exposures to certain toxic chemicals which include solvents, petrochemicals, pesticides, and specific industrial by-products (dioxins and polycyclic aromatic hydrocarbons) can play a role in cancer during childhood and perhaps years down the line (Pogoda, 1997).

Acknowledgment

This chapter was contributed by Petra Luber. Petra is an MPH candidate at the University of California School of Public Health.

References

American Academy of Pediatrics (AAP) (1997a). *Child health issues for the second session of the 106th congress*. Washington, DC: Department of Federal Affairs. Available at: <http://www.aap.org/advocacy/washing/chieh/htm>.

American Academy of Pediatrics (AAP) (1997b). Breastfeeding and the use of human milk. Work group on breastfeeding. *Pediatrics*, *100*(1997), 1035–1039.

American Cancer Society (ACS) (2007). *Cancer facts and figures 2007*. Atlanta, GA: American Cancer Society. Retrieved December 26, 2007, from <http://www.cancer.org/downloads/STT/CAFF2007PWSecured.pdf>.

Avol, E. L., Gauderman, W. J., Tan, S. M., London, S. J., & Peters, J. M. (2001). Respiratory effects of relocating to areas of differing air pollution levels. *American Journal of Respiratory and Critical Care Medicine*, *164*, 2067–2072. [Abstract/Free Full Text]

Berry, M., Lioy, P. J., Gelperin, K., Buckler, G., & Klotz, J. (1991). Accumulated exposure to ozone and measurement of health effects in children and counselors at two summer camps. *Environmental Research*, *54*, 135–150.

Butte, N. F., Garza, C., Smith, E. O., et al. (1984). Human milk intake and growth in exclusively breast-fed infants. *Journal of Pediatrics*, *104*, 187–195.

Cohen-Hubal, E. A., Sheldon, L. S., Burke, J. M., McCurdy, T. R., Berry, M. R., Rigas, M. L., et al. (2000a). Children's exposure assessment: A review of factors influencing children's exposure, and the data available to characterize and assess that exposure. *Environmental Health Perspectives*, *108*(6), 475–486.

Dunnill, M. S. (1962). *Thorax*, *17*, 329 cited in: Altman PL, Dittmer DS. Respiration and Circulation. Bethesda, MD: Federation of American Societies for Experimental Biology; 1971.

Eskenazi, B., Bradman, A., & Castriona, R. (1999). Exposure of children to organophosphate pesticides and their potential adverse health effects. *Environmental Health Perspectives*, *107*(Suppl. 3), 409–419.

EWG (2005). *Body Burden – The pollution in newborns – a benchmark investigation of industrial chemicals, pollutants and pesticides in umbilical cord blood*. Retrieved July 14, 2005 from <www.ewg.org/reports/bodyburden2/execsumm.php>.

Gauderman, W. J., Gilliland, G. F., Vora, H., et al. (2002). Association between air pollution and lung function growth in southern California children: Results from a second cohort. *American Journal of Respiratory and Critical Care Medicine*, *166*, 76–84.

Gouveia-Vigeant, T., & Tickner, J. (2003). Toxic chemicals and childhood cancer: A review of the evidence Lowell Center for Sustainable Production: sustainableproduction.org

Greene (2007). *Raising Baby Green: The Earth-Friendly Guide to Pregnancy, Childbirth, and Baby Care*. Alan Greene with Jeanette Pavini, Theresa Foy DiGeronimo. ISBN: 978-0-7879-9622-2

Gurunathan, S., Robson, M., Freeman, N., et al. (1998). Accumulation of chloropyrifos on residential surfaces and toys accessible to children. *Environmental Health Perspectives*, *106*(1), 9–16.

Jacobson, J. L., Jacobson, S. W., & Humphrey, H. E. B (1990). Effects of in utero exposure to polychlorinated biphenyls and related contaminants on cognitive functioning in young children. *Journal of Pediatrics*, *116*, 38–45.

Jensen, A. A. (1991). Transfer of chemical contaminants into human milk. In A. A. Jensen, & S. A. Sorach (Eds.), *Chemical contaminants in human milk* (pp. 9–19). Boca Raton (FL): CRC Press.

Kinney, P. L., Ware, J. H., Spengler, J. D., Dockery, D. W., Speizer, F. E., & Ferris, B. G., Jr. (1989). Short-term pulmonary function change in association with ozone levels. *The American Review of Respiratory Disease*, *139*, 56–61.

Lackmann, G. M. (2006). Human milk, environmental toxins and pollution of our infants: Disturbing findings during the first six months of life. *International Journal of Biomedical Science*, *2*(2), 178–183.

McConnell, R., Berhane, K., Gilliland, F., et al. (2002). Asthma in exercising children exposed to ozone: A cohort study. *Lancet*, *359*, 386–391. [CrossRef][Web of Science] [Medline].

Myers, G. J., & Davidson, P. W. (2000). Does methylmercury have a role in causing developmental disabilities in children? *Environmental Health Perspectives*, *108*(3), 413–419.

NAS (2000). *National Academy of Sciences. Clearing the air: Asthma and indoor air exposures*. Washington, DC: National Academy Press.

NCI (2008). *National Cancer Institute fact sheet on childhood cancer*. <http://www.cancer.gov/cancertopics/factsheet/Sites-Types/childhood>.

Nishioka, M. G., Burkholder, H. M., Brinkman, M. C., et al. (1999). Distribution of 2,4-dihlorophenoxyacetic acid in floor dust throughout homes following homeowner and commercial lawn application: Quantitative effects of children, pets, and shoes. *Environmental Science & Technology*, *33*, 1359–1365.

Pogoda, J. M., & Preston-Martin, S. (1997, November). Household pesticides and risk of pediatric brain tumors. *Environmental Health Perspectives*, *105*(11), 1214–1220.

Schwartz, J. (2004, April). Air pollution and children's health. *Pediatrics*, *113*(4 Suppl), 1037–1043.

Smith, D. (1999). Worldwide trends in DDT levels in human breast milk. *International Journal of Epidemiology*, *28*, 179–188.

Spektor, D. M., Mao, J., He, D., Thurston, G. D., Hayes, C., & Lippmann, M. (1991). Effects of single and multi day ozone exposures on respiratory function in active normal children. *Environmental Research*, *55*, 107–122.

US EPA (1997a). *Office of Research and Development strategy for research on risks to children*. Washington, DC: Office of Research and Development, Science Council Review Draft.

Visnjevac, V., & Mikov, M. (1986). Smoking and carboxyhaemoglobin concentrations in mothers and their newborn infants. *Human Toxicology*, *5*, 175–177.

Ware, J. H., Spengler, J. D., Neas, L. M., et al. (1993). Respiratory and irritant health effects of ambient volatile organic compounds. The Kanawha County Health Study. *American Journal of Epidemiology*, *137*, 1287 –1301.

WHO (1995). *Lead. Environmental health criteria* (Vol. 165) Geneva: World Health Organization.

Woodruff, T., Axelrad, D., & Kyle, A. (2000). *America's children and the environment: A first view of available measures*. Washington, DC: US Environmental Protection Agency (Report No. EPA 240-R-00-006).

Woodruff, T., Axelrad, D., Kyle, A., Nweke, O., & Miller, G. (2003). *America's children and the environment: Measures of contaminants, body burdens, and illnesses*. Washington, DC: US Environmental Protection Agency (Report No. EPA 240-R-03-001). Available at: <http://www.epa.gov/envirohealth/children>.

19 Health Care Costs and Corporate Accountability

Chapter Contents

Pollution-related health care results in large payments by insurance companies. About 30,000 hospital admissions and emergency room visits over a period of 2 years in California could have been prevented if federal clean air standards had been met (Romley et al., 2010). Fewer hospital admissions and emergency room visits translate to lower expenses for insurance companies and public health care payers such as Medicaid and Medicare. Between 2005 and 2007, private health insurers spent about $55 million, whereas Medicaid and Medicare spent about $27 million and $100 million, respectively, on air pollution-related hospital care in California (Romley et al., 2010). What this means is that the potential exists for health care payers to spend substantially less on preventable medical visits and pharmaceuticals through the reduction of pollutants (Romley et al., 2010). In the meantime, efforts are being made to recover health care costs from the entities responsible for the pollution.

19.1 Corporate Accountability to Medicare and Medicaid for Health Care Costs

19.1.1 Medicare Secondary Payer Act

Corporate entities, and not taxpayers, are starting to be held accountable for injuries caused by toxic chemical mishandling that result in short- and long-term health effects. In the past, medical care for toxic exposure rendered to the elderly and

Risks of Hazardous Wastes. DOI: 10.1016/B978-1-4377-7842-7.00019-2

disabled has been paid by taxpayers through Medicare, health insurance for the elderly and disabled. Despite attempts at corporate accountability through the judicial process, taxpayers have been footing the bill. As an effort to reduce health care costs, the Medicare Secondary Payer (MSP) Act was enacted in 1980. It ensures that Medicare does not pay for services and items that the corporate entities releasing the pollutants have the responsibility for paying (Department of Health and Human Services, 2010). Additionally, MSP penalizes parties who fail to report liability settlements involving Medicare beneficiaries. Parties that do not do so will be forced to at least reimburse funds that were wrongly paid. Additionally, providers, physicians, and other suppliers can be fined up to $2,000 for 'knowingly, willfully, and repeatedly providing inaccurate information relating to the existence of other health insurance or coverage' (Department of Health and Human Services, 2010). Through MSP provisions, Medicare saves approximately $6 billion every year on claims filed by non-Medicare insurances that hold primary responsibility for payment (Department of Health and Human Services, 2010). The MSP program helps ensure that Medicare does not need to pay for claims corporate entities are responsible for and that parties who violate it will be penalized.

19.1.2 Recovery Audit Contractor Program

Medicare also utilizes the Recovery Audit Contractor (RAC) program for the recovery of improper payments. The goal of the RAC program is to recover money that never should have been paid by Medicare. RACs help enforce the MSP through the use of data mining software on filed claim data and through complex audits, ensuring that there is no unnecessary Medicare spending (Huntoon, 2009). New initiatives supported by President Obama will help apply current laws that make it mandatory to report settlements involving Medicare beneficiaries to Medicare for reimbursement for medical care rendered as a result of the toxic exposure. According to a recent Presidential Memorandum, the President declared that reclaiming funds associated with improper payments is an important component of taxpayer dollar protection and helps cut waste and fraud (The White House, 2010). Through the RAC program, contractors are given incentives to discover these improper payments and to conduct audits (Huntoon, 2009). Outside auditors also increase transparency and encourage accountability.

19.2 Case Study – US Government Civil Suit Against Monsanto

19.2.1 Introduction to the Lawsuit

Monsanto, a chemical manufacturing corporation based in St. Louis, produced polychlorinated biphenyls (PCBs) at its Anniston Plant in Alabama from the 1930s to the 1970s. PCBs are now banned due to concerns over various health effects, such as skin ailments, cancer, diabetes, learning disorders, and immune deficiencies (ATSDR, 2001; Centers, 2003). After the original lawsuit concerning PCB exposure settled in 2001, the government filed a lawsuit regarding the distribution of associated Medicare and Medicaid funds.

The lawsuit was initiated by the government because of the failure to notify and reimburse Medicare after the settlement of the first case. The PCB exposure case had involved many Anniston residents, some of whom received medical care paid by Medicare. The residents received $300 million to resolve claims, and the companies agreed to a $700 million settlement to resolve outstanding cases in the initial lawsuit regarding toxic chemicals dumped around Anniston (O'Reilly, 2009; SLBJ, 2009).

US v. Stricker was filed in December 2009 by the Department of Justice under MSP. The purpose of the suit was to recover funds paid by Medicare for medical care rendered to 907 residents who had claimed to be injured by PCBs (Reeves, 2010). Additionally, the government sought reimbursement for double damages on payments made on behalf of the residents involved in the Anniston PCB exposure using Medicare (Korris, 2010; O'Reilly, 2009).

Several defendants were named by the government, including Monsanto. Solutia was sued as well, as it had taken over Monsanto's plant in Anniston in 1997 after spinning off the main company (O'Reilly, 2009). Pharmacia and Pfizer were sued as companies acting as subsidiary or successor companies, respectively (Centers, 2003; SLBJ, 2009). Additionally, Travelers Indemnity Company and American International Group, Inc., the liability insurance companies involved, and the attorneys representing the residents were sued (O'Reilly, 2009; SLBJ, 2009).

19.2.2 The Complaint

The federal government complained that the two liability insurance carriers, AIG and Travelers Indemnity, neglected to determine whether any of the residents involved in the toxic exposure lawsuit received medical care paid by Medicare (O'Reilly, 2009). Under the Code of Federal Regulations, Monsanto, Solutia, and Pharmacia were required to reimburse the US and Medicare for conditional Medicare payments made on behalf of settlement claimants (O'Reilly, 2009).

19.2.3 Current Status of the Case

US v. Stricker continues to be litigated and has not yet been resolved. Depending on the outcome of the case, insurance companies can potentially be held liable to Medicare for health care costs of toxic exposure caused by corporate entities that they insure. The case has the potential to set precedence for MSP cases in the future. Although the United States has previously brought actions against claimants and their attorneys to recover conditional payments, this is the first case in which the government has attempted to enforce the MSP Act against the attorneys, law firms, companies, and insurance carriers (Reeves, 2010). With the current administration's initiatives, combined with MSP, the government may be on the path to reducing Medicare payments.

Acknowledgment

This work was contributed by Adrienne M. Gundry, JD, RN. Adrienne is a clinical measurement analyst at Cedars-Sinai Medical Center in Los Angeles, and is an MPH candidate at the University of California Los Angeles.

References

Agency for Toxic Substances & Disease Registry (ATSDR) (2001). *ToxFAQs for polychlorinate biphenyls (PCBs).* Retrieved February from <http://www.atsdr.cdc.gov/tfacts17.html#bookmark05>.

Centers, J. (2003). Deal Reached in PCB Trials. Anniston Star. Retrieved August 21 from <http://www.mindfully.org/Industry/2003/Monsanto-PCB-Litigation21aug03.htm>.

Department of Health and Human Services (2010). Medicare Secondary Payer Fact Sheet, ICN 006903. Centers for Medicare and Medicaid Services (CMS). Retrieved August from <https://www.cms.gov/MLNProducts/downloads/MSP_Fact_Sheet.pdf>.

Huntoon, L. R. (2009). Physicians beware: Medicare RAC attacks coming. *Journal of American Physicians and Surgeons*, *14*(1), 2–4.

Korris, S. (2010). Medicare Kicked Off Injury Suit Recovery in '09. Legal Newsline. Retrieved June 17 from <http://www.legalnewsline.com/news/227595-medicare-kicked-off-injury-suit-recovery-in-09?utm_source=feedburner&utm_medium=feed&utm_campaign=Feed%3A+LegalNewsLine+%28LegalNewsLine.com%29>.

O'Reilly, C. (2009). Monsanto, Law Firms, Sued for Shortchanging Medicare (Update 1). Bloomberg. Retrieved December 18 from <http://www.bloomberg.com/apps/news?pid=newsarchive&sid=a22CNdFILyww>.

Reeves, J. (2010). *Feds sue over $300M PCBs settlement in Alabama.* Associated Press. Retrieved March 2 from <http://finance.yahoo.com/news/Feds-sue-over-300M-PCBs-apf-2473067180.html?x=0>.

Romley, J. A., Hackbarth, A., & Goldman, D. P. (2010). *The impact of air quality on hospital spending*. Santa Monica, CA: RAND Corporation.

SLBJ (2009). Solutia, Monsanto, Pfizer sued over Medicare payments. *St. Louis Business Journal.* Retrieved December 21 from <http://stlouis.bizjournals.com/stlouis/stories/2009/12/21/daily7.html>.

The White House (2010, March). *Presidential memorandum regarding finding and recapturing improper payments.* Washington, DC: Office of the Press Secretary.

20 Health and Safety Standards

Chapter Contents

Air contaminant standards have been established since the 1970s to protect the safety, health, and welfare of people who work in areas that may have hazardous airborne contaminants. These standards are intended to protect not only workers but also family members, customers, and nearby communities. Several federal agencies, including the Occupational Safety and Health Administration (OSHA), the National Institute for Occupational Safety and Health (NIOSH), the American Conference of Governmental Industrial Hygienists (ACGIH), the Agency for Toxic Substances and Disease Registry (ATSDR), the Environmental Protection Agency (EPA), and the World Health Organization (WHO), have developed recommended health and safety standards to protect the worker and the surrounding community from air contaminants.

20.1 Occupational Safety and Health Administration

The United States Occupational Safety and Health Administration (OSHA) is an agency of the United States Department of Labor signed into existence by President Richard Nixon in 1970 under the Occupational Safety and Health Act. The intent of the organization is to enforce standards for the workplace to prevent work-related injuries, illnesses, and deaths.

Standards for approximately 500 chemicals have been designated by OSHA and can be found in Tables Z-1, Z-2, and Z-3 of the OSHA General Industry Air Contaminants Standard, reproduced in Appendices A–C of this book. Exposure to any of the substances listed in the tables is to be limited in accordance with the

Risks of Hazardous Wastes. DOI: 10.1016/B978-1-4377-7842-7.00020-9

requirements listed in the tables. OSHA sets enforceable permissible exposure limits (PELs) for airborne chemicals to protect workers against the health effects of exposure to hazardous substances. PELs are regulatory limits on the amount or concentration of a substance in the air. The regulatory standards are not limited to inhalation exposure but also include dermal exposure.

Appendix A provides PELs for an extensive list of chemicals. Most OSHA PELs are based on an 8-hour time-weighted average (TWA) (OSHA, 2006). As listed in Appendix B, a number of substances have an 'acceptable ceiling concentration,' PEL ceiling values that must not be exceeded at anytime within 8 hours, except for a time period not exceeding the maximum duration and concentration allowed indicated in the 'acceptable maximum peak above the acceptable ceiling concentration for an eight-hour shift.'

For example, suppose substance X has a 15 ppm TWA, 30 ppm ceiling, and 60 ppm peak. An employee is exposed to a concentration of this substance above 40 ppm but never above 60 ppm for only a maximum period of 10 minutes. For the remainder of the 8-hour shift, exposure cannot exceed 15 ppm so that the cumulative exposure does not exceed the weighted average of 15 ppm.

The cumulative exposure for an 8-hour work shift is computed as follows:

$$E = (C_a T_a + C_b T_b + \ldots C_n T_n)/8$$

where, E is the equivalent exposure for the working shift. C is the concentration during any period of time T where the concentration remains constant. T is the duration in hours of the exposure at the concentration C. The value of E shall not exceed the 8-hour time-weighted average specified in Subpart Z or 29 CFR Part 1910 for the substance involved.

To illustrate the use of the formula above, assume that Substance Z has an 8-hour time-weighted average limit of 100 ppm. Suppose an employee is subject to the following exposure:

- Two hours exposure at 150 ppm
- Two hours exposure at 75 ppm
- Four hours exposure at 50 ppm.

Substitution into the formula gives:

$$(2 \times 150 + 2 \times 75 + 4 \times 50) \div 8 = 81.25 \text{ ppm}$$

Since 81.25 ppm is less than 100 ppm, the 8-hour time-weighted average limit, the exposure is acceptable.

OSHA has also established Immediately Dangerous to Life and Health (IDLH) standards. IDLH is the concentration of an airborne chemical which may cause irreversible health effects or death. OSHA standards are broad enough to include oxygen-deficient circumstances that are free from air contaminants (Appendix C). Usually IDLH values are used to determine selection of proper breathing apparatus that must be available to workers in a given situation (NIOSH, 1994).

Exposure limits have been used by industrial hygienists for over 50 years as a means of protecting worker health. By implementing these standards as a part of an occupational health and safety program, they are a primary tool in disease prevention.

20.2 National Institute for Occupational Safety and Health

The National Institute for Occupational Safety and Health (NIOSH) is the United States federal agency responsible for conducting research and making recommendations for the prevention of work-related injury and illness. NIOSH is part of the Centers for Disease Control and Prevention (CDC) within the US Department of Health and Human Services.

NIOSH published the *Pocket Guide to Chemical Hazards* as a source of general industrial hygiene information for workers, employers, and occupational health professionals. The *Pocket Guide* presents information for 677 chemicals or substance groupings (e.g. manganese compounds, tellurium compounds, inorganic tin compounds, etc.) that are found in the work environment. The chemicals or substances include all substances for which the National Institute for Occupational Safety and Health (NIOSH) has recommended exposure limits (RELs) and recommended short-term exposure limits (STEL) (NIOSH, 2005). These chemical standards have been compiled in Appendix D.

For NIOSH RELs, 'TWA' indicates a time-weighted average concentration for up to a 10-hour work day during a 40-hour work week. A short-term exposure limit (STEL) is designated by 'ST' preceding the value; unless noted otherwise, the STEL is a 15-minute TWA exposure that should not be exceeded at any time during a workday. A ceiling REL is designated by 'C' preceding the value; unless noted otherwise, the ceiling value should not be exceeded at any time.

Acting under the authority of the Occupational Safety and Health Act of 1970 (29 USC Chapter 15) and the Federal Mine Safety and Health Act of 1977 (30 USC Chapter 22), NIOSH develops and periodically revises recommended exposure limits (RELs) for hazardous substances or conditions in the workplace. NIOSH also recommends appropriate preventive measures to reduce or eliminate the adverse health and safety effects of these hazards. To formulate these recommendations, NIOSH evaluates all known and available medical, biological, engineering, chemical, trade, and other information relevant to the hazard. These recommendations are then published and transmitted to OSHA and the Mine Safety and Health Administration (MSHA) for use in promulgating legal standards.

In 1974, NIOSH joined OSHA in developing an immediately dangerous to life and health standards (IDLH) for substances with existing PELs. This joint effort was labeled the Standards Completion Program (SCP) and involved the cooperative efforts of several contractors and personnel from various divisions within NIOSH and OSHA. The SCP developed 380 substance-specific draft IDLH standards with supporting documentation that contained technical information.

The purpose for establishing an IDLH value in the Standards Completion Program was to determine the airborne concentration from which a worker could escape without injury or irreversible health effects from an IDLH exposure in the event of the failure of personal protection equipment. The IDLH was considered a maximum concentration above which only a highly reliable breathing apparatus providing maximum worker protection should be permitted. In determining IDLH values, NIOSH considered the ability of a worker to escape without loss of life or irreversible health effects along with certain transient effects, such as severe eye or respiratory irritation, disorientation, and loss of coordination, which could prevent escape. As a safety margin, IDLH values are based on effects that might occur as a consequence of a 30-minute exposure (NIOSH, 2005). IDLH values can be found in Appendix E and Appendix G.

Three-hundred and eighty-seven substances were originally included in the SCP. IDLHs were not specifically determined for all of them. The published data at that time for 40 of these substances (e.g. DDT and triphenyl phosphate) showed no evidence that an acute exposure to high concentrations would impede escape or cause any irreversible health effects following a 30-minute exposure and the designation 'NO EVIDENCE' was used in the listing of IDLHs. For all of these substances, respirators were selected on the basis of assigned protection factors. For some (e.g. copper fume and tetryl), an assigned protection factor of 2,000 times the PEL was arbitrarily used to determine the concentration above which only the 'most protective' respirators were permitted. However, for most particulate substances for which evidence for establishing an IDLH did not exist (e.g. ferbam and oil mist), the use of an assigned protection factor of 2,000 would have resulted in the assignment of respirators at concentrations that were not likely to be encountered in the occupational environment. In addition, exposure concentrations greater than 500 times the PEL for many airborne particulates could result in exposures that would hamper vision. Therefore, it was decided as part of the SCP (and during the review and revision of the IDLHs) that for such particulate substances, only the 'most protective' respirators would be permitted for use in concentrations exceeding 500 times the PEL.

IDLHs could not be determined during the SCP for 22 substances (e.g. bromoform and calcium oxide) because of a lack of relevant toxicity data and therefore the designation 'UNKNOWN' was used in the IDLH listing. For most of these substances, the concentrations above which only the 'most protective' respirators were allowed were based arbitrarily on assigned protection factors that ranged from 10 to 2,000 times the PEL, depending on the substance. There were also ten substances (e.g. n-pentane and ethyl ether) for which it was determined only that the IDLHs were in excess of the lower explosive limits (LELs). Therefore, the LEL was selected as the IDLH with the designation 'LEL' added in the IDLH listing. For these substances, only the 'most protective' respirators were permitted above the LEL in the SCP draft technical standards. For 14 substances (e.g. beryllium and endrin), the IDLHs determined during the SCP were greater than the concentrations permitted based on assigned respiratory protection factors. In most instances the IDLHs for these substances were set at concentrations 2,000 times the PEL.

20.3 American Conference of Governmental Industrial Hygienists

ACGIH is an organization comprised of industrial hygienists and environmental health and safety experts. In 1941 the ACGIH established the Threshold Limit Values of Chemical Substances Committee. This group was charged with investigating, recommending, and annually reviewing exposure limits for chemical substances. It became a standing committee in 1944. Two years later, the organization adopted its first list of 148 exposure limits, then referred to as Maximum Allowable Concentrations. The term 'Threshold Limit Values' (TLVs) was introduced in 1956. The first *Documentation of the Threshold Limit Values* was published in 1962 and is now in its seventh edition. Today's list of TLVs includes 644 chemical substances and physical agents (Appendix F) (ACGIH, 2010).

In 1971, ACGIH developed the Hygienic Guide Series that established the IDLH values that were eventually used in the Standards Completion program. These values can be found in Appendix G.

20.4 Agency for Toxic Substances and Disease Registry

The Comprehensive Environmental Response, Compensation, and Liability Act (CERCLA) [42 U.S.C. 9604 et seq.], as amended by the Superfund Amendments and Reauthorization Act (SARA) [Pub. L. 99 499], requires that the Agency for Toxic Substances and Disease Registry (ATSDR) develop jointly with the US Environmental Protection Agency (EPA), in order of priority, a list of hazardous substances most commonly found at facilities on the CERCLA National Priorities List (NPL) (42 U.S.C. 9604(i)(2)); prepare toxicological profiles for each substance included on the priority list of hazardous substances, and to ascertain significant human exposure levels (SHELs) for hazardous substances in the environment, and the associated acute, subacute, and chronic health effects (42 U.S.C. 9604(i)(3)); and assure the initiation of a research program to fill identified data needs associated with the substances (42 U.S.C. 9604(i)(5)).

The ATSDR Minimal Risk Levels (MRLs) were developed as an initial response to the mandate. Following discussions with scientists within the Department of Health and Human Services (HHS) and the EPA, ATSDR chose to adopt a practice similar to that of the EPA's Reference Dose (RfD) and Reference Concentration (RfC) for deriving substance-specific health guidance levels for non-neoplastic endpoints. An MRL is an estimate of the daily human exposure to a hazardous substance that is likely to be without appreciable risk of adverse non-cancer health effects over a specified duration of exposure. These substance-specific estimates, which are intended to serve as screening levels, are used by ATSDR health assessors and other responders to identify contaminants and potential health effects that may be of concern at hazardous waste sites. It is important to note that MRLs are not intended to define cleanup or action levels for ATSDR or other Agencies (Appendix H).

MRLs are intended to serve as a screening tool to help public health professionals decide where to look more closely. They may also be viewed as a mechanism to identify those hazardous waste sites that are not expected to cause adverse health effects. Most MRLs contain some degree of uncertainty because of the lack of precise toxicological information on the people who might be most sensitive (e.g. infants, elderly, and nutritionally or immunologically compromised) to effects of hazardous substances. ATSDR uses a conservative (i.e. protective) approach to address these uncertainties consistent with the public health principle of prevention. Although human data are preferred, MRLs often must be based on animal studies because relevant human studies are lacking. In the absence of evidence to the contrary, ATSDR assumes that humans are more sensitive than animals to the effects of hazardous substances and that certain persons may be particularly sensitive. Thus the resulting MRL may be as much as a hundredfold below levels shown to be non-toxic in laboratory animals.

Proposed MRLs undergo a rigorous review process. They are reviewed by the Health Effects/MRL Workgroup within the Division of Toxicology and Environmental Medicine; an expert panel of external peer reviewers; the agency-wide MRL Workgroup, with participation from other federal agencies, including EPA; and are submitted for public comment through the toxicological profile public comment period. Each MRL is subject to change as new information becomes available concomitant with updating the toxicological profile of the substance. MRLs in the most recent toxicological profiles supersede previously published levels. To date, 139 inhalation MRLs, 230 oral MRLs and eight external radiation MRLs have been derived. A listing of the current published MRLs by route and duration of exposure is provided in Appendix H.

20.5 World Health Organization

The World Health Organization was established in 1948 as a specialized agency of the United Nations serving as the directing and coordinating authority for international health matters and public health. One of WHO's constitutional functions is to provide objective and reliable information and advice in the field of human health, a responsibility that it fulfills in part through its publication programs. Through its publications, the Organization seeks to support national health strategies and address the most pressing public health concerns. Recognizing the need of humans for clean air, in 1987 the WHO Regional Office for Europe published air quality guidelines for Europe, containing health risk assessments of 28 chemical air contaminants (Appendix I) (WHO, 2000).

20.6 Environmental Protection Agency

The Environmental Protection Agency (EPA) developed RSLs (Regional Screening Levels) for the Superfund/RCRA programs. RSLs are risk-based concentrations, derived from standardized equations combining exposure information assumptions with toxicity data. (Toxicity data are reproduced in Appendix K.) They are considered by the Agency to be protective for humans (including sensitive groups), over a lifetime. The RSLs have been reproduced in Appendix J (USEPA, 2010a).

When EPA Region 9 first came out with a Draft of the RSL tables in 1992, there was concern expressed by California EPA's Department of Toxic Substances and Control (DTSC) that, for some chemicals, the risk-based concentrations calculated using Cal-EPA toxicity values were 'significantly' more protective than the risk-based RSLs calculated by Region 9. At an interagency meeting comprised of mostly toxicologists, it was agreed that values that differed by a factor of four or more would be said to have 'significant' difference in risk-based RSLs. Although four was a somewhat arbitrary cutoff point, it reflects a consideration that the numbers are not very precise and at best are order-of-magnitude estimates of risk.

Cal-Modified RSLs are included for those chemicals where Cal-EPA values are 'significantly' more protective. The original list of Cal-Modified RSLs (cadmium, chromium 6, nickel, PAHs benzo(a)pyrene and benzo(k)fluoranthene, tetrachloroethylene [PCE] and lead which has been withdrawn) were based on exposure factors and modeling assumptions presented in California EPA's Preliminary Endangerment Assessment Guidance Manual (PEA 1994). Please note that any Cal-Modified RSLs that have been added to this original list after 1995 have been calculated using Cal-EPA toxicity values and Region 9 exposure methodology. In the State of California, Cal-Modified RSLs should be used as screening levels because they are more stringent than the Federal numbers (USEPA, 2010b).

20.7 EPA Sector Notebooks

The US Environmental Protection Agency has a series of profiles of certain industries. These notebooks are organized by air, land, and water pollutants. Each notebook is a comprehensive environmental profile of the specific industry. They contain industrial process information and pollutant release data. Also, contained in the notebooks are compliance and enforcement histories. Sector Notebooks were used to develop the matrix shown in Appendix L. Major chemicals relevant to each industry are marked with an X in the matrix (USEPA, 2010c).

Industry Sector Notebooks that were used include:

- Profile of the Agricultural Chemical, Pesticide and Fertilizer Industry (2000)
- Profile of the Fossil Fuel Electric Power Generation Industry (1997)
- Profile of the Iron and Steel Industry (1995)
- Profile of the Lumber and Wood Products Industry (1995)
- Profile of the Oil and Gas Extraction Industry (1999)
- Profile of the Petroleum Refining Industry (1995)
- Profile of the Pulp and Paper Industry, 2nd Edition (2002)
- Profile of the Stone, Clay, Glass and Concrete Industry (1995)
- Profile of the Textiles Industry (1997).

References

American Conference of Governmental Industrial Hygienists (ACGIH) (2010). About; <http://www.acgih.org/about/history.htm>.

National Institute for Occupational Safety and Health (NIOSH) (1994). Documentation for Immediately Dangerous to Life or Health Concentrations (IDLH), Introduction; <http://www.cdc.gov/niosh/idlh/idlhintr.html>.

National Institute for Occupational Safety and Health (NIOSH) (2005). *NIOSH Pocket Guide to Chemical Hazards*. <http://www.cdc.gov/niosh/npg/pgintrod.html#chemicalname>.

US Environmental Protection Agency (USEPA) (2010a). User's Guide and Background Technical Document for USEPA Region 9's Preliminary Remediation Goals (PRG) Table. EPA Region 9, San Francisco, CA, USA; <http://www.epa.gov/region9/superfund/prg/files/04usersguide.pdf>.

US Environmental Protection Agency (USEPA) (2010b). Regional Screening Levels (Formerly PRGs). USEPA Region 9, San Francisco, CA, USA; <http://www.epa.gov/region9/superfund/prg/index.html>.

US Environmental Protection Agency (USEPA) (2010c). Sector Notebooks. Compliance and Enforcement, USEPA; <http://www.epa.gov/compliance/resources/publications/assistance/sectors/notebooks/index.html#about>.

US Occupational Safety and Health Administration (OSHA) (2006). Regulations (Standards – 29 CFR), Part 1910: Occupational Safety and Health Standards; Subpart Z: Toxic and Hazardous Substances; Standard number 1910.1000, Air Contaminants; <http://www.osha.gov/pls/oshaweb/owadisp.show_document?p_table=STANDARDS&p_id=9991>.

World Health Organization (WHO) (2000). Air Quality Guidelines for Europe, Second Edition. Regional Office for Europe, Copenhagen; <http://www.euro.who.int/document/e71922.pdf>.

21 The Failures of Regulatory Agencies and Their Inefficiency in Introducing New Chemicals into Regulation

Chapter Contents

Safe exposure values established by US regulatory agencies are used domestically in both the private and public sectors as well as internationally by other regulators. While use of these values is widespread, many are outdated, and efforts to update and introduce new chemicals have been severely hindered by broken policies.

21.1 Outdated OSHA Values are not Safe

Since the creation of OSHA by the Occupational Safety and Health Act in 1970, fatal workplace injuries declined 50%, and workplace injury rates in the US have dropped by a similar percentage (OSHA, 2010). However, the rates are still 'substantial.' One study found that in 1992, 6,500 workers were killed, 13.2 million workers were injured and the US economy suffered a loss of $171 billion through work-related injury/death (OSHA, 2010). In 2000, OSHA launched an internal review to determine the root cause of these injuries and provide a recommendation for improving the Administration.

The National Advisory Committee on Occupational Safety and Health (NACOSH) found that the primary contributor to the large amount of workplace deaths and

Risks of Hazardous Wastes. DOI: 10.1016/B978-1-4377-7842-7.00021-0

injuries was OSHA's inadequate standards setting process. Indeed, many of OSHA's standards are outdated, and do not reflect current scientific studies. For example, OSHA's Permissible Exposure Limits for ethylbenzene and 2-butoxyethanol were established in 1974, benzene in 1987, toluene in 1989, and naphthalene in 1995 (Solomon, 2010). The large amount of time elapsed between the last updated PELs is not entirely due to lack of effort. In 1989, OSHA attempted to lower 212 existing PELs and introduce 164 PELs for previously unregulated toxic substances. Their efforts were met and overturned in the Court of Appeals decision in 1993 (*AFL-CIO v. OSHA* 965 F.2d 962 11th Cir., 1992) (OSHA, 1993). The values were reset to pre-1989 levels.

This was a major setback, especially given that many of the pre-1989 standards were in major need of updating. When OSHA was signed into existence in 1971, the agency adopted earlier limits set by the American Conference of Governmental Industrial Hygenists (ACGIH) in the 1950s and 1960s (Solomon, 2010). In effect, many of the 'modern' OSHA PELs are actually 50–60 years old.

OSHA has recognized that PELs need updating. David Michaels, the assistant secretary of labor for OSHA, affirmed at the 2010 ACGIH annual conference in Denver that, 'OSHA has not taken leadership on this issue, and we need to.' If NACOSH's earlier assessment is correct, then updating OSHA's PELs is a matter of life and death. In 2007, 5,488 US workers died from workplace-related injuries and 4.0 million reported non-fatal injuries at work (CDC, 2009; Risk and Insurance, 2010).

OSHA's sister organization, the National Institute for Occupational Safety and Health (NIOSH), was established under the CDC to provide research, information, and education on worker health. NIOSH does not set or enforce regulations for workplace contaminant levels, and therefore it has the luxury of setting standards that do not have to appease businesses or other government agencies. Unsurprisingly, NIOSH values are much lower than OSHA values (Risk and Insurance, 2010).

21.2 Problems with the EPA's IRIS

21.2.1 Overview of the EPA's Integrated Risk Information System

The EPA's database, known as IRIS (Integrated Risk Information System), is an internationally renowned source for associated chemical hazards, and is used by various regulatory agencies both domestically and abroad. The database provides four evaluations for each chemical listed: (1) hazard identification – an exposure can cause an 'increased incidence of an adverse health effect' and the 'nature and strength of the evidence of causation'; (2) quantitative dose–response assessments – (Reference Dose); (3) exposure assessment – 'intensity, frequency and duration of actual or hypothetical exposure of humans to hazard'; (4) risk characterization – finalizes 1–3 into a comprehensive utilitarian recommendation. The final risk assessment is then used to develop policy (EPA, 2010).

The IRIS system was created in 1985 for a different role than it currently holds – to 'develop consensus opinions within the agency about the health effects of chronic exposure to chemicals' (GAO, 2008). It was not intended to be used outside of the EPA, especially not for policy considerations. As mentioned before, the IRIS database is widely used and currently boasts over 9 million access queries per year. According to the GAO report (2008), 'a private sector risk-assessment expert has stated that the IRIS database has become the most important source of regulatory toxicity values for use across EPA's programs, and is also widely used across state programs and internationally.' The database's 540 chemicals are evaluated in several unique ways, including categories such as quantitative/qualitative and cancer/non-cancer. Categorizations of this type are rare and incredibly valuable in the regulatory world.

21.2.2 The Government Accountability Office's Issues with the EPA's IRIS (GAO, 2008)

The GAO released a scathing report in 2008, criticizing the EPA's methodology regarding adding chemicals to the IRIS database. The GAO report elucidates two major faults within the IRIS assessment process. First, the EPA is required to submit drafts of IRIS assessments to the Office of Management and Budget (OMB) for reviews which all too often end in rejection. Second, due to the structure of the IRIS assessment process delays can work synergistically to create a domino effect, stalling the process entirely. These two problems make the EPA's ability to process and finalize new IRIS assessments impotent.

The OMB review process, which began in 2005 under the Bush administration, is a means for outside agencies to reject EPA IRIS preliminary risk assessments if they feel financially threatened. To explain further, the IRIS database has become so essential in setting new standards for inter-agency safe contaminant levels that any new recommendations made in the IRIS database are almost guaranteed to be implemented by other agencies as the safety standard. This means that IRIS values that lower the safety standard can indirectly force other agencies to redesign filtration systems, protective clothing and equipment to meet the new standards, or begin the cleanup/remediation of older sites that do not meet the new standards. This of course will cost other agencies a great deal of money. The avoidance of these costs is the aim of the OMB review. Under the OMB review other governmental agencies such as the Department of Defense can evaluate the proposed IRIS values and if they believe they will suffer monetarily from the new threshold values, pass on approval. This recent OMB review requirement is a testament to how instrumental the IRIS values have become in setting new regulations, and how they for all intensive purposes are no longer internalized within the EPA.

This external interference of the OMB and other agencies has severely hampered the EPA in its duties. Despite its new unofficial role in setting external safety standards, the IRIS database is still desperately required by the EPA to provide assessments and data for EPA programs. For example, the Clean Air Act suffered after the OMB cut five IRIS assessments that 'for the first time addressed acute, rather than chronic exposure'; lower acute exposure standards pose a greater chance in outdating respirator and safety equipment as such used by the DOD and other agencies. Furthermore, the OMB can eliminate IRIS evaluations for esoteric reasons, undercutting EPA authority. The lack of transparency in the OMB decisions reduces the credibility of IRIS.

The IRIS system is susceptible to delays, such as those caused by OMB rejections. The effect of these delays has crippled the EPA's ability to finalize assessments, including those for dioxin, formaldehyde, trichloroethylene (TCE), naphthalene, and tetrachloroethylene (PCE). As of December 2007, there were 70 assessments in backlog. Forty-eight of these have been stalled for over 5 years, and of those, 12 have been in limbo for 9 years. Delays allow for the development of new scientific methodology and create a window for new studies and consequent discoveries. The EPA requires the incorporation of these new studies into the IRIS report. This results in a renewal of the evaluation process, causing further delays and allowing for more scientific progress to occur, causing more delays and exacerbating the cycle. The progression of scientific methodology also affects chemicals currently on the IRIS database. In 2003 the EPA disclosed that 287 of the 540 chemicals in IRIS needed to be updated with current studies. These updates entail changing exposure values and/or improving quantitative dose–response accuracy, in theory requiring more staff to compile new assessments. According to the GAO's analysis of the EPA's data, more staff do not aid the process (see Figure 21.1). Despite a four-fold increase in staff, the amount of finalized IRIS assessments actually decreased between 2003

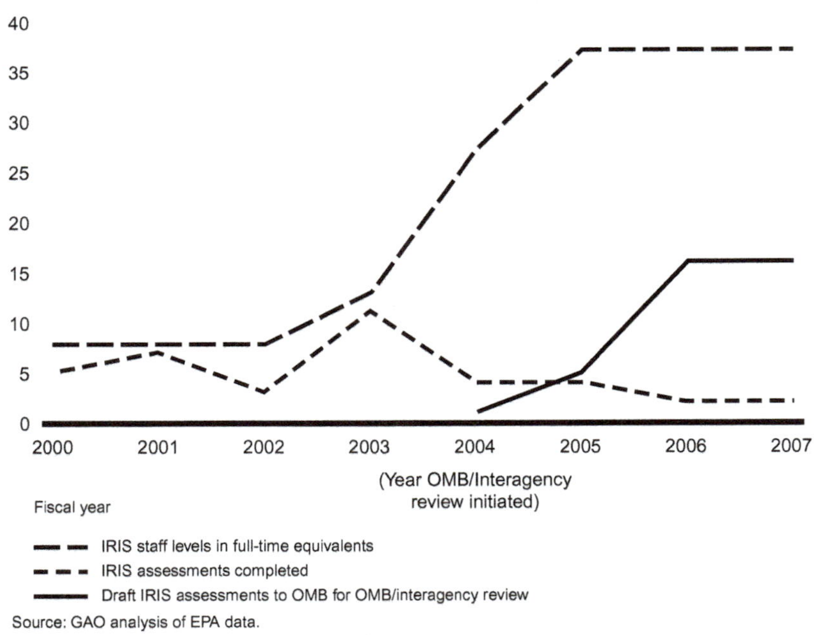

Figure 21.1 Number of completed IRIS assessments, draft assessments to OMB, and IRIS staff in full-time equivalents, fiscal years 2000–2007 (GAO, 2008).

and 2007, indicating methodology and not manpower was the limiting factor. Between the years 2006 and 2007, the EPA sent only four out of 32 IRIS assessments to completion (8%).

In a May 2009 memo, the EPA Administrator Lisa Jackson stated that the 'procedures formalized in an April 10, 2008 memorandum from the former Deputy Administrator, have reduced the transparency, timeliness and scientific integrity of the IRIS process.' The aforementioned April 10 memo supported the OMB review process. This change of procedure was due largely to the GAO report and the change of administration in the White House. In accordance with the May memo, the EPA's Office of Research and Development is currently developing a new IRIS process that will minimize external interference, only allowing input from other agencies in the form of technical comments from health scientists. The OMB's role will be significantly diminished if not eliminated.

The EPA has outlined nine new steps in creating an assessment of a chemical (EPA, 2010):

1. Federal Register announcement of EPA's IRIS agenda and call for scientific information from the public on the selected substances,
2. Search of the current scientific literature, a Federal Register announcement that the literature search is available on the IRIS internet site, and a call to submit additional scientific information on the substance,
3. Development of a draft Toxicological Review or other assessment document,
4. Internal peer consultation,
5. Internal Agency Review,
6. Science Consultation with other Federal agencies and White House offices,
7. External peer review and public comment,
8. Final internal Agency Review, Interagency Science Discussion and ORD management approval, and
9. Posting on the IRIS database.

Notice that these procedures exclude any review by the OMB.

21.3 Superfund Liability and Determination of Probable Responsible Parties

The criticism of the Superfund program centers on CERCLA's liability determination methods. Superfund liability is quite broad in its reach. Any entity that deposited any amount of contaminant at the site in question can be held liable. They are known as Probable Responsible Parties (PRPs). Superfund liability laws have fallen under criticism for largely two reasons. First, the laws (or lack thereof) at the time of disposal cannot be used to disclaim liability. In other words, if dumping large quantities of TCE into unlined pits was the legal means of disposal at the time, the disposing party can still be held accountable for cleanup costs. Even if the law at the time *required* that waste be disposed at a certain location and in a certain manner that today is considered unsafe, disposing parties can still be held liable. Critics say this practice unfairly punishes law-abiding corporations. Second, PRPs need not to have contributed the most pollution at a site to be held liable. All the EPA needs is for the party to be a contributor. The EPA uses this facet of CERCLA to act against the contributor with the most assets, not necessarily the most contaminating contributor, in order to reap the largest payout. The EPA will generally sue the wealthiest PRP, an international chemical company for instance. In turn, the chemical company will turn around and sue the other, smaller PRPs to recover the costs of cleanup. This method has been criticized because it generates excessive litigation, wasting time and money (Rahm, 1998).

Friction also arises between the EPA and municipalities when the local government has to bear the brunt of Superfund litigation. Typically this occurs when county or city landfills contaminated with household hazardous waste are designated Superfund sites. Because individual polluters are so numerous and untraceable, the city is the only party at which fingers can be pointed. As of 1997, roughly 20% of Superfund sites on the National Priorities List were county or municipal landfills, and about 30% of all Superfund sites included local governments as a PRP. Municipalities are funded by taxpayers; consequently, it is the taxpayers who will bear the brunt of municipal liability (Rahm, 1998).

21.4 Inadequacies within Hazardous Waste Legislation

21.4.1 Problems with the Toxic Substances Control Act

Congress created TSCA of 1976 to identify, analyze, and mitigate the 80,000 or so chemicals discharged into the environment by industry and manufacturing in the US (EPA, 2009b). Lawmakers designed the TSCA to address the existence and introduction of chemicals that 'present an unreasonable risk of injury to health or the environment.' Since its inception, however, the TSCA has not lived up to expectations, particularly concerning the analysis of chemicals.

The Act is a four-title document, with the bulk of the legislation residing in Title 1 – Control of Toxic Substances. Title 1 is the original document created in 1976, while Titles 2–4 were tacked on as amendments specifically to address asbestos (added 1986), radon (added 1988), and lead exposure (added 1992).

As stated in the Title 1, Section 1, the aims of the Act are three-fold: (1) to require manufacturers to provide the EPA with adequate data on their product chemicals, specifically regarding the health and environmental effects; (2) to award the USEPA with the authority to regulate these chemicals; and (3) to ensure that the USEPA does not use the authority to 'create unnecessary economic barriers to technological innovation,' while still performing its duty to verify that these innovations (chemicals) do not cause 'unreasonable risk' to the environment (TSCA). Therein lies a major fault with the Act. The TSCA places the burden of proving that a chemical causes an 'unreasonable risk' on the EPA itself, and not on the waste-producing companies. In other words, the EPA must experimentally demonstrate that the chemical in question will pose specific environmental and/or health risks before it can require companies to further test their product, or before the EPA can prohibit its introduction. This takes time, money, and manpower. Furthermore, even if a company conducts tests, it can take years for the EPA to collect and analyze data. With roughly 700 chemicals introduced each year, and around 80,000 left to analyze, the EPA is woefully behind on assessments.

Even if the EPA does manage to conduct testing, determines that a chemical poses a threat to public health, and moves to block the production of an existing toxic chemical, the agency still has to surmount a 'high legal threshold.' The EPA must prove that an 'unreasonable risk' is posed to the public by the chemical in question. In 1991, asbestos regulation was deterred in an appeals court because the EPA could not provide 'substantial evidence' that asbestos was harmful. According to the GAO, the EPA has only successfully controlled five existing chemicals since 1976 (GAO, 2008).

There is a portion of the TSCA that promotes industry testing and analysis of chemicals; however, this is for all effective purposes voluntary. Section 5 – Manufacturing and Processing Notices – of the document describes monitoring and enforcement of the toxic chemicals through the creation of premanufacture notices (PMN). If a chemical manufacturer intends to create or import a new chemical not on the TSCA's current inventory of chemicals, they must submit a PMN to the Administrator of the USEPA 90 days or more before introducing the chemical. The PMN contains company-compiled testing data on health effects (including epidemiological studies), ecological effects, physical and chemical properties, or environmental fate characteristics and (on sites under their control within the United States) exposure information. These notices then must be reviewed by the Administrator within a 90-day period. If, by the end of this period, the Administrator fails to challenge the chemical's introduction and the PMN form is completed correctly, the chemical is automatically allowed for manufacture or import. The wording contained in Section 5 of the TSCA is as follows:

> *If the Administrator finds that there is a reasonable basis to conclude that the manufacture, processing, distribution in commerce, use, or disposal of a chemical substance . . . presents or will present an unreasonable risk of injury to health or environment . . . the Administrator shall, before the expiration of the notification period . . . take the action . . . to protect against such risk.*

In short, the Administrator must provide substantial evidence that the chemical causes an unreasonable risk to the environment and impose sanctions if indeed that chemical is toxic within this 90-day period.

The TSCA places an unreasonable burden of proof upon the EPA if the agency attempts to ban or restrict a chemical. For the EPA to restrict a chemical, the agency must provide extensive testing and analysis, which it lacks the manpower and time to accomplish. Finally, if the EPA does restrict a chemical, the agency's decision will be fought in court and held to a precedent of 'substantial evidence.'

21.4.2 'Sham Recycling' and Problems with the RCRA's Definition of Solid Waste

To escape RCRA regulations, some hazardous waste facilities engage in what is known as 'sham recycling' (SR). SR is unregulated treatment, either illegal or legal, of hazardous waste (Comella, 1993). Recycling facilities do not have the extensive permitting process or requirements that treatment facilities must meet. Public meetings, various iterations of EPA applications and other EPA reviews required for the construction of a treatment plant can extend the time it takes to get a permit approved to four or more years. Once a permit is obtained, treatment facilities have to abide by federal regulations detailing cumbersome safety procedures, storage, and disposal methods. This onerous permitting process and the restrictions imposed once a permit is awarded drive some companies and businesses looking to turn a quick profit toward 'sham recycling.'

An example of 'sham recycling' can be found in the Louisiana-based company Marine Shale Processors (MSP). Throughout the 1980s, MSP was allowed to conditionally operate a hazardous waste incinerator for internal recycling. As long as the heat generated by the incinerator was used to recover energy for the facility, MSP could incinerate waste without a permit. The organization quickly violated these conditions, burning low-energy hazardous waste (like creosote sludge) that took more energy to incinerate than the process produced. The burn-refuse was sold as aggregate for road-filler. At the height of MSP's operations it was the largest hazardous waste incinerator in the US, burning over 240 hazardous materials.

By posing as a recycling operation, MSP avoided regulations normally imposed on incinerators of that size. These environmental safeguards required such things as a thorough post-incinerator waste analysis to ensure levels of contaminants were at a minimum, and safe storage/disposal of wastes in appropriately prepared landfills. At MSP, however, these safeguards were disregarded and hazardous waste seeped from the facility into surrounding soil and groundwater. After Louisiana regulators inspected the site in the late 1980s, MSP was charged with 55 violations of the RCRA, and five violations of the Clean Water Act along with a lawsuit in federal court.

The practice of sham recycling prompted the EPA to reexamine the RCRA and find a way to close the loophole that so many facilities were exploiting. The agency eventually determined the best fix would come in the form of a modification to the RCRA's definition of solid waste (DSW), which lies at the heart of the RCRA's hazardous waste regulations. For a material to be regulated as hazardous waste under the RCRA, it must be considered 'solid waste' (see Figure 1.1). This requirement naturally places a lot of weight on the DSW. Since 1987, contention over the definition itself has been rampant. The Federal DC Circuit Court of Appeals has issued seven decisions on the definition of solid waste in the past 20 years alone (Bergeson, 2010; Rubrecht, 2009).

In 2008, the EPA redefined the definition of solid waste to exclude many specific recycled materials from RCRA regulation – a move the EPA hoped would stymie industries skirting past the RCRA through a loophole in the recycling regulations. The EPA employed a two-part scheme to indirectly encourage recycling of *hazardous secondary materials* (EPA, 2008b). First, exceptions from the RCRA were added to federal regulations for many hazardous materials; however, only for those that were recycled through reclamation, i.e. hazardous materials that are reused by industries. Dubbed 'streamlining' by the EPA, it is in essence deregulation. The second step is to continue enforcing rules that attempt to prevent industries from dispelling hazardous waste into the environment. The hope is that by deregulating reclamation, and enforcing hazardous waste emission regulations, industries will be pinned. To explain another way, if a facility reclaims hazardous materials, theoretically there should be no waste dispelled. If the industry lies and discharges hazardous materials into the environment, regulatory agencies already monitoring for such discharges will detect them. As the EPA sees it, regulating reclaimed materials is redundant and unnecessary if discharge monitoring programs are effective.

Thirty days after the new definition was in place the Sierra Club produced a petition for review in the US Court of Appeals for District of Columbia Circuit (*Sierra Club v. EPA*, No. 09-1041), and requested the EPA reconsider the rule and refrain from implementation. The petition states that by removing the EPA's ability to regulate hazardous reclaimed waste, a significant threat to human and environmental health will be allowed.

The Sierra Club bases this statement largely on a study by the EPA in 2007 titled: *Assessment of Environmental Problems Associated with Recycling of Hazardous Secondary Materials* (EPA, 2007). In the study, which the Sierra Club claims was not exhaustive, 208 facilities that recycled hazardous materials caused 'environmental damage' in some form (Figure 21.2). *Environmental*

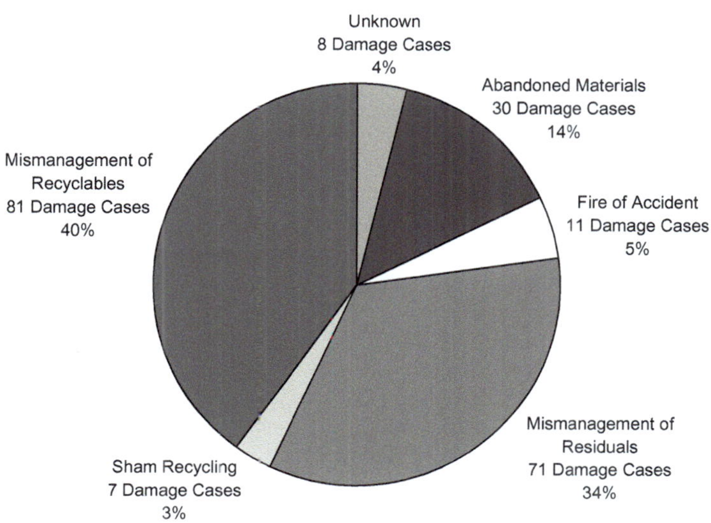

Figure 21.2 Primary causes of environmental damage from recycling processes (EPA, 2007).

damage involves the emission of hazardous waste into the environment in the form of spills, dumps, and other releases requiring cleanup that caused harm to humans or the environment; also, sites where hazardous materials were abandoned are included in the definition. Of these facilities, a vast majority (199) were specialized off-site recycling facilities which, according to the EPA's 2007 Environmental Assessment, are more likely to cause environmental damage.

Furthermore, it should be noted in Figure 4.2 that SR only accounted for 3% of the damage cases. These data contradict the EPA's reasoning for the redefinition of solid waste, which was drafted to cut down on SR. Ninety-seven percent of environmental damage is caused by a recycling-related mishap other than SR. If the EPA's 2007 report is an accurate measure of recycling faults, then the new DSW may cause more environmental damage than it will avert. In the EPA's own Regulatory Impact Analysis (EPA, 2008a), only an estimated additional 23,000 tons per year of hazardous waste will result from the new DSW rule (a 1.1% increase from 2005).

References

Bergeson, L. (2010). EPA Finally Amends Definition of Solid Waste. Chemical Processing. com. Retrieved August 25 from <http://www.chemicalprocessing.com/articles/2008/199.html>.

Centers for Disease Control and Prevention (CDC) (2009). About NIOSH. *NIOSH.* Retrieved June 26 from <http://www.cdc.gov/niosh/about.html>.

Comella, P. L. (1993). Understanding a sham: When is recycling, treatment? *Boston College Environmental Affairs Law Review*, *20*, 415–452.

Occupational Safety and Health Administration (OSHA) (1993). Air Contaminants, Final Rule, 29 CFR Part 1910, RIN 1218-AB26. <http://www.osha.gov/pls/oshaweb/owadisp.show_document?p_table=federal_register&p_id=13306>.

Occupational Safety and Health Administration (OSHA) (2010). Recommendations from the National Advisory Committee on Occupational Safety and Health (NACOSH) to the U.S. Department of Labor/OSHA and the U.S. Department of Health and Human Services/NIOSH. *National Advisory Committee on Occupational Safety and Health.* Retrieved June 8 from <http://www.osha.gov/dop/nacosh/nreport.html>.

Rahm, D. (1998). Controversial cleanup: Superfund and the implementation of U.S. hazardous waste policy. *Policy Studies Journal*, *26*(4), 719–734.

Rubrecht, G. L. (2009). Sierra Club Takes U.S. EPA to Court on Its Revisions to the Definition of Solid Waste. Jackson Kelly PLLC, Energy and Environment Monitor. Retrieved March 18 from <http://eem.jacksonkelly.com/2009/03/sierra-club-takes-us-epa-to-court-on-its-revisions-to-the-definition-of-solid-waste-.html>.

Solomon, G. (2010). OSHA Limits Don't Protect Gulf Coast Workers. The Huffington Post. Retrieved June 16 from <http://www.huffingtonpost.com/gina-solomon/osha-limits-dont-protect_b_614223.html>.

Toxic Substances Control Act (TSCA).

US Environmental Protection Agency (EPA) (2007). An Assessment of Environmental Problems Associated with Recycling of Hazardous Secondary Materials. Office of Solid

Waste. Retrieved January 11 from <http://www.regulations.gov/search/Regs/home.html#documentDetail?D=EPA-HQ-RCRA-2002-0031-0355>.

US Environmental Protection Agency (EPA) (2008a). USEPA's 2008 Final Rule Amendments to the Industrial Recycling Exclusions of the RCRA Definition of Solid Waste. Office of Solid Waste. Retrieved September 25 from <http://www.regulations.gov/search/Regs/home.html#documentDetail?R=0900006480728a11>.

US Environmental Protection Agency (EPA) (2008b). Introduction to the Definition of Solid Waste Final Rule. Office of Solid Waste. <www.epa.gov>.

US Environmental Protection Agency (EPA) (2009b). Testimony of Steve Owens, Assistant Administrator. Office of Prevention, Pesticides and Toxic Substances, Retrieved August 26, 2010 from <http://www.epa.gov/ocir/hearings/testimony/111_2009_2010/2009_1117_so.pdf>.

US Government Accountability Office (GAO) (2008, April 29). TOXIC CHEMICALS: EPA's New Assessment Process Will Increase Challenges EPA Faces in Evaluating and Regulating Chemicals, GAO-08-743T. Testimony Before the Committee on Environment and Public Works, US Senate.

22 Strategies for the Future – Waste Reduction and Recycling, Treatment Technologies, and Green Chemistry

Chapter Contents

22.1 Reduction and Recycling

Waste minimization is characterized as the reduction of waste from sources and the reuse of waste through recycling. The objective is to decrease the amount of hazardous waste bound for energy recovery, treatment, and disposal facilities. Therefore, hazardous waste minimization does not include treatment processes, such as those that alter the physical, chemical, or biological composition of waste. While hazardous waste minimization is not required by law, it is strongly encouraged through several acts of Congress.

In the 1984 amendments to the Resource Conservation and Recovery Act (RCRA), lawmakers modified national policy to make waste minimization the

Risks of Hazardous Wastes. DOI: 10.1016/B978-1-4377-7842-7.00022-2

preferred method for hazardous waste management. Facilities that deal with hazardous waste must have a working waste reduction program. With the passage of the Pollution Prevention Act (PPA) of 1990, the policy of waste prevention through reduction was extended outside of RCRA jurisdiction. The PPA stated that waste disposal and release should be 'employed only as a last resort,' and that hazardous waste should be 'prevented or reduced at the source whenever feasible' (US EPA, 2008b). Since that time, EPA Regions have pushed waste minimization programs on local industries and have received some successes as a result. In many cases, mercury and lead are targeted for reduction.

Programs have been established in an effort to reduce the effects of certain environmental pollutants generated by industry. The US EPA's National Waste Minimization Program has targeted 31 'Priority Chemicals' (PCs) for reduction from industrial products and wastestreams, based on their tendency to enter and persist in the environment, and the hazards they pose to human health (Table 22.1). EPA estimates reveal that after the implementation of waste reduction programs, total reported PC outputs by industry dropped 15% to 84.7 million pounds, from 2000 to 2004 (EPA, 2009).

22.1.1 Source Reduction

The most effective way to reduce waste outputs is to reduce inputs and to make processes more efficient. This is commonly called pollution prevention, or P2. This can be achieved through improvements such as: updating outdated and inefficient technologies, reformulating chemical products to exclude hazardous chemicals, switching to non-toxic materials used for manufacturing, and increasing efficiency of methods and protocols.

Hazardous waste reduction provides regulatory benefits as well. Companies that are able to reduce their hazardous waste outputs below 1,000 kg per month can move from a large-quantity generator (LQG) to a small-quantity generator (SQG) designation (please refer to Chapter 2). The definition of an LQG means that the facility generates more than 1,000 kg or more of RCRA hazardous waste in a month, generates or accumulates 1 kg of RCRA acute hazardous waste at any time, or generates or accumulates more than 100 kg of spill cleanup material contaminated with RCRA acute hazardous waste at any time. Benefits of becoming an SQG include permit exemptions and other deregulatory consequences, which can result in increased efficiency and cost-cutting for businesses.

Many large industries have volunteered to reduce specific hazardous waste associated with their manufacturing. In the EPA's mid-Atlantic region, 65 large organizations including the US Air Force, GE, Sony, Phillip Morris, and DuPont have pledged to the EPA that they will cut down on certain hazardous wastes (US EPA, 2010j). DuPont, for example, reduced its dioxin output by 63% at three US plants, and therefore saw the improvement of process yields. Other success stories using hazardous waste reduction include a General Electric plant reducing lead oxide waste by 25% and an Army depot reducing lead and cadmium waste by 99.5% and 57%, respectively.

Table 22.1 Priority Chemicals (EPA, 2009)

Chemical Name	**CASRN**
Organic chemicals and chemical compounds	
1,2,4-Trichlorobenzene	120-82-1
1,2,4,5-Tetrachlorobenzene	95-94-3
2,4,5-Trichlorophenol	95-95-4
4-Bromophenyl phenyl ether	101-55-3
Acenaphthene	83-32-9
Acenaphthylene	208-96-8
Anthracene	120-12-7
Benzo(g,h,i)perylene	191-24-2
Dibenzofuran	132-64-9
Dioxins/furans (considered one chemical on this list)	1746-01-6
Endosulfan, alpha & Endosulfan, beta (considered one chemical on this list)	959-98-8
	33213-65-9
Fluorene	86-73-7
Heptachlor & heptachlor epoxide (considered one chemical on this list)	76-44-8
	1024-57-3
Hexachlorobenzene	118-74-1
Hexachlorobutadiene	87-68-3
Hexachlorocyclohexane, gamma- (Lindane)	58-89-9
Hexachloroethane	67-72-1
Methoxychlor	72-43-5
Naphthalene	91-20-3
Pendimethalin	40487-42-1
Pentachlorobenzene	608-93-5
Pentachloronitrobenzene (Quintozene)	82-68-8
Pentachlorophenol	87-86-5
Phenanthrene	85-01-8
Polycyclic aromatic compounds (PACs)/PAH group (as defined in TRI)	
Polychlorinated biphenyls (PCBs)	1336-36-3
Pyrene	129-00-0
Trifluralin	1582-09-8
Metals and metal compounds	
Cadmium	7440-43-9
Lead	7439-92-1
Mercury	7439-97-6

In areas other than industry, such as biomedical research, waste minimization practices have also effectively decreased hazardous waste production. In a study by Rau et al. in 2000, the authors claimed that strong incentives such as environmental protection, regulatory compliance, liability avoidance, disposal cost avoidance, and improved community relations encouraged 90% of laboratories to adopt waste minimization programs. At these laboratories, substantial waste reduction and cost savings were reported due to these waste minimization programs.

22.1.2 Recycling

Recycling of materials is a general phrase used to describe processes that reuse waste products for materials or energy generation. If the waste is chemically unaltered, it can be reused for its original pre-waste purpose. Captured vapors emitted during dry cleaning operations, for example, can be condensed back into a liquid and reused for further cleaning. Additionally, some waste materials contain elements that can be salvaged for use in new products. Mercury in old cathode ray tube monitors, for example, can be reinstalled in fluorescent bulbs. The recycling and salvage of mercury in the United States has been so successful that the mining of mercury has all but ceased; recycled sources alone can now sustain the demand for mercury (US EPA, 2007).

However, recycling methods have not been embraced by many industries. As can be seen in Table 22.2, recycling is not the prominent endpoint for hazardous wastes. From 2007 data, the large-quantity generators (LQGs) of hazardous waste only recycle 7% (3,545,074 tons) of that which they produce. However, solvent recovery is the method used by most facilities (456 facilities or 23%), albeit at low quantities. Subterranean injection is the endpoint for by far the largest quantity of hazardous waste, though it is used by only a few, high-output facilities (US EPA, 2003; US EPA, 2008a).

Since 2001, the amount of hazardous waste that has been recovered by LQGs has increased somewhat; however, the portion of hazardous waste that is recovered has declined. This trend is due to the overall increase in hazardous waste produced.

Table 22.2 Waste Management Method Scale of Use, Adapted from US EPA (2008a)

Management Method	Tons Managed	Percentage of Quantity	No. of Facilities	Percentage of Facilities
Deepwell or underground injection	21,505,921	42.6	42	3
Other disposal	12,363,634	24.5	91	6.5
Aqueous organic treatment	3,106,828	6.2	67	4.8
Incineration	3,047,982	6	140	10
Landfill/surface impoundment	1,939,712	3.8	67	4.8
Aqueous inorganic treatment	1,879,946	3.7	170	12.2
Energy recovery	**1,764,693**	**3.5**	**91**	**6.5**
Other treatment	1,298,339	2.6	353	25.3
Metals recovery	**1,116,357**	**2.2**	**137**	**9.8**
Fuel blending	737,397	1.5	112	8
Stabilization	658,249	1.3	107	7.7
Sludge treatment	397,863	0.8	59	4.2
Other recovery	**335,093**	**0.7**	**65**	**4.7**
Solvents recovery	**328,931**	**0.7**	**456**	**32.7**
Land treatment/ application/farming	1,981	0	16	1.1

22.2 New Treatment Technologies

Hazardous waste treatment consists of manipulating biological, chemical, or physical attributes. However, current methods are not necessarily the best technology to treat hazardous wastes and there are still improvements to be made. Updated methods attempt not to cause major environmental damage or require much human intervention (Tedder and Pohland, 2000). Additionally, it has been shown that companies looking to site hazardous waste treatment facilities with these technologies tend to select areas that have little population density as the 'path of least resistance,' with little resources or organization to form opposition (Harris et al., 1985). To prevent the exposure of residents through more hazardous waste sites, minimization of the hazardous waste should be regarded as the most important hazardous waste management option. Although waste minimization should be the primary option, it may not always be the most feasible. The second-best option is to utilize new and improved technologies to diminish the need for hazardous waste treatment plants.

There have been numerous improvements in hazardous waste treatment technology. Some of the improvements are retrofits of traditional treatment, such as adding nanoparticles or other compounds to an existing treatment process. For example, soil washing and pump-and-treat can be aided by surfactants, or advanced oxidation processes can be aided by ozone. Other improvements include utilizing different compounds and organisms in bioremediation, such as using nanoparticles in photocatalytic treatment, reduction, and oxidation processes, or using *Escherichia coli* to degrade carbon tetrachloride in a biological system. These treatment technologies are a step in the right direction, and add to the current list of treatment technologies for hazardous waste listed in Title 40, Part 268 (Cha et al., 1998; Tedder and Pohland, 2000; US EPA, 2010i).

The following are selected treatment technologies for hazardous waste as detailed in Title 40, Part 268 (adapted from US EPA, 2010i):

- Venting of compressed gases into an absorbing or reacting media (i.e., solid or liquid) – venting can be accomplished through physical release utilizing valves/piping; physical penetration of the container; and/or penetration through detonation.
- Biodegradation of organics or non-metallic inorganics (i.e., degradable inorganics that contain the elements of phosphorus, nitrogen, and sulfur) in units operated under either aerobic or anaerobic conditions such that a surrogate compound or indicator parameter has been substantially reduced in concentration in the residuals (e.g., Total Organic Carbon can often be used as an indicator parameter for the biodegradation of many organic constituents that cannot be directly analyzed in wastewater residues).
- Carbon adsorption (granulated or powdered) of non-metallic inorganics, organo-metallics, and/or organic constituents, operated such that a surrogate compound or indicator parameter has not undergone breakthrough (e.g., Total Organic Carbon can often be used as an indicator parameter for the adsorption of many organic constituents that cannot be directly analyzed in wastewater residues). Breakthrough occurs when the carbon has become saturated with the constituent (or indicator parameter) and substantial change in adsorption rate associated with that constituent occurs.
- Chemical or electrolytic oxidation utilizing the following oxidation reagents (or waste reagents) or combinations of reagents: (1) Hypochlorite (e.g., bleach); (2) chlorine; (3) chlorine dioxide; (4) ozone or UV (ultraviolet light) assisted ozone; (5) peroxides;

(6) persulfates; (7) perchlorates; (8) permangantes; and/or (9) other oxidizing reagents of equivalent efficiency, performed in units operated such that a surrogate compound or indicator parameter has been substantially reduced in concentration in the residuals (e.g., Total Organic Carbon can often be used as an indicator parameter for the oxidation of many organic constituents that cannot be directly analyzed in wastewater residues). Chemical oxidation specifically includes what is commonly referred to as alkaline chlorination.

- Chemical reduction utilizing the following reducing reagents (or waste reagents) or combinations of reagents: (1) Sulfur dioxide; (2) sodium, potassium, or alkali salts or sulfites, bisulfites, metabisulfites, and polyethylene glycols (e.g., NaPEG and KPEG); (3) sodium hydrosulfide; (4) ferrous salts; and/or (5) other reducing reagents of equivalent efficiency, performed in units operated such that a surrogate compound or indicator parameter has been substantially reduced in concentration in the residuals (e.g., Total Organic Halogens can often be used as an indicator parameter for the reduction of many halogenated organic constituents that cannot be directly analyzed in wastewater residues). Chemical reduction is commonly used for the reduction of hexavalent chromium to the trivalent state.
- High temperature organic destruction technologies, such as combustion in incinerators, boilers, or industrial furnaces operated in accordance with the applicable requirements of 40 CFR part 264, subpart O, or 40 CFR part 265, subpart O, or 40 CFR part 266, subpart H, and in other units operated in accordance with applicable technical operating requirements; and certain non-combustive technologies, such as the Catalytic Extraction Process.
- Macroencapsulation with surface coating materials such as polymeric organics (e.g., resins and plastics) or with a jacket of inert inorganic materials to substantially reduce surface exposure to potential leaching media. Macroencapsulation specifically does not include any material that would be classified as a tank or container according to 40 CFR 260.10.
- Chemical precipitation of metals and other inorganics as insoluble precipitates of oxides, hydroxides, carbonates, sulfides, sulfates, chlorides, flourides, or phosphates. The following reagents (or waste reagents) are typically used alone or in combination: (1) Lime (i.e., containing oxides and/or hydroxides of calcium and/or magnesium; (2) caustic (i.e., sodium and/or potassium hydroxides; (3) soda ash (i.e., sodium carbonate); (4) sodium sulfide; (5) ferric sulfate or ferric chloride; (6) alum; or (7) sodium sulfate. Additional floculating, coagulation or similar reagents/processes that enhance sludge dewatering characteristics are not precluded from use.
- Recovery of acids or bases utilizing one or more of the following recovery technologies: (1) Distillation (i.e., thermal concentration); (2) ion exchange; (3) resin or solid adsorption; (4) reverse osmosis; and/or (5) incineration for the recovery of acid – Note: this does not preclude the use of other physical phase separation or concentration techniques such as decantation, filtration (including ultrafiltration), and centrifugation, when used in conjunction with the above listed recovery technologies.
- Steam stripping of organics from liquid wastes utilizing direct application of steam to the wastes operated such that liquid and vapor flow rates, as well as temperature and pressure ranges, have been optimized, monitored, and maintained. These operating parameters are dependent upon the design parameters of the unit, such as the number of separation stages and the internal column design, thus, resulting in a condensed extract high in organics that must undergo either incineration, reuse as a fuel, or other recovery/reuse and an extracted wastewater that must undergo further treatment as specified in the standard.
- Wet air oxidation performed in units operated such that a surrogate compound or indicator parameter has been substantially reduced in concentration in the residuals (e.g., Total Organic Carbon can often be used as an indicator parameter for the oxidation of many organic constituents that cannot be directly analyzed in wastewater residues).

Table 22.3 Principles of Green Chemistry

Prevention
Atom economy
Less hazardous chemical synthesis
Designing safer chemicals
Safer solvents and auxiliaries
Design for energy efficiency
Use of renewable feedstocks
Reduce derivatives
Catalysis
Design for degradation
Real-time analysis for pollution prevention
Inherently safer chemistry for accident prevention

Adapted from Anastas and Warner (1998).

22.3 Green Chemistry

22.3.1 Overview

Green Chemistry effectively reduces the production of hazardous waste while providing efficient and innovative products which can enhance the future of America. Green Chemistry is a process used in chemical development and, in the United States, is a trend supported by science, business, and cultural interest. Green Chemistry principles and practices must move from being a respected trend to being the mandated method for chemical creation and monitoring.

22.3.2 Purpose and Principles

Green Chemistry seeks to reduce pollution through prevention as chemical products are created by following 12 key principles (Table 22.3). Paul Anastas and John Warner introduced these principles in 1998 and they now function as the foundation of the field.

Anastas, the Director for the Center of Green Chemistry and Engineering at Yale University, describes the principles as a framework to 'optimize synthetic pathways and product design around minimum toxicity and material/energy inefficiency' (Anastas and Warner, 1998; US EPA, 2010g).

1. Prevention – It is better to prevent waste than to treat or clean up waste after it has been created.
2. Atom economy – Synthetic methods should be designed to maximize the incorporation of all materials used in the process into the final product.
3. Less hazardous chemical synthesis – Wherever practicable, synthetic methods should be designed to use and generate substances that possess little or no toxicity to human health and the environment.

4. Designing safer chemicals – Chemical products should be designed to effect their desired function while minimizing their toxicity.
5. Safer solvents and auxiliaries – The use of auxiliary substances (e.g., solvents, separation agents, etc.) should be made unnecessary wherever possible and innocuous when used.
6. Design for energy efficiency – Energy requirements of chemical processes should be recognized for their environmental and economic impacts and should be minimized. If possible, synthetic methods should be conducted at ambient temperature and pressure.
7. Use of renewable feedstocks – A raw material or feedstock should be renewable rather than depleting whenever technically and economically practicable.
8. Reduce derivatives – Unnecessary derivatization (use of blocking groups, protection/deprotection, temporary modification of physical/chemical processes) should be minimized or avoided if possible, because such steps require additional reagents and can generate waste.
9. Catalysis – Catalytic reagents (as selective as possible) are superior to stoichiometric reagents.
10. Design for degradation – Chemical products should be designed so that at the end of their function they break down into innocuous degradation products and do not persist in the environment.
11. Real-time analysis of pollution prevention – Analytical methodologies need to be further developed to allow for real-time, in-process monitoring and control prior to the formation of hazardous substances.
12. Inherently safer chemistry for accident prevention – Substances and the form of a substance used in a chemical process should be chosen to minimize the potential for chemical accidents, including releases, explosions, and fires.

22.3.3 Problems with the Toxic Substance Control Act of 1976

The United States regulates chemical use and production through EPA enforcement of the Toxic Substance Control Act (TSCA) of 1976, summarized in Table 22.4 (US EPA, 2010k). Limitations of the TSCA are numerous especially with regard to banning or regulating chemicals found to have significant health and safety risks. The EPA is acutely aware of the limitations, which were highlighted in testimony to the US Senate by John Stephenson of the Government Accountability Office in 2009 (USGAO, 2009). These problems have been detailed in the previous chapter, Failures of Regulatory Agencies, and will be briefly discussed below.

Below are a few of the key provisions of TSCA which restrict EPA's ability to protect America from toxic chemicals and therefore demonstrate why the principles of Green Chemistry must become the expectation of the law.

- Chemical Substance Inventory Creation in 1979: This is a list of chemicals currently manufactured or processed in the United States. Sixty thousand chemicals which were *automatically classified as safe and not subject to the regulations in TSCA*. The inventory now is composed of more than 84,000 chemicals (US EPA, 2010h).
- Chemical manufacturers/importers are exempt from providing information to the EPA about the health and environment effects of chemicals if (US EPA, 2010d):
  - chemicals are being used for research and development, or
  - less than 500 kg of the chemical is produced/imported at each plant site

- New chemical manufacturing requires a Premanufacture Notice (PMN) (US EPA, 2010e) to be filed with the EPA if it will be used for *commercial purposes*. The PMN requires manufactures to provide an *available* health and safety information; manufacturers are not required to test chemicals prior to their use.
    - *Materials exempt from PMN include:* '...naturally-occurring materials, products of incidental reactions, products of end-use reactions, mixtures (but not mixture components), impurities, byproducts, substances manufactured solely for export, nonisolated intermediates, and substances formed during the manufacture of an article (US EPA, 2010e).'
    - Manufacturers may designate information about chemicals as '*Confidential Business Information*' thus requiring the EPA to protect it from disclosure. Manufacturers have taken advantage of this provision and 95% of PMNs contain confidential business information. which may include details about the health and safety effects of chemicals (USGAO, 2009).
- The EPA must choose the least burdensome way to regulate existing chemicals and they must present information not only on the health and environmental effects of exposure, but the economic consequences of the regulation with consideration for the 'national economy, small business, technological innovation, the environment, and public health (USGAO, 2009).'

As a result of the TSCA, 60,000 chemicals were pronounced safe without proof of their safety, manufacturers were permitted to test chemicals without reporting the chemical or testing mechanism to the EPA, and by-products and products created during the manufacture of substances were made exempt from reporting requirements. Conversely, Green Chemistry requires consideration of substance safety from the initial choice of chemicals. It promotes accountability through understanding each step of the creation and degradation of the product. There is not room for being exempt from health and environmental considerations as a main focus is to understand the environmental impact of the product and minimize any toxic effect. Yet, Green Chemistry principles are not a mandate; motivation for approaching manufacturing from this perspective is individual.

Table 22.4 TSCA Overview (US EPA, 2010k)

Toxic Substance Control Act (TSCA) of 1976
Enforcement: EPA
TSCA Chemical Inventory Created: 1979
Number of chemicals assumed to be safe in 1979 inventory: 60,000
Number of chemicals in TSCA Chemical Inventory in 2010: 84,000
Regulates chemical substances: TSCA section 3(2)(A) 'any organic or inorganic substance of a particular molecular identity, including – (i) any combination of such substances occurring in whole or in part as a result of a chemical reaction or occurring in nature, and (ii) any element or uncombined radical.'
Does not regulate: food, pesticides, drugs, cosmetics, tobacco...
Chemicals exempt information requirement: Those used for research and development or manufactured/imported at less than 500 kg/plant site
Testing of new chemicals: Required if the EPA determines that there may be 'unreasonable risk of injury to health or the environment' (TSCA Sec 4(a)(1)(A)(i)). There is not a mandate for manufacturers to test chemicals.

22.3.4 Creating More Effective Legislation: Learning from the European Union and Canada

Both the European Union and Canada have enacted legislation to regulate toxic chemicals and promote environmental and health protection. REACH, or the Registration, Evaluation and Authorization of Chemicals, enacted by the European Union in 2007 *mandated the industry be responsible for the associated risks and safety precautions of their chemicals*. The REACH testing regulations applied to all chemicals – those currently in use and new chemicals. Similar to the United States TSCA, REACH requires reporting only after a company has created or imported a specified amount of the chemical. However, the limit (1 tonne) is per manufacturer, not each individual plant site (European Chemicals Agency, 2007).

The Canadian Environmental Protection Act (CEPA) passed in 1999 and also made testing requirements mandatory for all new and existing chemicals. By 2006, all existing chemicals had been reviewed and added to the list of toxic substances if appropriate. Any chemical determined to be toxic must meet preventive or control requirements which address each stage of the 'life cycle from the research and development stage through manufacture, use, storage, transport and ultimate disposal or recycling.' CEPA also lays out the steps for 'virtual elimination' of some bioaccumulative, persistent toxic substances so that any release into the environment remains below detection levels (Environment Canada, 2010a).

22.3.5 Change is Brewing in the United States: EPA Transparency and Proposed Legislation

A news release from the EPA in January 2010 announced the start to a new era of chemical reform by making some Confidential Business Information (CBI) available to the public (Kemery and Jones, 2010). Specifically, the names of chemicals on the public TSCA Inventory which have been identified as having significant health and safety risks have been unmasked. Manufacturers would frequently claim the name of the chemical as CBI while reporting significant health and safety risks thus leaving the public unable to protect themselves. Further changes in transparency are promised.

EPA Principles for TSCA reform have also been created. Six key areas are described as essential to reform in order to 'target chemicals of concern and promptly assess and regulate new and existing chemicals' (US EPA, 2010c; Table 20.5).

Further support for a more comprehensive approach to chemical management in the United States is gathering. In April 2010, new legislation was introduced: The Toxic Chemicals Safety Act of 2010 (H.R. 5820). Changes to the proposed legislation will likely be numerous, but currently it mandates testing and regulation of *new and existing chemicals* by manufacturers and promotion of Green Chemistry in development, safe alternatives and in workplace training (US House of Representatives, 2010).

The growing energy from government agencies and now lawmakers creates hope that the United States can improve chemical management beginning with substance creation through application and degradation. Transparency and awareness are a beginning, but successful change will require resources, dedication and continued

Table 22.5 Essential principles for chemical reform, adapted from US EPA (2010c)

Chemicals should be reviewed against safety standards that are based on sound science and reflect risk-based criteria protective of human health and the environment
Manufacturers should provide EPA with the necessary information to conclude that new and existing chemicals are safe and do not endanger public health or the environment
Risk management decisions should take into account sensitive subpopulations, cost, availability of substitutes and other relevant considerations
Manufacturers and EPA should assess and act on priority chemicals, both existing and new, in a timely manner
Green Chemistry should be encouraged and provisions assuring transparency and public access to information should be strengthened
EPA should be given a sustained source of funding for implementation

financial commitment. The next section explores groups, organizations and institutions that are creating resources and building support for Green Chemistry.

22.3.6 *Fortifying the Validity of Green Chemistry – Market Drivers and the Economic Impacts*

Consumers are a driving force behind Green Chemistry implementation. Shareholders have pushed companies to create strategies which protect health and the environment (Ambachtsheer et al., 2007). Research continues and education is growing to provide scientific validity to the principles of Green Chemistry. Climate change groups purport that Green Chemistry can reduce the effects of climate. The result is a society in support of clean, efficient and non-toxic practices; each builds further support for implementation of Green Chemistry.

As customer interest and demand for green products have increased, companies have responded with a plethora of products touting green and sustainable practices. The retail giant, Walmart, has changed to mainly recycled packaging, introduced a line of green cleaning supplies and is embarking on a campaign which will strongly encourage its suppliers to engage in sustainable activities (de Guzman, 2010). This is just one example from a multitude of companies choosing to provide green products because their customers request it.

Shareholders are realizing the importance of sustainability and environmentally sound practices. As the media attention to environmental hazards has increased, shareholders have wisely associated the value of their investment with the environmental responsibility of the business. If this were not enough motivation to promote environmental consideration by businesses, their leadership is also motivated because they can be individually held accountable for the actions performed under their supervision (Ambachtsheer et al., 2007).

Additionally, green chemistry can be used to decrease CO_2 production through the following methods described by Zecchini and Tundo (2010):

- creation of new products/processes/catalysts to reduce energy consumption and by-products/waste production

- development of products/processes/catalysts that don't use halogens or volatile organic solvents
- implementation of highly efficient processes
- use of CO2 as a C1 building block
- generation of H2, bio-hydrogen, from biomasses
- valorization of biomasses in bio-refineries
- syntheses of biodiesel from exhaust oils or from algae, bio-ethanol from lignin/cellulose biomasses
- creation of polymers from renewables
- use of novel materials for photovoltaic panels and fuel cells, and the improvement of battery capacity for electric vehicles
- improvement of catalytic converters for vehicles.

Realizing that the next generation of scientists must be trained in the principles of Green Chemistry, the EPA with the American Chemical Society (ACS) began developing education materials. ACS created a website dedicated to connecting educators to teaching resources for students in elementary school through graduate school. Textbooks, lab suggestions, lesson plans and networking opportunities are available through ACS. Further, ACS sponsors a biennial conference for chemistry teachers and professors (ACS: Chemistry For Life, 2010).

22.3.7 The Effect on Consumers

Ecolabels are created by manufacturers, non-profit organizations, government and other businesses and therefore are not created equal. They may represent an organization's sustainability efforts, the benign components of a product or other 'green' practices supported by the labeling company. A report by the World Resources Institute asked 66 questions of ecolabel companies (113 of 340 completed the survey) and found variation in certification requirements, accountability for meeting certification requirements and how 'green' is defined by each label (World Resource Institute and Big Room, Inc., 2010). The consumer must educate themselves about the validity of specific ecolabels.

Because there are no specific federal regulations regarding 'green' labeling, much has been done to assist people and businesses in identifying truly green products. For instance, when the federal government mandated its agencies to purchase environmentally preferred products, it provided guidelines for what this meant. They direct groups to programs and third-party organizations who have standards which are considered environmentally preferred (US EPA, 2010b).

The Federal Trade Commission created a guide for consumers, *'Sorting Out "Green" Advertising Claims'*. It explains many different components of labeling such as the numbers related to recycling, degradable and biodegradable, etc. (Federal Trade Commission, 1999) There are also many websites dedicated to assisting consumers seeking to understand specific ecolabels (Consumer Reports, 2010; Big Room, Inc., 2010).

Green Seal is a third-party certifying organization that evaluates products beginning with their extraction and continuing through manufacture and disposal (Green Seal, 2010). It is one of many organizations listed on the US EPA's Database of Environmental Information for Products and Services (US EPA, 2010a). As a third-party certifier, Green Seal meets criteria in verification, review, facilities inspection, testing protocols, transparency of information and more (Green Seal, 2010). There are other certifying organizations which also meet these criteria and because they focus on the lifecycle of the product, they provide certification to products that employ green chemistry principles.

PVC free: PVC production and degradation is harmful to health. When children's toys were required to be phthalate free in 2008 (Congressional Research Service, 2008), awareness of the harmful effects of PVC grew. Below are some products which are created without PVC:

- Medical products:
  - IV fluid bags – Research showed that the by-products of PVC such as DEHP and phthalates could leech from IV bags into their solutions. Hospitals have started to eliminate PVC IV bags and Catholic Healthcare West of California has made a complete conversion to PVC-free IV bags. B. Braun, their supplier, estimates that 840 tons of PVC containing waste is being eliminated through this conversion (B. Braun Medical Inc., 2005).
- Floor coverings:
  - When Kaiser Permanente requested an alternative to PVC containing carpet, C & A Floorcoverings Inc. created PVC-free carpet. Many other companies followed and eliminated PVC from their carpet and then extended the idea to chairs, shoes, medical devices and more (Walsh, 2005).
- Computers:
  - Large computer manufacturers have eliminated PVC almost entirely, created display technology that does not require mercury, removed brominated flame retardants and cadmium (Hewlett Packard, 2010; Jobs, 2007).

Windows: Window manufacturer Anderson has addressed the lifecycle of their products beginning with eliminating solvents in their wood treatment. Then, they use the wood fiber by-products of production to create their composite material and end the process by using 98% of what was initially put into manufacturing the windows. Anderson Windows bears the Green Seal certification (Anderson Corporation, 2007).

Superenzymes and catalysts: The use of new enzymes and biocatalysts has reduced the temperature needed for chemical reactions and decreased the need for solvents, and therefore resulted in great savings of energy and waste reduction. Pharmaceutical giant Pfizer estimates that biocatalysis for one of their medications will have saved over 200,000 metric tons of waste between 2007–2010 (Codexis, Inc., 2010; Pfizer, Inc., 2010).

Bioplastics: These materials are made from starches and without petroleum. They may not be biodegradable, but they do reduce toxicity (Montenegro, 2009). Bioplastics can be used in all of the same ways as PVC-based plastic. Electronic casings are one example of bioplastic implementation as Samsung introduced a phone in 2008 with bioplastic casing, no brominated flame retardants or PVC (Cadden, 2008).

22.3.8 Conclusion: Green Chemistry is Necessary

Consideration of the impact of any compound or chemical must be undertaken before it is created, tested or used. The impact of chemicals in our society provides impeccable evidence as the bioaccumulation, toxic environments, health conditions and deaths represent not preparing appropriately for chemical creation. It may seem ridiculous that humans could be careful in planning, scientific in implementation and ethical in response to both planning and implementation, but it must be attempted. Green Chemistry demands that this happens in the creation of chemicals and compounds and if this system of principles were enforced and expected, the scientific community would not only gain respect, but human life would be improved. Without Green Chemistry and accountability, the future is dim; we need to reduce the negative impact of chemicals in our society.

Acknowledgments

This section on green chemistry was contributed by Elizabeth Lytle, RN, BSN, PHN. Elizabeth currently works as the District Nurse for a K-8 school district in Northern California, and is an MPH candidate at the University of Los Angeles School of Public Health.

References

ACS: Chemistry For Life (2010). *ACS Green Chemistry Institute*. Retrieved August 10, 2010 from <http://portal.acs.org/portal/PublicWebSite/greenchemistry/>.

Ambachtsheer, J., Kron, J., Liroff, R. A., Little, T., & Massey, R. (2007). *Fiduciary guide to toxic chemical risk*. Falls Church, VA: Investor Environmental Health Network.

Anastas, P. T., & Warner, J. C. (1998). *Green chemistry: Theory and practice*. New York, NY: Oxford University Press.

Anderson Corporation (2007). *Environmental goals & progress*. <http://www.andersenwindows.com/servlet/Satellite/AW/Page/awGeneral-3/1200437179507#2>.

Big Room, Inc. (2010). *Ecolabel index*. <http://www.ecolabelindex.com/ecolabels/>.

B. Braun Medical Inc. (2005). *CHW switches to PVC/DEHP-free products to improve patient safety and protect the environment*. Retrieved November from <http://www.bbraunusa.com/index-78.html?B365AD2065B05CD0D716A989521C5E7B>.

Cadden, R. (2008). Samsung Using Cornstarch to Build Eco-Friendly Phone. Mobile Burn. Retrieved June 18 from <http://70.85.18.20/news.jsp?Id=4808>.

Cha, D. K., Sarr, D., Chiu, P. C., & Kim, D. W. (1998). Hazardous waste treatment technologies. *Literature Review*, *70*(4), 705–720.

Codexis, Inc. (2010). *Custom Biocatalysts*. <http://www.codexis.com/custom_biocatalysts>.

Congressional Research Service (2008). *H.R. 4040: Consumer Product Safety Improvement Act of 2008*.

Consumer Reports (2010). *Greener choices: Products for a better planet*. <http://www.greenerchoices.org/>.

de Guzman, D. (2010). Walmart's Green Goals Will Impact Supply Change. ICIS. Retrieved May 10 from <http://www.icis.com/Articles/2010/05/10/9356545/walmarts-green-goals-will-impact-chemical-supply-chain.html>.

Environment Canada (2010a). The Canadian Environmental Protection Act, 1999 and Virtual Elimination. Retrieved August from <http://www.ec.gc.ca/lcpe-cepa/default.asp?lang=En&n=D44ED61E-1>.

European Chemicals Agency (2007). *About REACH.* <http://guidance.echa.europa.eu/about_reach_en.htm>.

Federal Trade Commission (1999, April). FTC Facts for Consumers: Sorting out 'Green' Advertising Claims. Bureau of Consumer Protection, Office of Consumer and Business Education.

Harris, R. H., English, C. W., & Highland, J. H. (1985). Hazardous waste disposal: Emerging technologies and public policies to reduce public health risks. *Annual Review of Public Health*, *6*(1), 269–294.

Hewlett Packard (2010). *HP general specification for the environment (GSE) revision history.* <http://www.hp.com/hpinfo/globalcitizenship/environment/pdf/revisionhistory.pdf>.

Jobs, S. (2007). A Greener Apple. Apple, Inc. <http://www.apple.com/hotnews/agreenerapple/>.

Kemery, D., & Jones, E. (2010). EPA Increases Transparency on Chemical Risk Information: Action Part of Continued Comprehensive Reform of Toxic Substance Laws. United States Environmental Protection Agency (EPA) News Release. Retrieved January 21 from <http://yosemite.epa.gov/opa/admpress.nsf/bd4379a92ceceeac8525735900400c27/631cf22eb540c4db852576b2004eca47!OpenDocument>.

Montenegro, M. (2009, November 10). Bioplastics Man. *Seed Magazine.*

Pfizer, Inc. (2010). *Our green chemistry program: Commitments and actions.* <http://www.pfizer.com/responsibility/protecting_environment/green_chemistry_commitments.jsp>.

Tedder, D. W., & Pohland, F. G. (2000). *Emerging Technology in Hazardous Waste Management 8.* New York, NY: Kluwer Academic/Plenum Publishers.

US Environmental Protection Agency (EPA) (2003, July). National Analysis – The National Biennial RCRA Hazardous Waste Report (Based on 2001 Data), EPA530-R-03-007. Office of Solid Waste and Emergency Response Office.

US Environmental Protection Agency (EPA) (2007). Primary Mercury Mining: Historical and Current Trends. Commodity Mercury Stakeholder Meeting. Retrieved July 24 from <http://www.epa.gov/mercury/stocks/Primary%20Mercury%20Mining_7.23.07_FINAL.pdf>.

US Environmental Protection Agency (EPA) (2008a, November). National Analysis – The National Biennial RCRA Hazardous Waste Report (Based on 2007 Data), EPA530-R-08-012. Office of Solid Waste and Emergency Response.

US Environmental Protection Agency (EPA) (2008b). Wastes, Hazardous Waste, Waste Minimization Frequent Questions. Office of Resource Conservation and Recovery. Retrieved September from <http://www.epa.gov/epawaste/hazard/wastemin/minimize/faqs.htm#pp>.

US Environmental Protection Agency (EPA) (2009). Priority Chemicals – Wastes – Hazardous Waste – Waste Minimization. Office of Resource Conservation and Recovery. Retrieved May from <http://www.epa.gov/epawaste/hazard/wastemin/priority.htm>.

US Environmental Protection Agency (EPA) (2010a). Database of Environmental Information for Products and Services. Environmentally Preferable Purchasing Program. Retrieved September from <http://yosemite1.epa.gov/oppt/eppstand2.nsf>.

US Environmental Protection Agency (EPA) (2010b). Environmentally Preferred Purchasing: Frequently Asked Questions. Environmentally Preferable Purchasing Program. Retrieved May from <http://www.epa.gov/epp/pubs/about/faq.htm#c>.

US Environmental Protection Agency (EPA) (2010c). Essential Principles for Reform of Chemicals Management Legislation. Office of Pollution Prevention and Toxics. Retrieved April from <http://www.epa.gov/oppt/existingchemicals/pubs/principles.html>.

US Environmental Protection Agency (EPA) (2010d). General Information Gathering Authority. Chemical Information and Testing Branch. Retrieved April from <http://www.epa.gov/oppt/chemtest/pubs/sect8a.html>.

US Environmental Protection Agency (EPA) (2010e). Is Filing Necessary for My Chemical? Office of Pollution Prevention and Toxics. Retrieved August from <http://www.epa.gov/opptintr/newchems/pubs/whofiles.htm>.

US Environmental Protection Agency (EPA) (2010g). Twelve Principles of Green Chemistry. Industrial Chemistry Branch – Green Chemistry Program. Retrieved April from <http://www.epa.gov/gcc/pubs/principles.html>.

US Environmental Protection Agency (EPA) (2010h). What is the TSCA chemical substance inventory? Office of Pollution Prevention and Toxics. Retrieved August from <http://www.epa.gov/oppt/newchems/pubs/invntory.htm>.

US Environmental Protection Agency (EPA) (2010i). Title 40: Protection of the Environment, Part 268: Land Disposal Restrictions – Treatment Standards. Electronic Code of Federal Regulations (e-CFR), US Government Printing Office (GPO), Washington, D.C. Retrieved September from <http://ecfr.gpoaccess.gov/cgi/t/text/text-idx?c=ecfr&sid=6a376c0ac9477323ea42b6a8e796b54a&rgn=div5&view=text&node=40:26.0.1.1.3&idno=40#40:26.0.1.1.3.4.27.3>.

US Environmental Protection Agency (EPA) (2010j). Mid-Atlantic waste minimization – Region 3 national partnership for environmental priorities (NPEP). Land and Chemicals Division. Retrieved March from <http://www.epa.gov/reg3wcmd/wmmembers.htm>.

US Environmental Protection Agency (EPA) (2010k). *Summary of Toxic Substances Control Act of 1976*. Retrieved March from <http://www.epa.gov/lawsregs/laws/tsca.html>.

United States Government Accountability Office (GAO) (2009). Testimony Before the Committee on Environment and Public Works, U.S. Senate Chemical Regulation – Observations on Improving the Toxic Substances Control Act, GAO-10-292T. Retrieved December from <http://www.gao.gov/new.items/d10292t.pdf>.

US House of Representatives (2010, July 22). The Toxic Chemicals Safety Act of 2010 (H.R. 5820). Committee on Energy and Commerce.

Walsh, B. (2005). PVC-Free Agenda Makes Sense. Healthy Building Network. Retrieved July 11 from <http://www.healthybuilding.net/press/plastics_news_opition.html>.

World Resource Institute and Big Room, Inc. (2010). *2010 global ecolabel monitor – Towards transparency*.

Zecchini, F., & Tundo, P. (2010). Perspectives on chemistry and global climate change. *Chemistry International*, *32*(4), 8–10.

Appendix A: OSHA Permissible Exposure Limits (PEL)

CAS No.[c]	Chemical Name	ppm[a][1]	mg/m^3 [b][1]	Skin Designation
75-07-0	Acetaldehyde	200	360	
64-19-7	Acetic acid	10	25	
108-24-7	Acetic anhydride	5	20	
67-64-1	Acetone	1000	2400	
75-05-8	Acetonitrile	40	70	
53-96-3	2-Acetylaminofluorene			
79-27-6	Acetylene tetrabromide	1	14	
107-02-8	Acrolein	0.1	0.25	
79-06-1	Acrylamide		0.3	X
107-13-1	Acrylonitrile			
309-00-2	Aldrin		0.25	X
107-18-6	Allyl alcohol	2	5	X
107-05-1	Allyl chloride	1	3	
106-92-3	Allyl glycidyl ether	(C)10	(C)45	
2179-59-1	Allyl propyl disulfide	2	12	
1344-28-1	alpha-Alumina, total dust		15	
1344-28-1	alpha-Alumina, respirable fraction		5	
7429-90-5	Aluminum Metal, total dust (as Al)		15	
7429-90-5	Aluminum Metal, respirable fraction (as Al)		5	
92-67-1	4-Aminodiphenyl			
504-29-0	2-Aminopyridine	0.5	2	
7664-41-7	Ammonia	50	35	
7773-06-0	Ammonium sulfamate, total dust		15	
7773-06-0	Ammonium sulfamate, respirable fraction		5	
628-63-7	n-Amyl acetate	100	525	
626-38-0	sec-Amyl acetate	125	650	
62-53-3	Aniline and homologs	5	19	X
90-04-0	o-Anisidine		0.5	X
104-94-9	p-Anisidine		0.5	X
7440-36-0	Antimony and compounds (as Sb)		0.5	
86-88-4	1-Naphthylthiourea		0.3	
7440-38-2	Arsenic, inorganic (as As)			
	Arsenic, organic compounds (as As)		0.5	
7784-42-1	Arsine	0.05	0.2	
1332-21-4	Asbestos		(4)	
86-50-0	Azinphos-methyl		0.2	X
7440-39-3	Barium, soluble compounds (as Ba)		0.5	

(*Continued*)

Appendix A (Continued)

CAS No.[c]	Chemical Name	ppm[a][1]	mg/m³ [b][1]	Skin Designation
7727-43-7	Barium sulfate, total dust		15	
	Barium sulfate, respirable fraction		5	
17804-35-2	Benomyl, total dust		15	
17804-35-2	Benomyl, respirable fraction		5	
71-43-2	Benzene; see 1910.1028. See Appendix B for the limits applicable in the operations or sectors excluded in 1910.1028[d][1]			
94-36-0	Benzoyl peroxide		5	
100-44-7	Benzyl chloride	1	5	
7440-41-7	Beryllium and beryllium compounds (as Be)		[2]	
1304-82-1	Bismuth telluride, total dust		15	
1304-82-1	Bismuth telluride, respirable fraction		5	
1303-86-2	Boron oxide, total dust		15	
7637-07-2	Boron trifluoride	(C)1	(C)3	
7726-95-6	Bromine	0.1	0.7	
75-25-2	Bromoform	0.5	5	X
106-99-0	Butadiene	1 ppm/ 5 ppm STEL		
78-93-3	Methyl ethyl ketone	200	590	
111-76-2	2-Butoxyethanol	50	240	X
123-86-4	Butyl-acetate	150	710	
105-46-4	sec-Butyl acetate	200	950	
540-88-5	tert-Butyl acetate	200	950	
71-36-3	n-Butyl alcohol	100	300	
78-92-2	sec-Butyl alcohol	150	450	
75-65-0	tert-Butyl alcohol	100	300	
109-73-9	Butylamine	(C)5	(C)15	X
1189-85-1	tert-Butyl chromate (as CrO[3])			
2426-08-6	n-Butyl glycidyl ether	50	270	
109-79-5	n-Butyl mercaptan	10	35	
98-51-1	p-tert-Butyltoluene	10	60	
7440-43-9	Cadmium (as Cd)			
1317-65-3	Calcium carbonate, total dust		15	
1317-65-3	Calcium carbonate, respirable fraction		5	
1305-62-0	Calcium hydroxide, total dust		15	
1305-62-0	Calcium hydroxide, respirable fraction		5	
1305-78-8	Calcium oxide		5	
1344-95-2	Calcium silicate, total dust		15	
1344-95-2	Calcium silicate, respirable fraction		5	
7778-18-9	Calcium sulfate, total dust		15	
7778-18-9	Calcium sulfate, respirable fraction		5	
76-22-2	Camphor, synthetic		2	
63-25-2	Carbaryl		5	
1333-86-4	Carbon black		3.5	
124-38-9	Carbon dioxide	5000	9000	
75-15-0	Carbon disulfide		[2]	

(*Continued*)

CAS No.[c]	Chemical Name	ppm[a][1]	mg/m³ [b][1]	Skin Designation
630-08-0	Carbon monoxide	50	55	
56-23-5	Carbon tetrachloride		[2]	
9004-34-6	Cellulose, total dust		15	
9004-34-6	Cellulose, respirable fraction		5	
57-74-9	Chlordane		0.5	X
8001-35-2	Chlorinated camphene		0.5	X
	Chlorinated diphenyl oxide		0.5	
7782-50-5	Chlorine	(C)1	(C)3	
10049-04-4	Chlorine dioxide	0.1	0.3	
7790-91-2	Chlorine trifluoride	(C)0.1	(C)0.4	
107-20-0	Chloroacetaldehyde	(C)1	(C)3	
532-27-4	a-Chloroacetophenone	0.05	0.3	
108-90-7	Chlorobenzene	75	350	
2698-41-1	o-Chlorobenzylidene malononitrile	0.05	0.4	
74-97-5	Chlorobromomethane	200	1050	
53469-21-9	Chlorodiphenyl (42% Chlorine)		1	X
11097-69-1	Chlorodiphenyl (54% Chlorine)		0.5	X
67-66-3	Chloroform	(C)50	(C)240	
600-25-9	1-Chloro-1-nitropropane	20	100	
76-06-2	Chloropicrin	0.1	0.7	
126-99-8	beta-Chloroprene	25	90	X
1929-82-4	2-Chloro-6 (trichloromethyl) pyridine, total dust		15	
1929-82-4	2-Chloro-6 (trichloromethyl) pyridine, respirable fraction		5	
	Chromic acid and chromates (as CrO[3])		[2]	
7440-47-3	Chromium (II) compounds (as Cr)		0.5	
7440-47-3	Chromium (III) compounds (as Cr)		0.5	
	Chromium (VI) compounds, See 1910.1026[5]			
7440-47-3	Chromium metal and insol. salts (as Cr)		1	
2971-90-6	Clopidol, total dust		15	
2971-90-6	Clopidol, respirable fraction		5	
	Coal dust (less than 5% SiO[2]), respirable fraction		[3]	
	Coal dust (greater than or equal to 5% SiO[2]), respirable fraction		[3]	
65966-93-2	Coal tar pitch volatiles (benzene soluble fraction), anthracene, BaP, phenanthrene, acridine, chrysene, pyrene		0.2	
7440-48-4	Cobalt metal, dust, and fume (as Co)		0.1	
7440-50-8	Copper fume (as Cu)		0.1	
7440-50-8	Copper dusts and mists (as Cu)		1	
	Cotton dust [e][1], see 1910.1043		1	
136-78-7	Crag herbicide, total dust		15	
136-78-7	Crag herbicide, respirable fraction		5	
Varies	Cresol (all isomers)	5	22	X
4170-30-3	Crotonaldehyde	2	6	
98-82-8	Cumene	50	245	X
	Cyanides (as CN)		5	X

(Continued)

Appendix A (Continued)

CAS No.[c]	Chemical Name	ppm[(a)(1)]	mg/m³ (b)(1)	Skin Designation
110-82-7	Cyclohexane	300	1050	
108-93-0	Cyclohexanol	50	200	
108-94-1	Cyclohexanone	50	200	
110-83-8	Cyclohexene	300	1015	
542-92-7	Cyclopentadiene	75	200	
94-75-7	2,4-D (Dichlorophen-oxyacetic acid)		10	
17702-41-9	Decaborane	0.05	0.3	X
8065-48-3	Demeton		0.1	X
123-42-2	Diacetone alcohol	50	240	
334-88-3	Diazomethane	0.2	0.4	
19287-45-7	Diborane	0.1	0.1	
107-66-4	Dibutyl phosphate	1	5	
84-74-2	Dibutyl phthalate		5	
95-50-1	o-Dichlorobenzene	(C)50	(C)300	
106-46-7	p-Dichlorobenzene	75	450	
75-71-8	Dichlorodifluoromethane	1000	4950	
118-52-5	1,3-Dichloro-5,5-dimethylhydantoin		0.2	
50-29-3	Dichlorodiphenyltrichloroethane (DDT)		1	X
75-34-3	1,1-Dichloroethane	100	400	
540-59-0	1,2-Dichloroethylene	200	790	
111-44-4	Dichloroethyl ether	(C)15	(C)90	X
75-43-4	Dichloromonofluoromethane	1000	4200	
594-72-9	1,1-Dichloro-1-nitroethane	(C)10	(C)60	
76-14-2	Dichlorotetrafluoroethane	1000	7000	
62-73-7	Dichlorvos		1	X
102-54-5	Dicyclopentadienyl iron, total dust		15	
102-54-5	Dicyclopentadienyl iron, respirable fraction		5	
60-57-1	Dieldrin		0.25	X
109-89-7	Diethylamine	25	75	
100-37-8	2-Diethylaminoethanol	10	50	X
75-61-6	Difluorodibromomethane	100	860	
2238-07-5	Diglycidyl ether	(C)0.5	(C)2.8	
108-83-8	Diisobutyl ketone	50	290	
108-18-9	Diisopropylamine	5	20	X
127-19-5	Dimethyl acetamide	10	35	X
124-40-3	Dimethylamine	10	18	
121-69-7	N,N-Dimethylaniline	5	25	X
300-76-5	Dimethyl-1,2-dibromo-2,2-dichloroethyl phosphate		3	
68-12-2	Dimethylformamide	10	30	X
57-14-7	1,1-Dimethylhydrazine	0.5	1	X
131-11-3	Dimethylphthalate		5	
77-78-1	Dimethyl sulfate	1	5	X
528-29-0	o-Dinitrobenzene		1	X
99-65-0	m-Dinitrobenzene		1	X
100-25-4	p-Dinitrobenzene		1	X
534-52-1	Dinitro-o-cresol		0.2	X
25321-14-6	Dinitrotoluene		1.5	X

(Continued)

CAS No.[c]	Chemical Name	ppm[(a)(1)]	mg/m³ [(b)(1)]	Skin Designation
123-91-1	Dioxane	100	360	X
92-52-4	Diphenyl	0.2	1	
34590-94-8	Dipropylene glycol methyl ether	100	600	X
117-81-7	Di-sec octyl phthalate (Di-(2-ethylhexyl) phthalate)		5	
1302-74-5	Emery, total dust		15	
1302-74-5	Emery, respirable fraction		5	
72-20-8	Endrin		0.1	X
106-89-8	Epichlorohydrin	5	19	X
2104-64-5	EPN		0.5	X
141-43-5	Ethanolamine	3	6	
110-80-5	2-Ethoxyethanol (Cellosolve)	200	740	X
111-15-9	2-Ethoxyethyl acetate (Cellosolve acetate)	100	540	X
141-78-6	Ethyl acetate	400	1400	
140-88-5	Ethyl acrylate	25	100	X
64-17-5	Ethyl alcohol (Ethanol)	1000	1900	
75-04-7	Ethylamine	10	18	
541-85-5	5-Methyl-3-heptanone	25	130	
100-41-4	Ethyl benzene	100	435	
74-96-4	Ethyl bromide	200	890	
106-35-4	Ethyl butyl ketone (3-Heptanone)	50	230	
75-00-3	Ethyl chloride	1000	2600	
60-29-7	Ethyl ether	400	1200	
109-94-4	Ethyl formate	100	300	
75-08-1	Ethyl mercaptan	(C)10	(C)25	
78-10-4	Ethyl silicate	100	850	
107-07-3	Ethylene chlorohydrin	5	16	X
107-15-3	Ethylenediamine	10	25	
106-93-4	Ethylene dibromide		(2)	
107-06-2	Ethylene dichloride (1,2-Dichloroethane)		(2)	
628-96-6	Ethylene glycol dinitrate	(C)0.2	(C)1	X
100-74-3	N-Ethylmorpholine	20	94	X
14484-64-1	Ferbam, total dust		15	
12604-58-9	Ferrovanadium dust		1	
Varies	Fluorides (as F)		2.5	
7782-41-4	Fluorine	0.1	0.2	
75-69-4	Fluorotrichloromethane	1000	5600	
64-18-6	Formic acid	5	9	
98-01-1	Furfural	5	20	X
98-00-0	Furfuryl alcohol	50	200	
	Grain dust (oat, wheat, barley)		10	
56-81-5	Glycerin (mist), total dust		15	
56-81-5	Glycerin (mist), respirable fraction		5	
556-52-5	Glycidol	50	150	
7782-42-5	Graphite, natural, respirable fraction		(3)	
7440-44-0	Graphite, synthetic, total dust		15	
7440-44-0	Graphite, synthetic, respirable fraction		5	
13397-24-5	Gypsum, total dust		15	

(*Continued*)

Appendix A (Continued)

CAS No.[c]	Chemical Name	ppm[(a)(1)]	mg/m³ [(b)(1)]	Skin Designation
13397-24-5	Gypsum, respirable fraction		5	
7440-58-6	Hafnium		0.5	
76-44-8	Heptachlor		0.5	X
142-82-5	n-Heptane	500	2000	
67-72-1	Hexachloroethane	1	10	X
1335-87-1	Hexachloronaphthalene		0.2	X
110-54-3	n-Hexane	500	1800	
591-78-6	2-Hexanone	100	410	
108-10-1	Hexone	100	410	
108-84-9	sec-Hexyl acetate	50	300	
302-01-2	Hydrazine	1	1.3	X
10035-10-6	Hydrogen bromide	3	10	
7647-01-0	Hydrogen chloride	(C)5	(C)7	
74-90-8	Hydrogen cyanide	10	11	X
7664-39-3	Hydrogen fluoride		(2)	
7722-84-1	Hydrogen peroxide	1	1.4	
7783-07-5	Hydrogen selenide (as Se)	0.05	0.2	
7783-06-4	Hydrogen sulfide		(2)	
123-31-9	Hydroquinone		2	
7553-56-2	Iodine	(C)0.1	(C)1	
1309-37-1	Iron oxide fume		10	
123-92-2	Isomyl acetate	100	525	
123-51-3	Isomyl alcohol (primary)			
6032-29-7	Isomyl alcohol (secondary)	100	360	
110-19-0	Isobutyl acetate	150	700	
78-83-1	Isobutyl alcohol	100	300	
78-59-1	Isophorone	25	140	
108-21-4	Isopropyl acetate	250	950	
67-63-0	Isopropyl alcohol	400	980	
75-31-0	Isopropylamine	5	12	
108-20-3	Isopropyl ether	500	2100	
4016-14-2	Isopropyl glycidyl ether	50	240	
1332-58-7	Kaolin, total dust		15	
1332-58-7	Kaolin, respirable fraction		5	
463-51-4	Ketene	0.5	0.9	
1317-65-3	Limestone, total dust		15	
1317-65-3	Limestone, respirable fraction		5	
58-89-9	Lindane		0.5	X
7580-67-8	Lithium hydride		0.025	
68476-85-7	L.P.G. (Liquefied petroleum gas)	1000	1800	
546-93-0	Magnesite, total dust		15	
546-93-0	Magnesite, respirable fraction		5	
1309-48-4	Magnesium oxide fume, total particulate		15	
121-75-5	Malathion, total dust		15	X
108-31-6	Maleic anhydride	0.25	1	
7439-96-5	Manganese compounds (as Mn)		(C)5	
7439-96-5	Manganese fume (as Mn)		(C)5	

(Continued)

CAS No.[c]	Chemical Name	ppm[(a)(1)]	mg/m³ [(b)(1)]	Skin Designation
1317-65-3	Marble, total dust		15	
	Marble, respirable fraction		5	
7439-97-6	Mercury (aryl and inorganic) (as Hg)		(2)	
7439-97-6	Mercury (organo) alkyl compounds (as Hg)		(2)	
7439-97-6	Mercury (vapor) (as Hg)		(2)	
141-79-7	Mesityl oxide	25	100	
72-43-5	Methoxychlor, total dust		15	
109-86-4	Methyl cellosolve	25	80	X
110-49-6	Methyl cellosolve acetate	25	120	X
79-20-9	Methyl acetate	200	610	
74-99-7	Methyl acetylene	1000	1650	
59355-75-8	Methyl acetylene-propadiene mixture	1000	1800	
96-33-3	Methyl acrylate	10	35	X
109-87-5	Methylal	1000	3100	
67-56-1	Methyl alcohol	200	260	
74-89-5	Methylamine	10	12	
110-43-0	Methyl (n-amyl) ketone	100	465	
74-83-9	Methyl bromide	(C)20	(C)80	X
74-87-3	Methyl chloride		(2)	
71-55-6	Methyl chloroform	350	1900	
108-87-2	Methylcyclohexane	500	2000	
25639-42-3	Methylcyclohexanol	100	470	
583-60-8	o-Methylcyclohexanone	100	460	X
75-09-2	Methylene chloride		(2)	
107-31-3	Methyl formate	100	250	
60-34-4	Methyl hydrazine	(C)0.2	(C)0.35	X
74-88-4	Methyl iodide	5	28	X
110-12-3	Methyl isoamyl ketone	100	475	
108-11-2	Methyl isobutyl carbinol	25	100	X
624-83-9	Methyl isocyanate	0.02	0.05	X
74-93-1	Methyl mercaptan	(C)10	(C)20	
80-62-6	Methyl methacrylate	100	410	
98-83-9	alpha-Methyl styrene	(C)100	(C)480	
101-68-8	Methylene bisphenyl isocyanate	(C)0.02	(C)0.2	
7439-98-7	Molybdenum (as Mo), soluble compounds		5	
7439-98-7	Molybdenum (as Mo), soluble compounds, total dust		15	
100-61-8	Monomethyl aniline	2	9	X
110-91-8	Morpholine	20	70	X
8030-30-6	Naphtha (Coal tar)	100	400	
91-20-3	Naphthalene	10	50	
13463-39-3	Nickel carbonyl (as Ni)	0.001	0.007	
7440-02-0	Nickel, metal and insoluble compounds (as Ni)		1	
7440-02-0	Nickel, soluble compounds (as Ni)		1	
54-11-5	Nicotine		0.5	X
7697-37-2	Nitric acid	2	5	
10102-43-9	Nitric oxide	25	30	
100-01-6	p-Nitroaniline	1	6	X

(*Continued*)

Appendix A (Continued)

CAS No.[c]	Chemical Name	ppm[(a)(1)]	mg/m^3 [(b)(1)]	Skin Designation
98-95-3	Nitrobenzene	1	5	X
100-00-5	p-Nitrochlorobenzene		1	X
79-24-3	Nitroethane	100	310	
10102-44-0	Nitrogen dioxide	(C)5	(C)9	
7783-54-2	Nitrogen trifluoride	10	29	
55-63-0	Nitroglycerin	(C)0.2	(C)2	X
75-52-5	Nitromethane	100	250	
108-03-2	1-Nitropropane	25	90	
79-46-9	2-Nitropropane	25	90	
88-72-2	o-Nitrotoluene	5	30	X
99-08-1	m-Nitrotoluene	5	30	X
99-99-0	p-Nitrotoluene	5	30	X
2234-13-1	Octachloronaphthalene		0.1	X
111-65-9	Octane	500	2350	
8012-95-1	Oil mist, mineral		5	
20816-12-0	Osmium tetroxide (as Os)		0.002	
144-62-7	Oxalic acid		1	
7783-41-7	Oxygen difluoride	0.05	0.1	
10028-15-6	Ozone	0.1	0.2	
4685-14-7	Paraquat, respirable dust		0.5	X
56-38-2	Parathion		0.1	X
	Particulates not otherwise regulated, total dust (f)		15	
	Particulates not otherwise regulated, respirable fraction (f)[(f)(l)]		5	
19624-22-7	Pentaborane	0.005	0.01	
1321-64-8	Pentachloronaphthalene		0.5	X
87-86-5	Pentachlorophenol		0.5	X
115-77-5	Pentaerythritol			
	Total dust		15	
	Respirable fraction		5	
109-66-0	Pentane	1000	2950	
107-87-9	2-Pentanone	200	700	
127-18-4	Tetrachloroethylene		(2)	
594-42-3	Perchloromethyl mercaptan	0.1	0.8	
7616-94-6	Perchloryl fluoride	3	13.5	
8002-05-9	Petroleum distillates (Naphtha)	500	2000	
108-95-2	Phenol	5	19	X
106-50-3	p-Phenylene diamine		0.1	X
101-84-8	Phenyl ether, vapor	1	7	
8004-13-5	Phenyl ether-biphenyl mixture, vapor	1	7	
122-60-1	Phenyl glycidyl ether	10	60	
100-63-0	Phenylhydrazine	5	22	X
7786-34-7	Phosdrin		0.1	X
75-44-5	Phosgene	0.1	0.4	
7803-51-2	Phosphine	0.3	0.4	
7664-38-2	Phosphoric acid		1	
7723-14-0	Phosphorus (yellow)		0.1	

(Continued)

CAS No.[c]	Chemical Name	ppm[(a)(1)]	mg/m³ [(b)(1)]	Skin Designation
10026-13-8	Phosphorus pentachloride		1	
1314-80-3	Phosphorus pentasulfide		1	
7719-12-2	Phosphorus trichloride	0.5	3	
85-44-9	Phthalic anhydride	2	12	
1918-02-1	Picloram, total dust		15	
1918-02-1	Picloram, respirable fraction		5	
88-89-1	Picric acid		0.1	X
83-26-1	Pindone (2-pivalyl-1,3-indandione)		0.1	
26499-65-0	Plaster of paris, total dust		15	
	Plaster of paris, respirable fraction		5	
7440-06-4	Platinum (as Pt), soluble salts		0.002	
65997-15-1	Portland cement, total dust		15	
65997-15-1	Portland cement, respirable fraction		5	
74-98-6	Propane	1000	1800	
109-60-4	n-Propyl acetate	200	840	
71-23-8	n-Propyl alcohol	200	500	
627-13-4	n-Propyl nitrate	25	110	
78-87-5	Propylene dichloride	75	350	
75-55-8	Propylene imine	2	5	X
75-56-9	Propylene oxide	100	240	
8003-34-7	Pyrethrum		5	
110-86-1	Pyridine	5	15	
106-51-4	Quinone	0.1	0.4	
7440-16-6	Rhodium (metal fume and insoluble compounds, as Rh)		0.1	
7440-16-6	Rhodium (soluble compounds, as Rh)		0.001	
299-84-3	Ronnel		15	
83-79-4	Rotenone		5	
1309-37-1	Rouge, total dust		15	
1309-37-1	Rouge, respirable fraction		5	
7782-49-2	Selenium compounds		0.2	
7783-79-1	Selenium hexafluoride	0.05	0.4	
7440-21-3	Silicon, total dust		15	
7440-21-3	Silicon, respirable fraction		5	
409-21-2	Silicon carbide, total dust		15	
409-21-2	Silicon carbide, respirable fraction		5	
7440-22-4	Silver (metal and soluble compounds, as Ag)		0.01	
62-74-8	Sodium fluoroacetate		0.05	X
1310-73-2	Sodium hydroxide		2	
9005-25-8	Starch, total dust		15	
9005-25-8	Starch, respirable fraction		5	
7803-52-3	Stibine	0.1	0.5	
8052-41-3	Stoddard solvent	500	2900	
57-24-9	Strychnine		0.15	
100-42-5	Styrene		(2)	
57-50-1	Sucrose, total dust		15	
57-50-1	Sucrose, respirable fraction		5	
7446-09-5	Sulfur dioxide	5	13	

(*Continued*)

Appendix A (Continued)

CAS No.(c)	Chemical Name	ppm(a)(1)	mg/m^3 (b)(1)	Skin Designation
2551-62-4	Sulfur hexafluoride	1000	6000	
7664-93-9	Sulfuric acid		1	
10025-67-9	Sulfur monochloride	1	6	
5714-22-7	Sulfur pentafluoride	0.025	0.25	
2699-79-8	Sulfuryl fluoride	5	20	
93-76-5	2,4,5-T (2,4,5-tri-chlorophenoxyacetic acid)		10	
7440-25-7	Tantalum, metal and oxide dust as Ta		5	
3689-24-5	TEDP		0.2	X
13494-80-9	Tellurium and compounds (as Te)		0.1	
7783-80-4	Tellurium hexafluoride (as Te)	0.02	0.2	
3383-96-8	Temephos, total dust		15	
	Temephos, respirable fraction		5	
107-49-3	TEPP (Tetraethyl pyrophosphate)		0.05	X
26140-60-3	Terphenylis	(C)1	(C)9	
76-11-9	1,1,1,2-Tetrachloro-2, 2-difluoroethane	500	4170	
76-12-0	1,1,2,2-Tetrachloro-1, 2-difluoroethane	500	4170	
79-34-5	1,1,2,2-Tetrachloro-ethane	5	35	X
1335-88-2	Tetrachloronaphthalene		2	X
78-00-2	Tetraethyl lead (as Pb)		0.075	X
109-99-9	Tetrahydrofuran	200	590	
75-74-1	Tetramethyl lead (as Pb)		0.075	X
3333-52-6	Tetramethyl succinonitrile	0.5	3	X
509-14-8	Tetranitromethane	1	8	
479-45-8	Tetryl (2,4,6-trinitro-phenylmethyl-nitramine)		1.5	X
7440-28-0	Thallium, soluble compounds (as Tl)		0.1	X
96-69-5	4,4'-Thiobis (6-tert, Butyl-m-cresol)			
	Total dust		15	
	Respirable fraction		5	
137-26-8	Thiram		5	
7440-31-5	Tin, inorganic compounds (except oxides) (as Sn)		2	
7440-31-5	Tin, organic compounds (as Sn)		0.1	
13463-67-7	Titanium dioxide, total dust		15	
108-88-3	Toluene		(2)	
584-84-9	Toluene-2,4-diisocyanate	(C)0.02	(C)0.14	
95-53-4	o-Toluidine	5	22	X
126-73-8	Tributyl phosphate		5	
79-00-5	1,1,2-Trichloroethane	10	45	X
79-01-6	Trichloroethylene		(2)	
1321-65-9	Trichloronaphthalene	(5)	5	X
96-18-4	1,2,3-Trichloropropane	50	300	
76-13-1	1,1,2-Trichloro-1,2,2-trifluoroethane	1000	7600	
121-44-8	Triethylamine	25	100	
75-63-8	Trifluorobromomethane	1000	6100	
118-96-7	2,4,6-Trinitrotoluene		1.5	X
78-30-8	Triorthocresyl phosphate		0.1	
115-86-6	Triphenyl phosphate		3	
8006-64-2	Turpentine	100	560	
7440-61-1	Uranium (as U), soluble compounds		0.05	

(*Continued*)

CAS No.[c]	Chemical Name	ppm[a][1]	mg/m^3 [b][1]	Skin Designation
7440-61-1	Uranium (as U), insoluble compounds		0.25	
1314-62-1	Vanadium, respirable fraction		(C)0.5	
1314-62-1	Vanadium, fume		(C)0.1	
	Vegetable oil mist, total dust		15	
	Vegetable oil mist, respirable fraction		5	
25013-15-4	Vinyl toluene	100	480	
81-81-2	Warfarin		0.1	
1330-20-7	m-Xylenes	100	435	
95-47-6	o-Xylene	100	435	
106-42-3	p-Xylene	100	435	
1300-73-8	Xylidine	5	25	X
7440-65-5	Yttrium		1	
7646-85-7	Zinc chloride fume		1	
1314-13-2	Zinc oxide fume		5	
1314-13-2	Zinc oxide, total dust		15	
1314-13-2	Zinc oxide, respirable fraction		5	
557-05-1	Zinc stearate, total dust		15	
557-05-1	Zinc stearate, respirable fraction		5	
7440-67-7	Zirconium compounds (as Zr)		5	

[a] Parts of vapor or gas per million parts of contaminated air by volume at 25 degrees C and 760 torr.
[b] Milligrams of substance per cubic meter of air. When entry is in this column only, the value is exact; when listed with a ppm entry, it is approximate.
[c] The CAS number is for information only. Enforcement is based on the substance name. For an entry covering more than one metal compound measured as the metal, the CAS number for the metal is given – not CAS numbers for the individual compounds.
[d] The final benzene standard in 1910.1028 applies to all occupational exposures to benzene except in some circumstances the distribution and sale of fuels, sealed containers and pipelines, coke production, oil and gas drilling and production, natural gas processing, and the percentage exclusion for liquid mixtures; for the excepted subsegments, the benzene limits in Appendix B apply. See 1910.1028 for specific circumstances.
[e] This 8-hour TWA applies to respirable dust as measured by a vertical elutriator cotton dust sampler or equivalent instrument. The time-weighted average applies to the cotton waste processing operations of waste recycling (sorting, blending, cleaning and willowing) and garnetting. See also 1910.1043 for cotton dust limits applicable to other sectors.
[f] All inert or nuisance dusts, whether mineral, inorganic, or organic, not listed specifically by substance name are covered by the Particulates Not Otherwise Regulated (PNOR) limit which is the same as the inert or nuisance dust limit of OSHA's Table Z-3 for Mineral Dusts.
[1] The PELs are 8-hour TWAs unless otherwise noted; a (C) designation denotes a ceiling limit. They are to be determined from breathing-zone air samples.
[2] See OSHA Table Z-2, in Appendix B.
[3] See OSHA Table Z-3 Mineral Dusts.
[4] Varies with compound.
[5] See OSHA Table Z-2, Appendix B, for the exposure limits for any operations or sectors where the exposure limits in 1910.1026 are stayed or are otherwise not in effect.

Appendix B: OSHA Permissible Exposure Limits (PEL) Time-Weighted Averages

CAS No.	Chemical Name	8-hour TWA	Acceptable Ceiling Concentration	Acceptable Maximum Peak Above the Acceptable Ceiling Concentration for an 8-hr Shift	
				Concentration	Maximum Duration
71-43-2	Benzene[a]	10 ppm	25 ppm	50 ppm	10 minutes.
7440-41-7	Beryllium and beryllium compounds	2 μg/m^3	5 μg/m^3	25 μg/m^3	30 minutes.
1306-19-0	Cadmium fume[b]	0.1 mg/m^3	0.3 mg/m^3		
7440-43-9	Cadmium dust[b]	0.2 mg/m^3	0.6 mg/m^3		
75-15-0	Carbon disulfide	20 ppm	30 ppm	100 ppm	30 minutes.
56-23-5	Carbon tetrachloride	10 ppm	25 ppm	200 ppm	5 min. in any 3 hrs.
1333-82-0	Chromic acid and chromates[c]		1 mg/10 m^3		
106-93-4	Ethylene dibromide	20 ppm	30 ppm	50 ppm	5 minutes.
107-06-2	Ethylene dichloride	50 ppm	100 ppm	200 ppm	5 min. in any 3 hrs.
	Fluoride as dust	2.5 mg/m^3			
50-00-0	Formaldehyde				
7664-39-3	Hydrogen fluoride	3 ppm			
7783-06-4	Hydrogen sulfide		20 ppm	50 ppm	10 mins. once only if no other meas. exp. occurs.
7439-97-6	Mercury		1 mg/10 m^3		
74-87-3	Methyl chloride	100 ppm	200 ppm	300 ppm	5 mins. in any 3 hrs.
75-09-2	Methylene chloride				
	Organo (alkyl) mercury	0.01 mg/m^3	0.04 mg/m^3		

(Continued)

Appendix B (Continued)

CAS No.	Chemical Name	8-hour TWA	Acceptable Ceiling Concentration	Acceptable Maximum Peak Above the Acceptable Ceiling Concentration for an 8-hr Shift	
				Concentration	Maximum Duration
100-42-5	Styrene	100 ppm	200 ppm	600 ppm	5 mins. in any 3 hrs.
127-18-4	Tetrachloroethylene	100 ppm	200 ppm	300 ppm	5 mins. in any 3 hrs.
108-88-3	Toluene	200 ppm	300 ppm	500 ppm	10 minutes
79-01-6	Trichloroethylene	100 ppm	200 ppm	300 ppm	5 mins. in any 2 hrs.

Reference: 29 CFR Part Number: 1910 Part Title: Occupational Safety and Health Standards Subpart: Z Subpart Title: Toxic and Hazardous Substances Standard Number: 1910.1000 TABLE Z-2 Title: TABLE Z-2. Available HTTP://www.osha.gov/pls/oshaweb/owadisp.show_document?p_table=STANDARDS&p_id=9993

[a] This standard applies to the industry segments exempt from the 1 ppm 8-hour TWA and 5 ppm STEL of the benzene standard at 1910.1028.

[b] This standard applies to any operations or sectors for which the Cadmium standard, 1910.1027, is stayed or otherwise not in effect.

[c] This standard applies to any operations or sectors for which the exposures limit in the Chromium (VI) standard, Sec. 1910.1026, is stayed or is otherwise not in effect.

Appendix C: OSHA Immediately Dangerous to Life and Health (IDLH) Values

CAS No.	Chemical Name	IDLH (ppm)	STEL/Ceiling (ppm)
75-07-0	Acetaldehyde	2000^{++}	25^{+}
64-19-7	Acetic acid	50	15
108-24-7	Acetic anhydride	200	
67-64-1	Acetone	2500	1000
75-05-8	Acetonitrile	500	60
79-27-6	Acetylene tetrabromide	8	
107-02-8	Acrolein	2	0.3
79-06-1	Acrylamide	$60\ mg/m^{3++}$	
309-00-2	Aldrin	$25\ mg/m^{3++}$	
107-18-6	Allyl alcohol	20	4
107-05-1	Allyl chloride	250	2
106-92-3	Allyl glycidyl ether	50	10^{+}
2179-59-1	Allyl propyl disulfide		3
504-29-0	2-Aminopyridine	5	
7664-41-7	Ammonia	300	35
7773-06-0	Ammonium sulfamate	$1500\ mg/m^3$	
628-63-7	n-Amyl acetate	1000	
626-38-0	sec-Amyl acetate	1000	
628-63-7	Aniline and homologs	100^{++}	
62-53-3	o-Anisidine	$50\ mg/m^3$	
104-94-9	p-Anisidine	$50\ mg/m^3$	
7440-36-0	Antimony & compounds	$50\ mg/m^3$	
86-88-4	ANTU	$100\ mg/m^3$	
7784-42-1	Arsine	3^{++}	
86-50-0	Azinphos-methyl	$10\ mg/m^3$	
94-36-0	Benzoyl peroxide	$1500\ mg/m^3$	
100-44-7	Benzyl chloride	10	
1303-86-2	Boron oxide	$2000\ mg/m^3$	
7637-07-2	Boron trifluoride	25	1^{+}
7726-95-6	Bromine	3	0.2
75-25-2	Bromoform	850	
106-99-0	1,3 Butadiene	2000^{++}	
78-93-3	2-Butanone	3000	300
111-76-2	2-Butoxyethanol	700	

(Continued)

Appendix C (Continued)

CAS No.	Chemical Name	IDLH (ppm)	STEL/Ceiling (ppm)
123-86-4	Butyl-acetate	1700	200
105-46-4	sec-Butyl acetate	1700	
540-88-5	tert-Butly acetate	1500	
71-36-3	n-Butyl alcohol	1400	50^{+}
78-92-2	sec-Butyl alcohol	2000	
75-65-0	tert-Butyl alcohol	1600	
109-73-9	Butylamine	300	5^{+}
1189-85-1	tert-Butyl chromate	15 mg/m^3	0.1 mg/m^{3+}
2426-08-6	n-Butyl glycidyl ether	250	
109-79-5	n-Butyl mercaptan	500	
98-51-1	p-tert-Butyl-1-toluene	100	
1305-78-8	Calcium oxide	25 mg/m^3	
76-22-2	Camphor, synthetic	200 mg/m^3	19 mg/m^{3+}
63-25-2	Carbaryl	100 mg/m^3	
1333-86-4	Carbon black	1750 mg/m^3	
124-38-9	Carbon dioxide	40,000	30,000
75-15-0	Carbon disulfide	500	
630-08-0	Carbon monoxide	1200	
57-74-9	Chlorodane	100 mg/m^{3++}	
8001-35-2	Chlorinated camphene	200 mg/m3CA	
	Chlorinated diphenyl oxide	5 mg/m^3	
7782-50-5	Chlorine	10	1
10049-04-4	Chlorine dioxide	5	0.3
7790-91-2	Chlorine trifluoride	20	0.1^{+}
107-20-0	Chloroacetaldehyde	45	1^{+}
532-27-4	a-Chloroacetophenone	15 mg/m^3	
108-90-7	Chlorobenzene	1000	
2698-41-1	o-Chlorobenzylidene malononitrile	2 mg/m^3	0.05+
74-97-5	Chlorobromomethane	2000	
53469-21-9	Chlorodiphenyl (42% Cl)	5 mg/m^{3++}	
11097-69-1	Chlorobiphenyl (54% Cl)	5 mg/m^{3++}	
600-25-9	1-Chloro-1-nitro-propane	100	
76-06-2	Chloropicrin	2	
126-99-8	beta-Chloroprene	300	
1929-82-4	2-Chloro-6-trichloromethyl pyridine		20 mg/m^3
	Chromium (II) compounds (as Cr)	250 mg/m^3	
	Chromium (III) compounds (as Cr)	25 mg/m^3	
7440-47-3	Chromium metal and insoluble salts	250 mg/m^3	
65996-93-2	Coal tar pitch volatiles	80 mg/m^{3++}	
7440-48-4	Cobalt metal, dust, and fume (as Co)	20 mg/m^3	
1317-38-0	Copper fume (as Cu)	100 mg/m^3	
7440-50-8	Copper dusts and mists (as Cu)	100 mg/m^3	
	Cotton dust (raw)	100 mg/m^3	
136-78-7	Crag herbicide	500 mg/m^3	
(Varies)	Cresol, all isomers	250	
4170-30-3	Crotonaldehyde	50	
98-82-8	Cumene	900	
110-82-7	Cyclohexane	1300	
108-93-0	Cyclohexanol	400	

(Continued)

CAS No.	Chemical Name	IDLH (ppm)	STEL/Ceiling (ppm)
108-94-1	Cyclohexanone	700	
110-83-8	Cyclohexene	2000	
542-92-7	Cyclopentadiene	750	
94-75-7	2,4-D	100 mg/m^3	
17702-41-9	Decaborane	15 mg/m^3	0.15
8065-48-3	Demeton	10 V	
123-42-2	Diacetone alcohol	1800	
334-88-3	Diazomethane	2	
19287-45-7	Diborane	15	
107-66-4	Dibutyl phosphate	30	2
84-74-2	Dibutyl phthalate	4000 mg/m^3	
95-50-1	o-Dichlorobenzene	200	
106-46-7	p-Dichlorobenzene	150^{++}	
75-34-3	1,1-Dichloroethane	3000	
540-59-0	1,2-Dichloroethylene	1000	
111-44-4	Dichloroethyl ether	100^{++}	15
75-43-4	Dichloromonofluomethane	5000	
594-72-9	1,1-Dichloro-1-nitroethane	25	10^{+}
76-14-2	Dichlorotetrafluoroethane	15,000	
62-73-7	Dichlorvos	100 mg/m^3	
60-57-1	Dieldrin	50 mg/m^{3++}	
109-89-7	Diethylamine	200	15
100-37-8	2-Diethylaminoethanol	100	
75-61-6	Difluorodibromomethane	2000	
123639	Diglycidyl ether	10^{++}	0.5^{+}
108-83-8	Diisobutyl ketone	500	
108-18-9	Diisopropylamine	200	
127-19-5	Dimethyl acetamide	300	
124-40-3	Dimethylamine	500	15
121-69-7	Dimethylaniline	100	10
300-76-5	Dimethyl-1,2-dibromo-2,2-dichloroethyl phosphate	200 mg/m^3	
68-12-2	Dimethylformamide	500	
57-14-7	1,1-Dimethylhydrazine	15^{++}	
131-11-3	Dimethylphthalate	2000 mg/m^3	
77-78-1	Dimethyl sulfate	7b	
Varies	Dinitrobenzene (all isomers)	50 mg/m^3	
534-52-1	Dinitro-o-cresol	5 mg/m^3	
25321-14-6	Dinitrotoluene	50 mg/m^{3++}	
123-91-1	Dioxane	500^{++}	
92-52-4	Diphenyl	100 mg/m^3	
34590-94-8	Dipropylene glycol methyl ether	600	150
117-81-7	Di-sec octyl phthalate	10 mg/m^{3+}	
72-20-8	Endrin	2 mg/m^3	
106-89-8	Epichlorohydrin	75^{++}	
2104-64-5	EPN	5 mg/m^3	
141-43-5	Ethanolamine	30	6
110-80-5	2-Ethoxyethanol	500	
111-15-9	2-Ethoxyethyl acetate	500	
141-78-6	Ethyl acetate	2000	

(Continued)

Appendix C (Continued)

CAS No.	Chemical Name	IDLH (ppm)	STEL/Ceiling (ppm)
140-88-5	Ethyl acrylate	300	15
64-17-5	Ethyl alcohol	3300	
75-04-7	Ethylamine	600	15
100-41-4	Ethyl benzene	800	125
74-96-4	Ethyl bromide	2000	
106-35-4	Ethyl butyl ketone	1000	
75-00-3	Ethyl chloride	3800	
107-07-3	Ethylene chlorohydrin	7	
628-96-6	Ethylene glycol dinitrate	75 mg/m^3	0.2^+
60-29-7	Ethyl ether	1900	500
109-94-4	Ethyl formate	1500	
75-08-1	Ethyl mercaptan	500	10^+
78-10-4	Ethyl silicate	700	
107-15-3	Ethylenediamine	1000	
100-74-3	N-Ethylmorpholine	100	
14484-64-1	Ferbam (td)	800 mg/m^3	
12604-58-9	Ferrovanadium dust	500 mg/m^3	3 mg/m^{3+}
7782-41-4	Fluorine	25	2
75-69-4	Fluorotrichloromethane	2000	1000^+
64-18-6	Formic acid	30	10
98-01-1	Furfural	100	
98-00-0	Furfuryl alcohol	75	15
556-52-5	Glycidol	150	
7440-58-6	Hafnium	50 mg/m^3	
76-44-8	Heptachlor	35 mg/m^{3++}	
142-82-5	n-Heptane	750	500
67-72-1	Hexachloroethane	300^{++}	
1335-87-1	Hexachloronaphthalene	2 mg/m^3	
110-54-3	n-Hexane	1100	
591-78-6	2-Hexanone	1600	
108-10-1	Hexone	500	75
108-84-9	sec-Hexyl acetate	500	
302-01-2	Hydrazine	50	
10035-10-6	Hydrogen bromide	30	3^+
7647-01-0	Hydrogen chloride	50	5^+
74-90-8	Hydrogen cyanide	50	4.7^+
7722-84-1	Hydrogen peroxide	75	
7783-07-05	Hydrogen selenide (as Se)	1	
123-31-9	Hydroquinone	50 mg/m^3	
7553-56-2	Iodine	2	0.1^+
1309-37-1	Iron oxide fume	2500 mg/m^3	
123-92-2	Isoamyl acetate	1000	
6032-29-7	Isoamyl alcohol (primary)	500	125
123-51-3	Isoamyl alcohol (secondary)	500	125
110-19-0	Isobutyl acetate	1300	
78-83-1	Isobutyl alcohol	1600	
78-59-1	Isophorone	200	5^+
108-21-4	Isopropyl acetate	1800	310
67-63-0	Isopropyl alcohol	2000	500

(Continued)

CAS No.	Chemical Name	IDLH (ppm)	STEL/Ceiling (ppm)
75-31-0	Isopropylamine	750	10
108-20-3	Isopropyl ether	1400	310
4016-14-2	Isopropyl glycidyl ether	400	75
463-51-4	Ketene	5	1.5
58-89-9	Lindane	50 mg/m^3	
7580-67-8	Lithium hydride	0.5 mg/m^3	
68476-85-7	L.P.G. (Liquefied petroleum gas)	2000	
1309-48-4	Magnesium oxide fume (total particulate)	750 mg/m^3	
121-75-5	Malathion	250 mg/m^3	
108-31-6	Maleic anhydride	10 mg/m^3	
7439-96-5	Manganese compounds (as Mn)	500 mg/m^3	5 mg/m^{3+}
7439-96-5	Manganese fume (as Mn)	500 mg/m^3	5 mg/m^{3+}
141-79-7	Mesityl oxide	1400	25
72-43-5	Methoxychlor	5000 mg/m^{3+}	
109-86-4	Methyl cellosolve	200	
110-49-6	Methyl cellosolve acetate	200	
79-20-9	Methyl acetate	3100	250
74-99-7	Methyl acetylene	1700	
59355-75-8	Methyl acetylene propadiene mixture	3400	1250
96-33-3	Methyl acrylate	250	
109-87-5	Methylal	2200	
67-56-1	Methyl alcohol	6000	250
74-89-5	Methylamine	100	15
110-43-0	Methyl (n-amyl) ketone	800	
74-83-9	Methyl bromide	250^{++}	20^{+}
71-55-6	Methyl chloroform	700	450
108-87-2	Methylcyclohexane	1200	
25639-42-3	Methylcyclohexanol	500	
583-60-8	o-Methylcyclohexanone	600	75
101-68-8	Methylene bisphenyl isocyanate	75 mg/m^3	0.02^{+}
107-31-3	Methyl formate	4500	150
60-34-4	Methyl hydrazine	20^{++}	0.2^{+}
74-88-4	Methyl iodide	100^{++}	
108-11-2	Methyl isobutyl carbinol	400	40
624-83-9	Methyl isocyanate	3	
74-93-1	Methyl mercaptan	150	10^{+}
80-62-6	Methyl methacrylate	1000	
98-83-9	alpha-Methyl styrene	700	100^{+}
	Molybdenum (as Mo) Soluble compounds	1000 mg/m^3	
7439-98-7	Molybdenum (as Mo) Insoluble compounds	5000 mg/m^3	
100-61-8	Monomethyl aniline	100	
110-91-8	Morpholine	1400	
8030-30-6	Naphtha (Coal tar)	1000	
91-20-3	Naphthalene	250	15
13463-39-3	Nickel carbonyl (as Ni)	2^{++}	
7440-02-0	Nickel, metal and insoluble compounds (as Ni)	10 mg/m^{3++}	
54-11-5	Nicotine	5 mg/m^3	
7697-37-2	Nitric acid	25	4
10102-43-9	Nitric oxide	100	
100-01-6	p-Nitroaniline	300 mg/m^3	

(Continued)

Appendix C (Continued)

CAS No.	Chemical Name	IDLH (ppm)	STEL/Ceiling (ppm)
98-95-3	Nitrobenzene	200	
100-00-5	p-Nitrochlorobenzene	100 mg/m^{3+}	
79-24-3	Nitroethane	1000	
10102-44-0	Nitrogen dioxide	20	5
7783-54-2	Nitrogen trifluoride	1000	
55-63-0	Nitroglycerin	75 mg/m^3	0.2^+
75-52-5	Nitromethane	750	
108-03-2	1-Nitropropane	1000	
79-46-9	2-Nitropropane	100	
Varies	Nitrotoluene (all isomers)	200	
2234-13-1	Octachloronaphthalene	Unknown	0.3 mg/m^{3+}
111-65-9	Octane	1000	375
8012-95-1	Oil mist, mineral	2500 mg/m^3	10 mg/m^3
20816-12-0	Osmium tetroxide (as Os)	1 mg/m^3	0.0006
144-62-7	Oxalic acid	500 mg/m^3	2 mg/m^3
7783-41-7	Oxygen difluoride	0.5	0.05^+
10028-15-6	Ozone	5	0.1^+
1910-42-5	Paraquat	1 mg/m^3	
56-38-2	Parathion	10 mg/m^3	
19624-22-7	Pentaborane	1	0.015
1321-64-8	Pentachloronaphthalene	Unknown	
87-86-5	Pentachlorophenol	2.5 mg/m^3	
109-66-0	Pentane	1500	750
107-87-9	2-Pentanone	1500	250
594-42-3	Perchloromethyl mercaptan	10	
7616-94-6	Perchloryl fluoride	100	25^+
2228840	Petroleum distillates	1100	1480
108-95-2	Phenol	250	
106-50-3	p-Phenylene diamine	25 mg/m^3	
101-84-8	Phenyl ether, vapor	100	2
8004-13-5	Phenyl ether-biphenyl mixture, vapor	10	
122-60-1	Phenyl glycidyl ether	100^{++}	
100-63-0	Phenylhydrazine	15^{++}	
7786-34-7	Phosdrin	4	0.03
75-44-5	Phosgene	2	
7803-51-2	Phosphine	50	1
7664-38-2	Phosphoric acid	1000 mg/m^3	
7723-14-0	Phosphorus (yellow)	5 mg/m^3	
10026-13-8	Phosphorus pentachloride	70 mg/m^3	
1314-80-3	Phosphorus pentasulfide	250 mg/m^3	3 mg/m^3
2125683	Phosphorus trichloride	25	0.5
85-44-9	Phthalic anhydride	60 mg/m^3	
88-89-1	Picric acid	75 mg/m^3	
83-26-1	Pindone	100 mg/m^3	
	Platinum (as Pt) Soluble salts	4 mg/m^3	
65997-15-1	Portland cement	5000 mg/m^3	
74-98-6	Propane	2100	

(Continued)

CAS No.	Chemical Name	IDLH (ppm)	STEL/Ceiling (ppm)
109-60-4	n-Propyl acetate	1700	250
71-23-8	n-Propyl alcohol	800	250
78-87-5	Propylene dichloride	400^{++}	110
75-55-8	Propylene imine	100^{++}	
75-56-9	Propylene oxide	400^{++}	
627-13-4	n-Propyl nitrate	500	40
8003-34-7	Pyrethrum	5000 mg/m^3	
110-86-1	Pyridine	1000	
106-51-4	Quinone	100 mg/m^3	
7440-16-6	Rhodium, metal fume and insoluble compounds	100 mg/m^3	
	Rhodium, soluble compounds	2 mg/m^3	
299-84-3	Ronnel	300 mg/m^3	
83-79-4	Rotenone	2500 mg/m^3	
7782-49-2	Selenium compounds (as Se)	1 mg/m^3	
7783-79-1	Selenium hexafluoride (as Se)	2	
7440-22-4	Silver, metal and soluble compounds (as Ag)	10 mg/m^3	
62-74-8	Sodium fluoroacetate	2.5 mg/m^3	
1310-73-2	Sodium hydroxide	10 mg/m^3	2 mg/m^{3+}
7803-52-3	Stibine	5	
8052-41-3	Stoddard solvent	20,000 mg/m^3	
57-24-9	Strychnine	3 mg/m^3	
7446-09-5	Sulfur dioxide	100	5
7664-93-9	Sulfuric acid	15 mg/m^3	3 mg/m^{3+}
10025-67-9	Sulfur monochloride	5	1^+
5714-22-7	Sulfur pentafluoride	1	0.01^+
2699-79-8	Sulfuryl fluoride	200	10
93-76-5	2,4,5-T	250 mg/m^3	
7440-25-7	Tantalum, metal and oxide dust	2500 mg/m^3	
3689-24-5	TEDP	10 mg/m^3	
13494-80-9	Tellurium and compounds (as Te)	25 mg/m^3	
7783-80-4	Tellurium hexafluoride (as Te)	1	
107-49-3	TEPP	5 mg/m^3	
Varies	Terphenyls	500 mg/m^3	1
76-11-9	1,1,1,2-Tetrachloro-2,2-difluoroethane	2000	
76-12-0	1,1,2,2-Tetrachloro-1,2-difluoroethane	2000	
79-34-5	1,1,2,2-Tetrachloroethane	100^{++}	
1335-88-2	Tetrachloronaphthalene	Unknown	
78-00-2	Tetraethyl lead (as Pb)	40 mg/m^3	
109-99-9	Tetrahydrofuran	2000	250
75-74-1	Tetramethyl lead (as Pb)	40 mg/m^3	
3333-52-6	Tetramethyl succinonitrile	5	
509-14-8	Tetranitromethane	4	
479-45-8	Tetryl	750 mg/m^3	
	Thallium, soluble compounds (as Tl)	15 mg/m^3	
137-26-8	Thiram	100 mg/m^3	
7440-31-5	Tin, inorganic compounds (except oxides) (as Sn)	100 mg/m^3	
	Tin, organic compounds (as Sn)	25 mg/m^3	

(Continued)

Appendix C (Continued)

CAS No.	Chemical Name	IDLH (ppm)	STEL/Ceiling (ppm)
13463-67-7	Titanium dioxide	5000 mg/m^{3+}	
584-84-9	Toluene-2,4-diisocyanate	2.5^{++}	0.02^{+}
119-93-7	o-Toluidine	50^{++}	
126-73-8	Tributyl phosphate	30	
79-00-5	1,1,2-Trichloroethane	100^{++}	
1321-65-9	Trichloronaphthalene	Unknown	
96-18-4	1,2,3-Trichloropropane	100^{++}	
76-13-1	1,1,2-Trichloro-1,2,2-trifluoroethane	2000	1250
121-44-8	Triethylamine	200	3
75-63-8	Trifluorobromomethane	40,000	
118-96-7	2,4,6-Trinitrotoluene	500 mg/m^3	
78-30-8	Triorthocresyl phosphate	40 mg/m^3	
115-86-6	Triphenyl phosphate	1000 mg/m^3	
8006-64-2	Turpentine	800	
7440-61-1	Uranium (as U) Soluble compounds	10 mg/m^{3++}	
	Uranium (as U) Insoluble compounds	10 mg/m^{3++}	
1314-62-1	Vanadium	35 mg/m^3	0.5 mg/m^3
1314-62-1	Vanadium (fume)	35 mg/m^3	0.1 mg/m^3
25013-15-4	Vinyl toluene	400	100
81-81-2	Warfarin	100 mg/m^3	
Varies	Xylenes (o-, m-, p-isomers)	900	150
1300-73-8	Xylidine	50	
7440-65-5	Yttrium	500 mg/m^3	
7646-85-7	Zinc chloride fume	50 mg/m^3	
1314-13-2	Zinc oxide fume	500 mg/m^3	
1314-13-2	Zinc oxide	500 mg/m^3	
7440-67-7	Zirconium compounds (as Zr)	50 mg/m^3	10 mg/m^3

Reference: http://www.labsafety.com/refinfo/ezfacts/ezf232.htm
http://www.osha.gov/web/dep/chemicaldata/
Notes: $^{+}$ indicates a Ceiling Value
$^{++}$ indicates that the chemical is believed, by NIOSH, to be a potential carcinogen.

Appendix D: NIOSH Recommended Exposure Limits (REL)

CAS No.	Chemical Name	NIOSH REL TWA (ppm)	NIOSH REL TWA (mg/m^3)	NIOSH Ceiling REL (ppm)	NIOSH Ceiling REL (mg/m^3)	NIOSH REL STEL (ppm)	NIOSH REL STEL (mg/m^3)
107-18-6	AA	2 (skin)	5 (skin)			4 (skin)	10 (skin)
3383-96-8	Abate		10 (total) 5 (resp)				
64-19-7	Acetic acid	10	25			15	37
108-24-7	Acetic anhydride	5	20				
67-64-1	Acetone	250	590				
75-86-5	Acetone cyanohydrin			1 (15 min)	4 (15 min)		
75-05-8	Acetonitrile	20	34				
74-86-2	Acetylene			2500	2662		
50-78-2	Acetylsalicyclic acid		5				
107-02-8	Acrolein	0.1	0.25			0.3	0.8
79-06-1	Acrylamide		0.03 (skin)				
79-10-7	Acrylic acid	2 (skin)	6 (skin)				
107-13-1	Acrylonitrile	1 (15 min/skin)		10 (15 min/ skin)			
111-69-3	Adiponitrile	4	18				
309-00-2	Aldrin		0.25(skin)				
107-18-6	Allyl alcohol	2 (skin)	5 (skin)			4 (skin)	10 (skin)
107-05-1	Allyl chloride	1	3			2	6
2179-59-1	Allyl propyl disulfide	2	12			3	18
7429-90-5	Aluminum		10 (total) 5 (resp)				
	Aluminum (pyro powders and welding fumes, as Al)		5				
	Aluminum (soluble salts and alkyls, as Al)		2				
504-29-0	2-Aminopyridine	0.5	2				
61-82-5	Amitrole		0.2				
7664-41-7	Ammonia	25	18			35	27
12125-02-9	Ammonium chloride fume		10				20
7773-06-0	Ammonium sulfamate		10 (total) 5 (resp)				

628-63-7	n-Amyl acetate	100	525			
626-38-0	sec-Amyl acetate	125	650			
90-04-0	o-Anisidine		0.5 (skin)			
104-94-9	p-Anisidine		0.5 (skin)			
7440-36-0	Antimony		0.5			
86-88-4	ANTU		0.3			
7440-38-2	Arsenic (inorganic compounds, as As)				0.002 (15 min)	
7784-42-1	Arsine				0.002 (15min)	
8052-42-4	Asphalt fumes				5 mg/m^3 (15 min)	
1912-24-9	Atrazine		5			
86-50-0	Azinphos-methyl		0.2 (skin)			
10361-37-2	Barium chloride (as Ba)		0.5			
10022-31-8	Barium nitrate (as Ba)		0.5			
7727-43-7	Barium sulfate		10 (total) 5 (resp)			
17804-35-2	Benomyl		15 (total) 5 (resp)			
71-43-2	Benzene	0.1				1
108-98-5	Benzenethiol			0.1 (15 min)	0.5 (15 min)	
94-36-0	Benzoyl peroxide		5			
100-44-7	Benzyl chloride			1 (15 min)	5 (15 min)	
7440-41-7	Beryllium & beryllium compounds (as Be)		not to exceed 0.0005			
	Bismuth telluride, doped with Selenium sulfide (as Bi2Te3)		5			
1304-82-1	Bismuth telluride, undoped		10 (total) 5 (resp)			
1330-43-4	Borates, tetra, sodium salts (Anhydrous)		1			
1303-96-4	Borates, tetra, sodium salts (Decahydrate)		5			
12179-04-3	Borates, tetra, sodium salts (Pentahydrate)		1			
1303-86-2	Boron oxide		10			

(Continued)

Appendix D (Continued)

CAS No.	Chemical Name	NIOSH REL TWA (ppm)	NIOSH REL TWA (mg/m^3)	NIOSH Ceiling REL (ppm)	NIOSH Ceiling REL (mg/m^3)	NIOSH REL STEL (ppm)	NIOSH REL STEL (mg/m^3)
10294-33-4	Boron tribromide			1	10		
7637-07-2	Boron trifluoride			1	3		
314-40-9	Bromacil	1	10				
7726-95-6	Bromine	0.1	0.7			0.3	2
7789-30-2	Bromine pentafluoride	0.1	0.7				
75-25-2	Bromoform	0.5 (skin)	5 (skin)				
106-97-8	n-Butane	800	1900				
78-93-3	2-Butanone	200	590			300	885
111-76-2	2-Butoxyethanol	5 (skin)	24 (skin)				
112-07-2	2-Butoxyethanol acetate	5	33				
123-86-4	Butyl acetate	150	710			200	950
105-46-4	sec-Butyl acetate	200	950				
540-88-5	tert-Butyl acetate	200	950				
141-32-2	Butyl acrylate	10	55				
71-36-3	n-Butyl alcohol			50 (skin)	150 (skin)		
78-92-2	sec-Butyl alcohol	100	305			150	455
75-65-0	tert-Butyl alcohol	100	300			150	450
109-73-9	Butylamine			5 (skin)	15 (skin)		
1189-85-1	tert-Butyl chromate (as CrO3)		0.001 mg Cr(VI)/m^3				
2426-08-6	n-Butyl glycidyl ether			5.6 (15 min)	30 (15 min)		
138-22-7	n-Butyl lactate	5	25				
109-79-5	n-Butyl mercaptan			0.5 (15 min)	1.8 (15 min)		
89-72-5	o-sec-Butylphenol	5 (skin)	30 (skin)				
98-51-1	p-tert-Butyltoluene	10	60			20	120
109-74-0	n-Butyronitrile	8	22				
7778-44-1	Calcium arsenate (as As)				0.0002 (15 min)		
1317-65-3	Calcium carbonate		10 (total 5) (resp)				
156-62-7	Calcium cyanamide		0.5				

1305-62-0	Calcium hydroxide		5				
1305-78-8	Calcium oxide		2				
1344-95-2	Calcium silicate		10 (total 5 (resp)				
7778-18-9	Calcium sulfate		10 (total 5 (resp)				
76-22-2	Camphor (synthetic)		2				
105-60-2	Caprolactam	0.22 (vapor)	1 (dust) 1 (vapor)			0.66 (vapor)	3 (dust) 3 (vapor)
2425-06-1	Captafol		0.1 (skin)				
133-06-2	Captan		5				
63-25-2	Carbaryl		5				
1563-66-2	Carbofuran		0.1				
1333-86-4	Carbon black		3.5 (0.1 mg PAHs/m^3. . .carbon black in the presence of PAHs)				
124-38-9	Carbon dioxide	5000	9000			30000	54000
75-15-0	Carbon disulfide	1 (skin)	3 (skin)			10 (skin)	30 (skin)
630-08-0	Carbon monoxide	35	40	200	229		
558-13-4	Carbon tetrabromide	0.1	1.4			0.3	4
56-23-5	Carbon tetrachloride					2 (60 min)	12.6 (60 min)
353-50-4	Carbonyl fluoride	2	5			5	15
120-80-9	Catechol	5 (skin)	20 (skin)				
9004-34-6	Cellulose		10 (total 5) (resp)				
21351-79-1	Cesium hydroxide		2				
57-74-9	Chlordane		0.5 (skin)				
	Chlorinated diphenyl oxide		0.5				
7782-50-5	Chlorine			0.5 (15 min)	1.45 (15 min)		
10049-04-4	Chlorine dioxide	0.1	0.3			0.3	0.9
7790-91-2	Chlorine trifluoride			0.1	0.4		
107-20-0	Chloroacetaldehyde			1	3		
532-27-4	alpha-Chloroacetophenone	0.05	0.3				
79-04-9	Chloroacetyl chloride	0.05	0.2				
2698-41-1	o-Chlorobenzylidene malononitrile			0.05 (skin)	0.4 (skin)		

(Continued)

Appendix D (Continued)

CAS No.	Chemical Name	NIOSH REL TWA (ppm)	NIOSH REL TWA (mg/m^3)	NIOSH Ceiling REL (ppm)	NIOSH Ceiling REL (mg/m^3)	NIOSH REL STEL (ppm)	NIOSH REL STEL (mg/m^3)
74-97-5	Chlorobromomethane	200	1050				
75-45-6	Chlorodifluoromethane	1000	3500			1250	4375
53469-21-9	Chlorodiphenyl (42% chlorine)		0.001				
11097-69-1	Chlorodiphenyl (54% chlorine)		0.001				
67-66-3	Chloroform					2 (60 min)	9.78 (60 min)
600-25-9	1-Chloro-1-nitropropane	2	10				
76-15-3	Chloropentafluoroethane	1000	6320				
76-06-2	Chloropicrin	0.1	0.7				
126-99-8	beta-Chloroprene			1 (15 min)	3.6 (15 min)		
2039-87-4	o-Chlorostyrene	50	285			75	428
95-49-8	o-Chlorotoluene	50	250			75	375
1929-82-4	2-Chloro-6-trichloromethyl pyridine		10 (total) 5 (resp)				20 (total)
2921-88-2	Chlorpyrifos		0.2 (skin)			0.6 (skin)	
1333-82-0	Chromic acid and chromates		0.001				
	Chromium(II) compounds (as Cr)		0.5				
	Chromium(III) compounds (as Cr)		0.5				
7440-47-3	Chromium metal		0.5				
14977-61-8	Chromyl chloride		0.001 mg Cr (VI)/m^3				
2971-90-6	Clopidol		10 (total) 5 (resp)				20 (total)
	Coal dust		2.4 (resp, $<5\%$ SiO_2) 10 (resp, greater than or equal to 5% SiO_2)				
65996-93-2	Coal tar pitch volatiles		0.1 (cyclohexane-extractable fraction)				

10210-68-1	Cobalt carbonyl (as Co)		0.1				
16842-03-8	Cobalt hydrocarbonyl (as Co)		0.1				
7440-48-4	Cobalt metal, dust, and fume (as Co)		0.05				
	Coke oven emissions		0.2 (benzene-soluble fraction)				
7440-50-8	Copper (dusts and mists, as Cu)		1				
1317-38-0	Copper fume (as Cu)		0.1				
	Cotton dust (raw)		<0.2				
136-78-7	Crag herbicide		10 (total) 5 (resp)				
108-39-4	m-Cresol	2.3	10				
95-48-7	o-Cresol	2.3	10				
106-44-5	p-Cresol	2.3	10				
4170-30-3	Crotonaldehyde	2	6				
299-86-5	Crufomate		5				20
98-82-8	Cumene	50 (skin)	245 (skin)				
420-04-2	Cyanamide		2				
460-19-5	Cyanogen	10	20				
506-77-4	Cyanogen chloride			0.3	0.6		
110-82-7	Cyclohexane	300	1050				
1569-69-3	Cyclohexanethiol			0.5 (15 min)	2.4 (15 min)		
108-93-0	Cyclohexanol	50 (skin)	200 (skin)				
108-94-1	Cyclohexanone	25 (skin)	100 (skin)				
110-83-8	Cyclohexene	300	1015				
108-91-8	Cyclohexylamine	10	40				
121-82-4	Cyclonite		1.5 (skin)				3 (skin)
542-92-7	Cyclopentadiene	75	200				
287-92-3	Cyclopentane	600	1720				
13121-70-5	Cyhexatin		5				
94-75-7	2,4-D		10				
50-29-3	DDT		0.5				
17702-41-9	Decaborane	0.05 (skin)	0.3 (skin)			0.15 (skin)	0.9 (skin)
143-10-2	1-Decanethiol			0.5 (15 min)	3.6 (15 min)		

(Continued)

Appendix D (Continued)

CAS No.	Chemical Name	NIOSH REL TWA (ppm)	NIOSH REL TWA (mg/m^3)	NIOSH Ceiling REL (ppm)	NIOSH Ceiling REL (mg/m^3)	NIOSH REL STEL (ppm)	NIOSH REL STEL (mg/m^3)
8065-48-3	Demeton		0.1 (skin)				
123-42-2	Diacetone alcohol	50	240				
615-05-4	2,4-Diaminoanisole (and its salts)	(minimize occupational exposure)					
333-41-5	Diazinon		0.1 (skin)				
334-88-3	Diazomethane	0.2	0.4				
19287-45-7	Diborane	0.1	0.1				
102-81-8	2-N-Dibutylaminoethanol	2 (skin)	14 (skin)				
128-37-0	2,6-Di-tert-butyl-p-cresol		10				
107-66-4	Dibutyl phosphate	1	5			2	10
84-74-2	Dibutyl phthalate		5				
7572-29-4	Dichloroacetylene			0.1	0.4		
95-50-1	o-Dichlorobenzene			50	300		
75-71-8	Dichlorodifluoromethane	1000	4950				
118-52-5	1,3-Dichloro-5, 5-dimethylhydantoin		0.2				0.4
75-34-3	1,1-Dichloroethane	100	400				
540-59-0	1,2-Dichloroethylene	200	790				
111-44-4	Dichloroethyl ether	5 (skin)	30 (skin)			10 (skin)	60 (skin)
75-43-4	Dichloromonofluoromethane	10	40				
594-72-9	1,1-Dichloro-1-nitroethane	2	10				
542-75-6	1,3-Dichloropropene	1 (skin)	5 (skin)				
75-99-0	2,2-Dichloropropionic acid	1	6				
76-14-2	Dichlorotetrafluoroethane	1000	7000				
62-73-7	Dichlorvos		1(skin)				
141-66-2	Dicrotophos		0.25 (skin)				
77-73-6	Dicyclopentadiene	5	30				
102-54-5	Dicyclopentadienyl iron		10 (total) 5 (resp)				

60-57-1	Dieldrin		0.25 (skin)				
111-42-2	Diethanolamine	3	15				
109-89-7	Diethylamine	10	30			25	75
100-37-8	2-Diethylaminoethanol	10 (skin)	50 (skin)				
111-40-0	Diethylenetriamine	1 (skin)	4 (skin)				
96-22-0	Diethyl ketone	200	705				
84-66-2	Diethyl phthalate		5				
75-61-6	Difluorodibromomethane	100	860				
2238-07-5	Diglycidyl ether	0.1	0.5				
108-83-8	Diisobutyl ketone	25	150				
108-18-9	Diisopropylamine	5 (skin)	20 (skin)				
127-19-5	Dimethyl acetamide	10 (skin)	35 (skin)				
124-40-3	Dimethylamine	10	18				
121-69-7	N, N-Dimethylaniline	5 (skin)	25 (skin)			10 (skin)	50 (skin)
300-76-5	Dimethyl-1,2-dibromo-2,2-dichlorethyl phosphate		3 (skin)				
68-12-2	Dimethylformamide	10 (skin)	30 (skin)				
57-14-7	1,1-Dimethylhydrazine			0.06 (2 hr)	0.15 (2 hr)		
131-11-3	Dimethylphthalate		5				
77-78-1	Dimethyl sulfate	0.1 (skin)	0.5 (skin)				
148-01-6	Dinitolmide		5				
99-65-0	m-Dinitrobenzene		1 (skin)				
528-29-0	o-Dinitrobenzene		1 (skin)				
100-25-4	p-Dinitrobenzene		1 (skin)				
534-52-1	Dinitro-o-cresol		0.2 (skin)				
25321-14-6	Dinitrotoluene		1.5 (skin)				
117-81-7	Di-sec octyl phthalate		5				10
123-91-1	Dioxane			1 (30 min)	3.6 (30 min)		
78-34-2	Dioxathion		0.2 (skin)				
92-52-4	Diphenyl	0.2	1				
122-39-4	Diphenylamine		10				
34590-94-8	Dipropylene glycol methyl ether	100 (skin)	600 (skin)			150 (skin)	900 (skin)

(Continued)

Appendix D (Continued)

CAS No.	Chemical Name	NIOSH REL TWA (ppm)	NIOSH REL TWA (mg/m^3)	NIOSH Ceiling REL (ppm)	NIOSH Ceiling REL (mg/m^3)	NIOSH REL STEL (ppm)	NIOSH REL STEL (mg/m^3)
123-19-3	Dipropyl ketone	50	235				
85-00-7	Diquat		0.5				
97-77-8	Disulfiram		2				
298-04-4	Disulfoton		0.1 (skin)				
330-54-1	Diuron		10				
1321-74-0	Divinyl benzene	10	50				
112-55-0	1-Dodecanethiol			0.5 (15 min)	4.1 (15 min)		
115-29-7	Endosulfan		0.1 (skin)				
72-20-8	Endrin		0.1 (skin)				
13838-16-9	Enflurane			2 (60 min)	15.1 (60 min)		
106-89-8	Epichlorohydrin	5 (skin)	19 (skin)				
2104-64-5	EPN		0.5 (skin)				
141-43-5	Ethanolamine	3	8			6	15
563-12-2	Ethion		0.4 (skin)				
110-80-5	2-Ethoxyethanol	0.5 (skin)	1.8 (skin)				
111-15-9	2-Ethoxyethyl acetate	0.5 (skin)	2.7 (skin)				
141-78-6	Ethyl acetate	400	1400				
64-17-5	Ethyl alcohol	1000	1900				
75-04-7	Ethylamine	10	18				
100-41-4	Ethyl benzene	100	435			125	545
106-35-4	Ethyl butyl ketone	50	230				
75-00-3	Ethyl chloride	(handle with caution in the workplace)					
107-07-3	Ethylene chlorohydrin			1 (skin)	3 (skin)		
107-15-3	Ethylenediamine	10	25				
106-93-4	Ethylene dibromide	0.045 (15 min)		0.13 (15 min)			
107-06-2	Ethylene dichloride	1	4			2	8
628-96-6	Ethylene glycol dinitrate						0.1 (skin)

151-56-4	Ethyleneimine	<0.1 (10 min)	0.18 (10 min)	5 (10 min)	9 (10 min)		
75-21-8	Ethylene oxide	<0.1 (10 min)	0.18 (10 min)	5 (10 min)	9 (10 min)		
109-94-4	Ethyl formate	100	300				
16219-75-3	Ethylidene norbornene			5	25		
75-08-1	Ethyl mercaptan			0.5 (15 min)	1.3 (15 min)		
100-74-3	N-Ethylmorpholine	5 (skin)	23 (skin)				
78-10-4	Ethyl silicate	10	85				
22224-92-6	Fenamiphos		0.1 (skin)				
115-90-2	Fensulfothion		0.1				
14484-64-1	Ferbam		10				
12604-58-9	Ferrovanadium dust		1				3
	Fibrous glass dust		5 (total) 3 fibers/cm^3 (fiber diameter <or = to 3.5 um)				
7782-41-4	Fluorine	0.1	0.2				
75-69-4	Fluorotrichloromethane			1000	5600		
406-90-6	Fluoroxene			2 (60 min)	10.3 (60 min)		
944-22-9	Fonofos		0.1 (skin)				
50-00-0	Formaldehyde	0.016 (15 min)		0.1 (15 min)			
	Formalin (as formaldehyde)	0.016 (15 min)		0.1 (15 min)			
75-12-7	Formamide	10 (skin)	15 (skin)				
64-18-6	Formic acid	5	9				
98-00-0	Furfuryl alcohol	10 (skin)	40 (skin)			15 (skin)	60 (skin)
7782-65-2	Germanium tetrahydride	0.2	0.6				
111-30-8	Glutaraldehyde			0.2	0.8		
556-52-5	Glycidol	25	75				
107-16-4	Glycolonitrile			2 (15 min)	5 (15 min)		
	Grain dust (oat, wheat, barley)		4				
7782-42-5	Graphite (natural)		2.5 (resp)				
7440-44-0	Graphite (synthetic)		15 (total) 5 (resp)				
13397-24-5	Gypsum		10 (total 5 (resp)				
7440-58-6	Hafnium		0.5				
151-67-7	Halothane			2 (60 min)	16.2 (60 min)		

(Continued)

Appendix D (Continued)

CAS No.	Chemical Name	NIOSH REL TWA (ppm)	NIOSH REL TWA (mg/m^3)	NIOSH Ceiling REL (ppm)	NIOSH Ceiling REL (mg/m^3)	NIOSH REL STEL (ppm)	NIOSH REL STEL (mg/m^3)
76-44-8	Heptachlor		0.5 (skin)				
142-82-5	n-Heptane	85 (15 min)	350 (15 min)	440 (15 min)	1800 (15 min)		
1639-09-4	1-Heptanethiol			0.5 (15 min)	2.7 (15 min)		
87-68-3	Hexachlorobutadiene	0.02 (skin)	0.24 (skin)				
77-47-4	Hexachlorocyclopentadiene	0.01	0.1				
67-72-1	Hexachloroethane	1 (skin)	10 (skin)				
1335-87-1	Hexachloronaphthalene		0.2 (skin)				
2917-26-2	1-Hexadecanethiol			0.5 (15 min)	5.3 (15 min)		
684-16-2	Hexafluoroacetone	0.1 (skin)	0.7 (skin)				
822-06-0	Hexamethylene diisocyanate	0.005 (10 min)	0.035 (10 min)	0.020 (10 min)	0.140 (10 min)		
110-54-3	n-Hexane	50	180				
	Hexane isomers (excluding n-Hexane)	100 (15 min)	350 (15 min)	510 (15 min)	1800 (15 min)		
111-31-9	n-Hexanethiol			0.5 (15 min)	2.7 (15 min)		
591-78-6	2-Hexanone	1	4				
108-10-1	Hexone	50	205			75	300
108-84-9	sec-Hexyl acetate	50	300				
107-41-5	Hexylene glycol			25	125		
302-01-2	Hydrazine			0.03 (2 hour)	0.04 (2 hour)		
61788-32-7	Hydrogenated terphenyls	0.5	5				
10035-10-6	Hydrogen bromide			3	10		
7647-01-0	Hydrogen chloride			5	7		
74-90-8	Hydrogen cyanide					4.7 (skin)	5 (skin)
7664-39-3	Hydrogen fluoride	3 (15 min)	2.5 (15 min)	6 (15 min)	5 (15 min)		
7722-84-1	Hydrogen peroxide	1	1.4				
7783-07-5	Hydrogen selenide	0.05	0.2				
7664-93-9	Hydrogen sulfide			10 (10 min)	15 (10 min)		

123-31-9	Hydroquinone				2 (15 min)		
999-61-1	2-Hydroxypropyl acrylate	0.5 (skin)	3 (skin)				
95-13-6	Indene	10	45				
7440-74-6	Indium		0.1				
7553-56-2	Iodine			0.1	1		
75-47-8	Iodoform	0.6	10				
1309-37-1	Iron oxide dust and fume		5				
13463-40-6	Iron pentacarbonyl (as Fe)	0.1	0.23			0.2	0.45
	Iron salts (soluble, as Fe)		1				
123-92-2	Isoamyl acetate	100	525				
6032-29-7	Isoamyl alcohol (secondary)	100	360			125	450
123-51-3	Isoamyl alcohol (primary)	100	360			125	450
75-28-5	Isobutane	800	1900				
110-19-0	Isobutyl acetate	150	700				
78-83-1	Isobutyl alcohol	50	150				
78-82-0	Isobutyronitrile	8	22				
26952-21-6	Isooctyl alcohol	50 (skin)	270 (skin)				
78-59-1	Isophorone	4	23				
4098-71-9	Isophorone diisocyanate	0.005 (skin)	0.045 (skin)			0.02 (skin)	0.180 (skin)
67-63-0	Isopropyl alcohol	400	980			500	1225
768-52-5	N-Isopropylaniline	2 (skin)	10 (skin)				
108-20-3	Isopropyl ether	500	2100				
4016-14-2	Isopropyl glycidyl ether			50 (15 min)	240 (15 min)		
1332-58-7	Kaolin		10 (total) 5 (resp)				
143-50-0	Kepone		0.001				
8008-20-6	Kerosene		100				
463-51-4	Ketene	0.5	0.9			1.5	3
7439-92-1	Lead		0.050				
1317-65-3	Limestone		10 (total) 5 (resp)				
58-89-9	Lindane		0.5 (skin)				
7580-67-8	Lithium hydride		0.025				
68476-85-7	L.P.G. (Liquefied Petroleum Gas)	1000	1800				

(Continued)

Appendix D (Continued)

CAS No.	Chemical Name	NIOSH REL TWA (ppm)	NIOSH REL TWA (mg/m^3)	NIOSH Ceiling REL (ppm)	NIOSH Ceiling REL (mg/m^3)	NIOSH REL STEL (ppm)	NIOSH REL STEL (mg/m^3)
546-93-0	Magnesite		10 (total) 5 (resp)				
121-75-5	Malathion		10 (skin)				
108-31-6	Maleic anhydride	0.25	1				
109-77-3	Malononitrile	3	8				
7439-96-5	Manganese compounds (as Mn)		1				3
7439-96-5	Manganese fume (as Mn)		1				3
12079-65-1	Manganese cyclopentadienyl tricarbonyl (as Mn)		0.1 (skin)				
1317-65-3	Marble		10 (total) 5 (resp)				
7439-97-6	Mercury compounds [except (organo) alkyls] (as Hg)		0.05 (vapor/skin)		0.1 (skin)		
7439-97-6	Mercury (organo) alkyl compounds (as Hg)		0.01 (skin)				0.03 (skin)
141-79-7	Mesityl oxide	10	40				
79-41-4	Methacrylic acid	20 (skin)	70 (skin)				
16752-77-5	Methomyl		2.5				
76-38-0	Methoxyflurane			2 (60 min)	13.5 (60 min)		
150-76-5	4-Methoxyphenol		5				
79-20-9	Methyl acetate	200	610			250	760
74-99-7	Methyl acetylene	1000	1650				
59355-75-8	Methyl acetylene-propadiene mixture	1000	1800			1250	2250
96-33-3	Methyl acrylate	10 (skin)	35 (skin)				
126-98-7	Methylacrylonitrile	1 (skin)	3 (skin)				
109-87-5	Methylal	1000	3100				
67-56-1	Methyl alcohol	200 (skin)	260 (skin)			250 (skin)	325 (skin)

74-89-5	Methylamine	10	12				
109-86-4	Methyl cellosolve	0.1 (skin)	0.3 (skin)				
110-49-6	Methyl cellosolve acetate	0.1 (skin)	0.5 (skin)				
71-55-6	Methyl chloroform			350 (15 min)	1900 (15 min)		
137-05-3	Methyl-2-cyanoacrylate	2	8			4	16
108-87-2	Methylcyclohexane	400	1600				
25639-42-3	Methylcyclohexanol	50	235				
583-60-8	o-Methylcyclohexanone	50 (skin)	230 (skin)			75 (skin)	345 (skin)
12108-13-3	Methyl cyclopentadienyl manganese tricarbonyl (as Mn)		0.2 (skin)				
8022-00-2	Methyl demeton		0.5 (skin)				
101-68-8	Methylene bisphenyl isocyanate	0.005 (10 min)	0.05 (10 min)	0.02 (10 min)	0.20 (10 min)		
101-14-4	4, 4′-Methylenebis (2-chloroaniline)		0.003 (skin)				
5124-30-1	Methylene bis (4-cyclohexylisocyanate)			0.01	0.11		
1338-23-4	Methyl ethyl ketone peroxide			0.2	1.5		
107-31-3	Methyl formate	100	250			150	375
541-85-5	5-Methyl-3-heptanone	25	130				
60-34-4	Methyl hydrazine			0.04 (2 hour)	0.08 (2 hour)		
74-88-4	Methyl iodide	2 (skin)	10 (skin)				
110-12-3	Methyl isoamyl ketone	50	240				
108-11-2	Methyl isobutyl carbinol	25 (skin)	100 (skin)			40 (skin)	165 (skin)
624-83-9	Methyl isocyanate	0.02 (skin)	0.05 (skin)				
563-80-4	Methyl isopropyl ketone	200	705				
74-93-1	Methyl mercaptan			0.5 (15 min)	1 (15 min)		
80-62-6	Methyl methacrylate	100	410				
110-43-0	Methyl (n-amyl) ketone	100	465				
298-00-0	Methyl parathion		0.2 (skin)				
681-84-5	Methyl silicate	1	6				

(Continued)

Appendix D (Continued)

CAS No.	Chemical Name	NIOSH REL TWA (ppm)	NIOSH REL TWA (mg/m^3)	NIOSH Ceiling REL (ppm)	NIOSH Ceiling REL (mg/m^3)	NIOSH REL STEL (ppm)	NIOSH REL STEL (mg/m^3)
98-83-9	alpha-Methyl styrene	50	240			100	485
21087-64-9	Metribuzin		5				
12001-26-2	Mica (containing less than 1% quartz)		3 (resp)				
	Mineral wool fiber		3 fibers/cm^3 (fiber diameter < or = to 3.5 μm)				
6923-22-4	Monocrotophos		0.25				
100-61-8	Monomethyl aniline	0.5 (skin)	2 (skin)				
110-91-8	Morpholine	20 (skin)	70 (skin)			30 (skin)	105 (skin)
8030-30-6	Naphtha (coal tar)	100	400				
91-20-3	Naphthalene	10	50			15	75
3173-72-6	Naphthalene diisocyanate	0.005	0.040	0.020	0.170		
13463-39-3	Nickel carbonyl	0.001	0.007				
7440-02-0	Nickel metal and other compounds (as Ni)		0.015				
54-11-5	Nicotine		0.5 (skin)				
7697-37-2	Nitric acid	2	5			4	10
10102-43-9	Nitric oxide	25	30				
100-01-6	p-Nitroaniline		3 (skin)				
98-95-3	Nitrobenzene	1 (skin)	5 (skin)				
79-24-3	Nitroethane	100	310				
10102-44-0	Nitrogen dioxide					1	1.8
7783-54-2	Nitrogen trifluoride	10	29				
55-63-0	Nitroglycerine						0.1 (skin)
108-03-2	1-Nitropropane	25	90				
99-08-1	m-Nitrotoluene	2 (skin)	11 (skin)				
88-72-2	o-Nitrotoluene	2 (skin)	11 (skin)				
99-99-0	p-Nitrotoluene	2 (skin)	11 (skin)				

10024-97-2	Nitrous oxide	25	46				
111-84-2	Nonane	200	1050				
1455-21-6	1-Nonanethiol			0.5 (15 min)	3.3 (15 min)		
2234-13-1	Octachloronaphthalene		0.1 (skin)				0.3 (skin)
2885-00-9	1-Octadecanethiol			0.5 (15 min)	5.9 (15 min)		
111-65-9	Octane	75 (15 min)	350 (15 min)	385 (15 min)	1800 (15 min)		
111-88-6	1-Octanethiol			0.5 (15 min)	3.0 (15 min)		
8012-95-1	Oil mist (mineral)		5				10
20816-12-0	Osmium tetroxide	0.0002	0.002			0.0006	0.006
144-62-7	Oxalic acid		1				2
7783-41-7	Oxygen difluoride			0.05	0.1		
10028-15-6	Ozone			0.1	0.2		
8002-74-2	Paraffin wax fume		2				
1910-42-5	Paraquat		0.1 (resp/skin)				
56-38-2	Parathion		0.05 (skin)				
19624-22-7	Pentaborane	0.005	0.01			0.15	0.03
76-01-7	Pentachloroethane	(handle with caution in the workplace)					
1321-64-8	Pentachloronaphthalene		0.5 (skin)				
87-86-5	Pentachlorophenol		0.5 (skin)				
115-77-5	Pentaerythritol		10 (total) 5 (resp)				
109-66-0	n-Pentane	120 (15 min)	350 (15 min)	610 (15 min)	1800 (15 min)		
110-66-7	1-Pentanethiol			0.5 (15 min)	2.1 (15 min)		
107-87-9	2-Pentanone	150	530				
594-42-3	Perchloromethyl mercaptan	0.1	0.8				
7616-94-6	Perchloryl fluoride	3	14			6	28
93763-70-3	Perlite		10 (total) 5 (resp)				
8002-05-9	Petroleum distillates (naphtha)		350 (15 min)		1800 (15 min)		
108-95-2	Phenol	5 (15 min)	19 (15 min)	15.6 (15 min)	60 (15 min)		
92-84-2	Phenothiazine		5 (skin)				
106-50-3	p-Phenylene diamine		0.1 (skin)				

(Continued)

Appendix D (Continued)

CAS No.	Chemical Name	NIOSH REL TWA (ppm)	NIOSH REL TWA (mg/m^3)	NIOSH Ceiling REL (ppm)	NIOSH Ceiling REL (mg/m^3)	NIOSH REL STEL (ppm)	NIOSH REL STEL (mg/m^3)
8004-13-5	Phenyl ether-biphenyl mixture (vapor)	1	7				
101-84-8	Phenyl ether (vapor)	1	7				
122-60-1	Phenyl glycidyl ether			1 (15 min)	6 (15 min)		
100-63-0	Phenylhydrazine			0.14 (2 hour/ skin)	0.6 (2 hour/ skin)		
638-21-1	Phenylphosphine			0.05	0.25		
298-02-2	Phorate		0.05 (skin)				0.2 (skin)
7786-34-7	Phosdrin	0.01 (skin)	0.1 (skin)			0.03 (skin)	0.3 (skin)
75-44-5	Phosgene	0.1 (15 min)	0.4 (15 min)	0.2 (15 min)	0.8 (15 min)		
7803-51-2	Phosphine	0.3	0.4			1	1
7664-38-2	Phosphoric acid		1				1
10025-87-3	Phosphorus oxychloride	0.1	0.6			0.5	3
10026-13-8	Phosphorus pentachloride		1				
1314-80-3	Phosphorus pentasulfide		1				3
7719-12-2	Phosphorus trichloride	0.2	1.5			0.5	3
7723-14-0	Phosphorus (yellow)		0.1				
85-44-9	Phthalic anhydride	1	6				
626-17-5	m-Phthalodinitrile		5				
88-89-1	Picric acid		0.1 (skin)				0.3 (skin)
83-26-1	Pindone		0.1				
142-64-3	Piperazine dihydrochloride		5				
26499-65-0	Plaster of Paris		10 (total) 5 (resp)				
7440-06-4	Platinum		1				
7440-06-4	Platinum (soluble salts, as Pt)		0.002				
65997-15-1	Portland cement		10 (total) 5 (resp)				

151-50-8	Potassium cyanide (as CN)			4.7 (10 min)	5 (10 min)		
1310-58-3	Potassium hydroxide				2		
74-98-6	Propane	1000	1800				
107-03-9	1-Propanethiol			0.5 (15 min)	1.6 (15 min)		
107-19-7	Propargyl alcohol	1 (skin)	2 (skin)				
79-09-4	Propionic acid	10	30			15	45
107-12-0	Propionitrile	6	14				
114-26-1	Propoxur		0.5				
109-60-4	n-Propyl acetate	200	840			250	1050
71-23-8	n-Propyl alcohol	200 (skin)	500 (skin)			250 (skin)	625 (skin)
6423-43-4	Propylene glycol dinitrate	0.05 (skin)	0.3 (skin)				
107-98-2	Propylene glycol monomethyl ether	100	360			150	540
75-55-8	Propylene imine	2 (skin)	5 (skin)				
627-13-4	n-Propyl nitrate	25	105			40	170
8003-34-7	Pyrethrum		5				
110-86-1	Pyridine	5	15				
106-51-4	Quinone	0.1	0.4				
108-46-3	Resorcinol	10	45			20	90
7440-16-6	Rhodium (metal fume and insoluble compounds, as Rh)		0.1				
	Rhodium (soluble compounds, as Rh)		0.001				
299-84-3	Ronnel		10				
	Rosin core solder, pyrolysis products (as formaldehyde)		0.1				
83-79-4	Rotenone		5				
7782-49-2	Selenium		0.2				
7783-79-1	Selenium hexafluoride	0.05					
7631-86-9	Silica, amorphous		6				
14808-60-7	Silica, crystalline (as respirable dust)		0.05				

(*Continued*)

Appendix D (Continued)

CAS No.	Chemical Name	NIOSH REL TWA (ppm)	NIOSH REL TWA (mg/m^3)	NIOSH Ceiling REL (ppm)	NIOSH Ceiling REL (mg/m^3)	NIOSH REL STEL (ppm)	NIOSH REL STEL (mg/m^3)
7440-21-3	Silicon		10 (total) 5 (resp)				
409-21-2	Silicon carbide		10 (total 5 (resp)				
7803-62-5	Silicon tetrahydride	5	7				
7440-22-4	Silver (metal dust and soluble compounds, as Ag)		0.01				
	Soapstone (containing less than 1% quartz)		6 (total) 3 (resp)				
15096-52-3	Sodium aluminum fluoride (as F)		2.5				
26628-22-8	Sodium azide			0.1 (as HN_3 on skin)	0.3 (as NaN_3 on skin)		
7631-90-5	Sodium bisulfite		5				
143-33-9	Sodium cyanide (as CN)			4.7 (10 min)	5 (10 min)		
7681-49-4	Sodium fluoride (as F)		2.5				
62-74-8	Sodium fluoroacetate		0.05 (skin)				0.15 (skin)
1310-73-2	Sodium hydroxide				2		
7681-57-4	Sodium metabisulfite		5				
9005-25-8	Starch		10 (total) 5 (resp)				
7803-52-3	Stibine	0.1	0.5				
8052-41-3	Stoddard solvent		350 (15 min)		1800 (15 min)		
57-24-9	Strychnine		0.15				
100-42-5	Styrene	50	215			100	425
1395-21-7	Subtilisins						0.00006 (60 min)
110-61-2	Succinonitrile	6	20				
57-50-1	Sucrose		10 (total 5 (resp)				
7446-09-5	Sulfur dioxide	2	5			5	13
2551-62-4	Sulfur hexafluoride	1000	6000				

7664-93-9	Sulfuric acid		1				
10025-67-9	Sulfur monochloride			1	6		
5714-22-7	Sulfur pentafluoride			0.01	0.1		
7783-60-0	Sulfur tetrafluoride			0.1	0.4		
2699-79-8	Sulfuryl fluoride	5	20			10	40
35400-43-2	Sulprofos		1				
93-76-5	2,4,5-T		10				
14807-96-6	Talc (containing no asbestos and less than 1% quartz)		2 (resp)				
7440-25-7	Tantalum (metal and oxide dust, as Ta)		5				10
3689-24-5	TEDP		0.2 (skin)				
13494-80-9	Tellurium (as Te)		0.1				
7783-80-4	Tellurium hexafluoride (as Te)	0.02	0.2				
3383-96-8	Temephos		10 (total) 5 (resp)				
107-49-3	TEPP		0.05 (skin)				
92-06-8	m-Terphenyl			0.5	5		
84-15-1	o-Terphenyl			0.5	5		
92-94-4	p-Terphenyl			0.5	5		
76-11-9	1,1,1,2-Tetrachloro-2, 2-difluoroethane	500	4170				
76-12-0	1,1,2,2-Tetrachloro-1, 2-difluoroethane	500	4170				
630-20-6	1,1,1,2-Tetrachloroethane	(handle with caution in the workplace)					
79-34-5	1,1,2,2-Tetrachloroethane	1 (skin)	7 (skin)				
127-18-4	Tetrachloroethylene	(minimize occupational exposure)					
1335-88-2	Tetrachloronaphthalene		2 (skin)				
78-00-2	Tetraethyl lead (as Pb)		0.075 (skin)				
109-99-9	Tetrahydrofuran	200	590			250	735

(Continued)

Appendix D (Continued)

CAS No.	Chemical Name	NIOSH REL TWA (ppm)	NIOSH REL TWA (mg/m^3)	NIOSH Ceiling REL (ppm)	NIOSH Ceiling REL (mg/m^3)	NIOSH REL STEL (ppm)	NIOSH REL STEL (mg/m^3)
75-74-1	Tetramethyl lead (as Pb)		0.075 (skin)				
3333-52-6	Tetramethyl succinonitrile	0.5 (skin)	3 (skin)				
509-14-8	Tetranitromethane	1	8				
7722-88-5	Tetrasodium pyrophosphate		5				
479-45-8	Tetryl (2,4,6-Trinitro-phenylmethyl-nitramine)		1.5 (skin)				
	Thallium (soluble compounds, as Tl)		0.1 (skin)				
96-69-5	4,4′-Thiobis (6-tert-butyl-m-cresol)		10 (total) 5 (resp)				
68-11-1	Thioglycolic acid	1 (skin)	4 (skin)				
2125597	Thionyl chloride			1	5		
137-26-8	Thiram		5				
7440-31-5	Tin		2				
7440-31-5	Tin (organic compounds, as Sn)		0.1 (skin)				
21651-19-4	Tin(II) oxide (as Sn)		2				
18282-10-5	Tin(IV) oxide (as Sn)		2				
119-93-7	o-Tolidine				0.02 (60 min/ skin)		
108-88-3	Toluene	100	375			150	560
126-73-8	Tributyl phosphate	0.2	2.5				
76-03-9	Trichloroacetic acid	1	7				
120-82-1	1,2,4-Trichlorobenzene			5	40		
79-00-5	1,1,2-Trichloroethane	10 (skin)	45 (skin)				
79-01-6	Trichloroethylene		5 (skin)				
1321-65-9	Trichloronaphthalene		5 (skin)				
96-18-4	1,2,3-Trichloropropane	10 (skin)	60 (skin)				
76-13-1	1,1,2-Trichloro-1,2, 2-trifluoroethane	1000	7600			1250	9500

75-63-8	Trifluorobromomethane	1000	6100				
552-30-7	Trimellitic anhydride	0.005	0.04				
75-50-3	Trimethylamine	10	24			15	36
526-73-8	1,2,3-Trimethylbenzene	25	125				
95-63-6	1,2,4-Trimethylbenzene	25	125				
108-67-8	1,3,5-Trimethylbenzene	25	125				
121-45-9	Trimethyl phosphite	2	10				
118-96-7	2,4,6-Trinitrotoluene		0.5 (skin)				
78-30-8	Triorthocresyl phosphate		0.1 (skin)				
603-34-9	Triphenylamine		5				
115-86-6	Triphenyl phosphate		3				
7440-33-7	Tungsten		5				10
	Tungsten (soluble compounds, as W)		1				3
8006-64-2	Turpentine	100	560				
5332-52-5	1-Undecanethiol			0.5 (15 min)	3.9 (15 min)		
7440-61-1	Uranium (insoluble compounds, as U)		0.2				0.6
	Uranium (soluble compounds, as U)		0.05				
110-62-3	n-Valeraldehyde	50	175				
1314-62-1	Vanadium dust				0.05 mg V/m^3 (15 min)		
1314-62-1	Vanadium fume				0.05 mg V/m^3 (15 min)		
68956-68-3	Vegetable oil mist		10 (total) 5 (resp)				
108-05-4	Vinyl acetate			4 (15 min)	15 (15 min)		
106-87-6	Vinyl cyclohexene dioxide	10 (skin)	60 (skin)				
75-02-5	Vinyl fluoride	1		5			
75-38-7	Vinylidene fluoride	1		5			
25013-15-4	Vinyl toluene	100	480				
8032-32-4	VM & P Naphtha		350 (15 min)	1800 (15 min)			

(Continued)

Appendix D (Continued)

CAS No.	Chemical Name	NIOSH REL TWA (ppm)	NIOSH REL TWA (mg/m^3)	NIOSH Ceiling REL (ppm)	NIOSH Ceiling REL (mg/m^3)	NIOSH REL STEL (ppm)	NIOSH REL STEL (mg/m^3)
81-81-2	Warfarin		0.1				
	Wood dust		1				
108-38-3	m-Xylene	100	435			150	655
95-47-6	o-Xylene	100	435			150	655
106-42-3	p-Xylene	100	435			150	655
1477-55-0	m-Xylene-alpha, alpha′-diamine			0.1 (skin)			
1300-73-8	Xylidine	2 (skin)	10 (skin)				
7440-65-5	Yttrium		1				
7646-85-7	Zinc chloride fume		1				2
1314-13-2	Zinc oxide		5 (dust) 5 (fume)		15 (dust)		10 (fume)
557-05-1	Zinc stearate		10 (total) 5 (resp)				
7440-67-7	Zirconium compounds (as Zr)		5				10

Reference: http://www.cdc.gov/niosh/npg/npgname-a.html

Appendix E: NIOSH Immediately Dangerous to Life and Health (IDLH) Values

CAS No.	Chemical Name	Original IDLH	Revised IDLH
75-07-0	Acetaldehyde	10,000 ppm	2,000 ppm
64-19-7	Acetic acid	1,000 ppm	50 ppm
108-24-7	Acetic anhydride	1,000 ppm	200 ppm
67-64-1	Acetone	20,000 ppm	2,500 ppm [LEL]
75-05-8	Acetonitrile	4,000 ppm	500 ppm
79-27-6	Acetylene tetrabromide	10 ppm	8 ppm
107-02-8	Acrolein	5 ppm	2 ppm
79-06-1	Acrylamide	Unknown	60 mg/m^3
107-13-1	Acrylonitrile	500 ppm	85 ppm
309-00-2	Aldrin	100 mg/m^3	25 mg/m^3
107-18-6	Allyl alcohol	150 ppm	20 ppm
107-05-1	Allyl chloride	300 ppm	250 ppm
106-92-3	Allyl glycidyl ether	270 ppm	50 ppm
504-29-0	2-Aminopyridine	5 ppm	5 ppm [Unch]
7664-41-7	Ammonia	500 ppm	300 ppm
7773-06-0	Ammonium sulfamate	5,000 mg/m^3	1,500 mg/m^3
628-63-7	n-Amyl acetate	4,000 ppm	1,000 ppm
626-38-0	sec-Amyl acetate	9,000 ppm	1,000 ppm
62-53-3	Aniline	100 ppm	100 ppm [Unch]
90-04-0	o-Anisidine	50 mg/m^3	50 mg/m^3 [Unch]
104-94-9	p-Anisidine	50 mg/m^3	50 mg/m^3 [Unch]
7440-36-0 (Metal)	Antimony compounds (as Sb)	80 mg Sb/m^3	50 mg Sb/m^3
86-88-4	ANTU	100 mg/m^3	100 mg/m^3 [Unch]
7440-38-2 (Metal)	Arsenic (inorganic compounds, as As)	100 mg As/m^3	5 mg As/m^3
7784-42-1	Arsine	6 ppm	3 ppm
86-50-0	Azinphosmethyl	20 mg/m^3	10 mg/m^3
7440-39-3 (Metal)	Barium (soluble compounds, as Ba)	1,100 mg Ba/m^3	50 mg Ba/m^3
71-43-2	Benzene	3,000 ppm	500 ppm
94-36-0	Benzoyl peroxide	7,000 mg/m^3	1,500 mg/m^3
100447	Benzyl chloride	10 ppm	10 ppm [Unch]
7440-41-7 (Metal)	Beryllium compounds (as Be)	10 mg Be/m^3	4 mg Be/m^3
1303-86-2	Boron oxide	N.E.	2,000 mg/m^3
7637-07-2	Boron trifluoride	100 ppm	25 ppm
7726-95-6	Bromine	10 ppm	3 ppm
75-25-2	Bromoform	Unknown	850 ppm

(Continued)

Appendix E (Continued)

CAS No.	Chemical Name	Original IDLH	Revised IDLH
106-99-0	1,3-Butadiene	20,000 ppm [LEL]	2,000 ppm [LEL]
78933	2-Butanone	3,000 ppm	3,000 ppm [Unch]
111762	2-Butoxyethanol	700 ppm	700 ppm [Unch]
123864	n-Butyl acetate	10,000 ppm	1,700 ppm [LEL]
105464	sec-Butyl acetate	10,000 ppm	1,700 ppm [LEL]
540885	tert-Butyl acetate	10,000 ppm	1,500 ppm [LEL]
71363	n-Butyl alcohol	8,000 ppm	1,400 ppm [LEL]
78922	sec-Butyl alcohol	10,000 ppm	2,000 ppm
75650	tert-Butyl alcohol	8,000 ppm	1,600 ppm
109739	n-Butylamine	2,000 ppm	300 ppm
1189851	tert-Butyl chromate	30 mg/m^3 (as CrO_3)	15 mg Cr(VI)/m^3
2426086	n-Butyl glycidyl ether	3,500 ppm	250 ppm
109795	n-Butyl mercaptan	2,500 ppm	500 ppm
98511	p-tert-Butyltoluene	1,000 ppm	100 ppm
7440439 (Metal)	Cadmium dust (as Cd)	50 mg Cd/m^3	9 mg Cd/m^3
7440439 (Metal)	Cadmium fume (as Cd)	9 mg Cd/m^3	9 mg Cd/m^3 [Unch]
7778441	Calcium arsenate (as As)	100 mg As/m^3	5 mg As/m^3
1305788	Calcium oxide	Unknown	25 mg/m^3
76222	Camphor (synthetic)	200 mg/m^3	200 mg/m^3 [Unch]
63252	Carbaryl	600 mg/m^3	100 mg/m^3
1333864	Carbon black	N.E.	1,750 mg/m^3
124389	Carbon dioxide	50,000 ppm	40,000 ppm
75150	Carbon disulfide	500 ppm	500 ppm [Unch]
630080	Carbon monoxide	1,500 ppm	1,200 ppm
56235	Carbon tetrachloride	300 ppm	200 ppm
57749	Chlordane	500 mg/m^3	100 mg/m^3
8001352	Chlorinated camphene	200 mg/m^3	200 mg/m^3 [Unch]
31242930	Chlorinated diphenyl oxide	Unknown	5 mg/m^3
7782505	Chlorine	30 ppm	10 ppm
	Chlorine dioxide	10 ppm	5 ppm
7790912	Chlorine trifluoride	20 ppm	20 ppm [Unch]
107200	Chloroacetaldehyde	100 ppm	45 ppm
532274	alpha-Chloroacetophenone	100 mg/m^3	15 mg/m^3
108907	Chlorobenzene	2,400 ppm	1,000 ppm
2698411	o-Chlorobenzylidene malononitrile	2 mg/m^3	2 mg/m^3 [Unch]
74975	Chlorobromomethane	5,000 ppm	2,000 ppm
53469219	Chlorodiphenyl (42% chlorine)	10 mg/m^3	5 mg/m^3
11097691	Chlorodiphenyl (54% chlorine)	5 mg/m^3	5 mg/m^3 [Unch]
67663	Chloroform	1,000 ppm	500 ppm
600259	1-Chloro-1-nitropropane	2,000 ppm	100 ppm
76062	Chloropicrin	4 ppm	2 ppm
126998	beta-Chloroprene	400 ppm	300 ppm
1333-82-0 (CrO_3)	Chromic acid and chromates	30 mg/m^3 (as CrO_3)	15 mg Cr(VI)/m^3
Varies	Chromium (II) compounds [as Cr(II)]	N.E.	250 mg Cr(II)/m^3
Varies	Chromium (III) compounds [as Cr(III)]	N.E.	25 mg Cr(III)/m^3

(Continued)

CAS No.	Chemical Name	Original IDLH	Revised IDLH
7440473	Chromium metal (as Cr)	N.E.	250 mg Cr/m^3
65996932	Coal tar pitch volatiles	700 mg/m^3	80 mg/m^3
7440484 (Metal)	Cobalt metal, dust and fume (as Co)	20 mg Co/m^3	20 mg Co/m^3 [Unch]
7440508 (Metal)	Copper (dusts and mists, as Cu)	N.E.	100 mg Cu/m^3
1317380 (CuO)	Copper fume (as Cu)	N.E.	100 mg Cu/m^3
None	Cotton dust (raw)	N.E.	100 mg/m^3
136787	Crag (r) herbicide	5,000 mg/m^3	500 mg/m^3
95487 (oisomer), 108394 (misomer), 106445 (pisomer)	Cresol (o, m, p isomers)	250 ppm	250 ppm [Unch]
123739 (transisomer)	Crotonaldehyde	400 ppm	50 ppm
98828	Cumene	8,000 ppm	900 ppm [LEL]
Varies	Cyanides (as CN)	50 mg/m^3 (as CN)	25 mg/m^3 (as CN)
110827	Cyclohexane	10,000 ppm	1,300 ppm [LEL]
108930	Cyclohexanol	3,500 ppm	400 ppm
108941	Cyclohexanone	5,000 ppm	700 ppm
110838	Cyclohexene	10,000 ppm	2,000 ppm
542927	Cyclopentadiene	2,000 ppm	750 ppm
94757	2,4-D	500 mg/m^3	100 mg/m^3
50293	DDT	N.E.	500 mg/m^3
17702419	Decaborane	100 mg/m^3	15 mg/m^3
8065483	Demeton	20 mg/m^3	10 mg/m^3
123422	Diacetone alcohol	2,100 ppm	1,800 ppm [LEL]
334883	Diazomethane	2 ppm	2 ppm [Unch]
19287457	Diborane	40 ppm	15 ppm
107664	Dibutyl phosphate	125 ppm	30 ppm
84742	Dibutyl phthalate	9,300 mg/m^3	4,000 mg/m^3
95501	o-Dichlorobenzene	1,000 ppm	200 ppm
106467	p-Dichlorobenzene	1,000 ppm	150 ppm
75718	Dichlorodifluoromethane	50,000 ppm	15,000 ppm
118525	1,3-Dichloro 5,5-dimethylhydantoin	Unknown	5 mg/m^3
75343	1,1-Dichloroethane	4,000 ppm	3,000 ppm
540590	1,2-Dichloroethylene	4,000 ppm	1,000 ppm
111444	Dichloroethyl ether	250 ppm	100 ppm
75434	Dichloromonofluoromethane	50,000 ppm	5,000 ppm
594729	1,1-Dichloro 1-nitroethane	150 ppm	25 ppm
76142	Dichlorotetrafluoroethane	50,000 ppm	15,000 ppm
62737	Dichlorvos	200 mg/m^3	100 mg/m^3
60571	Dieldrin	450 mg/m^3	50 mg/m^3
109897	Diethylamine	2,000 ppm	200 ppm
100378	2-Diethylaminoethanol	500 ppm	100 ppm
75616	Difluorodibromomethane	2,500 ppm	2,000 ppm
2238075	Diglycidyl ether	25 ppm	10 ppm
108838	Diisobutyl ketone	2,000 ppm	500 ppm
108189	Diisopropylamine	1,000 ppm	200 ppm
127195	Dimethyl acetamide	400 ppm	300 ppm
124403	Dimethylamine	2,000 ppm	500 ppm
121697	N,N-Dimethylaniline	100 ppm	100 ppm [Unch]

(Continued)

Appendix E (Continued)

CAS No.	Chemical Name	Original IDLH	Revised IDLH
300765	Dimethyl 1,2-dibromo 2,2-dichlorethyl phosphate	1,800 mg/m³	200 mg/m³
68122	Dimethylformamide	3,500 ppm	500 ppm
57147	1,1-Dimethylhydrazine	50 ppm	15 ppm
131113	Dimethylphthalate	9,300 mg/m³	2,000 mg/m³
77781	Dimethyl sulfate	10 ppm	7 ppm
528290 (oisomer), 99650 (misomer), 100254 (pisomer)	Dinitrobenzene (o, m, p isomers)	200 mg/m³	50 mg/m³
534521	Dinitroocresol	5 mg/m³	5 mg/m³ [Unch]
25321146	Dinitrotoluene	200 mg/m³	50 mg/m³
117817	Di sec-octyl phthalate	Unknown	5,000 mg/m³
123911	Dioxane	2,000 ppm	500 ppm
92524	Diphenyl	300 mg/m³	100 mg/m³
34590948	Dipropylene glycol methyl ether	Unknown	600 ppm
72208	Endrin	2,000 mg/m³	2 mg/m³
106898	Epichlorohydrin	250 ppm	75 ppm
2104645	EPN	50 mg/m³	5 mg/m³
141435	Ethanolamine	1,000 ppm	30 ppm
110805	2-Ethoxyethanol	6,000 ppm	500 ppm
111159	2-Ethoxyethyl acetate	2,500 ppm	500 ppm
141786	Ethyl acetate	10,000 ppm	2,000 ppm [LEL]
140885	Ethyl acrylate	2,000 ppm	300 ppm
64175	Ethyl alcohol	15,000 ppm	3,300 ppm [LEL]
75047	Ethylamine	4,000 ppm	600 ppm
100414	Ethyl benzene	2,000 ppm	800 ppm [LEL]
74964	Ethyl bromide	3,500 ppm	2,000 ppm
106354	Ethyl butyl ketone	3,000 ppm	1,000 ppm
75003	Ethyl chloride	20,000 ppm	3,800 ppm [LEL]
107073	Ethylene chlorohydrin	10 ppm	7 ppm
107153	Ethylenediamine	2,000 ppm	1,000 ppm
106934	Ethylene dibromide	400 ppm	100 ppm
107062	Ethylene dichloride	1,000 ppm	50 ppm
628966	Ethylene glycol dinitrate	500 mg/m³	75 mg/m³
151564	Ethyleneimine	100 ppm	100 ppm [Unch]
75218	Ethylene oxide	800 ppm	800 ppm [Unch]
60297	Ethyl ether	19,000 ppm [LEL]	1,900 ppm [LEL]
109944	Ethyl formate	8,000 ppm	1,500 ppm
75081	Ethyl mercaptan	2,500 ppm	500 ppm
100743	N-Ethylmorpholine	2,000 ppm	100 ppm
78104	Ethyl silicate	1,000 ppm	700 ppm
14484641	Ferbam	N.E.	800 mg/m³
12604589	Ferrovanadium dust	N.E.	500 mg/m³
Varies	Fluorides (as F)	500 mg F/m³	250 mg F/m³
7782414	Fluorine	25 ppm	25 ppm [Unch]
75694	Fluorotrichloromethane	10,000 ppm	2,000 ppm
50000	Formaldehyde	30 ppm	20 ppm
64186	Formic acid	30 ppm	30 ppm [Unch]
98011	Furfural	250 ppm	100 ppm
98000	Furfuryl alcohol	250 ppm	75 ppm

(Continued)

CAS No.	Chemical Name	Original IDLH	Revised IDLH
556525	Glycidol	500 ppm	150 ppm
7782425	Graphite (natural)	N.E.	1,250 mg/m^3
7440586 (Metal)	Hafnium compounds (as Hf)	Unknown	50 mg Hf/m^3
7440586 (Metal)	Heptachlor	700 mg/m^3	35 mg/m^3
142825	n-Heptane	5,000 ppm	750 ppm
67721	Hexachloroethane	300 ppm	300 ppm [Unch]
1335871	Hexachloronaphthalene	2 mg/m^3	2 mg/m^3 [Unch]
110543	n-Hexane	5,000 ppm	1,100 ppm [LEL]
591786	2-Hexanone	5,000 ppm	1,600 ppm
108101	Hexone	3,000 ppm	500 ppm
108849	sec Hexyl acetate	4,000 ppm	500 ppm
302012	Hydrazine	80 ppm	50 ppm
10035106	Hydrogen bromide	50 ppm	30 ppm
7647010	Hydrogen chloride	100 ppm	50 ppm
74908	Hydrogen cyanide	50 ppm	50 ppm [Unch]
7664393	Hydrogen fluoride (as F)	30 ppm	30 ppm [Unch]
7722841	Hydrogen peroxide	75 ppm	75 ppm [Unch]
7783075	Hydrogen selenide (as Se)	2 ppm	1 ppm
7783064	Hydrogen sulfide	300 ppm	100 ppm
123319	Hydroquinone	Unknown	50 mg/m^3
7553562	Iodine	10 ppm	2 ppm
1309371	Iron oxide dust and fume (as Fe)	N.E.	2,500 mg Fe/m^3
12392-2	Isoamyl acetate	3,000 ppm	1,000 ppm
123513	Isoamyl alcohol (primary)	10,000 ppm	500 ppm
528754	Isoamyl alcohol (secondary)	10,000 ppm	500 ppm
110190	Isobutyl acetate	7,500 ppm	1,300 ppm [LEL]
78831	Isobutyl alcohol	8,000 ppm	1,600 ppm
78591	Isophorone	800 ppm	200 ppm
108214	Isopropyl acetate	16,000 ppm	1,800 ppm
67630	Isopropyl alcohol	12,000 ppm	2,000 ppm [LEL]
75310	Isopropylamine	4,000 ppm	750 ppm
108203	Isopropyl ether	10,000 ppm	1,400 ppm [LEL]
4016142	Isopropyl glycidyl ether	1,000 ppm	400 ppm
463514	Ketene	Unknown	5 ppm
7439921	Lead compounds (as Pb)	700 mg Pb/m^3	100 mg Pb/m^3
58899	Lindane	1,000 mg/m^3	50 mg/m^3
7580678	Lithium hydride	55 mg/m^3	0.5 mg/m^3
68476857	L.P.G.	19,000 ppm [LEL]	2,000 ppm [LEL]
1309484	Magnesium oxide fume	N.E.	750 mg/m^3
121755	Malathion	5,000 mg/m^3	250 mg/m^3
108316	Maleic anhydride	Unknown	10 mg/m^3
7439965	Manganese compounds (as Mn)	N.E.	500 mg Mn/m^3
7439976	Mercury compounds [except (organo) alkyls, as Hg]	28 mg Hg/m^3	10 mg Hg/m^3
Varies	Mercury (organo) alkyl compounds (as Hg)	10 mg Hg/m^3	2 mg Hg/m^3
141797	Mesityl oxide	5,000 ppm	1,400 ppm [LEL]
72435	Methoxychlor	N.E.	5,000 mg/m^3
79209	Methyl acetate	10,000 ppm	3,100 ppm [LEL]
74997	Methyl acetylene	15,000 ppm [LEL]	1,700 ppm [LEL]

(Continued)

Appendix E (Continued)

CAS No.	Chemical Name	Original IDLH	Revised IDLH
59355758	Methyl acetylenepropadiene mixture	15,000 ppm	3,400 ppm [LEL]
96333	Methyl acrylate	1,000 ppm	250 ppm
109875	Methylal	15,000 ppm [LEL]	2,200 ppm [LEL]
67561	Methyl alcohol	25,000 ppm	6,000 ppm
74895	Methylamine	100 ppm	100 ppm [Unch]
110430	Methyl (namyl) ketone	4,000 ppm	800 ppm
74839	Methyl bromide	2,000 ppm	250 ppm
109864	Methyl cellosolve (r)	2,000 ppm	200 ppm
110496	Methyl cellosolve (r) acetate	4,000 ppm	200 ppm
74873	Methyl chloride	10,000 ppm	2,000 ppm
71556	Methyl chloroform	1,000 ppm	700 ppm
108872	Methylcyclohexane	10,000 ppm	1,200 ppm [LEL]
25639423	Methylcyclohexanol	10,000 ppm	500 ppm
583608	o-Methylcyclohexanone	2,500 ppm	600 ppm
101688	Methylene bisphenyl isocyanate	100 mg/m^3	75 mg/m^3
75092	Methylene chloride	5,000 ppm	2,300 ppm
107313	Methyl formate	5,000 ppm	4,500 ppm
541855	5-Methyl 3-heptanone	3,000 ppm	100 ppm
60344	Methyl hydrazine	50 ppm	20 ppm
74884	Methyl iodide	800 ppm	100 ppm
108112	Methyl isobutyl carbinol	2,000 ppm	400 ppm
624839	Methyl isocyanate	20 ppm	3 ppm
74931	Methyl mercaptan	400 ppm	150 ppm
80626	Methyl methacrylate	4,000 ppm	1,000 ppm
98839	Methyl styrene	5,000 ppm	700 ppm
12001262	Mica	N.E.	1,500 mg/m^3
7439987	Molybdenum (insoluble compounds, as Mo)	N.E.	5,000 mg Mo/m^3
Varies	Molybdenum (soluble compounds, as Mo)	N.E.	1,000 mg Mo/m^3
100618	Monomethyl aniline	100 ppm	100 ppm [Unch]
110918	Morpholine	8,000 ppm	1,400 ppm [LEL]
8030306	Naphtha (coal tar)	10,000 ppm [LEL]	1,000 ppm [LEL]
91203	Naphthalene	500 ppm	250 ppm
13463393	Nickel carbonyl (as Ni)	7 ppm	2 ppm
7440020	Nickel metal and other compounds (as Ni)	N.E.	10 mg Ni/m^3
54115	Nicotine	35 mg/m^3	5 mg/m^3
7697372	Nitric acid	100 ppm	25 ppm
10102439	Nitric oxide	100 ppm	100 ppm [Unch]
100016	p-Nitroaniline	300 mg/m^3	300 mg/m^3 [Unch]
98953	Nitrobenzene	200 ppm	200 ppm [Unch]
100005	p-Nitrochlorobenzene	1,000 mg/m^3	100 mg/m^3
79243	Nitroethane	1,000 ppm	1,000 ppm [Unch]
10102440	Nitrogen dioxide	50 ppm	20 ppm
7783542	Nitrogen trifluoride	2,000 ppm	1,000 ppm
55630	Nitroglycerine	500 mg/m^3	75 mg/m^3
75525	Nitromethane	1,000 ppm	750 ppm
108032	1-Nitropropane	2,300 ppm	1,000 ppm

(Continued)

CAS No.	Chemical Name	Original IDLH	Revised IDLH
79469	2-Nitropropane	2,300 ppm	100 ppm
88722	Nitrotoluene (oisomer)	200 ppm	200 ppm [Unch]
99081	Nitrotoluene (misomer)	200 ppm	200 ppm [Unch]
99990	Nitrotoluene (p isomers)	200 ppm	200 ppm [Unch]
2234131	Octachloronaphthalene	Unknown	Unknown [Unch]
111659	Octane	5,000 ppm	1,000 ppm [LEL]
8012951	Oil mist (mineral)	N.E.	2,500 mg/m^3
20816120	Osmium tetroxide (as Os)	1 mg Os/m^3	1 mg Os/m^3 [Unch]
144627	Oxalic acid	500 mg/m^3	500 mg/m^3 [Unch]
7783417	Oxygen difluoride	0.5 ppm	0.5 ppm [Unch]
10028156	Ozone	10 ppm	5 ppm
1910425	Paraquat	1.5 mg/m^3	1 mg/m^3
56382	Parathion	20 mg/m^3	10 mg/m^3
19624227	Pentaborane	3 ppm	1 ppm
1321648	Pentachloronaphthalene	Unknown	Unknown [Unch]
87865	Pentachlorophenol	150 mg/m^3	2.5 mg/m^3
109660	n-Pentane	15,000 ppm [LEL]	1,500 ppm [LEL]
107879	2-Pentanone	5,000 ppm	1,500 ppm
594423	Perchloromethyl mercaptan	10 ppm	10 ppm [Unch]
7616946	Perchloryl fluoride	385 ppm	100 ppm
8002059	Petroleum distillates (naphtha)	10,000 ppm	1,100 ppm [LEL]
108952	Phenol	250 ppm	250 ppm [Unch]
106503	p-Phenylene diamine	Unknown	25 mg/m^3
101848	Phenyl ether (vapor)	N.E.	100 ppm
8004135	Phenyl etherbiphenyl mixture (vapor)	N.E.	10 ppm
122601	Phenyl glycidyl ether	Unknown	100 ppm
100630	Phenylhydrazine	295 ppm	15 ppm
7786347	Phosdrin	4 ppm	4 ppm [Unch]
75445	Phosgene	2 ppm	2 ppm [Unch]
7803512	Phosphine	200 ppm	50 ppm
7664382	Phosphoric acid	10,000 mg/m^3	1,000 mg/m^3
7723140	Phosphorus (yellow)	N.E.	5 mg/m^3
10026138	Phosphorus pentachloride	200 mg/m^3	70 mg/m^3
1314803	Phosphorus pentasulfide	750 mg/m^3	250 mg/m^3
7719122	Phosphorus trichloride	50 ppm	25 ppm
85449	Phthalic anhydride	10,000 mg/m^3	60 mg/m^3
88891	Picric acid	100 mg/m^3	75 mg/m^3
83261	Pindone	200 mg/m^3	100 mg/m^3
Varies	Platinum (soluble salts, as Pt)	N.E.	4 mg Pt/m^3
65997151	Portland cement	N.E.	5,000 mg/m^3
74986	Propane	20,000 ppm [LEL]	2,100 ppm [LEL]
109604	n-Propyl acetate	8,000 ppm	1,700 ppm
71238	n-Propyl alcohol	4,000 ppm	800 ppm
78875	Propylene dichloride	2,000 ppm	400 ppm
75558	Propylene imine	500 ppm	100 ppm
75569	Propylene oxide	2,000 ppm	400 ppm
627134	n-Propyl nitrate	2,000 ppm	500 ppm
8003347	Pyrethrum	5,000 mg/m^3	5,000 mg/m^3 [Unch]
110861	Pyridine	3,600 ppm	1,000 ppm

(Continued)

Appendix E (Continued)

CAS No.	Chemical Name	Original IDLH	Revised IDLH
106514	Quinone	300 mg/m^3	100 mg/m^3
7440166	Rhodium (metal fume and insoluble compounds, as Rh)	N.E.	100 mg Rh/m^3
Varies	Rhodium (soluble compounds, as Rh)	N.E.	2 mg Rh/m^3
299843	Ronnel	5,000 mg/m^3	300 mg/m^3
83794	Rotenone	Unknown	2,500 mg/m^3
7782492	Selenium compounds (as Se)	Unknown	1 mg Se/m^3
7783791	Selenium hexafluoride	5 ppm	2 ppm
7631869	Silica, amorphous	N.E.	3,000 mg/m^3
	Silica, crystalline (respirable dust)	N.E.	
14808607	cristobalite/tridymite:		25 mg/m^3
14808607	quartz/tripoli:		50 mg/m^3
7440224	Silver (metal dust and soluble compounds, as Ag)	N.E.	10 mg Ag/m^3
None	Soapstone	N.E.	3,000 mg/m^3
62748	Sodium fluoroacetate	5 mg/m^3	2.5 mg/m^3
1310732	Sodium hydroxide	250 mg/m^3	10 mg/m^3
7803523	Stibine	40 ppm	5 ppm
8052413	Stoddard solvent	29,500 mg/m^3	20,000 mg/m^3
57249	Strychnine	3 mg/m^3	3 mg/m^3 [Unch]
100425	Styrene	5,000 ppm	700 ppm
7446095	Sulfur dioxide	100 ppm	100 ppm [Unch]
7664939	Sulfuric acid	80 mg/m^3	15 mg/m^3
10025679	Sulfur monochloride	10 ppm	5 ppm
5714227	Sulfur pentafluoride	1 ppm	1 ppm [Unch]
2699798	Sulfuryl fluoride	1,000 ppm	200 ppm
93765	2,4,5-T	Unknown	250 mg/m^3
14807966	Talc	N.E.	1,000 mg/m^3
7440257	Tantalum (metal and oxide dust, as Ta)	N.E.	2,500 mg Ta/m^3
89245	TEDP	35 mg/m^3	10 mg/m^3
13494809	Tellurium compounds (as Te)	N.E.	25 mg Te/m^3
7783804	Tellurium hexafluoride	1 ppm	1 ppm [Unch]
107493	TEPP	10 mg/m^3	5 mg/m^3
26140603	Terphenyl (o, m, p isomers)	Unknown	500 mg/m^3
76119	1,1,1,2-Tetrachloro 2,2-difluoroethane	15,000 ppm	2,000 ppm
76120	1,1,2,2-Tetrachloro 1,2-difluoroethane	15,000 ppm	2,000 ppm
79345	1,1,2,2-Tetrachloroethane	150 ppm	100 ppm
127184	Tetrachloroethylene	500 ppm	150 ppm
1335882	Tetrachloronaphthalene	Unknown	Unknown [Unch]
78002	Tetraethyl lead (as Pb)	40 mg Pb/m^3	40 mg Pb/m^3 [Unch]
109999	Tetrahydrofuran	20,000 ppm [LEL]	2,000 ppm [LEL]
75741	Tetramethyl lead (as Pb)	40 mg Pb/m^3	40 mg Pb/m^3 [Unch]
3333526	Tetramethyl succinonitrile	5 ppm	5 ppm [Unch]
509148	Tetranitromethane	5 ppm	4 ppm

(Continued)

CAS No.	Chemical Name	Original IDLH	Revised IDLH
479458	Tetryl	N.E.	750 mg/m^3
Varies	Thallium (soluble compounds, as Tl)	20 mg Tl/m^3	15 mg Tl/m^3
137268	Thiram	1,500 mg/m^3	100 mg/m^3
7440315	Tin (inorganic compounds, as Sn)	400 mg Sn/m^3	100 mg Sn/m^3
Varies	Tin (organic compounds, as Sn)	Unknown	25 mg Sn/m^3
13463677	Titanium dioxide	N.E.	5,000 mg/m^3
108883	Toluene	2,000 ppm	500 ppm
584849	Toluene 2,4-diisocyanate	10 ppm	2.5 ppm
95534	o-Toluidine	100 ppm	50 ppm
126738	Tributyl phosphate	125 ppm	30 ppm
79005	1,1,2-Trichloroethane	500 ppm	100 ppm
79016	Trichloroethylene	1,000 ppm	1,000 ppm [Unch]
1321659	Trichloronaphthalene	Unknown	Unknown [Unch]
96184	1,2,3-Trichloropropane	1,000 ppm	100 ppm
76131	1,1,2-Trichloro 1,2,2-trifluoroethane	4,500 ppm	2,000 ppm
121448	Triethylamine	1,000 ppm	200 ppm
75638	Trifluorobromomethane	50,000 ppm	40,000 ppm
118967	2,4,6-Trinitrotoluene	1,000 mg/m^3	500 mg/m^3
78308	Triorthocresyl phosphate	40 mg/m^3	40 mg/m^3 [Unch]
115866	Triphenyl phosphate	N.E.	1,000 mg/m^3
8006642	Turpentine	1,500 ppm	800 ppm
7440611	Uranium (insoluble compounds, as U)	30 mg U/m^3	10 mg U/m^3
Varies	Uranium (soluble compounds, as U)	20 mg U/m^3	10 mg U/m^3
1314621	Vanadium dust	70 mg/m^3 (as V_2O_5)	35 mg V/m^3
1314621	Vanadium fume	70 mg/m^3 (as V_2O_5)	35 mg V/m^3
25013154	Vinyl toluene	5,000 ppm	400 ppm
81812	Warfarin	350 mg/m^3	100 mg/m^3
95476	Xylene (oisomer)	1,000 ppm	900 ppm
108383	Xylene (misomer)	1,000 ppm	900 ppm
106423	Xylene (p isomers)	1,000 ppm	900 ppm
1300738	Xylidine	150 ppm	50 ppm
7440655	Yttrium compounds (as Y)	N.E.	500 mg Y/m^3
7646857	Zinc chloride fume	4,800 mg/m^3	50 mg/m^3
1314132	Zinc oxide	2,500 mg/m^3	500 mg/m^3
7440677	Zirconium compounds (as Zr)	500 mg Zr/m^3	25 mg Zr/m^3

Reference: http://www.cdc.gov/niosh/idlh/intridl4.html

Appendix F: ACGIH Threshold Limit Value (TLV)

CAS No.	Chemical Name	TWA		STEL	
		ppm	mg/m^3	ppm	mg/m^3
75-07-0	Acetaldehyde			25*	
64-19-7	Acetic acid	10		15	
108-24-7	Acetic anhydride	5			
67-64-1	Acetone	500		750	
75-86-5	Acetone cyanohydrin				5*
75-05-8	Acetonitrile	20			
98-86-2	Acetophenone	10			
58-78-2	Acetylsalicylic acid		5		
107-02-8	Acrolein			0.1*	
79-06-1	Acrylamide		0.03		
79-10-7	Acrylic acid	2			
107-13-1	Acrylonitrile	2			
124-04-9	Adipic acid		5		
111-69-3	Adiponitrile	2			
15972-60-8	Alachlor		1		
309-00-2	Aldrin		0.05		
	Aliphatic hydrocarbon gases	1000			
	Alkanes				
107-18-6	Allyl alcohol	0.5			
107-05-1	Allyl chloride	1		2	
106-92-3	Allyl glycidyl ether	1			
2179-59-1	Allyl propyl disulfide	0.5			
7429-90-5	Aluminum, metal		1		
504-29-0	2-Aminopyridine	0.5			
61-82-5	Amitrole		0.2		
7664-41-7	Ammonia	25		35	
12125-02-9	Ammonium chloride fume		10		20
3825-26-1	Ammonium perfluorooctanoate		0.01		
7773-06-0	Ammonium sulfamate		10		
994-05-8	tert-Amyl methyl ether	20			
62-53-3	Aniline	2			
90-04-0	o-Anisidine		0.5		
104-94-9	p-Anisidine		0.5		
7440-36-0	Antimony & compounds		0.5		
7803-52-3	Antimony hydride	0.1			
86-88-4	ANTU		0.3		

(Continued)

Appendix F (Continued)

CAS No.	Chemical Name	TWA		STEL	
		ppm	mg/m^3	ppm	mg/m^3
7440-38-2	Arsenic		0.01		
7784-42-1	Arsine	0.005			
1332-21-4	Abestos, all forms	0.1 (f/cc)			
8052-42-4	Asphalt (Bitumen) fume		0.5		
1912-24-9	Atrazine		5		
86-50-0	Azinphos-methyl		0.2		
7440-39-3	Barium, soluble compounds		0.5		
7727-43-7	Barium sulfate		10		
17804-35-2	Benomyl		1		
71-43-2	Benzene	0.5		2.5	
98-07-7	Benzotrichloride			0.1	
98-88-4	Benzoyl chloride			0.5	
94-36-0	Benzoyl peroxide		5		
140-11-4	Benzyl acetate	10			
100-44-7	Benzyl chloride	1			
7440-41-7	Beryllium		5E-05		
92-52-4	Biphenyl	0.2			
3033-62-3	Bis (2-dimethylaminoethyl) ether	0.05		0.15	
1304-82-1	Bismuth telluride (undoped)		10		
varies	Borate compounds (inorganic)		2		6
1303-86-2	Boron oxide		10		
10294-33-4	Boron tribromide			1	
7637-07-2	Boron trifluoride			1	
314-40-9	Bromacil		10		
7726-95-6	Bromine	0.1		0.2	
7789-30-2	Bromine pentafluoride	0.1			
75-25-2	Bromoform	0.5			
106-94-5	1-Bromopropane	10			
106-99-0	1,3 Butadiene	2			
71-36-3	n-Butanol	20			
75-65-0	sec-Butanol	100			
75-65-0	tert-Butanol	100			
varies	Butenes (all isomers)	250			
111-76-2	2-Butoxyethanol (EGBE)	20			
112-07-2	2-Butoxyethyl acetate	20			
123-86-4	n-Butyl acetate	150		200	
105-46-4	sec-Butyl acetate	200			
540-88-5	tert-Butyl acetate	200			
141-32-2	n-Butyl acrylate	2			
78-92-2	sec-Butyl alcohol	100			
109-73-9	n-Butylamine			5	
128-37-0	Butylated hydroxytoluene (BHT)		2		
1189-85-1	tert-Butyl chromate, as CrO_3				0.1
2426-08-6	n-Butyl glycidyl ether (BGE)	3			
138-22-7	n-Butyl lactate	5			
109-79-5	n-Butyl mercaptan	0.5			
89-72-5	o-sec-Butylphenol	5			

(Continued)

CAS No.	Chemical Name	TWA		STEL	
		ppm	mg/m^3	ppm	mg/m^3
98-51-1	p-tert-Butyl toluene	1			
7440-43-9	Cadmium		0.01		
13765-19-0	Calcium chromate		0.001		
156-62-7	Calcium cyanamide		0.5		
1305-62-0	Calcium hydroxide		5		
1305-78-8	Calcium oxide		2		
1344-95-2	Calcium silicate		10		
7778-18-9	Calcium sulfate		10		
76-22-2	Camphor, synthetic	2		3	
105-60-2	Caprolactam		5		
2425-06-1	Captafol		0.1		
133-06-2	Captan		5		
63-25-2	Carbaryl		0.5		
1563-66-2	Carbofuran		0.1		
1333-86-4	Carbon black		3.5		
124-38-9	Carbon dioxide	5000		30,000	
75-15-0	Carbon disulfide	1			
630-08-0	Carbon monoxide	25			
558-13-4	Carbon tetrabromide	0.1		0.3	
56-23-5	Carbon tetrachloride	5		10	
353-50-4	Carbonyl fluoride	2		5	
120-80-9	Catechol	5			
9004-34-6	Cellulose		10		
21351-79-1	Cesium hydroxide		2		
57-74-9	Chlorodane		0.5		
8001-35-2	Chlorinated camphene		0.5		1
31242-93-0	o-Chlorinated diphenyl oxide		0.5		
7782-50-5	Chlorine	0.5		1	
10049-04-4	Chlorine dioxide	0.1		0.3	
7790-91-2	Chlorine trifluoride			0.1	
107-20-0	Chloroacetaldehyde			1	
78-95-5	Chloroacetone			1	
532-27-4	2-Chloroacetophenone	0.05			
79-04-9	Chloroacetyl chloride	0.05		0.15	
108-90-7	Chlorobenzene	10			
2698-41-1	o-Chlorobenzylidene malononitrile			0.05	
74-97-5	Chlorobromomethane	200			
75-45-6	Chlorodifluoromethane	1000			
53469-21-9	Chlorodiphenyl (42% Cl)		1		
11097-69-1	Chlorobiphenyl (54% Cl)		0.5		
67-66-3	Chloroform	10			
542-88-1	bis(Chloromethyl) ether	0.001			
600-25-9	1-Chloro-1-nitro-propane	2			
76-15-3	Chloropentafluoroethane	1000			
76-06-2	Chloropicrin	0.1			
127-00-4	1-Chloro-2-propanol	1			
126-99-8	beta-Chloroprene	10			
598-78-7	2-Chloropropionic acid	0.1			

(Continued)

Appendix F (Continued)

CAS No.	Chemical Name	TWA		STEL	
		ppm	mg/m³	ppm	mg/m³
2039-87-4	o-Chlorostyrene	50		75	
95-49-8	o-Chlorotoluene	50			
2921-88-2	Chlorpyrifos		0.1		
	Chromite ore processing (Chromate) (as Cr)		0.05		
1929-82-4	2-Chloro-6-trichloromethyl n-pyridine		10		
7440-47-3	Chromium (III) compounds		0.5		
14977-61-8	Chromyl chloride	0.025			
5392-40-5	Citral	5			
2971-90-6	Clopidol		10		
	Coal dust (Anthracite)		0.4		
	Coal dust (Bituminous)		0.9		
65996-93-2	Coal tar pitch volatiles		0.2		
7440-48-4	Cobalt and inorganic comounds		0.02		
10210-68-1	Cobalt cerbonyl (as Co)				
16842-03-8	Cobalt hydrocarbonyl		0.1		
1317-38-0	Copper fume		0.2		
7440-50-8	Copper dusts and mists		1		
	Cotton dust		0.1		
56-72-4	Coumaphos		0.05		
Varies	Cresol, all isomers		20		
4170-30-3	Crotonaldehyde			0.3	
299-86-5	Crufomate		5		
98-82-8	Cumene	50			
420-04-2	Cyanamide		2		
460-19-5	Cyanogen	10			
506-77-4	Cyanogen chloride			0.3	
110-82-7	Cyclohexane	100			
108-93-0	Cyclohexanol	50			
108-94-1	Cyclohexanone	20		50	
110-83-8	Cyclohexene	300			
108-91-8	Cyclohexylamine	10			
12-82-4	Cyclonite		0.5		
542-92-7	Cyclopentadiene	75			
287-92-3	Cyclopentane	600			
13121-70-5	Cyhexatin		5		
94-75-7	2,4-D		10		
50-29-3	DDT		1		
17702-41-9	Decaborane	0.05		0.15	
8065-48-3	Demeton		0.05		
919-86-8	Demeton-S-methyl		0.05		
123-42-2	Diacetone alcohol	50			
333-41-5	Diazinon		0.01		
334-88-3	Diazomethane	0.2			
19287-45-7	Diborane	0.1			
102-81-8	2-N-Dibutylaminoethanol	0.5			
2528-36-1	Dibutyl phenyl phosphate	0.3			
84-74-2	Dibutyl phthalate		5		

(*Continued*)

CAS No.	Chemical Name	TWA		STEL	
		ppm	mg/m³	ppm	mg/m³
107-66-4	Dibutyl phosphate		5		
79-43-6	Dichloroacetic acid	0.5			
7572-29-4	Dichloroacetylene			0.1	
95-50-1	o-Dichlorobenzene	25		50	
106-46-7	p-Dichlorobenzene	10			
764-41-0	1,4-Dichloro-2-butene	0.005			
75-71-8	Dichlorodifluoromethane	1000			
118-52-5	1,3-Dichloro-5,5-dimethyl hydantoin		0.2		0.4
75-34-3	1,1-Dichloroethane	100			
540-59-0	1,2-Dichloroethylene	200			
111-44-4	Dichloroethyl ether	5		10	
75-43-4	Dichlorofluoromethane	10			
75-09-2	Dichloromethane	50			
594-72-9	1,1-Dichloro-1-nitroethane	2			
542-75-6	1,3-Dichloropropene	1			
75-99-0	2,2-Dichloropropionic acid		5		
76-14-2	Dichlorotetrafluoroethane	1000			
62-73-7	Dichlorvos (DDVP)		0.1		
141-66-2	Dicrotophos		0.05		
77-73-6	Dicyclopentadiene	5			
102-54-5	Dicyclopentadienyl iron		10		
60-57-1	Dieldrin		0.1		
varies	Diesel fuel		100		
111-42-2	Diethanolamine		1		
109-89-7	Diethylamine	5		15	
100-37-8	2-Diethylaminoethanol	2			
111-40-0	Diethylene triamine	1			
117-81-7	Di(2-ethylhexyl)phthalate (DEHP)		5		
96-22-0	Diethyl ketone	200		300	
84-66-2	Diethyl phthalate		5		
75-61-6	Difluorodibromomethane	100			
123639	Diglycidyl ether	0.01			
108-83-8	Diisobutyl ketone	25			
108-18-9	Diisopropylamine	5			
127-19-5	N,N-Dimethylacetamide	10			
124-40-3	Dimethylamine	5			
121-69-7	Dimethylaniline	5		10	
79-44-7	Dimethyl carbamoyl chloride	0.005			
624-92-0	Dimethyl disulfide	0.5			
14857-34-2	Dimethylethoxysilane	0.5		1.5	
68-12-2	Dimethylformamide	10			
57-14-7	1,1-Dimethylhydrazine	0.01			
131-11-3	Dimethyl phthalate		5		
77-78-1	Dimethyl sulfate	0.1			
75-18-3	Dimethyl sulfide	10			
varies	Dinitrobenzene (all isomers)	0.15			
534-52-1	Dinitro-o-cresol		0.2		
148-01-6	3,5-Dinitro-o-toluamide		1		
25321-14-6	Dinitrotoluene		0.2		

(Continued)

Appendix F (Continued)

CAS No.	Chemical Name	TWA		STEL	
		ppm	mg/m³	ppm	mg/m³
123-91-1	1,4-Dioxane	20			
78-34-2	Dioxathion		0.1		
646-06-0	1,3-Dioxolane	20			
122-39-4	Diphenylamine		10		
123-19-3	Dipropyl ketone	50			
97-77-8	Disulfiram		2		
298-04-4	Disulfoton		0.05		
330-54-1	Diuron		10		
1321-74-0	Divinyl benzene	10			
112-55-0	Dodecyl mercaptan	0.1			
115-29-7	Endosulfan		0.1		
72-20-8	Endrin		0.1		
13838-16-9	Enflurane	75			
106-89-8	Epichlorohydrin	0.5			
2104-64-5	EPN		0.1		
64-17-5	Ethanol			1000	
141-43-5	Ethanolamine	3		6	
563-12-2	Ethion		0.05		
110-80-5	2-Ethoxyethanol	5			
111-15-9	2-Ethoxyethyl acetate	5			
141-78-6	Ethyl acetate	400			
140-88-5	Ethyl acrylate	5		15	
75-04-7	Ethylamine	5		15	
541-85-5	Ethyl amyl ketone	10			
100-41-4	Ethyl benzene	100		125	
74-96-4	Ethyl bromide	5			
637-92-3	Ethyl tert-butyl ether (ETBE)	5			
106-35-4	Ethyl butyl ketone	50		75	
75-00-3	Ethyl chloride	100			
7085-85-0	Ethyl cyanoacrylate	0.2			
74-85-1	Ethylene	200			
107-07-3	Ethylene chlorohydrin			1	
107-15-3	Ethylenediamine	10			
107-06-2	Ethylene dichloride	10			
107-21-1	Ethylene glycol		100		
628-96-6	Ethylene glycol dinitrate (EGDN)	0.05			
75-21-8	Ethylene oxide	1			
151-56-4	Ethyleneimine	0.05		0.1	
60-29-7	Ethyl ether	400		500	
109-94-4	Ethyl formate	100			
149-57-5	2-Ethylhexanoic acid		5		
16219-7-3	Ethylidene norbornene			5	
75-08-1	Ethyl mercaptan	0.5			
100-74-3	N-Ethylmorpholine	5			
78-10-4	Ethyl silicate	10			
22224-92-6	Fenamiphos		0.05		
115-90-2	Fensulfothion		0.01		
55-38-9	Fenthion		0.05		

(*Continued*)

CAS No.	Chemical Name	TWA		STEL	
		ppm	mg/m³	ppm	mg/m³
14484-64-1	Ferbam (td)		5		
12604-58-9	Ferrovanadium dust		1		3
	Flour dust		0.5		
varies	Fluorides (as F)		2.5		
7782-41-4	Fluorine	1		2	
944-22-9	Fonofos		0.01		
50-00-0	Formaldehyde			0.3	
75-12-7	Formamide	10			
64-18-6	Formic acid	5		10	
98-01-1	Furfural	2			
98-00-0	Furfuryl alcohol	10		15	
1303-00-0	Gallium arsenide		0.0003		
86290-81-5	Gasoline	300		500	
7782-65-2	Germanium tetrahydride	0.2			
111-30-8	Glutaraldehyde			0.05	
56-81-5	Glycerin mist		10		
556-52-5	Glycidol	2			
107-22-2	Glyoxal		0.1		
	Grain dust (oat, wheat, barley)		4		
7782-42-5	Graphite (all forms except fibers)		2		
7440-58-6	Hafnium		0.5		
151-67-7	Halothane	50			
76-44-8	Heptachlor		0.05		
varies	Heptane (all isomers)	400		500	
118-74-1	Hexachlorobenzene		0.002		
87-68-3	Hexachlorobutadiene	0.02			
77-47-4	Hexachlorocyclopentadiene	0.01			
67-72-1	Hexachloroethane	1			
1335-87-1	Hexachloronaphthalene		0.2		
684-16-2	Hexafluoroacetone	0.1			
116-15-4	Hexafluoropropylene	0.1			
varies	Hexahydrophthalic anhydride (all isomers)				0.005
822-06-0	Hexamethylene diisocyanate	0.005			
110-54-3	n-Hexane	50			
varies	Hexane (other isomers)	500		1000	
124-09-4	1,6-Hexanediamine	0.5			
592-41-6	1-Hexene	50			
108-84-9	sec-Hexyl acetate	50			
107-41-5	Hexylene glycol			25	
302-01-2	Hydrazine	0.01			
61788-32-7	Hydrogenated terphenyls (nonirradiated)	0.5			
10035-10-6	Hydrogen bromide			2	
7647-01-0	Hydrogen chloride			2	
74-90-8	Hydrogen cyanide			4.7	
varies	Cyanide salts			5	
7664-39-3	Hydrogen fluoride	0.5		2	
7722-84-1	Hydrogen peroxide	1			
2148909	Hydrogen selenide (as Se)	0.05			
7783-06-4	Hydrogen sulfide	1		5	

(Continued)

Appendix F (Continued)

CAS No.	Chemical Name	TWA		STEL	
		ppm	mg/m^3	ppm	mg/m^3
123-31-9	Hydroquinone		1		
999-61-1	2-Hydroxypropyl acrylate	0.5			
95-13-6	Indene	5			
7440-74-6	Indium (and compounds)		0.1		
7553-56-2	Iodine	0.01		0.1	
7553-56-2	Iodides	0.01			
75-47-8	Iodoform	0.6			
1309-37-1	Iron oxide (Fe_2O_3)		5		
13463-40-6	Iron pentacarbonyl	0.1		0.2	
varies	Iron salts (soluble, as Fe)		1		
123-51-3	Isoamyl alcohol	100		125	
78-83-1	Isobutanol	50			
110-19-0	Isobutyl acetate	150			
542-56-3	Isobutyl nitrite			1	
26952-21-6	Isooctyl alcohol	50			
78-59-1	Isophorone			5	
4098-71-9	Isophorone diisocyanate	0.005			
109-59-1	2-Isopropoxyethanol	25			
108-21-4	Isopropyl acetate	100		200	
75-31-0	Isopropylamine	5		10	
768-52-5	N-Isopropylaniline	2			
108-20-3	Isopropyl ether	250		310	
4016-14-2	Isopropyl glycidyl ether (IGE)	50		75	
1332-58-7	Kaolin		2		
8008-20-6	Kerosene	200			
463-51-4	Ketene	0.5		1.5	
7439-92-1	Lead		0.05		
58-89-9	Lindane		0.5		
7580-67-8	Lithium hydride		0.025		
68476-85-7	L.P.G. (Liquefied petroleum gas)	1000			
1309-48-4	Magnesium oxide		10		
121-75-5	Malathion		1		
108-31-6	Maleic anhydride	0.1			
7439-96-5	Manganese compounds (as Mn)		0.2		
12079-65-1	Manganese cyclopentadienyl tricarbonyl		0.01		
7439-97-6	Mercury alkyl compounds	0.01		0.03	
141-79-7	Mesityl oxide	15		25	
79-41-4	Methacrylic acid	20			
74-82-8	Methane	1000			
67-56-1	Methanol	200		250	
16752-77-5	Methomyl		2.5		
72-43-5	Methoxychlor		10		
109-86-4	2-Methoxyethanol (EGME)	0.1			
110-49-6	2-Methoxyethyl acetate (EGMEA)	0.1			
34590-94-8	(2-Methoxymethylethoxy)propanol (DPGME)	100		150	
150-76-5	4-Methoxyphenol		5		
107-98-2	1-Methoxy-2-propanol (PGME)	100		150	
79-20-9	Methyl acetate	200		250	

(*Continued*)

CAS No.	Chemical Name	TWA		STEL	
		ppm	mg/m³	ppm	mg/m³
74-99-7	Methyl acetylene	1000			
59355-75-8	Methyl acetylene propadiene mixture	1000		1250	
96-33-3	Methyl acrylate	2			
126-98-7	Methylacrylonitrile	1			
109-87-5	Methylal	1000			
74-89-5	Methylamine	5		15	
110-43-0	Methyl (n-amyl) ketone	50			
100-61-8	N-Methyl aniline	0.5			
74-83-9	Methyl bromide	1			
1634-04-4	Methyl tert-butyl ether (MTBE)	50			
591-78-6	Methyl n-butyl ketone	5		10	
74-87-3	Methyl chloride	50		100	
71-55-6	Methyl chloroform	350		450	
137-05-3	Methyl 2-cyanoacrylate	0.2			
108-87-2	Methylcyclohexane	400			
25639-42-3	Methylcyclohexanol	50			
583-60-8	o-Methylcyclohexanone	50		75	
1208-13-3	2-Methylcyclopentadienyl manganese tricarbonyl		0.2		
8022-00-2	Methyl demeton		0.05		
101-68-8	Methylene bisphenyl isocyanate	0.005			
101-14-4	4,4′-Methylene bis(2-chloroaniline)	0.01			
5124-30-1	Methylene bis(4-cyclohexylisocyanate)	0.005			
101-77-9	4,4′-Methylene dianiline	0.1			
78-93-3	Methyl ethyl ketone (MEK)	200	300		
1338-23-4	Methyl ethyl ketone peroxide			0.2	
107-31-3	Methyl formate	100		150	
60-34-4	Methyl hydrazine	0.01			
74-88-4	Methyl iodide	2			
110-12-3	Methyl isoamyl ketone	50			
108-11-2	Methyl isobutyl carbinol	25		40	
108-10-1	Methyl isobutyl ketone	20		75	
624-83-9	Methyl isocyanate	0.02			
563-80-4	Methyl isopropyl ketone	200			
74-93-1	Methyl mercaptan	0.5			
80-62-6	Methyl methacrylate	50		100	
90-12-0	1-Methyl naphthalene	0.5			
91-57-6	2-Methyl naphthalene	0.5			
298-00-0	Methyl parathion		0.02		
107-87-9	Methyl propyl ketone			150	
681-84-5	Methyl silicate	1			
98-83-9	alpha-Methyl styrene	10			
78-94-4	Methyl vinyl ketone			0.2	
21087-64-9	Metribuzin		5		
7786-34-7	Mevinphos		0.01		
12001-26-2	Mica		3		
7439-98-7	Molybdenum (as Mo) Soluble compounds		0.5		
7439-98-7	Molybdenum (as Mo) Metal and insoluble compounds		10, 3		

(Continued)

Appendix F (Continued)

CAS No.	Chemical Name	TWA		STEL	
		ppm	mg/m^3	ppm	mg/m^3
79-11-8	Monochloroacetic acid	0.5			
6923-22-4	Monocrotophos		0.05		
110-91-8	Morpholine	20			
300-76-5	Naled		0.1		
91-20-3	Naphthalene	10		15	
8006-14-2	Natural gas	1000			
9006-04-6	Natural rubber latex		0.0001		
7440-02-0	Nickel – Ni (elemental)		1.5		
7440-02-0	Nickel – Ni (soluble inorganic compounds)		0.1		
7440-02-0	Nickel – Ni (insoluble inorganic compounds)		0.2		
12035-72-2	Nickel – Ni (nickel subsulfide) (as Ni)		0.1		
13463-39-3	Nickel carbonyl (as Ni)	0.05			
54-11-5	Nicotine		0.5		
1929-82-4	Nitrapyrin		10		20
7697-37-2	Nitric acid	2		4	
10102-43-9	Nitric oxide	25			
100-01-6	p-Nitroaniline		3		
98-95-3	Nitrobenzene	1			
100-00-5	p-Nitrochlorobenzene	0.1			
79-24-3	Nitroethane	100			
10102-44-0	Nitrogen dioxide	3		5	
7783-54-2	Nitrogen trifluoride	10			
55-63-0	Nitroglycerin	0.05			
75-52-5	Nitromethane	20			
108-03-2	1-Nitropropane	25			
79-46-9	2-Nitropropane	10			
varies	Nitrotoluene (all isomers)	2			
99-55-8	5-Nitro-o-toluidine		1		
10024-97-2	Nitrous oxide	50			
111-84-2	Nonane (all isomers)	200			
2234-13-1	Octachloronaphthalene		0.1		0.3
111-65-9	Octane (all isomers)	300			
20816-12-0	Osmium tetroxide (as Os)	0.0002		0.006	
144-62-7	Oxalic acid		1		2
80-51-3	p,p′-Oxybis(benzenesulfonyl hydrazide)		0.1		
7783-41-7	Oxygen difluoride		0.05		
10028-15-6	Ozone (heavy work)	0.05			
10028-15-6	Ozone (moderate work)	0.08			
10028-15-6	Ozone (light work)	0.1			
10028-15-6	Ozone (heavy, moderate, or light workloads ≤2 hrs)	0.2			
8002-74-2	Paraffin wax fume		2		
4685-14-7	Paraquat		0.5		
4685-14-7	Paraquat		0.1		
56-38-2	Parathion		0.5		
19624-22-7	Pentaborane	0.005		0.015	
1321-64-8	Pentachloronaphthalene		0.5		

(Continued)

CAS No.	Chemical Name	TWA		STEL	
		ppm	mg/m^3	ppm	mg/m^3
82-68-8	Pentachloronitrobenzene		0.5		
87-86-5	Pentachlorophenol		0.5		
115-77-5	Pentaerythritol		10		
varies	Pentane (all isomers)	600			
varies	Pentyl acetate (all isomers)	50		100	
594-42-3	Perchloromethyl mercaptan	0.1			
7616-94-6	Perchloryl fluoride	3		6	
19430-93-4	Perfluorobutyl ethylene	100			
382-21-8	Perfluoroisobutylene			0.01	
	Persulfates (as persulfate)		0.1		
108-95-2	Phenol	5			
92-84-2	Phenothiazine		5		
95-54-5	o-Phenylenediamine		0.1		
108-45-2	m-Phenylenediamine		0.1		
106-50-3	p-Phenylenediamine		0.1		
101-84-8	Phenyl ether, vapor	1		2	
122-60-1	Phenyl glycidyl ether (PGE)	0.1			
100-63-0	Phenylhydrazine	0.1			
108-98-5	Phenyl mercaptan	0.1			
638-21-1	Phenylphosphine			0.05	
298-02-2	Phorate		0.05		
75-44-5	Phosgene	0.1			
7803-51-2	Phosphine	0.3		1	
7664-38-2	Phosphoric acid		1		3
7723-14-0	Phosphorus (yellow)		0.1		
10025-87-3	Phosphorous oxychloride	0.1			
10026-13-8	Phosphorus pentachloride	0.1			
1314-80-3	Phosphorus pentasulfide		1		3
2125683	Phosphorus trichloride	0.2		0.5	
85-44-9	Phthalic anhydride	1			
626-17-5	m-Phthalodinitrile		5		
1918-02-1	Picloram		10		
88-89-1	Picric acid		0.1		
83-26-1	Pindone		0.1		
142-64-3	Piperazine dihydrochloride		5		
7440-06-4	Platinum (metal)		1		
7440-06-4	Platinum (soluble salts)		0.002		
9002-86-2	Polyvinyl chloride (PVC)		1		
65997-15-1	Portland cement		1		
1310-58-3	Potassium hydroxide				2
74-98-6	Propane	1000			
71-23-8	n-Propanol	100			
67-63-0	2-Propanol	200		400	
107-19-7	Propargyl alcohol	1			
57-57-8	beta-Propiolactone	0.5			
123-38-6	Propionaldehyde	20			
79-09-4	Propionic acid	10			
114-26-1	Propoxur		0.5		
109-60-4	n-Propyl acetate	200		250	

(Continued)

Appendix F (Continued)

CAS No.	Chemical Name	TWA		STEL	
		ppm	mg/m³	ppm	mg/m³
115-07-1	Propylene	500			
78-87-5	Propylene dichloride	10			
6423-43-4	Propylene glycol dinitrate	0.05			
75-56-9	Propylene oxide	2			
75-55-8	Propyleneimine	0.2		0.4	
627-13-4	n-Propyl nitrate	25		40	
8003-34-7	Pyrethrum		5		
110-86-1	Pyridine	1			
106-51-4	Quinone	0.1			
108-46-3	Resorcinol	10		20	
7440-16-6	Rhodium (metal fume and insoluble compounds)		1		
7440-16-6	Rhodium (soluble compounds)		0.01		
299-84-3	Ronnel		5		
83-79-4	Rotenone		5		
7782-49-2	Selenium compounds (as Se)		0.2		
7783-79-1	Selenium hexafluoride (as Se)	0.05			
136-78-7	Sesone		10		
varies	Silica (crystalline)		0.025		
409-21-2	Silicon carbide (nonfibrous)		10		
409-21-2	Silicon carbide (nonfibrous)		3		
7803-62-5	Silicon tetrahydride	5			
7440-22-4	Silver (metal, dust, fume) (as Ag)		0.1		
7440-22-4	Silver (soluble compounds) (as Ag)		0.01		
7631-90-5	Sodium bisulfite		5		
62-74-8	Sodium fluoroacetate		0.05		
1310-73-2	Sodium hydroxide				2
7681-57-4	Sodium metabisulfite		5		
9005-25-8	Starch		10		
	Stearates		10		
8052-41-3	Stoddard solvent	100			
7789-06-2	Strontium chromate		0.0005		
57-24-9	Strychnine		0.15		
100-42-5	Styrene (monomer)	20		40	
1395-21-7	Subtilisins (as crystalline active enzyme)				6E-05
57-50-1	Sucrose		10		
74222-97-2	Sulfometuron methyl		5		
3689-24-5	Sulfotepp (TEDP)		0.1		
7446095	Sulfur dioxide			0.25	
2551-62-4	Sulfur hexafluoride	1000			
7664-93-9	Sulfuric acid		0.2		
10025-67-9	Sulfur monochloride			1	
5714-22-7	Sulfur pentafluoride			0.01	
7783-60-0	Sulfur tetrafluoride			0.1	
2699-79-8	Sulfuryl fluoride	5		10	
35400-43-2	Sulprofos		0.1		
93-76-5	2,4,5-T		10		
14807-96-6	Talc		2		
13494-80-9	Tellurium and compounds (as Te)		0.1		

(Continued)

CAS No.	Chemical Name	TWA		STEL	
		ppm	mg/m³	ppm	mg/m³
7783-80-4	Tellurium hexafluoride (as Te)	0.02			
3383-96-8	Temephos		1		
13071-79-9	Terbufos		0.01		
100-21-0	Terephthalic acid		10		
26140-60-3	Terphenyls				5
79-27-6	1,1,2,2-Tetrabromoethane	0.1			
76-11-9	1,1,1,2-Tetrachloro-2,2-difluoroethane	100			
76-12-0	1,1,2,2-Tetrachloro-1,2-difluoroethane	50			
79-34-5	1,1,2,2-Tetrachloroethane	1			
127-18-4	Tetrachloroethylene	25		100	
1335-88-2	Tetrachloronaphthalene		2		
78-00-2	Tetraethyl lead (as Pb)		0.1		
107-49-3	Tetraethyl pyrophosphate (TEPP)		0.01		
116-14-3	Tetrafluoroethylene	2			
109-99-9	Tetrahydrofuran	50		100	
124-64-1	Tetrakis (hydroxymethyl) phosphonium chloride		2		
55566-30-8	Tetrakis (hydroxymethyl) phosphonium sulfate		2		
75-74-1	Tetramethyl lead (as Pb)		0.15		
3333-52-6	Tetramethyl succinonitrile	0.5			
509-14-8	Tetranitromethane	0.005			
7440-28-0	Thallium, and compounds (as Tl)		0.02		
96-69-5	4,4′-Thiobis (6-tert, Butyl-m-cresol) (td)		10		
68-1-1	Thioglycolic acid	1			
7719-09-7	Thionyl chloride			0.2	
137-26-8	Thiram		0.05		
7440-31-5	Tin, oxide and inorganic compounds, except tin hydride (as Sn)		2		
7440-31-5	Tin, metal (as Sn)		2		
7440-31-5	Tin, organic compounds (as Sn)		0.1		0.2
13463-67-7	Titanium dioxide		10		
108-88-3	Toluene	20			
584-84-9	Toluene-2,4	0.005		0.02	
91-08-7	or 2,6-diisocyanate (or as a mixture)	0.005		0.02	
95-53-4	o-Toluidine	2			
108-44-1	m-Toluidine	2			
106-49-0	p-Toluidine	2			
126-73-8	Tributyl phosphate	0.2			
76-03-9	Trichloroacetic acid	1			
120-82-1	1,2,4-Trichlorobenzene			5	
79-00-5	1,1,2-Trichloroethane	10			
79-01-6	Trichloroethylene	10		25	
75-69-4	Trichlorofluoromethane			1000	
1321-65-9	Trichloronaphthalene		5		
96-18-4	1,2,3-Trichloropropane	10			
76-13-1	1,1,2-Trichloro-1,2,2- trifluoroethane	1000		1250	
52-68-6	Trichlorphon		1		
10-71-6	Triethanolamine		5		
121-44-8	Triethylamine	1		3	

(Continued)

Appendix F (Continued)

CAS No.	Chemical Name	TWA		STEL	
		ppm	mg/m^3	ppm	mg/m^3
75-63-8	Trifluorobromomethane	1000			
2451-62-9	1,3,5-Triglycidyl-s-triazinetrione		0.05		
552-30-7	Trimellitic anhydride		0.0005		0.002
75-50-3	Trimethylamine	5		15	
25551-13-7	Trimethyl benzene	25			
121-45-9	Trimethyl phosphite	2			
118-96-7	2,4,6-Trinitrotoluene		0.1		
78-30-8	Triorthocresyl phosphate		0.1		
115-86-6	Triphenyl phosphate		3		
7440-33-7	Tungsten (metal and insoluble compounds) (as W)		5		10
7440-33-7	Tungsten (should be compounds) (as W)		1		3
varies	Turpentine (and selected monoterpenes)	20			
7440-61-1	Uranium (soluble and insoluble compounds) (as U)		0.2		0.6
110-62-3	n-Valeraldehyde	50			
1314-62-1	Vanadium pentoxide (as V)		0.05		
108-05-4	Vinyl acetate	10		15	
593-60-2	Vinyl bromide	0.5			
75-01-4	Vinyl chloride	1			
100-40-3	4-Vinyl cyclohexane	0.1			
106-87-6	Vinyl cyclohexene dioxide	0.1			
75-02-5	Vinyl fluoride	1			
88-12-0	N-Vinyl-2-pyrrolidone	0.05			
75-35-4	Vinylidene chloride	5			
75-38-7	Vinylidene fluoride	500			
25013-15-4	Vinyl toluene	50		100	
81-81-2	Warfarin		0.1		
	Wood dusts (western red cedar)		0.5		
	Wood dusts (all other species)		1		
Varies	Xylenes (o-, m-, p-isomers)	100		150	
1477-55-0	m-Xylene alpha,alpha′-diamine				0.1
1300-73-8	Xylidine (mixed isomers)	0.5			
7440-65-5	Yttrium (and compounds) (as Y)		1		
7646-85-7	Zinc chloride fume		1		2
varies	Zinc chromates (as Cr)		0.01		
1314-13-2	Zinc oxide		2		10
7440-67-7	Zirconium compounds (as Zr)		5		10

* Ceiling-based value.
Reference: ACGHI TLVs and BEIs Handbook 2010.

Appendix G: ACGIH Immediately Dangerous to Life and Health (IDLH) Values

CAS No.	Chemical Name	ppm	mg/m^3
75-07-0	Acetaldehyde	10000	
64-19-7	Acetic acid	1000	
108-24-7	Acetic anhydride	1000	
67-64-1	Acetone	20000	
75-05-8	Acetonitrile	4000	
79-27-6	Acetylene tetrabromide	10	
107-02-8	Acrolein	5	
79-06-1	Acrylamide	unknown	
107-13-1	Acrylonitrile	500	
309-00-2	Aldrin		100
107-18-6	Allyl alcohol	150	
107-05-1	Allyl chloride	300	
106-92-3	Allyl glycidyl ether	270	
504-29-0	2-Aminopyridine	5	
7664-41-7	Ammonia	500	
7773-06-0	Ammonium sulfamate		5000
628-63-7	n-Amyl acetate	4000	
626-38-0	sec-Amyl acetate	9000	
62-53-3	Aniline	100	
90-04-0	o-Anisidine		50
104-94-9	p-Anisidine		50
7440-36-0 (Metal)	Antimony compounds (as Sb)		80 Sb/m^3
86-88-4	ANTU		100
7440-38-2 (Metal)	Arsenic (inorganic compounds, as As)		100 As/m^3
7784-42-1	Arsine	6	
86-50-0	Azinphosmethyl		20
7440-39-3 (Metal)	Barium (soluble compounds, as Ba)		1100 mg Ba/m^3
71-43-2	Benzene	3000	
94-36-0	Benzoyl peroxide		7000
100447	Benzyl chloride	10	
7440-41-7 (Metal)	Beryllium compounds (as Be)		10 mg Be/m^3
1303-86-2	Boron oxide	no evidence	
7637-07-2	Boron trifluoride	100	
7726-95-6	Bromine	10	
75-25-2	Bromoform	unknown	
106-99-0	1,3-Butadiene	20000 (LEL)	
78933	2-Butanone	3000	
111762	2-Butoxyethanol	700	

(Continued)

Appendix G (Continued)

CAS No.	Chemical Name	ppm	mg/m³
123864	n-Butyl acetate	10000	
105464	sec-Butyl acetate	10000	
540885	tert-Butyl acetate	10000	
71363	n-Butyl alcohol	8000	
78922	sec-Butyl alcohol	10000	
75650	tert-Butyl alcohol	8000	
109739	n-Butylamine	2000	
1189851	tert-Butyl chromate		30 mg CrO_3/m^3
2426086	n-Butyl glycidyl ether	3500	
109795	n-Butyl mercaptan	2500	
98511	p-tert-Butyltoluene	1000	
7440439 (Metal)	Cadmium dust (as Cd)		50 mg Cd/m^3
7440439 (Metal)	Cadmium fume (as Cd)		
7778441	Calcium arsenate (as As)		100 mg As/m^3
1305788	Calcium oxide	unknown	
76222	Camphor (synthetic)		200
63252	Carbaryl		600
1333864	Carbon black	no evidence	
124389	Carbon dioxide	50000	
75150	Carbon disulfide	500	
630080	Carbon monoxide	1500	
56235	Carbon tetrachloride	300	
57749	Chlordane		500
8001352	Chlorinated camphene		200
31242930	Chlorinated diphenyl oxide	unknown	
7782505	Chlorine	30	
	Chlorine dioxide	unknown	
7790912	Chlorine trifluoride	20	
107200	Chloroacetaldehyde	100	
532274	alpha-Chloroacetophenone		100
108907	Chlorobenzene	2400	
2698411	o-Chlorobenzylidene malononitrile		2
74975	Chlorobromomethane	5000	
53469219	Chlorodiphenyl (42% chlorine)		10
11097691	Chlorodiphenyl (54% chlorine)		
67663	Chloroform	1000	
600259	1-Chloro-1-nitropropane	2000	
76062	Chloropicrin	4	
126998	beta-Chloroprene	400	
1333-82-0 (CrO_3)	Chromic acid and chromates		30 (as CrO_3)
VARIES	Chromium (II) compounds [as Cr(II)]	no evidence	
VARIES	Chromium (III) compounds [as Cr(III)]	no evidence	
7440473	Chromium metal (as Cr)	no evidence	
65996932	Coal tar pitch volatiles		700
7440484 (Metal)	Cobalt metal, dust and fume (as Co)		20 mg Co/m^3
7440508 (Metal)	Copper (dusts and mists, as Cu)	no evidence	
1317380 (CuO)	Copper fume (as Cu)	no evidence	
NONE	Cotton dust (raw)	no evidence	
136787	Crag (r) herbicide		5000
	Cresol (o, m, p isomers)	250	

(Continued)

CAS No.	Chemical Name	ppm	mg/m³
95487 (oisomer), 108394 (misomer), 106445 (pisomer)			
123739 (transisomer)	Crotonaldehyde	400	
98828	Cumene	8000	
VARIES	Cyanides (as CN)		50
110827	Cyclohexane	10000	
108930	Cyclohexanol	3500	
108941	Cyclohexanone	5000	
110838	Cyclohexene	10000	
542927	Cyclopentadiene	2000	
94757	2,4-D		500
50293	DDT	no evidence	
17702419	Decaborane		100
8065483	Demeton		20
123422	Diacetone alcohol	2100	
334883	Diazomethane	2	
19287457	Diborane	40	
107664	Dibutyl phosphate	125	
84742	Dibutyl phthalate		9300
95501	o-Dichlorobenzene	1000	
106467	p-Dichlorobenzene	1000	
75718	Dichlorodifluoromethane	50000	
118525	1,3-Dichloro 5,5-dimethylhydantoin	unknown	
75343	1,1-Dichloroethane	4000	
540590	1,2-Dichloroethylene	4000	
111444	Dichloroethyl ether	250	
75434	Dichloromonofluoromethane	50000	
594729	1,1-Dichloro 1-nitroethane	150	
76142	Dichlorotetrafluoroethane	50000	
62737	Dichlorvos		200
60571	Dieldrin		450
109897	Diethylamine	2000	
100378	2-Diethylaminoethanol	500	
75616	Difluorodibromomethane	2500	
2238075	Diglycidyl ether	25	
108838	Diisobutyl ketone	2000	
108189	Diisopropylamine	1000	
127195	Dimethyl acetamide	400	
124403	Dimethylamine	2000	
121697	N,N-Dimethylaniline	100	
300765	Dimethyl 1,2-dibromo 2,2-dichlorethyl phosphate		1800
68122	Dimethylformamide	3500	
57147	1,1-Dimethylhydrazine	50	
131113	Dimethylphthalate		9300
77781	Dimethyl sulfate	10	
528290 (oisomer), 99650 (misomer), 100254 (pisomer)	Dinitrobenzene (o, m, p isomers)		200

(Continued)

Appendix G (Continued)

CAS No.	Chemical Name	ppm	mg/m^3
534521	Dinitro-o-cresol		5
25321146	Dinitrotoluene		200
117817	Di sec-octyl phthalate	unknown	
123911	Dioxane	2000	
92524	Diphenyl		300
34590948	Dipropylene glycol methyl ether	unknown	
72208	Endrin		2000
106898	Epichlorohydrin	250	
2104645	EPN		50
141435	Ethanolamine	1000	
110805	2-Ethoxyethanol	6000	
111159	2-Ethoxyethyl acetate	2500	
141786	Ethyl acetate	10000	
140885	Ethyl acrylate	2000	
64175	Ethyl alcohol	15000	
75047	Ethylamine	4000	
100414	Ethyl benzene	2000	
74964	Ethyl bromide	3500	
106354	Ethyl butyl ketone	3000	
75003	Ethyl chloride	20000	
107073	Ethylene chlorohydrin	10	
107153	Ethylenediamine	2000	
106934	Ethylene dibromide	400	
107062	Ethylene dichloride	1000	
628966	Ethylene glycol dinitrate		500
151564	Ethyleneimine	100	
75218	Ethylene oxide	800	
60297	Ethyl ether	19000 (LEL)	
109944	Ethyl formate	8000	
75081	Ethyl mercaptan	2500	
100743	N-Ethylmorpholine	2000	
78104	Ethyl silicate	1000	
14484641	Ferbam	no evidence	
12604589	Ferrovanadium dust	no evidence	
VARIES	Fluorides (as F)		500 mg F/m^3
7782414	Fluorine	25	
75694	Fluorotrichloromethane	10000	
50000	Formaldehyde	30	
64186	Formic acid	30	
98011	Furfural	250	
98000	Furfuryl alcohol	250	
556525	Glycidol	500	
7782425	Graphite (natural)	no evidence	
7440586 (Metal)	Hafnium compounds (as Hf)	unknown	
7440586 (Metal)	Heptachlor		700
142825	n-Heptane	5000	
67721	Hexachloroethane	300	
1335871	Hexachloronaphthalene		2
110543	n-Hexane	5000	

(Continued)

CAS No.	Chemical Name	ppm	mg/m^3
591786	2-Hexanone	5000	
108101	Hexone	3000	
108849	sec Hexyl acetate	4000	
302012	Hydrazine	80	
10035106	Hydrogen bromide	50	
7647010	Hydrogen chloride	100	
74908	Hydrogen cyanide	50	
7664393	Hydrogen fluoride (as F)	30	
7722841	Hydrogen peroxide	75	
7783075	Hydrogen selenide (as Se)	2	
7783064	Hydrogen sulfide	300	
123319	Hydroquinone	unknown	
7553562	Iodine	10	
1309371	Iron oxide dust and fume (as Fe)	no evidence	
12392-2	Isoamyl acetate	3000	
123513	Isoamyl alcohol (primary)	10000 (LEL)	
528754	Isoamyl alcohol (secondary)	10000 (LEL)	
110190	Isobutyl acetate	7500	
78831	Isobutyl alcohol	8000	
78591	Isophorone	800	
108214	Isopropyl acetate	16000	
67630	Isopropyl alcohol	12000	
75310	Isopropylamine	4000	
108203	Isopropyl ether	10000	
4016142	Isopropyl glycidyl ether	1000	
463514	Ketene	unknown	
7439921	Lead compounds (as Pb)		700 mg Pb/m^3
58899	Lindane	1000	
7580678	Lithium hydride		55
68476857	L.P.G.	19000 (LEL)	
1309484	Magnesium oxide fume	no evidence	
121755	Malathion		5000
108316	Maleic anhydride	unknown	
7439965	Manganese compounds (as Mn)	no evidence	
7439976	Mercury compounds [except (organo) alkyls, as Hg]		28 mg Hg/m^3
Varies	Mercury (organo) alkyl compounds (as Hg)		10 mg Hg/m^3
141797	Mesityl oxide	5000	
72435	Methoxychlor	no evidence	
79209	Methyl acetate	10000	
74997	Methyl acetylene	15000 (LEL)	
59355758	Methyl acetylenepropadiene mixture	15000	
96333	Methyl acrylate	1000	
109875	Methylal	15000 (LEL)	
67561	Methyl alcohol	25000	
74895	Methylamine	100	
110430	Methyl (namyl) ketone	4000	
74839	Methyl bromide	2000	
109864	Methyl cellosolve (r)	2000	
110496	Methyl cellosolve (r) acetate	4000	

(Continued)

Appendix G (Continued)

CAS No.	Chemical Name	ppm	mg/m^3
74873	Methyl chloride	10000	
71556	Methyl chloroform	1000	
108872	Methylcyclohexane	10000	
25639423	Methylcyclohexanol	10000	
583608	o-Methylcyclohexanone	2500	
101688	Methylene bisphenyl isocyanate		100
75092	Methylene chloride	5000	
107313	Methyl formate	5000	
541855	5-Methyl 3-heptanone		3000
60344	Methyl hydrazine	50	
74884	Methyl iodide	800	
108112	Methyl isobutyl carbinol	2000	
624839	Methyl isocyanate	20	
74931	Methyl mercaptan	400	
80626	Methyl methacrylate	4000	
98839	Methyl styrene	5000	
12001262	Mica	no evidence	
7439987	Molybdenum (insoluble compounds, as Mo)	no evidence	
Varies	Molybdenum (soluble compounds, as Mo)	no evidence	
100618	Monomethyl aniline	100	
110918	Morpholine	8000	
8030306	Naphtha (coal tar)	10000 (LEL)	
91203	Naphthalene	500	
13463393	Nickel carbonyl (as Ni)	7	
7440020	Nickel metal and other compounds (as Ni)	no evidence	
54115	Nicotine		35
7697372	Nitric acid	100	
10102439	Nitric oxide	100	
100016	p-Nitroaniline		300
98953	Nitrobenzene	200	
100005	p-Nitrochlorobenzene		1000
79243	Nitroethane	1000	
10102440	Nitrogen dioxide	50	
7783542	Nitrogen trifluoride	2000	
55630	Nitroglycerine		500
75525	Nitromethane	1000	
108032	1-Nitropropane	2300	
79469	2-Nitropropane	2300	
88722	Nitrotoluene (oisomer)	200	
99081	Nitrotoluene (misomer)	200	
99990	Nitrotoluene (p isomers)	200	
2234131	Octachloronaphthalene	unknown	
111659	Octane	5000	
8012951	Oil mist (mineral)	no evidence	
20816120	Osmium tetroxide (as Os)		1 mg Os/m^3
144627	Oxalic acid		500
7783417	Oxygen difluoride	0.5	
10028156	Ozone	10	
1910425	Paraquat		1.5

(Continued)

CAS No.	Chemical Name	ppm	mg/m^3
56382	Parathion		20
19624227	Pentaborane	3	
1321648	Pentachloronaphthalene	unknown	
87865	Pentachlorophenol		150
109660	n-Pentane	15000 (LEL)	
107879	2-Pentanone	5000	
594423	Perchloromethyl mercaptan	10	
7616946	Perchloryl fluoride	385	
8002059	Petroleum distillates (naphtha)	10000	
108952	Phenol	250	
106503	p-Phenylene diamine	unknown	
101848	Phenyl ether (vapor)	no evidence	
8004135	Phenyl etherbiphenyl mixture (vapor)	no evidence	
122601	Phenyl glycidyl ether	unknown	
100630	Phenylhydrazine	295	
7786347	Phosdrin	4	
75445	Phosgene	2	
7803512	Phosphine	200	
7664382	Phosphoric acid		10000
7723140	Phosphorus (yellow)	no evidence	
10026138	Phosphorus pentachloride		200
1314803	Phosphorus pentasulfide		750
7719122	Phosphorus trichloride	50	
85449	Phthalic anhydride		10000
88891	Picric acid		100
83261	Pindone		200
Varies	Platinum (soluble salts, as Pt)	no evidence	
65997151	Portland cement	no evidence	
74986	Propane	20000 (LEL)	
109604	n-Propyl acetate	8000	
71238	n-Propyl alcohol	4000	
78875	Propylene dichloride	2000	
75558	Propylene imine	500	
75569	Propylene oxide	2000	
627134	n-Propyl nitrate	2000	
8003347	Pyrethrum		5000
110861	Pyridine	3600	
106514	Quinone		300
7440166	Rhodium (metal fume and insoluble compounds, as Rh)	no evidence	
Varies	Rhodium (soluble compounds, as Rh)	no evidence	
299843	Ronnel		5000
83794	Rotenone	unknown	
7782492	Selenium compounds (as Se)	unknown	
7783791	Selenium hexafluoride	5	
7631869	Silica, amorphous	no evidence	
	Silica, crystalline (respirable dust) cristobalite/tridymite	no evidence	
14808607	Silica, crystalline (respirable dust) quartz/tripoli	no evidence	

(Continued)

Appendix G (Continued)

CAS No.	Chemical Name	ppm	mg/m^3
7440224	Silver (metal dust as Ag)	no evidence	
	Silver (soluble compounds as Ag)	no evidence	
none assigned	Soapstone	no evidence	
62748	Sodium fluoroacetate		5
1310732	Sodium hydroxide		250
7803523	Stibine	40	
8052413	Stoddard solvent		29500
57249	Strychnine		3
100425	Styrene	5000	
7446095	Sulfur dioxide	100	
7664939	Sulfuric acid		80
10025679	Sulfur monochloride	10	
5714227	Sulfur pentafluoride	1	
2699798	Sulfuryl fluoride	1000	
93765	2,4,5-T	unknown	
14807966	Talc	no evidence	
7440257	Tantalum (metal and oxide dust, as Ta)	no evidence	
89245	TEDP		35
13494809	Tellurium compounds (as Te)	no evidence	
7783804	Tellurium hexafluoride	1	
107493	TEPP		10
26140603	Terphenyl (o, m, p isomers)	unknown	
76119	1,1,1,2-Tetrachloro 2,2-difluoroethane	15000	
76120	1,1,2,2-Tetrachloro 1,2-difluoroethane	15000	
79345	1,1,2,2-Tetrachloroethane	150	
127184	Tetrachloroethylene	500	
1335882	Tetrachloronaphthalene		2
78002	Tetraethyl lead (as Pb)		40 mg Pb/m^3
109999	Tetrahydrofuran	20000 (LEL)	
75741	Tetramethyl lead (as Pb)		40 mg Pb/m^3
3333526	Tetramethyl succinonitrile	5	
509148	Tetranitromethane	5	
479458	Tetryl	no evidence	
Varies	Thallium (soluble compounds, as Tl)		20 mg Tl/m^3
137268	Thiram		1500
7440315	Tin (inorganic compounds, as Sn)		400 mg Sn/m^3
Varies	Tin (organic compounds, as Sn)	unknown	
13463677	Titanium dioxide	no evidence	
108883	Toluene	2000	
584849	Toluene 2,4-diisocyanate	10	
95534	o-Toluidine	100	
126738	Tributyl phosphate	125	
79005	1,1,2-Trichloroethane	500	
79016	Trichloroethylene	1000	
1321659	Trichloronaphthalene	unknown	
96184	1,2,3-Trichloropropane	1000	
76131	1,1,2-Trichloro 1,2,2-trifluoroethane	4500	
121448	Triethylamine	1000	
75638	Trifluorobromomethane	50000	
118967	2,4,6-Trinitrotoluene	no evidence	

(*Continued*)

CAS No.	Chemical Name	ppm	mg/m^3
78308	Triorthocresyl phosphate		40
115866	Triphenyl phosphate	no evidence	
8006642	Turpentine	1500	
7440611	Uranium (insoluble compounds, as U)		30 mg U/m^3
Varies	Uranium (soluble compounds, as U)		20 mg U/m^3
1314621	Vanadium dust		70
1314621	Vanadium fume		70
25013154	Vinyl toluene		5000
81812	Warfarin		350
95476	Xylene (oisomer)		1000
108383	Xylene (misomer)		1000
106423	Xylene (p isomers)		1000
1300738	Xylidine	150	
7440655	Yttrium compounds (as Y)	no evidence	
7646857	Zinc chloride fume		4800
1314132	Zinc oxide	no evidence	
7440677	Zirconium compounds (as Zr)		500 mg Zr/m^3

Reference: http://www.cdc.gov/niosh/idlh/intridl4.html

Appendix H: ATSDR Minimal Risk Levels (MRLs) December 2006

CAS No.	Chemical Name	Route	Duration	MRL	Factors	Endpoint	Draft/Final	Cover Date
000083-32-9	ACENAPHTHENE	Oral	Int.	0.6 mg/kg/day	300	Hepatic	Final	August-95
000067-64-1	ACETONE	Inh.	Acute	26 ppm	9	Neurol.	Final	May-94
			Int.	13 ppm	100	Neurol.		
			Chr.	13 ppm	100	Neurol.		
		Oral	Int.	2 mg/kg/day	100	Hemato.		
000107-02-8	ACROLEIN	Inh.	Acute	0.003 ppm	100	Resp.	Draft	August-07
			Int.	0.00004 ppm	300	Resp.		
		Oral	Int.	0.004 mg/kg/day	100	Gastro.		
000079-06-1	ACRYLAMIDE	Oral	Acute	0.02 mg/kg/day	100	Repro.	Draft	September-09
			Int.	0.002 mg/kg/day	100	Neurol.		
000107-13-1	ACRYLONITRILE	Inh.	Acute	0.1 ppm	10	Neurol.	Final	December-90
		Oral	Acute	0.1 mg/kg/day	100	Develop.		
			Int.	0.01 mg/kg/day	1000	Repro.		
			Chr.	0.04 mg/kg/day	100	Hemato.		
000309-00-2	ALDRIN	Oral	Acute	0.002 mg/kg/day	1000	Develop.	Final	September-02
			Chr.	0.00003 mg/kg/day	1000	Hepatic		
007429-90-5	ALUMINUM	Oral	Int.	1.0 mg/kg/day	30	Neurol.	Draft	September-07
			Chr.	1.0 mg/kg/day	90	Neurol.		
007440-35-9	AMERICIUM	Rad.	Acute	4 mSv/yr	3	Develop.	Final	October-04
			Chr.	1 mSv/yr	3	Other		
007664-41-7	AMMONIA	Inh.	Acute	1.7 ppm	30	Resp.	Final	October-04
			Chr.	0.1 ppm	30	Resp.		
000120-12-7	ANTHRACENE	Oral	Int.	10 mg/kg/day	100	Hepatic	Final	August-95
007440-38-2	ARSENIC	Oral	Acute*	0.005 mg/kg/day	10	Gastro.	Draft	August-07
	*provisional		Chr.	0.0003 mg/kg/day	3	Dermal		
001912-24-9	ATRAZINE	Oral	Acute	0.01 mg/kg/day	100	Body Wt.	Final	September-03
			Int.	0.003 mg/kg/day	300	Repro.		
007440-39-3	BARIUM, SOLUBLE SALTS	Oral	Int.	0.2 mg/kg/day	100	Renal	Draft	August-07
			Chr.	0.2 mg/kg/day	100	Renal		
000071-43-2	BENZENE	Inh.	Acute	0.009 ppm	300	Immuno.	Draft	August-07
			Int.	0.006 ppm	300	Immuno.		
			Chr.	0.003 ppm	10	Immuno.		

007440-41-7	BERYLLIUM	Oral	Chr.	0.002 mg/kg/day	300	Gastro.	Final	September-02
000542-88-1	BIS(CHLOROMETHYL) ETHER	Inh.	Int.	0.0003 ppm	100	Resp.	Final	December-89
000111-44-4	BIS(2 CHLOROETHYL) ETHER	Inh.	Int.	0.02 ppm	1000	Body Wt.	Final	December-89
007440-42-8	BORON	Inh.	Acute	0.01 mg/m^3	30	Resp.	Draft	September-07
		Oral	Acute	0.2 mg/kg/day	100	Develop.		
			Int.	0.2 mg/kg/day	66	Develop.		
000075-27-4	BROMODICHLOROMETHANE	Oral	Acute	0.04 mg/kg/day	1000	Hepatic	Final	December-89
			Chr.	0.02 mg/kg/day	1000	Renal		
000075-25-2	BROMOFORM	Oral	Acute	0.7 mg/kg/day	100	Hepatic	Final	September-05
			Int.	0.2 mg/kg/day	300	Hepatic		
			Chr.	0.02 mg/kg/day	3000	Hepatic		
000074-83-9	BROMOMETHANE	Inh.	Acute	0.05 ppm	100	Neurol.	Final	September-92
			Int.	0.05 ppm	100	Neurol.		
			Chr.	0.005 ppm	100	Neurol.		
		Oral	Int.	0.003 mg/kg/day	100	Gastro.		
000106-99-1	BUTADIENE, 1,3-	Inh.	Acute	0.1 ppm	90	Develop.	Draft	September-09
007440-43-9	CADMIUM	Inh.	Acute	0.00003 mg/m^3	300	Resp.	Draft	September-08
			Chr.	0.00001 mg/m^3	9	Renal		
		Oral	Acute	0.0005 mg/kg/day	100	Musculo.		
			Chr.	0.0001 mg/kg/day	3	Renal		
000075-15-0	CARBON DISULFIDE	Inh.	Chr.	0.3 ppm	30	Neurol.	Final	August-96
		Oral	Acute	0.01 mg/kg/day	300	Hepatic		
000056-23-5	CARBON TETRACHLORIDE	Inh.	Int.	0.03 ppm	30	Hepatic	Final	September-05
			Chr.	0.03 ppm	30	Hepatic		
		Oral	Acute	0.02 mg/kg/day	300	Hepatic		
			Int.	0.007 mg/kg/day	100	Hepatic		
007440-46-2	CESIUM	Rad.	Acute	4 mSv	3	Develop.	Final	October-04
			Chr.	1 mSv/yr	3	Other		
000057-74-9	CHLORDANE	Inh.	Int.	0.0002 mg/m^3	100	Hepatic	Final	May-94
			Chr.	0.00002 mg/m^3	1000	Hepatic		
		Oral	Acute	0.001 mg/kg/day	1000	Develop.		
			Int.	0.0006 mg/kg/day	100	Hepatic		
			Chr.	0.0006 mg/kg/day	100	Hepatic		

(Continued)

Appendix H (Continued)

CAS No.	Chemical Name	Route	Duration	MRL	Factors	Endpoint	Draft/Final	Cover Date
000143-50-0	CHLORDECONE	Oral	Acute	0.01 mg/kg/day	100	Neurol.	Final	August-95
			Int.	0.0005 mg/kg/day	100	Renal		
			Chr.	0.0005 mg/kg/day	100	Renal		
000470-90-6	CHLORFENVINPHOS	Oral	Acute	0.002 mg/kg/day	1000	Neurol.	Final	September-97
			Int.	0.002 mg/kg/day	1000	Immuno.		
			Chr.	0.0007 mg/kg/day	1000	Neurol.		
010049-04-4	CHLORINE DIOXIDE	Inh.	Int.	0.001 ppm	300	Resp.	Final	October-04
007758-19-2	CHLORITE	Oral	Int.	0.1 mg/kg/day	30	Neurol.	Final	October-04
000108-90-7	CHLOROBENZENE	Oral	Int.	0.4 mg/kg/day	100	Hepatic	Final	December-90
000124-48-1	CHLORODIBROMOMETHANE	Oral	Acute	0.1 mg/kg/day	300	Hepatic	Final	September-05
			Chr.	0.09 mg/kg/day	300	Hepatic		
000075-00-3	CHLOROETHANE	Inh.	Acute	15 ppm	100	Develop.	Final	December-98
000067-66-3	CHLOROFORM	Inh.	Acute	0.1 ppm	30	Hepatic	Final	September-97
			Int.	0.05 ppm	300	Hepatic		
			Chr.	0.02 ppm	100	Hepatic		
		Oral	Acute	0.3 mg/kg/day	100	Hepatic		
			Int.	0.1 mg/kg/day	100	Hepatic		
			Chr.	0.01 mg/kg/day	1000	Hepatic		
000074-87-3	CHLOROMETHANE	Inh.	Acute	0.5 ppm	100	Neurol.	Final	December-98
			Int.	0.2 ppm	300	Hepatic		
			Chr.	0.05 ppm	1000	Neurol.		
002921-88-2	CHLORPYRIFOS	Oral	Acute	0.003 mg/kg/day	10	Neurol.	Final	September-97
			Int.	0.003 mg/kg/day	10	Neurol.		
			Chr.	0.001 mg/kg/day	100	Neurol.		
016065-83-1	CHROMIUM (III), SOLUBLE PARTICULATES	Inh.	Int.	0.0001 mg/m^3	300	Resp.	Draft	September-08
016065-83-1	CHROMIUM (III), INSOLUBLE PARTICULATES	Inh	Int.	0.005 mg/m^3	90	Resp.	Draft	September-08
018540-29-9	CHROMIUM (VI)	Oral	Int.	0.005 mg/kg/day	100	Hemato.	Draft	September-08
			Chr.	0.001 mg/kg/day	100	Gastro.		
018540-29-9	CHROMIUM (VI), AEROSOL MISTS	Inh.	Int.	0.000005 mg/m^3	100	Resp.	Draft	September-08
			Chr.	0.000005 mg/m^3	100	Resp.		

018540-29-9	CHROMIUM (VI), PARTICULATES	Inh.	Int.	0.0003 mg/m^3	30	Resp.	Draft	September-08
007440-48-4	COBALT	Inh.	Chr.	0.0001 mg/m^3	10	Resp.	Final	October-04
		Oral	Int.	0.01 mg/kg/day	100	Hemato.		
		Rad.	Acute	4 mSv	3	Develop.		
			Chr.	1 mSv/yr	3	Other		
007440-50-8	COPPER	Oral	Acute	0.01 mg/kg/day	3	Gastro.	Final	October-04
			Int.	0.01 mg/kg/day	3	Gastro.		
001319-77-3	CRESOLS	Oral	Int.	0.1 mg/kg/day	100	Resp.	Draft	September-08
000143-33-9	CYANIDE, SODIUM	Oral	Int.	0.05 mg/kg/day	100	Repro.	Final	July-06
002691-41-0	CYCLOTETRAMETHYLENE TETRANITRAMINE (HMX)	Oral	Acute	0.1 mg/kg/day	1000	Neurol.	Final	September-97
			Int.	0.05 mg/kg/day	1000	Hepatic		
000121-82-4	CYCLOTRIMETHYLENETRINITRAMINE (RDX)	Oral	Acute	0.06 mg/kg/day	100	Neurol.	Final	June-95
			Int.	0.03 mg/kg/day	300	Repro.		
068085-85-8	CYHALOTHRIN	Oral	Acute	0.01 mg/kg/day	100	Gastro.	Final	September-03
			Int.	0.01 mg/kg/day	100	Gastro.		
052315-07-8	CYPERMETHRIN	Oral	Acute	0.02 mg/kg/day	100	Neurol.	Final	September-03
000050-29-3	DDT, P,P′-	Oral	Acute	0.0005 mg/kg/day	1000	Develop.	Final	September-02
			Int.	0.0005 mg/kg/day	100	Hepatic		
000117-81-7	DI(2-ETHYLHEXYL)PHTHALATE	Oral	Int.	0.1 mg/kg/day	100	Repro.	Final	September-02
			Chr.	0.06 mg/kg/day	100	Repro.		
000084-74-2	DI-N-BUTYL PHTHALATE	Oral	Acute	0.5 mg/kg/day	100	Develop.	Final	September-01
000117-84-0	DI-N-OCTYL PHTHALATE	Oral	Acute	3 mg/kg/day	300	Hepatic	Final	September-97
			Int.	0.4 mg/kg/day	100	Hepatic		
000333-41-5	DIAZINON	Inh.	Int.	0.01 mg/m^3	30	Neurol.	Draft	September-08
		Oral	Acute	0.006 mg/kg/day	100	Neurol.		
			Int.	0.002 mg/kg/day	100	Neurol.		
			Chr.	0.0007 mg/kg/day	100	Neurol.		
000095-50-1	DICHLOROBENZENE, 1,2-	Oral	Acute	0.7 mg/kg/day	100	Hepatic	Final	July-06
			Int.	0.6 mg/kg/day	100	Hepatic		
			Chr.	0.3 mg/kg/day	100	Renal		

(Continued)

Appendix H (Continued)

CAS No.	Chemical Name	Route	Duration	MRL	Factors	Endpoint	Draft/Final	Cover Date
000541-73-1	DICHLOROBENZENE, 1,3-	Oral	Acute	0.4 mg/kg/day	100	Hepatic	Final	July-06
			Int.	0.02 mg/kg/day	100	Endocr.		
000106-46-7	DICHLOROBENZENE, 1,4-	Inh.	Acute	2 ppm	10	Ocular	Final	July-06
			Int.	0.2 ppm	100	Hepatic		
			Chr.	0.01 ppm	30	Resp.		
		Oral	Int.	0.07 mg/kg/day	100	Hepatic		
			Chr.	0.07 mg/kg/day	100	Hepatic		
000062-73-7	DICHLORVOS	Inh.	Acute	0.002 ppm	100	Neurol.	Final	September-97
			Int.	0.0003 ppm	100	Neurol.		
			Chr.	0.00006 ppm	100	Neurol.		
		Oral	Acute	0.004 mg/kg/day	1000	Neurol.		
			Int.	0.003 mg/kg/day	10	Neurol.		
			Chr.	0.0005 mg/kg/day	100	Neurol.		
000060-57-1	DIELDRIN	Oral	Int.	0.0001 mg/kg/day	100	Neurol.	Final	September-02
			Chr.	0.00005 mg/kg/day	100	Hepatic		
000084-66-2	DIETHYL PHTHALATE	Oral	Acute	7 mg/kg/day	300	Repro.	Final	June-95
			Int.	6 mg/kg/day	300	Hepatic		
001445-75-6	DIISOPROPYL METHYLPHOSPHONATE (DIMP)	Oral	Int.	0.8 mg/kg/day	100	Hemato.	Final	October-98
			Chr.	0.6 mg/kg/day	100	Hemato.		
000075-60-5	DIMETHYLARSINIC ACID (DMA)	Oral	Chr.	0.02 mg/kg/day	100	Renal	Final	August-07
000298-04-4	DISULFOTON	Inh.	Acute	0.006 mg/m^3	30	Neurol.	Final	August-95
			Int.	0.0002 mg/m^3	30	Neurol.		
		Oral	Acute	0.001 mg/kg/day	100	Neurol.		
			Int.	0.00009 mg/kg/day	100	Develop.		
			Chr.	0.00006 mg/kg/day	1000	Neurol.		
000115-29-7	ENDOSULFAN	Oral	Int.	0.005 mg/kg/day	100	Immuno.	Final	September-00
			Chr.	0.002 mg/kg/day	100	Hepatic		
000072-20-8	ENDRIN	Oral	Int.	0.002 mg/kg/day	100	Neurol.	Final	August-96
			Chr.	0.0003 mg/kg/day	100	Neurol.		

000563-12-2	ETHION	Oral	Acute	0.002 mg/kg/day	30	Neurol.	Final	September-00
			Int.	0.002 mg/kg/day	30	Neurol.		
			Chr.	0.0004 mg/kg/day	150	Neurol.		
000100-41-4	ETHYLBENZENE	Inh.	Acute	10 ppm	30	Neurol.	Draft	September-07
			Int.	0.7 ppm	300	Neurol.		
			Chr.	0.3 ppm	300	Renal		
		Oral	Int.	0.5 mg/kg/day	100	Renal		
000107-21-1	ETHYLENE GLYCOL	Inh.	Acute	2 mg/m^3	10	Resp.	Draft	September-07
		Oral	Acute	0.8 mg/kg/day	100	Develop.		
			Int.	0.8 mg/kg/day	100	Develop.		
000075-21-8	ETHYLENE OXIDE	Inh.	Int.	0.09 ppm	100	Renal	Final	December-90
000206-44-0	FLUORANTHENE	Oral	Int.	0.4 mg/kg/day	300	Hepatic	Final	August-95
000086-73-7	FLUORENE	Oral	Int.	0.4 mg/kg/day	300	Hepatic	Final	August-95
007681-49-4	FLUORIDE, SODIUM	Oral	Chr.	0.05 mg/kg/day	3	Musculo.	Final	September-03
007782-41-4	FLUORINE	Inh.	Acute	0.01 ppm	10	Resp.	Final	September-03
000050-00-0	FORMALDEHYDE	Inh.	Acute	0.04 ppm	9	Resp.	Final	July-99
			Int.	0.03 ppm	30	Resp.		
			Chr.	0.008 ppm	30	Resp.		
		Oral	Int.	0.3 mg/kg/day	100	Gastro.		
			Chr.	0.2 mg/kg/day	100	Gastro.		
068476-30-2	FUEL OIL NO. 2	Inh.	Acute	0.02 mg/m^3	1000	Neurol.	Final	June-95
000086-50-0	GUTHION	Inh.	Acute	0.02 mg/m^3	30	Neurol.	Draft	September-08
			Int.	0.01 mg/m^3	30	Neurol.		
			Chr.	0.01 mg/m^3	30	Neurol.		
		Oral	Acute	0.01 mg/kg/day	100	Neurol.		
			Int.	0.003 mg/kg/day	100	Neurol.		
			Chr.	0.003 mg/kg/day	100	Neurol.		
000076-44-8	HEPTACHLOR	Oral	Acute	0.0006 mg/kg/day	3000	Repro.	Final	August-07
			Int.	0.0001 mg/kg/day	300	Immuno.		
000118-74-1	HEXACHLOROBENZENE	Oral	Acute	0.008 mg/kg/day	300	Develop.	Final	September-02
			Int.	0.0001 mg/kg/day	90	Repro.		
			Chr.	0.00005 mg/kg/day	300	Develop.		
000087-68-3	HEXACHLOROBUTADIENE	Oral	Int.	0.0002 mg/kg/day	1000	Renal	Final	May-94

(Continued)

Appendix H (Continued)

CAS No.	Chemical Name	Route	Duration	MRL	Factors	Endpoint	Draft/Final	Cover Date
000319-84-6	HEXACHLOROCYCLOHEXANE, ALPHA-	Oral	Chr.	0.008 mg/kg/day	100	Hepatic	Final	September-05
000319-85-7	HEXACHLOROCYCLOHEXANE, BETA-	Oral	Acute	0.05 mg/kg/day	100	Neurol.	Final	September-05
			Int.	0.0006 mg/kg/day	300	Hepatic		
000058-89-9	HEXACHLOROCYCLOHEXANE, GAMMA-	Oral	Acute	0.003 mg/kg/day	300	Develop.	Final	September-05
			Int.	0.00001 mg/kg/day	1000	Immuno.		
000077-47-4	HEXACHLOROCYCLOPENTADIENE	Inh.	Int.	0.01 ppm	30	Resp.	Final	July-99
			Chr.	0.0002 ppm	90	Resp.		
		Oral	Int.	0.1 mg/kg/day	100	Renal		
000067-72-1	HEXACHLOROETHANE	Inh.	Acute	6 ppm	30	Neurol.	Final	September-97
			Int.	6 ppm	30	Neurol.		
		Oral	Acute	1 mg/kg/day	100	Hepatic		
			Int.	0.01 mg/kg/day	100	Hepatic		
000822-06-0	HEXAMETHYLENE DIISOCYANATE	Inh.	Int.	0.00003 ppm	30	Resp.	Final	October-98
			Chr.	0.00001 ppm	90	Resp.		
000110-54-3	HEXANE, N-	Inh.	Chr.	0.6 ppm	100	Neurol.	Final	July-99
000302-01-2	HYDRAZINE	Inh.	Int.	0.004 ppm	300	Hepatic	Final	September-97
007664-39-3	HYDROGEN FLUORIDE	Inh.	Acute	0.02 ppm	30	Resp.	Final	September-03
007783-06-4	HYDROGEN SULFIDE	Inh.	Acute	0.07 ppm	27	Resp.	Final	July-06
			Int.	0.02 ppm	30	Resp.		
007553-56-2	IODIDE	Oral	Acute	0.01 mg/kg/day	1	Endocr.	Final	October-04
			Chr.	0.01 mg/kg/day	1	Endocr.		
HZ1800-45-T	IONIZING RADIATION, N.O.S.	Rad.	Acute	4 mSv	3	Neurol.	Final	September-99
			Chr.	1 mSv/yr	3	Other		
000078-59-1	ISOPHORONE	Oral	Int.	3 mg/kg/day	100	Other	Final	December-89
			Chr.	0.2 mg/kg/day	1000	Hepatic		
050815-00-4	JP-4	Inh.	Int.	9 mg/m^3	300	Hepatic	Final	June-95
HZ0600-26-T	JP-5/JP-8	Inh.	Int.	3 mg/m^3	300	Hepatic	Final	October-98
HZ0600-22-T	JP-7	Inh.	Chr.	0.3 mg/m^3	300	Hepatic	Final	June-95
008008-20-6	KEROSENE	Inh.	Int.	0.01 mg/m^3	1000	Hepatic	Final	June-95

000121-75-5	MALATHION	Inh.	Acute	0.2 mg/m^3	100	Neurol.	Final	September-03
			Int.	0.02 mg/m^3	1000	Resp.		
		Oral	Int.	0.02 mg/kg/day	10	Neurol.		
			Chr.	0.02 mg/kg/day	100	Neurol.		
007439-96-5	MANGANESE, RESPIRABLE	Inh.	Chr.	0.0003 mg/m^3	100	Neurol.	Draft	September-08
007487-94-7	MERCURIC CHLORIDE	Oral	Acute	0.007 mg/kg/day	100	Renal	Final	March-99
			Int.	0.002 mg/kg/day	100	Renal		
007439-97-6	MERCURY	Inh.	Chr.	0.0002 mg/m^3	30	Neurol.	Final	March-99
000072-43-5	METHOXYCHLOR	Oral	Int.	0.005 mg/kg/day	1000	Repro.	Draft	September-00
000298-00-0	METHYL PARATHION	Oral	Int.	0.0007 mg/kg/day	300	Neurol.	Final	September-01
			Chr.	0.0003 mg/kg/day	100	Hemato.		
001634-04-4	METHYL-T-BUTYL ETHER	Inh.	Acute	2 ppm	100	Neurol.	Final	August-96
			Int.	0.7 ppm	100	Neurol.		
			Chr.	0.7 ppm	100	Renal		
		Oral	Acute	0.4 mg/kg/day	100	Neurol.		
			Int.	0.3 mg/kg/day	300	Hepatic		
000075-09-2	METHYLENE CHLORIDE	Inh.	Acute	0.6 ppm	100	Neurol.	Final	September-00
			Int.	0.3 ppm	90	Hepatic		
			Chr.	0.3 ppm	30	Hepatic		
		Oral	Acute	0.2 mg/kg/day	100	Neurol.		
			Chr.	0.06 mg/kg/day	100	Hepatic		
022967-92-6	METHYLMERCURY	Oral	Chr.	0.0003 mg/kg/day	4	Develop.	Final	March-99
002385-85-5	MIREX	Oral	Chr.	0.0008 mg/kg/day	100	Hepatic	Final	August-95
000124-58-3	MONOMETHYLARSONIC ACID (MMA)	Oral	Int.	0.1 mg/kg/day	100	Gastro.	Final	August-07
			Chr.	0.01 mg/kg/day	100	Renal		
000621-64-7	N-NITROSODI-N-PROPYLAMINE	Oral	Acute	0.095 mg/kg/day	100	Hepatic	Final	December-89
000091-20-3	NAPHTHALENE	Inh.	Chr.	0.0007 ppm	300	Resp.	Final	September-05
		Oral	Acute	0.6 mg/kg/day	90	Neurol.		
			Int.	0.6 mg/kg/day	90	Neurol.		
007440-02-0	NICKEL	Inh.	Int.	0.0002 mg/m^3	30	Resp.	Final	September-05
			Chr.	0.00009 mg/m^3	30	Resp.		
000087-86-5	PENTACHLOROPHENOL	Oral	Acute	0.005 mg/kg/day	1000	Develop.	Final	September-01
			Int.	0.001 mg/kg/day	1000	Repro.		
			Chr.	0.001 mg/kg/day	1000	Endocr.		

(Continued)

Appendix H (Continued)

CAS No.	Chemical Name	Route	Duration	MRL	Factors	Endpoint	Draft/Final	Cover Date
007778-74-7	PERCHLORATES	Oral	Chr.	0.0007 mg/kg/day	10	Endocr.	Draft	August-08
052645-53-1	PERMETHRIN	Oral	Acute	0.3 mg/kg/day	100	Neurol.	Final	September-03
			Int.	0.2 mg/kg/day	100	Neurol.		
000108-95-2	PHENOL	Oral	Acute	1 mg/kg/day	100	Body Wt.	Final	September-08
007723-14-0	PHOSPHORUS, WHITE	Inh.	Acute	0.02 mg/m^3	30	Resp.	Final	September-97
		Oral	Int.	0.0002 mg/kg/day	100	Repro.		
036355-01-8	POLYBROMINATED BIPHENYLS (PBBs)	Oral	Acute	0.01 mg/kg/day	100	Endocr.	Final	October-04
032534-81-9	POLYBROMINATED DIPHENYL ETHERS (PBDEs), LOWER BROMINATED	Inh	Int.	0.006 mg/m^3	90	Endocr.	Final	October-04
032536-52-0		Oral	Acute	0.03 mg/kg/day	30	Endocr.		
			Int.	0.007 mg/kg/day	300	Hepatic		
001163-19-5	PBDE, DECABROMINATED	Oral	Int.	10 mg/kg/day	100	Develop.	Final	November-00
011097-69-1	POLYCHLORINATED BIPHENYLS (PCBs) (Aroclor 1254)	Oral	Int.	0.03 µg/kg/day	300	Neurol.	Final	November-00
			Chr.	0.02 µg/kg/day	300	Immuno.		
006423-43-4	PROPYLENE GLYCOL DINITRATE	Inh.	Acute	0.003 ppm	10	Neurol.	Final	June-95
			Int.	0.00004 ppm	1000	Hemato.		
			Chr.	0.00004 ppm	1000	Hemato.		
000057-55-6	PROPYLENE GLYCOL	Inh.	Int.	0.009 ppm	1000	Resp.	Final	September-97
HZ0900-26-T	REFRACTORY CERAMIC FIBERS	Inh.	Chr.	0.03 fibers/cc	30	Resp.	Final	October-04
007782-49-2	SELENIUM	Oral	Chr.	0.005 mg/kg/day	3	Dermal	Final	September-03
007440-24-6	STRONTIUM	Oral	Int.	2 mg/kg/day	30	Musculo.	Final	October-04
000100-42-5	STYRENE	Inh.	Acute	2 ppm	10	Neurol.	Draft	September-07
			Chr.	0.2 ppm	100	Neurol.		
		Oral	Acute	0.1 mg/kg/day	1000	Neurol.		
007446-09-5	SULFUR DIOXIDE	Inh.	Acute	0.01 ppm	9	Resp.	Final	December-98
000505-60-2	SULFUR MUSTARD	Inh.	Acute	0.0007 mg/m^3	30	Ocular	Final	September-03
			Int.	0.00002 mg/m^3	30	Ocular		
		Oral	Acute	0.5 µg/kg/day	1000	Develop.		
			Int.	0.07 µg/kg/day	300	Gastro.		

000078-51-3	TRISBUTOXYETHYL PHOSPHATE (TBEP)	Oral	Acute	4.8 mg/kg/day	100	Body Wt.	Draft	September-09
			Int.	0.2 mg/kg/day	100	Hepatic		
000126-73-8	TRIBUTYL PHOSPHATE (TnBP)	Oral	Acute	1.1 mg/kg/day	100	Body Wt.	Draft	September-09
			Int.	0.02 mg/kg/day	100	Renal		
			Chr.	0.02 mg/kg/day	100	Renal		
000115-96-8	TRIS (2-CHLOROETHYL) PHOSPHATE (TCEP)	Oral	Int.	0.6 mg/kg/day	100	Neurol.	Draft	September-09
			Chr.	0.3 mg/kg/day	100	Renal		
013674-87-8	TRIS (1,3-DICHLORO-2-PROPYL) PHOSPHATE (TDCP)	Oral	Int.	0.05 mg/kg/day	100	Renal	Draft	September-09
			Chr.	0.02 mg/kg/day	100	Renal		
000127-18-4	TETRACHLOROETHYLENE	Inh.	Acute	0.2 ppm	10	Neurol.	Final	September-97
			Chr.	0.04 ppm	100	Neurol.		
		Oral	Acute	0.05 mg/kg/day	100	Develop.		
007440-31-5	TIN, INORGANIC	Oral	Int.	0.3 mg/kg/day	100	Hemato.	Final	September-05
000683-18-1	TIN, DIBUTYL-	Oral	Int.	0.005 mg/kg/day	1000	Immunol.	Final	September-05
000056-36-9	TIN, TRIBUTYL-	Oral	Int.	0.0003 mg/kg/day	100	Immunol.	Final	September-05
			Chr.	0.0003 mg/kg/day	100	Immunol.		
007550-45-0	TITANIUM TETRACHLORIDE	Inh.	Int.	0.01 mg/m^3	90	Resp.	Final	September-97
			Chr.	0.0001 mg/m^3	90	Resp.		
000108-88-3	TOLUENE	Inh.	Acute	1 ppm	10	Neurol.	Final	September-00
			Chr.	0.08 ppm	100	Neurol.		
		Oral	Acute	0.8 mg/kg/day	300	Neurol.		
			Int.	0.02 mg/kg/day	300	Neurol.		
008001-35-2	TOXAPHENE	Oral	Acute	0.005 mg/kg/day	1000	Hepatic	Final	August-96
			Int.	0.001 mg/kg/day	300	Hepatic		
000079-01-6	TRICHLOROETHYLENE	Inh.	Acute	2 ppm	30	Neurol.	Final	September-97
			Int.	0.1 ppm	300	Neurol.		
		Oral	Acute	0.2 mg/kg/day	300	Develop.		
HZ1800-90-T	URANIUM, HIGHLY SOLUBLE SALTS	Inh.	Int.	0.0004 mg/m^3	90	Renal	Final	September-99
			Chr.	0.0003 mg/m^3	30	Renal		
		Oral	Int.	0.002 mg/kg/day	30	Renal		

(*Continued*)

Appendix H (Continued)

CAS No.	Chemical Name	Route	Duration	MRL	Factors	Endpoint	Draft/Final	Cover Date
HZ1800-92-T	URANIUM, INSOLUBLE COMPOUNDS	Inh.	Int.	0.008 mg/m^3	30	Renal	Final	September-99
007440-62-2	VANADIUM	Inh.	Acute	0.0008 mg/m^3	90	Resp.	Draft	September-09
			Chr.	0.0001 mg/m^3	30	Resp.		
		Oral	Int.	0.003 mg/kg/day	100	Hemato.		
000108-05-4	VINYL ACETATE	Inh.	Int.	0.01 ppm	100	Resp.	Final	July-92
000075-01-4	VINYL CHLORIDE	Inh.	Acute	0.5 ppm	30	Develop.	Final	September-06
			Int.	0.03 ppm	30	Hepatic		
		Oral	Chr.	0.003 mg/kg/day	30	Hepatic		
001330-20-7	XYLENES, MIXED	Inh.	Acute	2 ppm	30	Neurol.	Final	August-07
			Int.	0.6 ppm	90	Neurol.		
			Chr.	0.05 ppm	300	Neurol.		
		Oral	Acute	1 mg/kg/day	100	Neurol.		
			Int.	0.4 mg/kg/day	1000	Neurol.		
			Chr.	0.2 mg/kg/day	1000	Neurol.		
007440-66-6	ZINC	Oral	Int.	0.3 mg/kg/day	3	Hemato.	Final	September-05
			Chr.	0.3 mg/kg/day	3	Hemato.		
000090-12-0	1-METHYLNAPHTHALENE	Oral	Chr.	0.07 mg/kg/day	1000	Resp.	Final	September-05
000091-57-6	2-METHYLNAPHTHALENE	Oral	Chr.	0.04 mg/kg/day	100	Resp.	Final	September-05
000075-35-4	1,1-DICHLOROETHENE	Inh.	Int.	0.02 ppm	100	Hepatic	Final	May-94
		Oral	Chr.	0.009 mg/kg/day	1000	Hepatic		
000057-14-7	1,1-DIMETHYLHYDRAZINE	Inh.	Int.	0.0002 ppm	300	Hepatic	Final	September-97
000071-55-6	1,1,1-TRICHLOROETHANE	Inh.	Acute	2 ppm	100	Neurol.	Final	July-06
			Int.	0.7 ppm	100	Neurol.		
		Oral	Int.	20 mg/kg/day	100	Body Wt.		
000079-00-5	1,1,2-TRICHLOROETHANE	Oral	Acute	0.3 mg/kg/day	100	Neurol.	Final	December-89
			Int.	0.04 mg/kg/day	100	Hepatic		
000079-34-5	1,1,2,2-TETRACHLOROETHANE	Oral	Int.	0.5 mg/kg/day	100	Hepatic	Draft	September-08
000096-12-8	1,2-DIBROMO-3-CHLOROPROPANE	Inh.	Int.	0.0002 ppm	100	Repro.	Final	September-92
		Oral	Int.	0.002 mg/kg/day	1000	Repro.		
000107-06-2	1,2-DICHLOROETHANE	Inh.	Chr.	0.6 ppm	90	Hepatic	Final	September-01
		Oral	Int.	0.2 mg/kg/day	300	Renal		

000156-59-2	1,2-DICHLOROETHENE, CIS-	Oral	Acute	1 mg/kg/day	100	Hemato.	Final	August-96
			Int.	0.3 mg/kg/day	100	Hemato.		
000156-60-5	1,2-DICHLOROETHENE, TRANS-	Inh.	Acute	0.2 ppm	1000	Hepatic	Final	August-96
			Int.	0.2 ppm	1000	Hepatic		
		Oral	Int.	0.2 mg/kg/day	100	Hepatic		
000078-87-5	1,2-DICHLOROPROPANE	Inh.	Acute	0.05 ppm	1000	Resp.	Final	December-89
			Int.	0.007 ppm	1000	Resp.		
		Oral	Acute	0.1 mg/kg/day	1000	Neurol.		
			Int.	0.07 mg/kg/day	1000	Hemato.		
			Chr.	0.09 mg/kg/day	1000	Hepatic		
000540 73 8	1,2-DIMETHYLHYDRAZINE	Oral	Int.	0.0008 mg/kg/day	1000	Hepatic	Final	September-97
000096-18-4	1,2,3-TRICHLOROPROPANE	Inh.	Acute	0.0003 ppm	100	Resp.	Final	September-92
		Oral	Int.	0.08 mg/kg/day	100	Hepatic		
000542-75-6	1,3-DICHLOROPROPENE	Inh.	Int.	0.008 ppm	30	Resp.	Draft	September-08
			Chr.	0.007 ppm	30	Resp.		
		Oral	Int.	0.04 mg/kg/day	100	Gastro.		
			Chr.	0.03 mg/kg/day	100	Gastro.		
000099-65-0	1,3-DINITROBENZENE	Oral	Acute	0.008 mg/kg/day	100	Repro.	Final	June-95
			Int.	0.0005 mg/kg/day	1000	Hemato.		
000123-91-1	1,4-DIOXANE	Inh.	Acute	2 ppm	30	Ocular	Draft	September-07
			Int.	1 ppm	30	Hepatic		
			Chr.	1 ppm	30	Hepatic		
		Oral	Acute	4 mg/kg/day	100	Resp.		
			Int.	0.6 mg/kg/day	100	Hepatic		
			Chr.	0.1 mg/kg/day	100	Hepatic		
000111-76-2	2-BUTOXYETHANOL	Inh.	Acute	6 ppm	9	Hemato.	Final	October-98
			Int.	3 ppm	9	Hemato.		
			Chr.	0.2 ppm	3	Hemato.		
		Oral	Acute	0.4 mg/kg/day	90	Hemato.		
			Int.	0.07 mg/kg/day	1000	Hepatic		
000078-88-6	2,3-DICHLOROPROPENE	Inh.	Acute	0.002 ppm	90	Resp.	Final	September-08
057117-31-4	2,3,4,7,8-PENTACHLORODIBENZOFURAN	Oral	Acute	0.001 ug/kg/day	3000	Immuno.	Final	May-94

(*Continued*)

Appendix H (Continued)

CAS No.	Chemical Name	Route	Duration	MRL	Factors	Endpoint	Draft/Final	Cover Date
			Int.	0.00003 ug/kg/day	3000	Hepatic		
001746-01-6	2,3,7,8-TETRACHLORODIBENZO-P-DIOXIN	Oral	Acute	0.0002 ug/kg/day	21	Immuno.	Final	December-98
			Int.	0.00002 ug/kg/day	30	Lymphor.		
			Chr.	0.000001 ug/kg/day	90	Develop.		
000120-83-2	2,4-DICHLOROPHENOL	Oral	Int.	0.003 mg/kg/day	100	Immuno.	Final	July-99
000051-28-5	2,4-DINITROPHENOL	Oral	Acute	0.01 mg/kg/day	100	Body Wt.	Final	August-95
000121-14-2	2,4-DINITROTOLUENE	Oral	Acute	0.05 mg/kg/day	100	Neurol.	Final	December-98
			Chr.	0.002 mg/kg/day	100	Hemato.		
000118-96-7	2,4,6-TRINITROTOLUENE	Oral	Int.	0.0005 mg/kg/day	1000	Hepatic	Final	June-95
000606-20-2	2,6-DINITROTOLUENE	Oral	Int.	0.004 mg/kg/day	1000	Hemato.	Final	December-98
000106-48-9	4-CHLOROPHENOL	Oral	Acute	0.01 mg/kg/day	100	Hepatic	Final	July-99
000101-14-4	4,4′-METHYLENEBIS(2-CHLOROANILINE)	Oral	Chr.	0.003 mg/kg/day	3000	Hepatic	Final	May-94
000101-77-9	4,4′-METHYLENEDIANILINE	Oral	Acute	0.2 mg/kg/day	300	Hepatic	Final	October-98
			Int.	0.08 mg/kg/day	100	Hepatic		
000534-52-1	4,6-DINITRO-O-CRESOL	Oral	Acute	0.004 mg/kg/day	100	Neurol.	Final	August-95
			Int.	0.004 mg/kg/day	100	Neurol.		

Reference: http://www.atsdr.cdc.gov/mrls/pdfs/atsdr_mrls_december_2009.pdf

Appendix I: WHO Air Quality Guidelines

CAS No.	Chemical Name	Time-Weighted Average (TWA)	Averaging Time
7440-43-9	Cadmium	5 ng/m^3	annual
75-15-0	Carbon disulfide	100 μg/m^3	24 hr
630-08-0	Carbon monoxide	100 mg/m^3	15 min
		60 mg/m^3	30 min
		30 mg/m^3	1 hr
		10 mg/m^3	8 hr
107-06-2	1,2-Dichloroethane	0.7 mg/m^3	24 hr
75-09-2	Dichloromethane	3 mg/m^3	1 week
50-00-0	Formaldehyde	0.1 mg/m^3	30 min
7783-06-4	Hydrogen sulfide	150 μg/m^3	24 hr
7439-92-1	Lead	0.5 μg/m^3	annual
7439-96-5	Manganese	0.15 μg/m^3	annual
7439-97-6	Mecury	1 μg/m^3	annual
10102-44-0	Nitrogen dioxide	200 μg/m^3	1 hr
		40 μg/m^3	annual
10028-15-6	Ozone	100 μg/m^3	8 hr
	Particulate matter (2.5)	25 μg/m^3	24 hr
		10 μg/m^3	annual
	Particulate matter (10)	50 μg/m^3	24 hr
		20 μg/m^3	annual
100-42-5	Styrene	0.26 mg/m^3	1 week
7446-09-5	Sulfur dioxide	500 μg/m^3	10 min
		20 μg/m^3	24 hr
		50 μg/m^3	annual
127-18-4	Tetrachloroethylene	0.25 mg/m^3	annual
108-88-3	Toulene	0.26 mg/m^3	1 week
1314-62-1	Vanadium	1 μg/m^3	24 hr

References: http://www.euro.who.int/document/e71922.pdf
http://www.euro.who.int/__data/assets/pdf_file/0005/74732/E71922.pdf
http://whqlibdoc.who.int/hq/2006/WHO_SDE_PHE_OEH_06.02_eng.pdf

Appendix J: EPA Regional Screening Levels (RSLs)

CAS No.	Chemical Name	Regional Screening Levels (RSLs)									
		'Direct Contact Exposure Pathways'									
		Residential Soil (mg/kg)		Industrial Soil (mg/kg)		Residential Air (μg/m³)		Industrial Air (μg/m³)		Tap Water (μg/l)	
1596-84-5	ALAR	2.70E+01	c	9.60E+01	c	4.80E−01	c	2.40E+00	c	3.70E+00	c
30560-19-1	Acephate	5.60E+01	c**	2.00E+02	c*					7.70E+00	c*
75-07-0	Acetaldehyde	1.00E+01	c**	5.20E+01	c**	1.10E+00	c**	5.60E+00	c**	2.20E+00	c**
34256-82-1	Acetochlor	1.20E+03	n	1.20E+04	n					7.30E+02	n
67-64-1	Acetone	6.10E+04	n	6.30E+05	nms	3.20E+04	n	1.40E+05	n	2.20E+04	n
75-86-5	Acetone Cyanohydrin	2.00E+02	n	2.10E+03	n	6.30E+01	n	2.60E+02	n	5.80E+01	n
75-05-8	Acetonitrile	8.70E+02	n	3.70E+03	n	6.30E+01	n	2.60E+02	n	1.30E+02	n
98-86-2	Acetophenone	7.80E+03	ns	1.00E+05	nms					3.70E+03	n
53-96-3	Acetylaminofluorene, 2-	1.30E−01	c	4.50E−01	c	1.90E−03	c	9.40E−03	c	1.80E−02	c
107-02-8	Acrolein	1.50E−01	n	6.50E−01	n	2.10E−02	n	8.80E−02	n	4.20E−02	n
79-06-1	Acrylamide	2.30E−01	c	3.40E+00	c	9.60E−03	c	1.20E−01	c	4.30E−02	c
79-10-7	Acrylic Acid	3.00E+04	n	2.90E+05	nm	1.00E+00	n	4.40E+00	n	1.80E+04	n
107-13-1	Acrylonitrile	2.40E−01	c*	1.20E+00	c*	3.60E−02	c*	1.80E−01	c*	4.50E−02	c*
111-69-3	Adiponitrile	8.50E+06	nm	3.60E+07	nm	6.30E+00	n	2.60E+01	n		
15972-60-8	Alachlor	8.70E+00	c*	3.10E+01	c					1.20E+00	c
116-06-3	Aldicarb	6.10E+01	n	6.20E+02	n					3.70E+01	n
1646-88-4	Aldicarb Sulfone	6.10E+01	n	6.20E+02	n					3.70E+01	n
309-00-2	Aldrin	2.90E−02	c*	1.00E−01	c	5.00E−04	c	2.50E−03	c	4.00E−03	c
74223-64-6	Ally	1.50E+04	n	1.50E+05	nm					9.10E+03	n
107-18-6	Allyl Alcohol	3.00E+02	n	3.10E+03	n	1.00E−01	n	4.40E−01	n	1.80E+02	n
107-05-1	Allyl Chloride	6.80E−01	c**	3.40E+00	c**	4.10E−01	c**	2.00E+00	c**	6.50E−01	c**
7429-90-5	Aluminum	7.70E+04	n	9.90E+05	nm	5.20E+00	n	2.20E+01	n	3.70E+04	n
20859-73-8	Aluminum Phosphide	3.10E+01	n	4.10E+02	n					1.50E+01	n
67485-29-4	Amdro	1.80E+01	n	1.80E+02	n					1.10E+01	n
834-12-8	Ametryn	5.50E+02	n	5.50E+03	n					3.30E+02	n
92-67-1	Aminobiphenyl, 4-	2.30E−02	c	8.20E−02	c	4.10E−04	c	2.00E−03	c	3.20E−03	c
591-27-5	Aminophenol, m-	4.90E+03	n	4.90E+04	n					2.90E+03	n

123-30-8	Aminophenol, p-	1.20E+03	n	1.20E+04	n					7.30E+02	n
33089-61-1	Amitraz	1.50E+02	n	1.50E+03	n					9.10E+01	n
7664-41-7	Ammonia					1.00E+02	n	4.40E+02	n		
7790-98-9	Ammonium Perchlorate	5.50E+01	n	7.20E+02	n					2.60E+01	n
7773-06-0	Ammonium Sulfamate	1.60E+04	n	2.00E+05	nm					7.30E+03	n
62-53-3	Aniline	8.50E+01	c**	3.00E+02	c*	1.00E+00	n	4.40E+00	n	1.20E+01	c*
7440-36-0	Antimony (metallic)	3.10E+01	n	4.10E+02	n					1.50E+01	n
1314-60-9	Antimony Pentoxide	3.90E+01	n	5.10E+02	n					1.80E+01	n
11071-15-1	Antimony Potassium Tartrate	7.00E+01	n	9.20E+02	n					3.30E+01	n
1332-81-6	Antimony Tetroxide	3.10E+01	n	4.10E+02	n					1.50E+01	n
1309-64-4	Antimony Trioxide	2.80E+05	nm	1.20E+06	nm	2.10E−01	n	8.80E−01	n		
74115-24-5	Apollo	7.90E+02	n	8.00E+03	n					4.70E+02	n
140-57-8	Aramite	1.90E+01	c	6.90E+01	c	3.40E−01	c	1.70E+00	c	2.70E+00	c
7440-38-2	Arsenic, Inorganic	3.90E−01	c*	1.60E+00	c	5.70E−04	c*	2.90E−03	c*	4.50E−02	c
7784-42-1	Arsine	2.70E−01	n	3.60E+00	n	5.20E−02	n	2.20E−01	n	1.30E−01	n
76578-14-8	Assure	5.50E+02	n	5.50E+03	n					3.30E+02	n
3337-71-1	Asulam	3.10E+03	n	3.10E+04	n					1.80E+03	n
1912-24-9	Atrazine	2.10E+00	c	7.50E+00	c					2.90E−01	c
492-80-8	Auramine	7.30E−01	c	3.30E+00	c	9.70E−03	c	4.90E−02	c	7.60E−02	c
65195-55-3	Avermectin B1	2.40E+01	n	2.50E+02	n					1.50E+01	n
103-33-3	Azobenzene	5.10E+00	c	2.30E+01	c	7.80E−02	c	4.00E−01	c	1.20E−01	c
7440-39-3	Barium	1.50E+04	n	1.90E+05	nm	5.20E−01	n	2.20E+00	n	7.30E+03	n
114-26-1	Baygon	2.40E+02	n	2.50E+03	n					1.50E+02	n
43121-43-3	Bayleton	1.80E+03	n	1.80E+04	n					1.10E+03	n
68359-37-5	Baythroid	1.50E+03	n	1.50E+04	n					9.10E+02	n
1861-40-1	Benefin	1.80E+04	n	1.80E+05	nm					1.10E+04	n
17804-35-2	Benomyl	3.10E+03	n	3.10E+04	n					1.80E+03	n
25057-89-0	Bentazon	1.80E+03	n	1.80E+04	n					1.10E+03	n
100-52-7	Benzaldehyde	7.80E+03	ns	1.00E+05	nms					3.70E+03	n
71-43-2	Benzene	1.10E+00	c*	5.40E+00	c*	3.10E−01	c	1.60E+00	c*	4.10E−01	c
108-98-5	Benzenethiol	7.80E−01	n	1.00E+01	n					3.70E−01	n
92-87-5	Benzidine	5.00E−04	c	7.50E−03	c	1.40E−05	c	1.80E−04	c	9.40E−05	c
65-85-0	Benzoic Acid	2.40E+05	nm	2.50E+06	nm					1.50E+05	n

(Continued)

Appendix J (Continued)

CAS No.	Chemical Name	Regional Screening Levels (RSLs) 'Direct Contact Exposure Pathways'									
		Residential Soil (mg/kg)		Industrial Soil (mg/kg)		Residential Air (μg/m^3)		Industrial Air (μg/m^3)		Tap Water (μg/l)	
98-07-7	Benzotrichloride	4.90E−02	c	2.20E−01	c					5.20E−03	c
100-51-6	Benzyl Alcohol	6.10E+03	n	6.20E+04	n					3.70E+03	n
100-44-7	Benzyl Chloride	1.00E+00	c*	4.90E+00	c*	5.00E−02	c*	2.50E−01	c*	7.90E−02	c*
7440-41-7	Beryllium and compounds	1.60E+02	n	2.00E+03	n	1.00E−03	c*	5.10E−03	c*	7.30E+01	n
141-66-2	Bidrin	6.10E+00	n	6.20E+01	n					3.70E+00	n
42576-02-3	Bifenox	5.50E+02	n	5.50E+03	n					3.30E+02	n
82657-04-3	Biphenthrin	9.20E+02	n	9.20E+03	n					5.50E+02	n
92-52-4	Biphenyl, 1,1′-	3.90E+03	ns	5.10E+04	ns					1.80E+03	n
108-60-1	Bis(2-chloro-1-methylethyl) ether	4.60E+00	c	2.20E+01	c	2.40E−01	c	1.20E+00	c	3.20E−01	c
111-91-1	Bis(2-chloroethoxy)methane	1.80E+02	n	1.80E+03	n					1.10E+02	n
111-44-4	Bis(2-chloroethyl)ether	2.10E−01	c	1.00E+00	c	7.40E−03	c	3.70E−02	c	1.20E−02	c
117-81-7	Bis(2-ethylhexyl)phthalate	3.50E+01	c*	1.20E+02	c	1.00E+00	c	5.10E+00	c	4.80E+00	c
542-88-1	Bis(chloromethyl)ether	7.70E−05	c	3.90E−04	c	3.90E−05	c	2.00E−04	c	6.20E−05	c
80-05-7	Bisphenol A	3.10E+03	n	3.10E+04	n					1.80E+03	n
7440-42-8	Boron and Borates only	1.60E+04	n	2.00E+05	nm	2.10E+01	n	8.80E+01	n	7.30E+03	n
7637-07-2	Boron Trifluoride	3.10E+03	n	4.10E+04	n	1.40E+01	n	5.70E+01	n	1.50E+03	n
15541-45-4	Bromate	9.10E−01	c	4.10E+00	c					9.60E−02	c
107-04-0	Bromo-2-chloroethane, 1-	2.40E−02	c	1.20E−01	c	4.10E−03	c	2.00E−02	c	6.50E−03	c
108-86-1	Bromobenzene	3.00E+02	n	1.80E+03	ns	6.30E+01	n	2.60E+02	n	8.80E+01	n
75-27-4	Bromodichloromethane	2.70E−01	c	1.40E+00	c	6.60E−02	c	3.30E−01	c	1.20E−01	c
75-25-2	Bromoform	6.10E+01	c*	2.20E+02	c*	2.20E+00	c	1.10E+01	c	8.50E+00	c*
74-83-9	Bromomethane	7.30E+00	n	3.20E+01	n	5.20E+00	n	2.20E+01	n	8.70E+00	n
2104-96-3	Bromophos	3.10E+02	n	3.10E+03	n					1.80E+02	n
1689-84-5	Bromoxynil	1.20E+03	n	1.20E+04	n					7.30E+02	n
1689-99-2	Bromoxynil Octanoate	1.20E+03	n	1.20E+04	n					7.30E+02	n
106-99-0	Butadiene, 1,3-	5.40E−02	c*	2.60E−01	c*	8.10E−02	c*	4.10E−01	c*	1.80E−02	c
71-36-3	Butanol, N-	6.10E+03	n	6.20E+04	n					3.70E+03	n

85-68-7	Butyl Benzyl Phthlate	2.60E+02	c*	9.10E+02	c					3.50E+01	c
78-92-2	Butyl Alcohol, sec-	1.60E+05	nm	2.00E+06	nm	3.10E+04	n	1.30E+05	n	7.30E+04	n
2008-41-5	Butylate	3.10E+03	n	3.10E+04	n					1.80E+03	n
25013-16-5	Butylated Hydroxyanisole	3.20E+03	c	1.40E+04	c	4.30E+01	c	2.20E+02	c	3.40E+02	c
85-70-1	Butylphthalyl Butylglycolate	6.10E+04	n	6.20E+05	nm					3.70E+04	n
75-60-5	Cacodylic Acid	1.20E+03	n	1.20E+04	n					7.30E+02	n
7440-43-9	Cadmium (Diet)	7.00E+01	n	8.00E+02	n						
7440-43-9	Cadmium (Water)					1.40E−03	c**	6.80E−03	c**	1.80E+01	n
105-60-2	Caprolactam	3.10E+04	n	3.10E+05	nm					1.80E+04	n
2425-06-1	Captafol	3.20E+00	c*	1.10E+01	c	5.70E−02	c	2.90E−01	c	4.50E−01	c
133-06-2	Captan	2.10E+02	c*	7.50E+02	c	3.70E+00	c	1.90E+01	c	2.90E+01	c
63-25-2	Carbaryl	6.10E+03	n	6.20E+04	n					3.70E+03	n
1563-66-2	Carbofuran	3.10E+02	n	3.10E+03	n					1.80E+02	n
75-15-0	Carbon Disulfide	8.20E+02	ns	3.70E+03	ns	7.30E+02	n	3.10E+03	n	1.00E+03	n
56-23-5	Carbon Tetrachloride	6.10E−01	c	3.00E+00	c	4.10E−01	c	2.00E+00	c	4.40E−01	c
55285-14-8	Carbosulfan	6.10E+02	n	6.20E+03	n					3.70E+02	n
5234-68-4	Carboxin	6.10E+03	n	6.20E+04	n					3.70E+03	n
1306-38-3	Ceric Oxide	1.30E+06	nm	5.40E+06	nm	9.40E−01	n	3.90E+00	n		
302-17-0	Chloral Hydrate	6.10E+03	n	6.20E+04	n					3.70E+03	n
133-90-4	Chloramben	9.20E+02	n	9.20E+03	n					5.50E+02	n
118-75-2	Chloranil	1.20E+00	c	4.30E+00	c					1.70E−01	c
12789-03-6	Chlordane	1.60E+00	c*	6.50E+00	c*	2.40E−02	c*	1.20E−01	c*	1.90E−01	c*
143-50-0	Chlordecone (Kepone)	4.90E−02	c	1.70E−01	c	5.30E−04	c	2.70E−03	c	6.70E−03	c
470-90-6	Chlorfenvinphos	4.30E+01	n	4.30E+02	n					2.60E+01	n
90982-32-4	Chlorimuron, Ethyl-	1.20E+03	n	1.20E+04	n					7.30E+02	n
7782-50-5	Chlorine	7.50E+03	n	9.10E+04	n	1.50E−01	n	6.40E−01	n	3.70E+03	n
10049-04-4	Chlorine Dioxide	2.30E+03	n	3.00E+04	n	2.10E−01	n	8.80E−01	n	1.10E+03	n
7758-19-2	Chlorite (Sodium Salt)	2.30E+03	n	3.10E+04	n					1.10E+03	n
75-68-3	Chloro-1,1-difluoroethane, 1-	5.80E+04	ns	2.40E+05	nms	5.20E+04	n	2.20E+05	n	1.00E+05	n
126-99-8	Chloro-1,3-butadiene, 2-	8.40E+00	n	3.60E+01	n	7.30E+00	n	3.10E+01	n	1.40E+01	n
3165-93-3	Chloro-2-methylaniline HCl, 4-	1.10E+00	c	3.70E+00	c					1.50E−01	c
107-20-0	Chloroacetaldehyde, 2-	2.40E+00	c	1.10E+01	c					2.50E−01	c
79-11-8	Chloroacetic Acid	1.20E+02	n	1.20E+03	n					7.30E+01	n
532-27-4	Chloroacetophenone, 2-	4.30E+04	n	1.80E+05	nm	3.10E−02	n	1.30E−01	n		

(Continued)

Appendix J (Continued)

CAS No.	Chemical Name	Regional Screening Levels (RSLs) 'Direct Contact Exposure Pathways'				
		Residential Soil (mg/kg)	Industrial Soil (mg/kg)	Residential Air (μg/m³)	Industrial Air (μg/m³)	Tap Water (μg/l)
106-47-8	Chloroaniline, p-	2.40E+00 c	8.60E+00 c			3.40E−01 c
108-90-7	Chlorobenzene	2.90E+02 n	1.40E+03 ns	5.20E+01 n	2.20E+02 n	9.10E+01 n
510-15-6	Chlorobenzilate	4.40E+00 c	1.60E+01 c	7.80E−02 c	4.00E−01 c	6.10E−01 c
74-11-3	Chlorobenzoic Acid, p-	1.80E+03 n	1.80E+04 n			1.10E+03 n
98-56-6	Chlorobenzotrifluoride, 4-	2.10E+02 ns	2.30E+03 ns	3.10E+02 n	1.30E+03 n	9.30E+01 n
109-69-3	Chlorobutane, 1-	3.10E+03 ns	4.10E+04 ns			1.50E+03 n
75-45-6	Chlorodifluoromethane	5.30E+04 ns	2.20E+05 nms	5.20E+04 n	2.20E+05 n	1.00E+05 n
67-66-3	Chloroform	2.90E−01 c	1.50E+00 c	1.10E−01 c	5.30E−01 c	1.90E−01 c
74-87-3	Chloromethane	1.20E+02 n	5.00E+02 n	9.40E+01 n	3.90E+02 n	1.90E+02 n
107-30-2	Chloromethyl Methyl Ether	1.90E−02 c	9.40E−02 c	3.50E−03 c	1.80E−02 c	5.60E−03 c
91-58-7	Chloronaphthalene, Beta-	6.30E+03 ns	8.20E+04 ns			2.90E+03 n
88-73-3	Chloronitrobenzene, o-	1.60E+00 c	5.70E+00 c	1.00E−02 n	4.40E−02 n	2.20E−01 c
100-00-5	Chloronitrobenzene, p-	6.10E+01 n	2.70E+02 c**	6.30E−01 n	2.60E+00 n	1.10E+01 c**
95-57-8	Chlorophenol, 2-	3.90E+02 n	5.10E+03 n			1.80E+02 n
76-06-2	Chloropicrin	5.70E+05 nm	2.40E+06 nm	4.20E−01 n	1.80E+00 n	
1897-45-6	Chlorothalonil	1.60E+02 c**	5.60E+02 c*	2.70E+00 c	1.40E+01 c	2.20E+01 c*
95-49-8	Chlorotoluene, o-	1.60E+03 ns	2.00E+04 ns			7.30E+02 n
106-43-4	Chlorotoluene, p-	5.50E+03 ns	7.20E+04 ns			2.60E+03 n
54749-90-5	Chlorozotocin	2.70E−03 c	1.20E−02 c	3.50E−05 c	1.80E−04 c	2.80E−04 c
101-21-3	Chlorpropham	1.20E+04 n	1.20E+05 nm			7.30E+03 n
2921-88-2	Chlorpyrifos	1.80E+02 n	1.80E+03 n			1.10E+02 n
5598-13-0	Chlorpyrifos Methyl	6.10E+02 n	6.20E+03 n			3.70E+02 n
64902-72-3	Chlorsulfuron	3.10E+03 n	3.10E+04 n			1.80E+03 n
60238-56-4	Chlorthiophos	4.90E+01 n	4.90E+02 n			2.90E+01 n
16065-83-1	Chromium(III), Insoluble Salts	1.20E+05 nm	1.50E+06 nm			5.50E+04 n
18540-29-9	Chromium(VI)	2.90E−01 c	5.60E+00 c	1.10E−05 c	1.50E−04 c	4.30E−02 c
7440-47-3	Chromium, Total					

7440-48-4	Cobalt	2.30E+01	n	3.00E+02	n	2.70E−04	c*	1.40E−03	c*	1.10E+01	n
8007-45-2	Coke Oven Emissions					1.50E−03	c	2.00E−02	c		
7440-50-8	Copper	3.10E+03	n	4.10E+04	n					1.50E+03	n
108-39-4	Cresol, m-	3.10E+03	n	3.10E+04	n	6.30E+02	n	2.60E+03	n	1.80E+03	n
95-48-7	Cresol, o-	3.10E+03	n	3.10E+04	n	6.30E+02	n	2.60E+03	n	1.80E+03	n
106-44-5	Cresol, p-	3.10E+02	n	3.10E+03	n	6.30E+02	n	2.60E+03	n	1.80E+02	n
59-50-7	Cresol, p-chloro-m-	6.10E+03	n	6.20E+04	n					3.70E+03	n
1319-77-3	Cresols	7.50E+03	n	9.10E+04	ns	6.30E+02	n	2.60E+03	n	9.30E+02	n
123-73-9	Crotonaldehyde, trans-	3.40E−01	c	1.50E+00	c					3.50E−02	c
98-82-8	Cumene	2.10E+03	ns	1.10E+04	ns	4.20E+02	n	1.80E+03	n	6.80E+02	n
135-20-6	Cupferron	2.90E+00	c	1.30E+01	c	3.90E−02	c	1.90E−01	c	3.10E−01	c
21725-46-2	Cyanazine	5.80E−01	c	2.10E+00	c					8.00E−02	c
	Cyanides										
592-01-8	~Calcium Cyanide	3.10E+03	n	4.10E+04	n					1.50E+03	n
544-92-3	~Copper Cyanide	3.90E+02	n	5.10E+03	n					1.80E+02	n
57-12-5	~Cyanide (CN-)	1.60E+03	n	2.00E+04	n					7.30E+02	n
460-19-5	~Cyanogen	3.10E+03	ns	4.10E+04	ns					1.50E+03	n
506-68-3	~Cyanogen Bromide	7.00E+03	n	9.20E+04	n					3.30E+03	n
506-77-4	~Cyanogen Chloride	3.90E+03	n	5.10E+04	ns					1.80E+03	n
74-90-8	~Hydrogen Cyanide	1.90E+01	n	8.00E+01	n	3.10E+00	n	1.30E+01	n	6.20E+00	n
151-50-8	~Potassium Cyanide	3.90E+03	n	5.10E+04	n					1.80E+03	n
506-61-6	~Potassium Silver Cyanide	1.60E+04	n	2.00E+05	nm					7.30E+03	n
506-64-9	~Silver Cyanide	7.80E+03	n	1.00E+05	nm					3.70E+03	n
143-33-9	~Sodium Cyanide	3.10E+03	n	4.10E+04	n					1.50E+03	n
463-56-9	~Thiocyanate	1.60E+01	n	2.00E+02	n					7.30E+00	n
557-21-1	~Zinc Cyanide	3.90E+03	n	5.10E+04	n					1.80E+03	n
110-82-7	Cyclohexane	7.00E+03	ns	2.90E+04	ns	6.30E+03	n	2.60E+04	n	1.30E+04	n
87-84-3	Cyclohexane, 1,2,3,4,5-pentabromo-6-chloro-	2.10E+01	c	7.50E+01	c					2.90E+00	c
108-94-1	Cyclohexanone	3.10E+05	nm	3.10E+06	nm					1.80E+05	n
108-91-8	Cyclohexylamine	1.20E+04	n	1.20E+05	nm					7.30E+03	n
68085-85-8	Cyhalothrin/karate	3.10E+02	n	3.10E+03	n					1.80E+02	n
52315-07-8	Cypermethrin	6.10E+02	n	6.20E+03	n					3.70E+02	n
66215-27-8	Cyromazine	4.60E+02	n	4.60E+03	n					2.70E+02	n
72-54-8	DDD	2.00E+00	c	7.20E+00	c	3.50E−02	c	1.80E−01	c	2.80E−01	c

(Continued)

Appendix J (Continued)

CAS No.	Chemical Name	Regional Screening Levels (RSLs) 'Direct Contact Exposure Pathways'									
		Residential Soil (mg/kg)		Industrial Soil (mg/kg)		Residential Air (μg/m³)		Industrial Air (μg/m³)		Tap Water (μg/l)	
72-55-9	DDE, p,p′-	1.40E+00	c	5.10E+00	c	2.50E−02	c	1.30E−01	c	2.00E−01	c
50-29-3	DDT	1.70E+00	c*	7.00E+00	c*	2.50E−02	c	1.30E−01	c	2.00E−01	c*
1861-32-1	Dacthal	6.10E+02	n	6.20E+03	n					3.70E+02	n
75-99-0	Dalapon	1.80E+03	n	1.80E+04	n					1.10E+03	n
1163-19-5	Decabromodiphenyl ether, 2,2′,3,3′,4,4′,5,5′,6,6′- (BDE-209)	4.30E+02	n	2.50E+03	c**					9.60E+01	c**
8065-48-3	Demeton	2.40E+00	n	2.50E+01	n					1.50E+00	n
103-23-1	Di(2-ethylhexyl)adipate	4.00E+02	c*	1.40E+03	c					5.60E+01	c
2303-16-4	Diallate	8.00E+00	c	2.80E+01	c					1.10E+00	c
333-41-5	Diazinon	4.30E+01	n	4.30E+02	n					2.60E+01	n
96-12-8	Dibromo-3-chloropropane, 1,2-	5.40E−03	c	6.90E−02	c	1.60E−04	c	2.00E−03	c	3.20E−04	c
106-37-6	Dibromobenzene, 1,4-	6.10E+02	n	6.20E+03	n					3.70E+02	n
124-48-1	Dibromochloromethane	6.80E−01	c	3.30E+00	c	9.00E−02	c	4.50E−01	c	1.50E−01	c
106-93-4	Dibromoethane, 1,2-	3.40E−02	c	1.70E−01	c	4.10E−03	c	2.00E−02	c	6.50E−03	c
74-95-3	Dibromomethane (Methylene Bromide)	2.50E+01	n	1.10E+02	n	4.20E+00	n	1.80E+01	n	8.20E+00	n
84-74-2	Dibutyl Phthalate	6.10E+03	n	6.20E+04	n					3.70E+03	n
NA	Dibutyltin Compounds	1.80E+01	n	1.80E+02	n					1.10E+01	n
1918-00-9	Dicamba	1.80E+03	n	1.80E+04	n					1.10E+03	n
764-41-0	Dichloro-2-butene, 1,4-	6.50E−03	c	3.30E−02	c	5.80E−04	c	2.90E−03	c	1.20E−03	c
1476-11-5	Dichloro-2-butene, cis-1,4-	6.90E−03	c	3.50E−02	c	5.80E−04	c	2.90E−03	c	1.20E−03	c
110-57-6	Dichloro-2-butene, trans-1,4-	6.90E−03	c	3.50E−02	c	5.80E−04	c	2.90E−03	c	1.20E−03	c
79-43-6	Dichloroacetic Acid	9.70E+00	c*	3.40E+01	c*					1.30E+00	c
95-50-1	Dichlorobenzene, 1,2-	1.90E+03	ns	9.80E+03	ns	2.10E+02	n	8.80E+02	n	3.70E+02	n
106-46-7	Dichlorobenzene, 1,4-	2.40E+00	c	1.20E+01	c	2.20E−01	c	1.10E+00	c	4.30E−01	c
91-94-1	Dichlorobenzidine, 3,3′-	1.10E+00	c	3.80E+00	c	7.20E−03	c	3.60E−02	c	1.50E−01	c
90-98-2	Dichlorobenzophenone, 4,4′-	5.50E+02	n	5.50E+03	n					3.30E+02	n
75-71-8	Dichlorodifluoromethane	1.80E+02	n	7.80E+02	n	2.10E+02	n	8.80E+02	n	3.90E+02	n

75-34-3	Dichloroethane, 1,1-	3.30E+00	c	1.70E+01	c	1.50E+00	c	7.70E+00	c	2.40E+00	c
107-06-2	Dichloroethane, 1,2-	4.30E−01	c	2.20E+00	c	9.40E−02	c	4.70E−01	c	1.50E−01	c
75-35-4	Dichloroethylene, 1,1-	2.40E+02	n	1.10E+03	n	2.10E+02	n	8.80E+02	n	3.40E+02	n
540-59-0	Dichloroethylene, 1,2- (Mixed Isomers)	7.00E+02	n	9.20E+03	ns					3.30E+02	n
156-59-2	Dichloroethylene, 1,2-cis-	7.80E+02	n	1.00E+04	ns					3.70E+02	n
156-60-5	Dichloroethylene, 1,2-trans-	1.50E+02	n	6.90E+02	n	6.30E+01	n	2.60E+02	n	1.10E+02	n
120-83-2	Dichlorophenol, 2,4-	1.80E+02	n	1.80E+03	n					1.10E+02	n
94-75-7	Dichlorophenoxy Acetic Acid, 2,4-	6.90E+02	n	7.70E+03	n					3.70E+02	n
94-82-6	(Dichlorophenoxy)butyric Acid, 4-(2,4-)	4.90E+02	n	4.90E+03	n					2.90E+02	n
78-87-5	Dichloropropane, 1,2-	8.90E−01	c*	4.50E+00	c*	2.40E−01	c*	1.20E+00	c*	3.90E−01	c*
142-28-9	Dichloropropane, 1,3	1.60E+03	ns	2.00E+04	ns					7.30E+02	n
616-23-9	Dichloropropanol, 2,3-	1.80E+02	n	1.80E+03	n					1.10E+02	n
542-75-6	Dichloropropene, 1,3-	1.70E+00	c*	8.10E+00	c*	6.10E−01	c*	3.10E+00	c*	4.30E−01	c*
62-73-7	Dichlorvos	1.70E+00	c*	5.90E+00	c*	2.90E−02	c*	1.50E−01	c*	2.30E−01	c*
77-73-6	Dicyclopentadiene	2.70E+01	n	1.20E+02	n	7.30E+00	n	3.10E+01	n	1.40E+01	n
60-57-1	Dieldrin	3.00E−02	c	1.10E−01	c	5.30E−04	c	2.70E−03	c	4.20E−03	c
NA	Diesel Engine Exhaust					8.10E−03	c	4.10E−02	c		
111-42-2	Diethanolamine	4.30E+06	nm	1.80E+07	nm	3.10E+00	n	1.30E+01	n		
84-66-2	Diethyl Phthalate	4.90E+04	n	4.90E+05	nm					2.90E+04	n
112-34-5	Diethylene Glycol Monobutyl Ether	1.80E+03	n	1.80E+04	n	1.00E−01	n	4.40E−01	n	1.10E+03	n
111-90-0	Diethylene Glycol Monoethyl Ether	3.60E+03	n	3.60E+04	n	3.10E−01	n	1.30E+00	n	2.20E+03	n
617-84-5	Diethylformamide	6.10E+01	n	6.20E+02	n					3.70E+01	n
56-53-1	Diethylstilbestrol	1.40E−03	c	4.90E−03	c	2.40E−05	c	1.20E−04	c	1.90E−04	c
43222-48-6	Difenzoquat	4.90E+03	n	4.90E+04	n					2.90E+03	n
35367-38-5	Diflubenzuron	1.20E+03	n	1.20E+04	n					7.30E+02	n
75-37-6	Difluoroethane, 1,1-	5.20E+04	ns	2.20E+05	nms	4.20E+04	n	1.80E+05	n	8.30E+04	n
94-58-6	Dihydrosafrole	1.50E+01	c	6.50E+01	c	1.90E−01	c	9.40E−01	c	1.50E+00	c
108-20-3	Diisopropyl Ether	1.40E+03	n	5.80E+03	ns	4.20E+02	n	1.80E+03	n	8.30E+02	n
1445-75-6	Diisopropyl Methylphosphonate	6.30E+03	ns	8.20E+04	ns					2.90E+03	n
55290-64-7	Dimethipin	1.20E+03	n	1.20E+04	n					7.30E+02	n
60-51-5	Dimethoate	1.20E+01	n	1.20E+02	n					7.30E+00	n
119-90-4	Dimethoxybenzidine, 3,3′-	3.50E+01	c	1.20E+02	c					4.80E+00	c
756-79-6	Dimethyl Methylphosphonate	2.90E+02	c*	1.00E+03	c*					4.00E+01	c*

(Continued)

Appendix J (Continued)

CAS No.	Chemical Name	Regional Screening Levels (RSLs) 'Direct Contact Exposure Pathways'									
		Residential Soil (mg/kg)		Industrial Soil (mg/kg)		Residential Air (μg/m³)		Industrial Air (μg/m³)		Tap Water (μg/l)	
60-11-7	Dimethylamino Azobenzene [p-]	1.10E−01	c	3.70E−01	c	1.90E−03	c	9.40E−03	c	1.50E−02	c
21436-96-4	Dimethylaniline HCl, 2,4-	8.40E−01	c	3.00E+00	c					1.20E−01	c
95-68-1	Dimethylaniline, 2,4-	6.50E−01	c	2.30E+00	c					9.00E−02	c
121-69-7	Dimethylaniline, N,N-	1.60E+02	n	2.00E+03	ns					7.30E+01	n
119-93-7	Dimethylbenzidine, 3,3′-	4.40E−02	c	1.60E−01	c					6.10E−03	c
68-12-2	Dimethylformamide	6.10E+03	n	6.20E+04	n	3.10E+01	n	1.30E+02	n	3.70E+03	n
57-14-7	Dimethylhydrazine, 1,1-	6.10E+00	n	6.10E+01	n	2.10E−03	n	8.80E−03	n	3.70E+00	n
540-73-8	Dimethylhydrazine, 1,2-	8.80E−04	c	3.10E−03	c	1.50E−05	c	7.70E−05	c	1.20E−04	c
105-67-9	Dimethylphenol, 2,4-	1.20E+03	n	1.20E+04	n					7.30E+02	n
576-26-1	Dimethylphenol, 2,6-	3.70E+01	n	3.70E+02	n					2.20E+01	n
95-65-8	Dimethylphenol, 3,4-	6.10E+01	n	6.20E+02	n					3.70E+01	n
120-61-6	Dimethylterephthalate	7.80E+03	ns	1.00E+05	nms					3.70E+03	n
513-37-1	Dimethylvinylchloride	1.40E+01	c	6.40E+01	c	1.90E−01	c	9.40E−01	c	1.50E+00	c
534-52-1	Dinitro-o-cresol, 4,6-	4.90E+00	n	4.90E+01	n					2.90E+00	n
131-89-5	Dinitro-o-cyclohexyl Phenol, 4,6-	1.20E+02	n	1.20E+03	n					7.30E+01	n
528-29-0	Dinitrobenzene, 1,2-	6.10E+00	n	6.20E+01	n					3.70E+00	n
99-65-0	Dinitrobenzene, 1,3-	6.10E+00	n	6.20E+01	n					3.70E+00	n
100-25-4	Dinitrobenzene, 1,4-	6.10E+00	n	6.20E+01	n					3.70E+00	n
51-28-5	Dinitrophenol, 2,4-	1.20E+02	n	1.20E+03	n					7.30E+01	n
25321-14-6	Dinitrotoluene Mixture, 2,4/2,6-	7.10E−01	c	2.50E+00	c					9.90E−02	c
121-14-2	Dinitrotoluene, 2,4-	1.60E+00	c*	5.50E+00	c	2.70E−02	c	1.40E−01	c	2.20E−01	c
606-20-2	Dinitrotoluene, 2,6-	6.10E+01	n	6.20E+02	n					3.70E+01	n
35572-78-2	Dinitrotoluene, 2-Amino-4,6-	1.50E+02	n	2.00E+03	n					7.30E+01	n
19406-51-0	Dinitrotoluene, 4-Amino-2,6-	1.50E+02	n	1.90E+03	n					7.30E+01	n
88-85-7	Dinoseb	6.10E+01	n	6.20E+02	n					3.70E+01	n
123-91-1	Dioxane, 1,4-	4.40E+01	c	1.60E+02	c	3.20E−01	c	1.60E+00	c	6.10E+00	c
	Dioxins										

NA	~Hexachlorodibenzo-p-dioxin, Mixture	9.40E−05	c	3.90E−04	c	1.90E−06	c	9.40E−06	c	1.10E−05	c
1746-01-6	~TCDD, 2,3,7,8-	4.50E−06	c*	1.80E−05	c*	6.40E−08	c	3.20E−07	c	5.20E−07	c*
957-51-7	Diphenamid	1.80E+03	n	1.80E+04	n					1.10E+03	n
127-63-9	Diphenyl Sulfone	4.90E+01	n	4.90E+02	n					2.90E+01	n
122-39-4	Diphenylamine	1.50E+03	n	1.50E+04	n					9.10E+02	n
122-66-7	Diphenylhydrazine, 1,2-	6.10E−01	c	2.20E+00	c	1.10E−02	c	5.60E−02	c	8.40E−02	c
85-00-7	Diquat	1.30E+02	n	1.40E+03	n					8.00E+01	n
1937-37-7	Direct Black 38	6.60E−02	c	2.30E−01	c	1.20E−03	c	5.80E−03	c	9.10E−03	c
2602-46-2	Direct Blue 6	6.60E−02	c	2.30E−01	c	1.20E−03	c	5.80E−03	c	9.10E−03	c
16071-86-6	Direct Brown 95	7.20E−02	c	2.60E−01	c	1.30E−03	c	6.50E−03	c	1.00E−02	c
298-04-4	Disulfoton	2.40E+00	n	2.50E+01	n					1.50E+00	n
505-29-3	Dithiane, 1,4-	6.10E+02	n	6.20E+03	n					3.70E+02	n
330-54-1	Diuron	1.20E+02	n	1.20E+03	n					7.30E+01	n
2439-10-3	Dodine	2.40E+02	n	2.50E+03	n					1.50E+02	n
759-94-4	EPTC	2.00E+03	ns	2.60E+04	ns					9.10E+02	n
115-29-7	Endosulfan	3.70E+02	n	3.70E+03	n					2.20E+02	n
145-73-3	Endothall	1.20E+03	n	1.20E+04	n					7.30E+02	n
72-20-8	Endrin	1.80E+01	n	1.80E+02	n					1.10E+01	n
106-89-8	Epichlorohydrin	2.00E+01	n	8.80E+01	n	1.00E+00	n	4.40E+00	n	2.10E+00	n
106-88-7	Epoxybutane, 1,2-	1.70E+02	n	7.20E+02	n	2.10E+01	n	8.80E+01	n	4.20E+01	n
16672-87-0	Ethephon	3.10E+02	n	3.10E+03	n					1.80E+02	n
563-12-2	Ethion	3.10E+01	n	3.10E+02	n					1.80E+01	n
111-15-9	Ethoxyethanol Acetate, 2-	1.80E+04	n	1.80E+05	nm	3.10E+02	n	1.30E+03	n	1.10E+04	n
110-80-5	Ethoxyethanol, 2-	2.40E+04	n	2.50E+05	nm	2.10E+02	n	8.80E+02	n	1.50E+04	n
141-78-6	Ethyl Acetate	7.00E+04	ns	9.20E+05	nms					3.30E+04	n
140-88-5	Ethyl Acrylate	1.30E+01	c	6.00E+01	c					1.40E+00	c
75-00-3	Ethyl Chloride	1.50E+04	ns	6.10E+04	ns	1.00E+04	n	4.40E+04	n	2.10E+04	n
60-29-7	Ethyl Ether	1.60E+04	ns	2.00E+05	nms					7.30E+03	n
97-63-2	Ethyl Methacrylate	7.00E+03	ns	9.20E+04	ns					3.30E+03	n
2104-64-5	Ethyl-p-nitrophenyl Phosphonate	6.10E−01	n	6.20E+00	n					3.70E−01	n
100-41-4	Ethylbenzene	5.40E+00	c	2.70E+01	c	9.70E−01	c	4.90E+00	c	1.50E+00	c
109-78-4	Ethylene Cyanohydrin	1.80E+03	n	1.80E+04	n					1.10E+03	n
107-15-3	Ethylene Diamine	5.50E+03	n	5.50E+04	n					3.30E+03	n
107-21-1	Ethylene Glycol	1.20E+05	nm	1.20E+06	nm	4.20E+02	n	1.80E+03	n	7.30E+04	n

(*Continued*)

Appendix J (Continued)

CAS No.	Chemical Name	Regional Screening Levels (RSLs) 'Direct Contact Exposure Pathways'									
		Residential Soil (mg/kg)		Industrial Soil (mg/kg)		Residential Air (µg/m^3)		Industrial Air (µg/m^3)		Tap Water (µg/l)	
111-76-2	Ethylene Glycol Monobutyl Ether	6.10E+03	n	6.20E+04	n	1.70E+03	n	7.00E+03	n	3.70E+03	n
75-21-8	Ethylene Oxide	1.70E−01	c	8.30E−01	c	2.80E−02	c	1.40E−01	c	4.40E−02	c
96-45-7	Ethylene Thiourea	4.90E+00	n	3.80E+01	c**	1.90E−01	c	9.40E−01	c	1.50E+00	c**
151-56-4	Ethyleneimine	9.80E−03	c	4.40E−02	c	1.30E−04	c	6.50E−04	c	1.00E−03	c
84-72-0	Ethylphthalyl Ethyl Glycolate	1.80E+05	nm	1.80E+06	nm					1.10E+05	n
101200-48-0	Express	4.90E+02	n	4.90E+03	n					2.90E+02	n
22224-92-6	Fenamiphos	1.50E+01	n	1.50E+02	n					9.10E+00	n
39515-41-8	Fenpropathrin	1.50E+03	n	1.50E+04	n					9.10E+02	n
2164-17-2	Fluometuron	7.90E+02	n	8.00E+03	n					4.70E+02	n
16984-48-8	Fluoride	3.10E+03	n	4.10E+04	n	1.40E+01	n	5.70E+01	n	1.50E+03	n
7782-41-4	Fluorine (Soluble Fluoride)	4.70E+03	n	6.10E+04	n	1.40E+01	n	5.70E+01	n	2.20E+03	n
59756-60-4	Fluridone	4.90E+03	n	4.90E+04	n					2.90E+03	n
56425-91-3	Flurprimidol	1.20E+03	n	1.20E+04	n					7.30E+02	n
66332-96-5	Flutolanil	3.70E+03	n	3.70E+04	n					2.20E+03	n
69409-94-5	Fluvalinate	6.10E+02	n	6.20E+03	n					3.70E+02	n
133-07-3	Folpet	1.40E+02	c*	4.90E+02	c					1.90E+01	c
72178-02-0	Fomesafen	2.60E+00	c	9.10E+00	c					3.50E−01	c
944-22-9	Fonofos	1.20E+02	n	1.20E+03	n					7.30E+01	n
50-00-0	Formaldehyde	1.20E+04	n	1.20E+05	nm	1.90E−01	c*	9.40E−01	c*	7.30E+03	n
64-18-6	Formic Acid	1.20E+05	nm	1.20E+06	nm	3.10E+00	n	1.30E+01	n	7.30E+04	n
39148-24-8	Fosetyl-AL	1.80E+05	nm	1.80E+06	nm					1.10E+05	n
	Furans										
132-64-9	~Dibenzofuran	7.80E+01	n	1.00E+03	ns					3.70E+01	n
110-00-9	~Furan	7.80E+01	n	1.00E+03	n					3.70E+01	n
67-45-8	Furazolidone	1.30E−01	c	4.50E−01	c					1.80E−02	c
98-01-1	Furfural	1.80E+02	n	1.80E+03	n	5.20E+01	n	2.20E+02	n	1.10E+02	n
531-82-8	Furium	3.20E−01	c	1.10E+00	c	5.70E−03	c	2.90E−02	c	4.50E−02	c

60568-05-0	Furmecyclox	1.60E+01	c	5.70E+01	c	2.80E−01	c	1.40E+00	c	2.20E+00	c
77182-82-2	Glufosinate, Ammonium	2.40E+01	n	2.50E+02	n					1.50E+01	n
111-30-8	Glutaraldehyde	1.10E+05	nm	4.80E+05	nm	8.30E−02	n	3.50E−01	n		
765-34-4	Glycidyl	2.40E+01	n	2.50E+02	n	1.00E+00	n	4.40E+00	n	1.50E+01	n
1071-83-6	Glyphosate	6.10E+03	n	6.20E+04	n					3.70E+03	n
42874-03-3	Goal	1.80E+02	n	1.80E+03	n					1.10E+02	n
86-50-0	Guthion	1.80E+02	n	1.80E+03	n	1.00E+01	n	4.40E+01	n	1.10E+02	n
69806-40-2	Haloxyfop, Methyl	3.10E+00	n	3.10E+01	n					1.80E+00	n
79277-27-3	Harmony	7.90E+02	n	8.00E+03	n					4.70E+02	n
76-44-8	Heptachlor	1.10E−01	c	3.80E−01	c	1.90E−03	c	9.40E−03	c	1.50E−02	c
1024-57-3	Heptachlor Epoxide	5.30E−02	c*	1.90E−01	c*	9.40E−04	c	4.70E−03	c	7.40E−03	c*
87-82-1	Hexabromobenzene	1.20E+02	n	1.20E+03	n					7.30E+01	n
68631-49-2	Hexabromodiphenyl Ether, 2,2′,4,4′,5,5′-(BDE-153)	1.60E+01	n	2.00E+02	n					7.30E+00	n
118-74-1	Hexachlorobenzene	3.00E−01	c	1.10E+00	c	5.30E−03	c	2.70E−02	c	4.20E−02	c
87-68-3	Hexachlorobutadiene	6.20E+00	c**	2.20E+01	c*	1.10E−01	c	5.60E−01	c	8.60E−01	c*
319-84-6	Hexachlorocyclohexane, Alpha-	7.70E−02	c	2.70E−01	c	1.40E−03	c	6.80E−03	c	1.10E−02	c
319-85-7	Hexachlorocyclohexane, Beta-	2.70E−01	c	9.60E−01	c	4.60E−03	c	2.30E−02	c	3.70E−02	c
58-89-9	Hexachlorocyclohexane, Gamma- (Lindane)	5.20E−01	c*	2.10E+00	c	7.80E−03	c	4.00E−02	c	6.10E−02	c
608-73-1	Hexachlorocyclohexane, Technical	2.70E−01	c	9.60E−01	c	4.80E−03	c	2.40E−02	c	3.70E−02	c
77-47-4	Hexachlorocyclopentadiene	3.70E+02	n	3.70E+03	n	2.10E−01	n	8.80E−01	n	2.20E+02	n
67-72-1	Hexachloroethane	3.50E+01	c**	1.20E+02	c**	6.10E−01	c	3.10E+00	c	4.80E+00	c**
70-30-4	Hexachlorophene	1.80E+01	n	1.80E+02	n					1.10E+01	n
121-82-4	Hexahydro-1,3,5-trinitro-1,3,5-triazine (RDX)	5.50E+00	c*	2.40E+01	c					6.10E−01	c
822-06-0	Hexamethylene Diisocyanate, 1,6-	3.40E+00	n	1.40E+01	n	1.00E−02	n	4.40E−02	n	2.10E−02	n
110-54-3	Hexane, N-	5.70E+02	ns	2.60E+03	ns	7.30E+02	n	3.10E+03	n	8.80E+02	n
124-04-9	Hexanedioic Acid	1.20E+05	nm	1.20E+06	nm					7.30E+04	n
591-78-6	Hexanone, 2-	2.10E+02	n	1.40E+03	n	3.10E+01	n	1.30E+02	n	4.70E+01	n
51235-04-2	Hexazinone	2.00E+03	n	2.00E+04	n					1.20E+03	n
302-01-2	Hydrazine	2.10E−01	c	9.50E−01	c	5.00E−04	c*	2.50E−03	c*	2.20E−02	c
10034-93-2	Hydrazine Sulfate	2.10E−01	c	9.50E−01	c	5.00E−04	c	2.50E−03	c	2.20E−02	c
7647-01-0	Hydrogen Chloride	2.80E+07	nm	1.20E+08	nm	2.10E+01	n	8.80E+01	n		
7664-39-3	Hydrogen Fluoride	3.10E+03	n	4.10E+04	n	1.50E+01	n	6.10E+01	n	1.50E+03	n
7783-06-4	Hydrogen Sulfide	2.80E+06	nm	1.20E+07	nm	2.10E+00	n	8.80E+00	n		

(Continued)

Appendix J (Continued)

CAS No.	Chemical Name	Regional Screening Levels (RSLs) 'Direct Contact Exposure Pathways'									
		Residential Soil (mg/kg)		Industrial Soil (mg/kg)		Residential Air (μg/m^3)		Industrial Air (μg/m^3)		Tap Water (μg/l)	
123-31-9	Hydroquinone	8.10E+00	c	2.90E+01	c					1.10E+00	c
35554-44-0	Imazalil	7.90E+02	n	8.00E+03	n					4.70E+02	n
81335-37-7	Imazaquin	1.50E+04	n	1.50E+05	nm					9.10E+03	n
7553-56-2	Iodine	7.80E+02	n	1.00E+04	n					3.70E+02	n
36734-19-7	Iprodione	2.40E+03	n	2.50E+04	n					1.50E+03	n
7439-89-6	Iron	5.50E+04	n	7.20E+05	nm					2.60E+04	n
78-83-1	Isobutyl Alcohol	2.30E+04	ns	3.10E+05	nms					1.10E+04	n
78-59-1	Isophorone	5.10E+02	c*	1.80E+03	c*	2.10E+03	n	8.80E+03	n	7.10E+01	c
33820-53-0	Isopropalin	9.20E+02	n	9.20E+03	n					5.50E+02	n
67-63-0	Isopropanol	9.90E+09	nm	4.20E + 10	nm	7.30E+03	n	3.10E+04	n		
1832-54-8	Isopropyl Methyl Phosphonic Acid	6.10E+03	n	6.20E+04	n					3.70E+03	n
82558-50-7	Isoxaben	3.10E+03	n	3.10E+04	n					1.80E+03	n
NA	JP-7	4.30E+08	nm	1.80E+09	nm	3.10E+02	n	1.30E+03	n	6.30E+02	n
23950-58-5	Kerb	4.60E+03	n	4.60E+04	n					2.70E+03	n
77501-63-4	Lactofen	1.20E+02	n	1.20E+03	n					7.30E+01	n
	Lead Compounds										
301-04-2	~Lead Acetate	2.30E+00	c	1.00E+01	c	3.00E−02	c	1.50E−01	c	2.40E−01	c
7439-92-1	~Lead and Compounds	4.00E+02	n	8.00E+02	n						
1335-32-6	~Lead Subacetate	1.70E+01	c	7.50E+01	c	2.20E−01	c	1.10E+00	c	1.80E+00	c
78-00-2	~Tetraethyl Lead	6.10E−03	n	6.20E−02	n					3.70E−03	n
330-55-2	Linuron	1.20E+02	n	1.20E+03	n					7.30E+01	n
7439-93-2	Lithium	1.60E+02	n	2.00E+03	n					7.30E+01	n
7791-03-9	Lithium Perchlorate	5.50E+01	n	7.20E+02	n					2.60E+01	n
83055-99-6	Londax	1.20E+04	n	1.20E+05	nm					7.30E+03	n
94-74-6	MCPA	3.10E+01	n	3.10E+02	n					1.80E+01	n
94-81-5	MCPB	6.10E+02	n	6.20E+03	n					3.70E+02	n
93-65-2	MCPP	6.10E+01	n	6.20E+02	n					3.70E+01	n

121-75-5	Malathion	1.20E+03	n	1.20E+04	n					7.30E+02	n
108-31-6	Maleic Anhydride	6.10E+03	n	6.10E+04	n	7.30E−01	n	3.10E+00	n	3.70E+03	n
123-33-1	Maleic Hydrazide	3.10E+04	n	3.10E+05	nm					1.80E+04	n
109-77-3	Malononitrile	6.10E+00	n	6.20E+01	n					3.70E+00	n
8018-01-7	Mancozeb	1.80E+03	n	1.80E+04	n					1.10E+03	n
12427-38-2	Maneb	3.10E+02	n	3.10E+03	n					1.80E+02	n
7439-96-5	Manganese (Diet)										
7439-96-5	Manganese (Water)	1.80E+03	n	2.30E+04	n	5.20E−02	n	2.20E−01	n	8.80E+02	n
950-10-7	Mephosfolan	5.50E+00	n	5.50E+01	n					3.30E+00	n
24307-26-4	Mepiquat Chloride	1.80E+03	n	1.80E+04	n					1.10E+03	n
	Mercury Compounds										
7487-94-7	~Mercuric Chloride	2.30E+01	n	3.10E+02	n	3.10E−02	n	1.30E−01	n	1.10E+01	n
1344-48-5	~Mercuric Sulfide	2.30E+01	n	3.10E+02	n					1.10E+01	n
7439-97-6	~Mercury (Elemental)	5.60E+00	ns	3.40E+01	ns	3.10E−01	n	1.30E+00	n	5.70E−01	n
NA	~Mercury, Inorganic Salts	2.30E+01	n	3.10E+02	n					1.10E+01	n
22967-92-6	~Methyl Mercury	7.80E+00	n	1.00E+02	n					3.70E+00	n
62-38-4	~Phenylmercuric Acetate	4.90E+00	n	4.90E+01	n					2.90E+00	n
150-50-5	Merphos	1.80E+00	n	1.80E+01	n					1.10E+00	n
78-48-8	Merphos Oxide	1.80E+00	n	1.80E+01	n					1.10E+00	n
57837-19-1	Metalaxyl	3.70E+03	n	3.70E+04	n					2.20E+03	n
126-98-7	Methacrylonitrile	3.20E+00	n	1.80E+01	n	7.30E−01	n	3.10E+00	n	1.00E+00	n
10265-92-6	Methamidophos	3.10E+00	n	3.10E+01	n					1.80E+00	n
67-56-1	Methanol	3.10E+04	n	3.10E+05	nm	4.20E+03	n	1.80E+04	n	1.80E+04	n
950-37-8	Methidathion	6.10E+01	n	6.20E+02	n					3.70E+01	n
16752-77-5	Methomyl	1.50E+03	n	1.50E+04	n					9.10E+02	n
99-59-2	Methoxy-5-nitroaniline, 2-	9.90E+00	c	3.50E+01	c	1.70E−01	c	8.80E−01	c	1.40E+00	c
72-43-5	Methoxychlor	3.10E+02	n	3.10E+03	n					1.80E+02	n
110-49-6	Methoxyethanol Acetate, 2-	1.20E+02	n	1.20E+03	n	9.40E+01	n	3.90E+02	n	7.30E+01	n
109-86-4	Methoxyethanol, 2-	1.80E+02	n	1.80E+03	n	2.10E+01	n	8.80E+01	n	1.10E+02	n
79-20-9	Methyl Acetate	7.80E+04	ns	1.00E+06	nms					3.70E+04	n
96-33-3	Methyl Acrylate	2.30E+03	n	3.10E+04	ns					1.10E+03	n
78-93-3	Methyl Ethyl Ketone (2-butanone)	2.80E+04	n	2.00E+05	nms	5.20E+03	n	2.20E+04	n	7.10E+03	n
108-10-1	Methyl Isobutyl Ketone (4-methyl-2-pentanone)	5.30E+03	ns	5.30E+04	ns	3.10E+03	n	1.30E+04	n	2.00E+03	n

(*Continued*)

Appendix J (Continued)

CAS No.	Chemical Name	Regional Screening Levels (RSLs) 'Direct Contact Exposure Pathways'									
		Residential Soil (mg/kg)		Industrial Soil (mg/kg)		Residential Air (μg/m^3)		Industrial Air (μg/m^3)		Tap Water (μg/l)	
624-83-9	Methyl Isocyanate	1.40E+06	nm	6.00E+06	nm	1.00E+00	n	4.40E+00	n		
80-62-6	Methyl Methacrylate	4.80E+03	ns	2.10E+04	ns	7.30E+02	n	3.10E+03	n	1.40E+03	n
298-00-0	Methyl Parathion	1.50E+01	n	1.50E+02	n					9.10E+00	n
993-13-5	Methyl Phosphonic Acid	3.70E+03	n	3.70E+04	n					2.20E+03	n
25013-15-4	Methyl Styrene (Mixed Isomers)	2.50E+02	n	1.60E+03	ns	4.20E+01	n	1.80E+02	n	6.00E+01	n
66-27-3	Methyl Methanesulfonate	4.90E+00	c	1.70E+01	c	8.70E−02	c	4.40E−01	c	6.80E−01	c
1634-04-4	Methyl tert-Butyl Ether (MTBE)	4.30E+01	c	2.20E+02	c	9.40E+00	c	4.70E+01	c	1.20E+01	c
99-55-8	Methyl-5-Nitroaniline, 2-	1.50E+01	c	5.20E+01	c					2.00E+00	c
70-25-7	Methyl-N-nitro-N-nitrosoguanidine, N-	7.70E−02	c	3.40E−01	c	1.00E−03	c	5.10E−03	c	8.10E−03	c
636-21-5	Methylaniline Hydrochloride, 2-	3.70E+00	c	1.30E+01	c	6.60E−02	c	3.30E−01	c	5.20E−01	c
124-58-3	Methylarsonic Acid	6.10E+02	n	6.20E+03	n					3.70E+02	n
56-49-5	Methylcholanthrene, 3-	2.20E−02	c	7.80E−02	c	3.90E−04	c	1.90E−03	c	3.10E−03	c
75-09-2	Methylene Chloride	1.10E+01	c	5.30E+01	c	5.20E+00	c	2.60E+01	c	4.80E+00	c
101-14-4	Methylene-bis(2-chloroaniline), 4,4′-	1.20E+00	c	1.70E+01	c*	2.20E−03	c	2.90E−02	c	2.20E−01	c
101-61-1	Methylene-bis(N,N-dimethyl) Aniline, 4,4′-	1.10E+01	c	3.70E+01	c	1.90E−01	c	9.40E−01	c	1.50E+00	c
101-77-9	Methylenebisbenzenamine, 4,4′-	3.00E−01	c	1.10E+00	c	5.30E−03	c	2.70E−02	c	4.20E−02	c
101-68-8	Methylenediphenyl Diisocyanate	8.50E+05	nm	3.60E+06	nm	6.30E−01	n	2.60E+00	n		
98-83-9	Methylstyrene, Alpha-	5.50E+03	ns	7.20E+04	ns					2.60E+03	n
51218-45-2	Metolachlor	9.20E+03	n	9.20E+04	n					5.50E+03	n
21087-64-9	Metribuzin	1.50E+03	n	1.50E+04	n					9.10E+02	n
8012-95-1	Mineral Oils	2.30E+05	nm	3.10E+06	nm					1.10E+05	n
2385-85-5	Mirex	2.70E−02	c	9.60E−02	c	4.80E−04	c	2.40E−03	c	3.70E−03	c
2212-67-1	Molinate	1.20E+02	n	1.20E+03	n					7.30E+01	n
7439-98-7	Molybdenum	3.90E+02	n	5.10E+03	n					1.80E+02	n
10599-90-3	Monochloramine	7.80E+03	n	1.00E+05	nm					3.70E+03	n
100-61-8	Monomethylaniline	1.20E+02	n	1.20E+03	n					7.30E+01	n
74-31-7	N,N′-Diphenyl-1,4-benzenediamine	1.80E+01	n	1.80E+02	n					1.10E+01	n

300-76-5	Naled	1.20E+02	n	1.20E+03	n					7.30E+01	n
64724-95-6	Naphtha, High Flash Aromatic (HFAN)	2.30E+03	n	3.10E+04	n	1.00E+02	n	4.40E+02	n	1.80E+02	n
91-59-8	Naphthylamine, 2-	2.70E−01	c	9.60E−01	c					3.70E−02	c
15299-99-7	Napropamide	6.10E+03	n	6.20E+04	n					3.70E+03	n
13463-39-3	Nickel Carbonyl	3.70E+03	n	4.40E+04	n	5.20E−02	n	2.20E−01	n	1.80E+03	n
1313-99-1	Nickel Oxide	3.80E+03	n	4.70E+04	n	1.00E−01	n	4.40E−01	n	1.80E+03	n
NA	Nickel Refinery Dust	3.70E+03	n	4.40E+04	n	1.00E−02	c**	5.10E−02	c**	1.80E+03	n
7440-02-0	Nickel Soluble Salts	1.50E+03	n	2.00E+04	n	9.40E−03	c*	4.70E−02	c**	7.30E+02	n
12035-72-2	Nickel Subsulfide	3.80E−01	c	1.70E+00	c	5.10E−03	c*	2.60E−02	c**	4.00E−02	c
14797-55-8	Nitrate	1.30E+05	nm	1.60E+06	nm					5.80E+04	n
14797-65-0	Nitrite	7.80E+03	n	1.00E+05	nm					3.70E+03	n
88-74-4	Nitroaniline, 2-	6.10E+02	n	6.00E+03	n	5.20E−02	n	2.20E−01	n	3.70E+02	n
100-01-6	Nitroaniline, 4-	2.40E+01	c*	8.60E+01	c*	6.30E+00	n	2.60E+01	n	3.40E+00	c*
98-95-3	Nitrobenzene	4.80E+00	c*	2.40E+01	c*	6.10E−02	c	3.10E−01	c	1.20E−01	c
9004-70-0	Nitrocellulose	2.30E+08	nm	3.10E+09	nm					1.10E+08	n
67-20-9	Nitrofurantoin	4.30E+03	n	4.30E+04	n					2.60E+03	n
59-87-0	Nitrofurazone	3.70E−01	c	1.30E+00	c	6.60E−03	c	3.30E−02	c	5.20E−02	c
55-63-0	Nitroglycerin	6.10E+00	n	6.20E+01	n					3.70E+00	n
556-88-7	Nitroguanidine	6.10E+03	n	6.20E+04	n					3.70E+03	n
75-52-5	Nitromethane	4.90E+00	c*	2.50E+01	c*	2.70E−01	c*	1.40E+00	c*	5.40E−01	c*
79-46-9	Nitropropane, 2-	**1.30E−02**	c	6.40E−02	c	9.00E−04	c	4.50E−03	c	1.80E−03	c
759-73-9	Nitroso-N-ethylurea, N-	**1.80E−02**	c	6.40E−02	c	3.20E−04	c	1.60E−03	c	2.50E−03	c
684-93-5	Nitroso-N-methylurea, N-	**4.00E−03**	c	1.40E−02	c	7.20E−05	c	3.60E−04	c	5.60E−04	c
924-16-3	Nitroso-di-N-butylamine, N-	**8.70E−02**	c	4.00E−01	c	1.50E−03	c	7.70E−03	c	2.40E−03	c
621-64-7	Nitroso-di-N-propylamine, N-	**6.90E−02**	c	2.50E−01	c	1.20E−03	c	6.10E−03	c	9.60E−03	c
1116-54-7	Nitrosodiethanolamine, N-	**1.70E−01**	c	6.20E−01	c	3.00E−03	c	1.50E−02	c	2.40E−02	c
55-18-5	Nitrosodiethylamine, N-	**7.70E−04**	c	1.10E−02	c	2.20E−05	c	2.90E−04	c	1.40E−04	c
62-75-9	Nitrosodimethylamine, N-	**2.30E−03**	c	3.40E−02	c	6.90E−05	c	8.80E−04	c	4.20E−04	c
86-30-6	Nitrosodiphenylamine, N-	**9.90E+01**	c	3.50E+02	c	9.40E−01	c	4.70E+00	c	1.40E+01	c
10595-95-6	Nitrosomethylethylamine, N-	**2.20E−02**	c	7.80E−02	c	3.90E−04	c	1.90E−03	c	3.10E−03	c
59-89-2	Nitrosomorpholine [N-]	**7.20E−02**	c	2.60E−01	c	1.30E−03	c	6.50E−03	c	1.00E−02	c
100-75-4	Nitrosopiperidine [N-]	**5.20E−02**	c	1.80E−01	c	9.00E−04	c	4.50E−03	c	7.20E−03	c
930-55-2	Nitrosopyrrolidine, N-	**2.30E−01**	c	8.20E−01	c	4.00E−03	c	2.00E−02	c	3.20E−02	c

(Continued)

Appendix J (Continued)

CAS No.	Chemical Name	Regional Screening Levels (RSLs) 'Direct Contact Exposure Pathways'				
		Residential Soil (mg/kg)	Industrial Soil (mg/kg)	Residential Air ($\mu g/m^3$)	Industrial Air ($\mu g/m^3$)	Tap Water ($\mu g/l$)
99-08-1	Nitrotoluene, m-	**6.10E+00** n	6.20E+01 n			3.70E+00 n
88-72-2	Nitrotoluene, o-	2.90E+00 c*	1.30E+01 c*			3.10E−01 c
99-99-0	Nitrotoluene, p-	3.00E+01 c**	1.10E+02 c*			4.20E+00 c*
111-84-2	Nonane, n-	2.10E+01 ns	2.30E+02 ns	2.10E+02 n	8.80E+02 n	1.10E+01 n
27314-13-2	Norflurazon	2.40E+03 n	2.50E+04 n			1.50E+03 n
85509-19-9	Nustar	4.30E+01 n	4.30E+02 n			2.60E+01 n
32536-52-0	Octabromodiphenyl Ether	1.80E+02 n	1.80E+03 n			1.10E+02 n
2691-41-0	Octahydro-1,3,5,7-tetranitro-1,3,5,7-tetra (HMX)	3.80E+03 n	4.90E+04 n			1.80E+03 n
152-16-9	Octamethylpyrophosphoramide	1.20E+02 n	1.20E+03 n			7.30E+01 n
19044-88-3	Oryzalin	3.10E+03 n	3.10E+04 n			1.80E+03 n
19666-30-9	Oxadiazon	3.10E+02 n	3.10E+03 n			1.80E+02 n
23135-22-0	Oxamyl	1.50E+03 n	1.50E+04 n			9.10E+02 n
76738-62-0	Paclobutrazol	7.90E+02 n	8.00E+03 n			4.70E+02 n
1910-42-5	Paraquat Dichloride	2.70E+02 n	2.80E+03 n			1.60E+02 n
56-38-2	Parathion	3.70E+02 n	3.70E+03 n			2.20E+02 n
1114-71-2	Pebulate	3.10E+03 n	3.10E+04 n			1.80E+03 n
40487-42-1	Pendimethalin	2.40E+03 n	2.50E+04 n			1.50E+03 n
32534-81-9	Pentabromodiphenyl Ether	1.20E+02 n	1.20E+03 n			7.30E+01 n
60348-60-9	Pentabromodiphenyl Ether, 2,2′,4,4′,5- (BDE-99)	7.80E+00 n	1.00E+02 n			3.70E+00 n
608-93-5	Pentachlorobenzene	4.90E+01 n	4.90E+02 n			2.90E+01 n
76-01-7	Pentachloroethane	5.40E+00 c	1.90E+01 c			7.50E−01 c
82-68-8	Pentachloronitrobenzene	1.90E+00 c*	6.60E+00 c			2.60E−01 c
87-86-5	Pentachlorophenol	3.00E+00 c	9.00E+00 c	4.80E−01 c	2.40E+00 c	5.60E−01 c
109-66-0	Pentane, n-	8.70E+02 ns	3.70E+03 ns	1.00E+03 n	4.40E+03 n	2.10E+03 n
14797-73-0	Perchlorate and Perchlorate Salts	5.50E+01 n	7.20E+02 n			2.60E+01 n
52645-53-1	Permethrin	3.10E+03 n	3.10E+04 n			1.80E+03 n
62-44-2	Phenacetin	2.20E+02 c	7.80E+02 c	3.90E+00 c	1.90E+01 c	3.10E+01 c

13684-63-4	Phenmedipham	1.50E+04	n	1.50E+05	nm					9.10E+03	n
108-95-2	Phenol	1.80E+04	n	1.80E+05	nm	2.10E+02	n	8.80E+02	n	1.10E+04	n
108-45-2	Phenylenediamine, m-	3.70E+02	n	3.70E+03	n					2.20E+02	n
95-54-5	Phenylenediamine, o-	1.00E+01	c	3.70E+01	c					1.40E+00	c
106-50-3	Phenylenediamine, p-	1.20E+04	n	1.20E+05	nm					6.90E+03	n
90-43-7	Phenylphenol, 2-	2.50E+02	c	8.90E+02	c					3.50E+01	c
298-02-2	Phorate	1.20E+01	n	1.20E+02	n					7.30E+00	n
75-44-5	Phosgene	3.30E−01	n	1.40E+00	n	3.10E−01	n	1.30E+00	n		
732-11-6	Phosmet	1.20E+03	n	1.20E+04	n					7.30E+02	n
7803-51-2	Phosphine	2.30E+01	n	3.10E+02	n	3.10E−01	n	1.30E+00	n	1.10E+01	n
7664-38-2	Phosphoric Acid	1.40E+07	nm	6.00E+07	nm	1.00E+01	n	4.40E+01	n		
7723-14-0	Phosphorus, White	1.60E+00	n	2.00E+01	n					7.30E−01	n
100-21-0	Phthalic Acid, P-	6.10E+04	n	6.20E+05	nm					3.70E+04	n
85-44-9	Phthalic Anhydride	1.20E+05	nm	1.20E+06	nm	2.10E+01	n	8.80E+01	n	7.30E+04	n
1918-02-1	Picloram	4.30E+03	n	4.30E+04	n					2.60E+03	n
96-91-3	Picramic Acid (2-Amino-4,6-dinitrophenol)	6.10E+00	n	6.20E+01	n					3.70E+00	n
29232-93-7	Pirimiphos, Methyl	6.10E+02	n	6.20E+03	n					3.70E+02	n
59536-65-1	Polybrominated Biphenyls	1.60E−02	c*	5.70E−02	c*	2.80E−04	c	1.40E−03	c	2.20E−03	c
	Polychlorinated Biphenyls (PCBs)										
12674-11-2	~Aroclor 1016	3.90E+00	n	2.10E+01	c**	1.20E−01	c	6.10E−01	c	9.60E−01	c**
11104-28-2	~Aroclor 1221	1.40E−01	c	5.40E−01	c	4.30E−03	c	2.10E−02	c	6.80E−03	c
11141-16-5	~Aroclor 1232	1.40E−01	c	5.40E−01	c	4.30E−03	c	2.10E−02	c	6.80E−03	c
53469-21-9	~Aroclor 1242	2.20E−01	c	7.40E−01	c	4.30E−03	c	2.10E−02	c	3.40E−02	c
12672-29-6	~Aroclor 1248	2.20E−01	c	7.40E−01	c	4.30E−03	c	2.10E−02	c	3.40E−02	c
11097-69-1	~Aroclor 1254	2.20E−01	c**	7.40E−01	c*	4.30E−03	c	2.10E−02	c	3.40E−02	c*
11096-82-5	~Aroclor 1260	2.20E−01	c	7.40E−01	c	4.30E−03	c	2.10E−02	c	3.40E−02	c
39635-31-9	~Heptachlorobiphenyl, 2,3,3′,4,4′,5,5′- (PCB 189)	3.40E−02	c	1.10E−01	c	6.40E−04	c	3.20E−03	c	5.20E−03	c
52663-72-6	~Hexachlorobiphenyl, 2,3′,4,4′,5,5′- (PCB 167)	3.40E−01	c	1.10E+00	c	6.40E−03	c	3.20E−02	c	5.20E−02	c
69782-90-7	~Hexachlorobiphenyl, 2,3,3′,4,4′,5′- (PCB 157)	6.80E−03	c	2.30E−02	c	1.30E−04	c	6.50E−04	c	1.00E−03	c
38380-08-4	~Hexachlorobiphenyl, 2,3,3′,4,4′,5- (PCB 156)	6.80E−03	c	2.30E−02	c	1.30E−04	c	6.50E−04	c	1.00E−03	c
32774-16-6	~Hexachlorobiphenyl, 3,3′,4,4′,5,5′- (PCB 169)	3.40E−04	c	1.10E−03	c	6.40E−06	c	3.20E−05	c	5.20E−05	c
65510-44-3	~Pentachlorobiphenyl, 2′,3,4,4′,5- (PCB 123)	3.40E−02	c	1.10E−01	c	6.40E−04	c	3.20E−03	c	5.20E−03	c
31508-00-6	~Pentachlorobiphenyl, 2,3′,4,4′,5- (PCB 118)	3.40E−02	c	1.10E−01	c	6.40E−04	c	3.20E−03	c	5.20E−03	c

(*Continued*)

Appendix J (Continued)

CAS No.	Chemical Name	Regional Screening Levels (RSLs) 'Direct Contact Exposure Pathways'									
		Residential Soil (mg/kg)		Industrial Soil (mg/kg)		Residential Air (μg/m³)		Industrial Air (μg/m³)		Tap Water (μg/l)	
32598-14-4	~Pentachlorobiphenyl, 2,3,3′,4,4′- (PCB 105)	3.40E−02	c	1.10E−01	c	6.40E−04	c	3.20E−03	c	5.20E−03	c
74472-37-0	~Pentachlorobiphenyl, 2,3,4,4′,5- (PCB 114)	6.80E−04	c	2.30E−03	c	1.30E−04	c	6.50E−04	c	1.00E−04	c
57465-28-8	~Pentachlorobiphenyl, 3,3′,4,4′,5- (PCB 126)	3.40E−05	c	1.10E−04	c	6.40E−07	c	3.20E−06	c	5.20E−06	c
1336-36-3	~Polychlorinated Biphenyls (high risk)	2.20E−01	c	7.40E−01	c	4.30E−03	c	2.20E−02	c		
1336-36-3	~Polychlorinated Biphenyls (low risk)					2.40E−02	c	1.20E−01	c	1.70E−01	c
1336-36-3	~Polychlorinated Biphenyls (lowest risk)					1.20E−01	c	6.10E−01	c		
32598-13-3	~Tetrachlorobiphenyl, 3,3′,4,4′- (PCB 77)	3.40E−02	c	1.10E−01	c	6.40E−04	c	3.20E−03	c	5.20E−03	c
70362-50-4	~Tetrachlorobiphenyl, 3,4,4′,5- (PCB 81)	3.40E−02	c	1.10E−01	c	6.40E−04	c	3.20E−03	c	5.20E−03	c
9016-87-9	Polymeric Methylene Diphenyl Diisocyanate (PMDI)	8.50E+05	nm	3.60E+06	nm	6.30E−01	n	2.60E+00	n		
	Polynuclear Aromatic Hydrocarbons (PAHs)										
83-32-9	~Acenaphthene	3.40E+03	n	3.30E+04	n					2.20E+03	n
120-12-7	~Anthracene	1.70E+04	n	1.70E+05	nm					1.10E+04	n
56-55-3	~Benz[a]anthracene	1.50E−01	c	2.10E+00	c	8.70E−03	c	1.10E−01	c	2.90E−02	c
205-82-3	~Benzo(j)fluoranthene	5.30E−01	c	2.40E+00	c	2.20E−02	c	1.10E−01	c	5.60E−02	c
50-32-8	~Benzo[a]pyrene	1.50E−02	c	2.10E−01	c	8.70E−04	c	1.10E−02	c	2.90E−03	c
205-99-2	~Benzo[b]fluoranthene	1.50E−01	c	2.10E+00	c	8.70E−03	c	1.10E−01	c	2.90E−02	c
207-08-9	~Benzo[k]fluoranthene	1.50E+00	c	2.10E+01	c	8.70E−03	c	1.10E−01	c	2.90E−01	c
218-01-9	~Chrysene	1.50E+01	c	2.10E+02	c	8.70E−02	c	1.10E+00	c	2.90E+00	c
53-70-3	~Dibenz[a,h]anthracene	1.50E−02	c	2.10E−01	c	8.00E−04	c	1.00E−02	c	2.90E−03	c
192-65-4	~Dibenzo(a,e)pyrene	5.30E−02	c	2.40E−01	c	2.20E−03	c	1.10E−02	c	5.60E−03	c
57-97-6	~Dimethylbenz(a)anthracene, 7,12-	1.80E−03	c	6.20E−03	c	3.40E−05	c	1.70E−04	c	2.70E−04	c
206-44-0	~Fluoranthene	2.30E+03	n	2.20E+04	n					1.50E+03	n
86-73-7	~Fluorene	2.30E+03	n	2.20E+04	n					1.50E+03	n
193-39-5	~Indeno[1,2,3-cd]pyrene	1.50E−01	c	2.10E+00	c	8.70E−03	c	1.10E−01	c	2.90E−02	c
90-12-0	~Methylnaphthalene, 1-	2.20E+01	c	9.90E+01	c					2.30E+00	c
91-57-6	~Methylnaphthalene, 2-	3.10E+02	n	4.10E+03	ns					1.50E+02	n

91-20-3	~Naphthalene	3.60E+00	c*	1.80E+01	c*	7.20E−02	c*	3.60E−01	c*	1.40E−01	c*
57835-92-4	~Nitropyrene, 4-	5.30E−01	c	2.40E+00	c	2.20E−02	c	1.10E−01	c	5.60E−02	c
129-00-0	~Pyrene	1.70E+03	n	1.70E+04	n					1.10E+03	n
7778-74-7	Potassium Perchlorate	5.50E+01	n	7.20E+02	n					2.60E+01	n
67747-09-5	Prochloraz	3.20E+00	c	1.10E+01	c					4.50E−01	c
26399-36-0	Profluralin	3.70E+02	n	3.70E+03	n					2.20E+02	n
1610-18-0	Prometon	9.20E+02	n	9.20E+03	n					5.50E+02	n
7287-19-6	Prometryn	2.40E+02	n	2.50E+03	n					1.50E+02	n
1918-16-7	Propachlor	7.90E+02	n	8.00E+03	n					4.70E+02	n
709-98-8	Propanil	3.10E+02	n	3.10E+03	n					1.80E+02	n
2312-35-8	Propargite	1.20E+03	n	1.20E+04	n					7.30E+02	n
107-19-7	Propargyl Alcohol	1.20E+02	n	1.20E+03	n					7.30E+01	n
139-40-2	Propazine	1.20E+03	n	1.20E+04	n					7.30E+02	n
122-42-9	Propham	1.20E+03	n	1.20E+04	n					7.30E+02	n
60207-90-1	Propiconazole	7.90E+02	n	8.00E+03	n					4.70E+02	n
123-38-6	Propionaldehyde	8.00E+01	n	3.40E+02	n	8.30E+00	n	3.50E+01	n	1.70E+01	n
103-65-1	Propyl Benzene	3.40E+03	ns	2.10E+04	ns	1.00E+03	n	4.40E+03	n	1.30E+03	n
115-07-1	Propylene	4.30E+09	nm	1.80E + 10	nm	3.10E+03	n	1.30E+04	n		
57-55-6	Propylene Glycol	1.20E+06	nm	1.20E+07	nm					7.30E+05	n
6423-43-4	Propylene Glycol Dinitrate	5.70E+01	n	2.40E+02	n	2.80E−01	n	1.20E+00	n	5.70E−01	n
1569-02-4	Propylene Glycol Monoethyl Ether	4.30E+04	n	4.30E+05	nm					2.60E+04	n
107-98-2	Propylene Glycol Monomethyl Ether	4.30E+04	n	4.30E+05	nm	2.10E+03	n	8.80E+03	n	2.60E+04	n
75-56-9	Propylene Oxide	1.90E+00	c	8.80E+00	c	6.60E−01	c*	3.30E+00	c*	2.30E−01	c
81335-77-5	Pursuit	1.50E+04	n	1.50E+05	nm					9.10E+03	n
51630-58-1	Pydrin	1.50E+03	n	1.50E+04	n					9.10E+02	n
110-86-1	Pyridine	7.80E+01	n	1.00E+03	n					3.70E+01	n
13593-03-8	Quinalphos	3.10E+01	n	3.10E+02	n					1.80E+01	n
91-22-5	Quinoline	1.60E−01	c	5.70E−01	c					2.20E−02	c
NA	Refractory Ceramic Fibers	4.30E+07	nm	1.80E+08	nm	3.10E+01	n	1.30E+02	n		
10453-86-8	Resmethrin	1.80E+03	n	1.80E+04	n					1.10E+03	n
299-84-3	Ronnel	3.10E+03	n	3.10E+04	n					1.80E+03	n
83-79-4	Rotenone	2.40E+02	n	2.50E+03	n					1.50E+02	n
94-59-7	Safrole	2.20E+00	c	7.80E+00	c	3.90E−02	c	1.90E−01	c	3.10E−01	c

(Continued)

Appendix J (Continued)

CAS No.	Chemical Name	Regional Screening Levels (RSLs) 'Direct Contact Exposure Pathways' Residential Soil (mg/kg)		Industrial Soil (mg/kg)		Residential Air (μg/m³)		Industrial Air (μg/m³)		Tap Water (μg/l)	
78587-05-0	Savey	1.50E+03	n	1.50E+04	n					9.10E+02	n
7783-00-8	Selenious Acid	3.90E+02	n	5.10E+03	n					1.80E+02	n
7782-49-2	Selenium	3.90E+02	n	5.10E+03	n	2.10E+01	n	8.80E+01	n	1.80E+02	n
7446-34-6	Selenium Sulfide	3.90E+02	n	5.10E+03	n	2.10E+01	n	8.80E+01	n	1.80E+02	n
74051-80-2	Sethoxydim	5.50E+03	n	5.50E+04	n					3.30E+03	n
7631-86-9	Silica (crystalline, respirable)	4.30E+06	nm	1.80E+07	nm	3.10E+00	n	1.30E+01	n		
7440-22-4	Silver	3.90E+02	n	5.10E+03	n					1.80E+02	n
122-34-9	Simazine	4.00E+00	c*	1.40E+01	c					5.60E−01	c
62476-59-9	Sodium Acifluorfen	7.90E+02	n	8.00E+03	n					4.70E+02	n
26628-22-8	Sodium Azide	3.10E+02	n	4.10E+03	n					1.50E+02	n
148-18-5	Sodium Diethyldithiocarbamate	1.80E+00	c	6.40E+00	c					2.50E−01	c
7681-49-4	Sodium Fluoride	3.90E+03	n	5.10E+04	n	1.40E+01	n	5.70E+01	n	1.80E+03	n
62-74-8	Sodium Fluoroacetate	1.20E+00	n	1.20E+01	n					7.30E−01	n
13718-26-8	Sodium Metavanadate	7.80E+01	n	1.00E+03	n					3.70E+01	n
7601-89-0	Sodium Perchlorate	5.50E+01	n	7.20E+02	n					2.60E+01	n
961-11-5	Stirofos (Tetrachlorovinphos)	2.00E+01	c*	7.20E+01	c					2.80E+00	c
7440-24-6	Strontium, Stable	4.70E+04	n	6.10E+05	nm					2.20E+04	n
57-24-9	Strychnine	1.80E+01	n	1.80E+02	n					1.10E+01	n
100-42-5	Styrene	6.30E+03	ns	3.60E+04	ns	1.00E+03	n	4.40E+03	n	1.60E+03	n
80-07-9	Sulfonylbis(4-chlorobenzene), 1,1′-	4.90E+01	n	4.90E+02	n					2.90E+01	n
7664-93-9	Sulfuric Acid	1.40E+06	nm	6.00E+06	nm	1.00E+00	n	4.40E+00	n		
88671-89-0	Systhane	1.50E+03	n	1.50E+04	n					9.10E+02	n
21564-17-0	TCMTB	1.80E+03	n	1.80E+04	n					1.10E+03	n
34014-18-1	Tebuthiuron	4.30E+03	n	4.30E+04	n					2.60E+03	n
3383-96-8	Temephos	1.20E+03	n	1.20E+04	n					7.30E+02	n
5902-51-2	Terbacil	7.90E+02	n	8.00E+03	n					4.70E+02	n
13071-79-9	Terbufos	1.50E+00	n	1.50E+01	n					9.10E−01	n

886-50-0	Terbutryn	6.10E+01	n	6.20E+02	n					3.70E+01	n
5436-43-1	Tetrabromodiphenyl ether, 2,2′,4,4′- (BDE-47)	7.80E+00	n	1.00E+02	n					3.70E+00	n
95-94-3	Tetrachlorobenzene, 1,2,4,5-	1.80E+01	n	1.80E+02	n					1.10E+01	n
630-20-6	Tetrachloroethane, 1,1,1,2-	1.90E+00	c	9.30E+00	c	3.30E−01	c	1.70E+00	c	5.20E−01	c
79-34-5	Tetrachloroethane, 1,1,2,2-	5.60E−01	c	2.80E+00	c	4.20E−02	c	2.10E−01	c	6.70E−02	c
127-18-4	Tetrachloroethylene	5.50E−01	c	2.60E+00	c	4.10E−01	c	2.10E+00	c	1.10E−01	c
58-90-2	Tetrachlorophenol, 2,3,4,6-	1.80E+03	n	1.80E+04	n					1.10E+03	n
5216-25-1	Tetrachlorotoluene, p- alpha, alpha, alpha-	2.40E−02	c	8.60E−02	c					3.40E−03	c
3689-24-5	Tetraethyl Dithiopyrophosphate	3.10E+01	n	3.10E+02	n					1.80E+01	n
811-97-2	Tetrafluoroethane, 1,1,1,2-	1.10E+05	nms	4.60E+05	nms	8.30E+04	n	3.50E+05	n	1.70E+05	n
479-45-8	Tetryl (Trinitrophenylmethylnitramine)	2.40E+02	n	2.50E+03	n					1.50E+02	n
7440-28-0	Thallium (Soluble Salts)										
28249-77-6	Thiobencarb	6.10E+02	n	6.20E+03	n					3.70E+02	n
111-48-8	Thiodiglycol	5.40E+03	n	6.80E+04	n					2.60E+03	n
39196-18-4	Thiofanox	1.80E+01	n	1.80E+02	n					1.10E+01	n
23564-05-8	Thiophanate, Methyl	4.90E+03	n	4.90E+04	n					2.90E+03	n
137-26-8	Thiram	3.10E+02	n	3.10E+03	n					1.80E+02	n
7440-31-5	Tin	4.70E+04	n	6.10E+05	nm					2.20E+04	n
7550-45-0	Titanium Tetrachloride	1.40E+05	nm	6.00E+05	nm	1.00E−01	n	4.40E−01	n		
108-88-3	Toluene	5.00E+03	ns	4.50E+04	ns	5.20E+03	n	2.20E+04	n	2.30E+03	n
106-49-0	Toluidine, p-	2.60E+00	c	9.10E+00	c					3.50E−01	c
8001-35-2	Toxaphene	4.40E−01	c	1.60E+00	c	7.60E−03	c	3.80E−02	c	6.10E−02	c
66841-25-6	Tralomethrin	4.60E+02	n	4.60E+03	n					2.70E+02	n
688-73-3	Tri-n-butyltin	1.80E+01	n	1.80E+02	n					1.10E+01	n
2303-17-5	Triallate	7.90E+02	n	8.00E+03	n					4.70E+02	n
82097-50-5	Triasulfuron	6.10E+02	n	6.20E+03	n					3.70E+02	n
615-54-3	Tribromobenzene, 1,2,4-	3.10E+02	n	3.10E+03	n					1.80E+02	n
126-73-8	Tributyl Phosphate	5.30E+01	c	1.90E+02	c					7.30E+00	c
NA	Tributyltin Compounds	1.80E+01	n	1.80E+02	n					1.10E+01	n
56-35-9	Tributyltin Oxide	1.80E+01	n	1.80E+02	n					1.10E+01	n
76-13-1	Trichloro-1,2,2-trifluoroethane, 1,1,2-	4.30E+04	ns	1.80E+05	nms	3.10E+04	n	1.30E+05	n	5.90E+04	n
76-03-9	Trichloroacetic Acid										
33663-50-2	Trichloroaniline HCl, 2,4,6-	1.70E+01	c	5.90E+01	c					2.30E+00	c

(Continued)

Appendix J (Continued)

CAS No.	Chemical Name	Regional Screening Levels (RSLs) 'Direct Contact Exposure Pathways'									
		Residential Soil (mg/kg)		Industrial Soil (mg/kg)		Residential Air (μg/m³)		Industrial Air (μg/m³)		Tap Water (μg/l)	
634-93-5	Trichloroaniline, 2,4,6-	1.40E+01	c	5.10E+01	c					2.00E+00	c
87-61-6	Trichlorobenzene, 1,2,3-	4.90E+01	n	4.90E+02	ns					2.90E+01	n
120-82-1	Trichlorobenzene, 1,2,4-	2.20E+01	c**	9.90E+01	c**	2.10E+00	n	8.80E+00	n	2.30E+00	c**
71-55-6	Trichloroethane, 1,1,1-	8.70E+03	ns	3.80E+04	ns	5.20E+03	n	2.20E+04	n	9.10E+03	n
79-00-5	Trichloroethane, 1,1,2-	1.10E+00	c	5.30E+00	c	1.50E−01	c	7.70E−01	c	2.40E−01	c
79-01-6	Trichloroethylene	2.80E+00	c	1.40E+01	c	1.20E+00	c	6.10E+00	c	2.00E+00	c
75-69-4	Trichlorofluoromethane	7.90E+02	n	3.40E+03	ns	7.30E+02	n	3.10E+03	n	1.30E+03	n
95-95-4	Trichlorophenol, 2,4,5-	6.10E+03	n	6.20E+04	n					3.70E+03	n
88-06-2	Trichlorophenol, 2,4,6-	4.40E+01	c**	1.60E+02	c**	7.80E−01	c	4.00E+00	c	6.10E+00	c**
93-76-5	Trichlorophenoxyacetic Acid, 2,4,5-	6.10E+02	n	6.20E+03	n					3.70E+02	n
93-72-1	Trichlorophenoxypropionic Acid, 2,4,5-	4.90E+02	n	4.90E+03	n					2.90E+02	n
598-77-6	Trichloropropane, 1,1,2-	3.90E+02	n	5.10E+03	ns					1.80E+02	n
96-18-4	Trichloropropane, 1,2,3-	5.00E−03	c	9.50E−02	c	3.10E−01	n	1.30E+00	n	7.20E−04	c
96-19-5	Trichloropropene, 1,2,3-	7.80E−01	n	3.30E+00	n	3.10E−01	n	1.30E+00	n	6.20E−01	n
58138-08-2	Tridiphane	1.80E+02	n	1.80E+03	n					1.10E+02	n
121-44-8	Triethylamine	1.20E+02	n	5.20E+02	n	7.30E+00	n	3.10E+01	n	1.50E+01	n
1582-09-8	Trifluralin	6.30E+01	c**	2.20E+02	c*					8.70E+00	c*
512-56-1	Trimethyl Phosphate	1.30E+01	c	4.70E+01	c					1.80E+00	c
95-63-6	Trimethylbenzene, 1,2,4-	6.20E+01	n	2.60E+02	ns	7.30E+00	n	3.10E+01	n	1.50E+01	n
108-67-8	Trimethylbenzene, 1,3,5-	7.80E+02	ns	1.00E+04	ns					3.70E+02	n
99-35-4	Trinitrobenzene, 1,3,5-	2.20E+03	n	2.70E+04	n					1.10E+03	n
118-96-7	Trinitrotoluene, 2,4,6-	1.90E+01	c**	7.90E+01	c**					2.20E+00	c**
791-28-6	Triphenylphosphine Oxide	1.20E+03	n	1.20E+04	n					7.30E+02	n
13674-87-8	Tris(1,3-Dichloro-2-propyl) Phosphate	1.20E+03	n	1.20E+04	n					7.30E+02	n
115-96-8	Tris(2-chloroethyl)phosphate	2.40E+01	c*	8.60E+01	c*					3.40E+00	c*
78-42-2	Tris(2-ethylhexyl)phosphate	1.50E+02	c*	5.40E+02	c					2.10E+01	c
NA	Uranium (Soluble Salts)	2.30E+02	n	3.10E+03	n	3.10E−01	n	1.30E+00	n	1.10E+02	n

51-79-6	Urethane	6.40E−01	c	2.90E+00	c	8.40E−03	c	4.20E−02	c	6.70E−02	c
1314-62-1	Vanadium Pentoxide	4.00E+02	c**	2.00E+03	c**	2.90E−04	c*	1.50E−03	c*	3.30E+02	n
36907-42-3	Vanadium Sulfate	1.60E+03	n	2.00E+04	n					7.30E+02	n
NA	Vanadium and Compounds	3.90E+02	n	5.20E+03	n					1.80E+02	n
7440-62-2	Vanadium, Metallic	5.50E+00	n	7.20E+01	n	1.00E−01	n	4.40E−01	n	2.60E+00	n
1929-77-7	Vernolate	6.10E+01	n	6.20E+02	n					3.70E+01	n
50471-44-8	Vinclozolin	1.50E+03	n	1.50E+04	n					9.10E+02	n
108-05-4	Vinyl Acetate	9.70E+02	n	4.10E+03	ns	2.10E+02	n	8.80E+02	n	4.10E+02	n
593-60-2	Vinyl Bromide	1.10E−01	c*	5.60E−01	c*	7.60E−02	c*	3.80E−01	c*	1.50E−01	c*
75-01-4	Vinyl Chloride	6.00E−02	c	1.70E+00	c	1.60E−01	c	2.80E+00	c	1.60E−02	c
81 81 2	Warfarin	1.80E+01	n	1.80E+02	n					1.10E+01	n
1330-20-7	Xylene, Mixture	6.30E+02	ns	2.70E+03	ns	1.00E+02	n	4.40E+02	n	2.00E+02	n
106-42-3	Xylene, p-	3.40E+03	ns	1.70E+04	ns	7.30E+02	n	3.10E+03	n	1.20E+03	n
108-38-3	Xylene, m-	3.40E+03	ns	1.70E+04	ns	7.30E+02	n	3.10E+03	n	1.20E+03	n
95-47-6	Xylene, o-	3.80E+03	ns	1.90E+04	ns	7.30E+02	n	3.10E+03	n	1.20E+03	n
7440-66-6	Zinc (Metallic)	2.30E+04	n	3.10E+05	nm					1.10E+04	n
1314-84-7	Zinc Phosphide	2.30E+01	n	3.10E+02	n					1.10E+01	n
12122-67-7	Zineb	3.10E+03	n	3.10E+04	n					1.80E+03	n

Key: I = IRIS; P = PPRTV; A = ATSDR; C = Cal EPA; X = PPRTV Appendix; H = HEAST; J = New Jersey; E = Environmental Criteria and Assessment Office; S = see user guide Section 5; L = see user guide on lead; M = mutagen; V = volatile; F = see FAQ #29; c = cancer; * = where: n SL < 100X c SL; ** = where n SL < 10X c SL; n = noncancer; m = concentration may exceed ceiling limit (see user guide); s = concentration may exceed Csat (see user guide); SSL values are based on DAF = 1.

Reference: REGIONAL SCREENING LEVELS (Formerly PRGs) Version 9 May 2010 Stanford J. Smucker Ph.D., EPA Region IX Technical Support Section (SFD-8-4) 75 Hawthorne Street, San Francisco, CA 94105-3901.

Appendix K: Toxicity and Chemical-Specific Information

CAS No.	Chemical Name	SFO (mg/kg-day)$^{-1}$		IUR (μg/m^3)$^{-1}$		RfD$_o$ (mg/kg-day)		RfC$_i$ (mg/m^3)		VOC	Muta-gen	GIABS	ABS	C$_{sat}$ (mg/kg)
1596-84-5	~ALAR	1.8E−02	C	5.1E−06	C2	1.5E−01	I	J	K	L	M	1	0.1	0.2
30560-19-1	Acephate	8.7E−03	I			4.0E−03	I					1	0.1	
75-07-0	Acetaldehyde			2.2E−06	I			9.0E−03	I	V		1		1.07E+05
34256-82-1	Acetochlor					2.0E−02	I					1	0.1	
67-64-1	Acetone					9.0E−01	I	3.1E+01	A	V		1		1.14E+05
75-86-5	Acetone Cyanohydrin					3.0E−03	P	6.0E−02	P	V		1		1.06E+05
75-05-8	Acetonitrile							6.0E−02	I	V		1		1.28E+05
98-86-2	Acetophenone					1.0E−01	I			V		1		2.52E+03
53-96-3	Acetylaminofluorene, 2-	3.8E+00	C	1.3E−03	C							1	0.1	
107-02-8	Acrolein					5.0E−04	I	2.0E−05	I	V		1		2.27E+04
79-06-1	Acrylamide	5.0E−01	I	1.0E−04	I	2.0E−03	I	6.0E−03	I		M	1	0.1	
79-10-7	Acrylic Acid					5.0E−01	I	1.0E−03	I			1	0.1	
107-13-1	Acrylonitrile	5.4E−01	I	6.8E−05	I	4.0E−02	A	2.0E−03	I	V		1		1.13E+04
111-69-3	Adiponitrile							6.0E−03	P			1	0.1	
15972-60-8	Alachlor	5.6E−02	C			1.0E−02	I					1	0.1	
116-06-3	Aldicarb					1.0E−03	I					1	0.1	
1646-88-4	Aldicarb Sulfone					1.0E−03	I					1	0.1	
309-00-2	Aldrin	1.7E+01	I	4.9E−03	I	3.0E−05	I					1	0.1	
74223-64-6	Ally					2.5E−01	I					1	0.1	
107-18-6	Allyl Alcohol					5.0E−03	I	1.0E−04	X			1	0.1	
107-05-1	Allyl Chloride	2.1E−02	C	6.0E−06	C			1.0E−03	I	V		1		1.42E+03
7429-90-5	Aluminum					1.0E+00	P	5.0E−03	P			1		
20859-73-8	Aluminum Phosphide					4.0E−04	I					1		
67485-29-4	Amdro					3.0E−04	I					1	0.1	
834-12-8	Ametryn					9.0E−03	I					1	0.1	
92-67-1	Aminobiphenyl, 4-	2.1E+01	C	6.0E−03	C							1	0.1	
591-27-5	Aminophenol, m-					8.0E−02	P					1	0.1	
123-30-8	Aminophenol, p-					2.0E−02	P					1	0.1	
33089-61-1	Amitraz					2.5E−03	I					1	0.1	
7664-41-7	Ammonia							1.0E−01	I			1		
7790-98-9	Ammonium Perchlorate					7.0E−04	I					1		

7773-06-0	Ammonium Sulfamate					2.0E−01	I					1		
62-53-3	Aniline	5.7E−03	I	1.6E−06	C	7.0E−03	P	1.0E−03	I			1	0.1	
7440-36-0	Antimony (metallic)					4.0E−04	I					0.15		
1314-60-9	Antimony Pentoxide					5.0E−04	H					0.15		
11071-15-1	Antimony Potassium Tartrate					9.0E−04	H					0.15		
1332-81-6	Antimony Tetroxide					4.0E−04	H					0.15		
1309-64-4	Antimony Trioxide							2.0E−04	I			0.15		
74115-24-5	Apollo					1.3E−02	I					1	0.1	
140-57-8	Aramite	2.5E−02	I	7.1E−06	I	5.0E−02	H					1	0.1	
7440-38-2	Arsenic, Inorganic	1.5E+00	I	4.3E−03	I	3.0E−04	I	1.5E−05	C			1	0.03	
7784-42-1	Arsine					3.5E−06	C	5.0E−05	I			1		
76578-14-8	Assure					9.0E−03	I					1	0.1	
3337-71-1	Asulam					5.0E−02	I					1	0.1	
1912-24-9	Atrazine	2.3E−01	C			3.5E−02	I					1	0.1	
492-80-8	Auramine	8.8E−01	C	2.5E−04	C									
65195-55-3	Avermectin B1					4.0E−04	I					1	0.1	
103-33-3	Azobenzene	1.1E−01	I	3.1E−05	I					V		1		
7440-39-3	Barium					2.0E−01	I	5.0E−04	H			0.07		
114-26-1	Baygon					4.0E−03	I					1	0.1	
43121-43-3	Bayleton					3.0E−02	I					1	0.1	
68359-37-5	Baythroid					2.5E−02	I					1	0.1	
1861-40-1	Benefin					3.0E−01	I					1	0.1	
17804-35-2	Benomyl					5.0E−02	I					1	0.1	
25057-89-0	Bentazon					3.0E−02	I					1	0.1	
100-52-7	Benzaldehyde					1.0E−01	I			V		1		1.16E+03
71-43-2	Benzene	5.5E−02	I	7.8E−06	I	4.0E−03	I	3.0E−02	I	V		1		1.82E+03
108-98-5	Benzenethiol					1.0E−05	H			V		1		1.26E+03
92-87-5	Benzidine	2.3E+02	I	6.7E−02	I	3.0E−03	I				M	1	0.1	
65-85-0	Benzoic Acid					4.0E+00	I					1	0.1	
98-07-7	Benzotrichloride	1.3E+01	I							V		1		3.24E+02
100-51-6	Benzyl Alcohol					1.0E−01	P					1	0.1	
100-44-7	Benzyl Chloride	1.7E−01	I	4.9E−05	C	2.0E−03	P	1.0E−03	P	V		1		1.46E+03

(Continued)

Appendix K (Continued)

CAS No.	Chemical Name	SFO (mg/kg-day)$^{-1}$		IUR (μg/m^3)$^{-1}$		RfD$_o$ (mg/kg-day)		RfC$_i$ (mg/m^3)		VOC	Muta-gen	GIABS	ABS	C$_{sat}$ (mg/kg)
7440-41-7	Beryllium and compounds			2.4E−03	I	2.0E−03	I	2.0E−05	I			0.007		
141-66-2	Bidrin					1.0E−04	I					1	0.1	
42576-02-3	Bifenox					9.0E−03	P					1	0.1	
82657-04-3	Biphenthrin					1.5E−02	I					1	0.1	
92-52-4	Biphenyl, 1,1′-					5.0E−02	I			V		1		2.14E+02
108-60-1	Bis(2-chloro-1-methylethyl)ether	7.0E−02	H	1.0E−05	H	4.0E−02	I			V		1		1.02E+03
111-91-1	Bis(2-chloroethoxy)methane					3.0E−03	P					1	0.1	
111-44-4	Bis(2-chloroethyl)ether	1.1E+00	I	3.3E−04	I					V		1		5.05E+03
117-81-7	Bis(2-ethylhexyl)phthalate	1.4E−02	I	2.4E−06	C	2.0E−02	I					1	0.1	
542-88-1	Bis(chloromethyl)ether	2.2E+02	I	6.2E−02	I					V		1		4.22E+03
80-05-7	Bisphenol A					5.0E−02	I					1	0.1	
7440-42-8	Boron and Borates only					2.0E−01	I	2.0E−02	H			1		
7637-07-2	Boron Trifluoride					4.0E−02	C	1.3E−02	C			1		
15541-45-4	Bromate	7.0E−01	I			4.0E−03	I					1		
107-04-0	Bromo-2-chloroethane, 1-	2.0E+00	X	6.0E−04	X					V		1		2.38E+03
108-86-1	Bromobenzene					8.0E−03	I	6.0E−02	I	V		1		6.79E+02
75-27-4	Bromodichloromethane	6.2E−02	I	3.7E−05	C	2.0E−02	I			V		1		9.31E+02
75-25-2	Bromoform	7.9E−03	I	1.1E−06	I	2.0E−02	I					1	0.1	
74-83-9	Bromomethane					1.4E−03	I	5.0E−03	I	V		1		3.59E+03
2104-96-3	Bromophos					5.0E−03	H					1	0.1	
1689-84-5	Bromoxynil					2.0E−02	I					1	0.1	
1689-99-2	Bromoxynil Octanoate					2.0E−02	I					1	0.1	
106-99-0	Butadiene, 1,3-	3.4E+00	C	3.0E−05	I			2.0E−03	I	V		1		6.67E+02
71-36-3	Butanol, N-					1.0E−01	I					1	0.1	
85-68-7	Butyl Benzyl Phthlate	1.9E−03	P			2.0E−01	I					1	0.1	
78-92-2	Butyl Alcohol, sec-					2.0E+00	P	3.0E+01	P					
2008-41-5	Butylate					5.0E−02	I					1	0.1	
25013-16-5	Butylated Hydroxyanisole	2.0E−04	C	5.7E−08	C									
85-70-1	Butylphthalyl Butylglycolate					1.0E+00	I					1	0.1	

75-60-5	Cacodylic Acid					2.0E−02	A				1	0.1	
7440-43-9	Cadmium (Diet)			1.8E−03	I	1.0E−03	I	1.0E−05	A		0.025	0.001	
7440-43-9	Cadmium (Water)			1.8E−03	I	5.0E−04	I	1.0E−05	A		0.05	0.001	
105-60-2	Caprolactam					5.0E−01	I				1	0.1	
2425-06-1	Captafol	1.5E−01	C	4.3E−05	C	2.0E−03	I				1	0.1	
133-06-2	Captan	2.3E−03	C	6.6E−07	C	1.3E−01	I				1	0.1	
63-25-2	Carbaryl					1.0E−01	I				1	0.1	
1563-66-2	Carbofuran					5.0E−03	I				1	0.1	
75-15-0	Carbon Disulfide					1.0E−01	I	7.0E−01	I	V	1		7.38E+02
56-23-5	Carbon Tetrachloride	7.0E−02	I	6.0E−06	I	4.0E−03	I	1.0E−01	I	V	1		4.58E+02
55285-14-8	Carbosulfan					1.0E−02	I				1	0.1	
5234-68-4	Carboxin					1.0E−01	I				1	0.1	
1306-38-3	Ceric Oxide							9.0E−04	I		1		
302-17-0	Chloral Hydrate					1.0E−01	I				1	0.1	
133-90-4	Chloramben					1.5E−02	I				1	0.1	
118-75-2	Chloranil	4.0E−01	H								1	0.1	
12789-03-6	Chlordane	3.5E−01	I	1.0E−04	I	5.0E−04	I	7.0E−04	I		1	0.04	
143-50-0	Chlordecone (Kepone)	1.0E+01	I	4.6E−03	C	3.0E−04	I				1	0.1	
470-90-6	Chlorfenvinphos					7.0E−04	A				1	0.1	
90982-32-4	Chlorimuron, Ethyl-					2.0E−02	I				1	0.1	
7782-50-5	Chlorine					1.0E−01	I	1.5E−04	A		1		
10049-04-4	Chlorine Dioxide					3.0E−02	I	2.0E−04	I		1		
7758-19-2	Chlorite (Sodium Salt)					3.0E−02	I				1		
75-68-3	Chloro-1,1-difluoroethane, 1-							5.0E+01	I	V	1		1.15E+03
126-99-8	Chloro-1,3-butadiene, 2-					2.0E−02	H	7.0E−03	H	V	1		7.52E+02
3165-93-3	Chloro-2-methylaniline HCl, 4-	4.6E−01	H								1	0.1	
107-20-0	Chloroacetaldehyde, 2-	2.7E−01	X										
79-11-8	Chloroacetic Acid					2.0E−03	H				1	0.1	
532-27-4	Chloroacetophenone, 2-							3.0E−05	I		1	0.1	
106-47-8	Chloroaniline, p-	2.0E−01	P			4.0E−03	I				1	0.1	
108-90-7	Chlorobenzene					2.0E−02	I	5.0E−02	P	V	1		7.61E+02
510-15-6	Chlorobenzilate	1.1E−01	C	3.1E−05	C	2.0E−02	I				1	0.1	

(Continued)

Appendix K (Continued)

CAS No.	Chemical Name	SFO (mg/kg-day)$^{-1}$		IUR (μg/m^3)$^{-1}$		RfD$_o$ (mg/kg-day)		RfC$_i$ (mg/m^3)		VOC	Muta-gen	GIABS	ABS	C$_{sat}$ (mg/kg)
74-11-3	Chlorobenzoic Acid, p-					3.0E−02	X					1	0.1	
98-56-6	Chlorobenzotrifluoride, 4-					3.0E−03	P	3.0E−01	P	V		1		1.17E+02
109-69-3	Chlorobutane, 1-					4.0E−02	P			V		1		7.28E+02
75-45-6	Chlorodifluoromethane							5.0E+01	I	V		1		1.68E+03
67-66-3	Chloroform	3.1E−02	C	2.3E−05	I	1.0E−02	I	9.8E−02	A	V		1		2.54E+03
74-87-3	Chloromethane							9.0E−02	I	V		1		1.32E+03
107-30-2	Chloromethyl Methyl Ether	2.4E+00	C	6.9E−04	C					V		1		2.58E+04
91-58-7	Chloronaphthalene, Beta-					8.0E−02	I			V		1		1.75E+02
88-73-3	Chloronitrobenzene, o-	3.0E−01	P			3.0E−03	P	1.0E−05	X			1	0.1	
100-00-5	Chloronitrobenzene, p-	6.3E−03	P			1.0E−03	P	6.0E−04	P			1	0.1	
95-57-8	Chlorophenol, 2-					5.0E−03	I			V		1		2.19E+04
76-06-2	Chloropicrin							4.0E−04	C					
1897-45-6	Chlorothalonil	3.1E−03	C	8.9E−07	C	1.5E−02	I					1	0.1	
95-49-8	Chlorotoluene, o-					2.0E−02	I			V		1		9.07E+02
106-43-4	Chlorotoluene, p-					7.0E−02	P			V		1		2.53E+02
54749-90-5	Chlorozotocin	2.4E+02	C	6.9E−02	C									
101-21-3	Chlorpropham					2.0E−01	I					1	0.1	
2921-88-2	Chlorpyrifos					3.0E−03	I					1	0.1	
5598-13-0	Chlorpyrifos Methyl					1.0E−02	H					1	0.1	
64902-72-3	Chlorsulfuron					5.0E−02	I					1	0.1	
60238-56-4	Chlorthiophos					8.0E−04	H					1	0.1	
16065-83-1	Chromium(III), Insoluble Salts					1.5E+00	I					0.013		
18540-29-9	Chromium(VI)	5.0E−01	J	8.4E−02	I	3.0E−03	I	1.0E−04	I		M	0.025		
7440-47-3	Chromium, Total											0.013		
7440-48-4	Cobalt			9.0E−03	P	3.0E−04	P	6.0E−06	P			1		
8007-45-2	Coke Oven Emissions			6.2E−04	I						M	1	0.1	
7440-50-8	Copper					4.0E−02	H					1		
108-39-4	Cresol, m-					5.0E−02	I	6.0E−01	C			1	0.1	
95-48-7	Cresol, o-					5.0E−02	I	6.0E−01	C			1	0.1	
106-44-5	Cresol, p-					5.0E−03	H	6.0E−01	C			1	0.1	

59-50-7	Cresol, p-chloro-m-					1.0E−01	X				1	0.1	
1319-77-3	Cresols					1.0E−01	A	6.0E−01	C	V	1		5.02E+04
123-73-9	Crotonaldehyde, trans-	1.9E+00	H							V	1		1.66E+04
98-82-8	Cumene					1.0E−01	I	4.0E−01	I	V	1		2.68E+02
135-20-6	Cupferron	2.2E−01	C	6.3E−05	C								
21725-46-2	Cyanazine	8.4E−01	H			2.0E−03	H				1	0.1	
	Cyanides												
592-01-8	~Calcium Cyanide					4.0E−02	I				1		
544-92-3	~Copper Cyanide					5.0E−03	I				1		
57-12-5	~Cyanide (CN-)					2.0E−02	I			V	1		1.00E+07
460-19-5	~Cyanogen					4.0E−02	I			V	1		1.45E+03
506-68-3	~Cyanogen Bromide					9.0E−02	I			V	1		1.03E+05
506-77-4	~Cyanogen Chloride					5.0E−02	I			V	1		4.29E+03
74-90-8	~Hydrogen Cyanide					2.0E−02	I	3.0E−03	I	V	1		1.18E+05
151-50-8	~Potassium Cyanide					5.0E−02	I				1		
506-61-6	~Potassium Silver Cyanide					2.0E−01	I				0.04		
506-64-9	~Silver Cyanide					1.0E−01	I				0.04		
143-33-9	~Sodium Cyanide					4.0E−02	I				1		
463-56-9	~Thiocyanate					2.0E−04	P			V	1		4.56E+03
557-21-1	~Zinc Cyanide					5.0E−02	I				1		
110-82-7	Cyclohexane							6.0E+00	I	V	1		1.17E+02
87-84-3	Cyclohexane, 1,2,3,4,5-pentabromo-6-chloro-	2.3E−02	H								1	0.1	
108-94-1	Cyclohexanone					5.0E+00	I				1	0.1	
108-91-8	Cyclohexylamine					2.0E−01	I				1	0.1	
68085-85-8	Cyhalothrin/karate					5.0E−03	I				1	0.1	
52315-07-8	Cypermethrin					1.0E−02	I				1	0.1	
66215-27-8	Cyromazine					7.5E−03	I				1	0.1	
72-54-8	DDD	2.4E−01	I	6.9E−05	C						1	0.1	
72-55-9	DDE, p,p′-	3.4E−01	I	9.7E−05	C						1	0.1	
50-29-3	DDT	3.4E−01	I	9.7E−05	I	5.0E−04	I				1	0.03	
1861-32-1	Dacthal					1.0E−02	I				1	0.1	

(Continued)

Appendix K (Continued)

CAS No.	Chemical Name	SFO $(mg/kg\text{-}day)^{-1}$		IUR $(\mu g/m^3)^{-1}$		RfD_o (mg/kg-day)		RfC_i (mg/m^3)		VOC	Mutagen	GIABS	ABS	C_{sat} (mg/kg)
75-99-0	Dalapon					3.0E−02	I					1	0.1	
1163-19-5	Decabromodiphenyl ether, 2,2′,3,3′,4,4′,5,5′,6,6′- (BDE−209)	7.0E−04	I			7.0E−03	I					1	0.1	
8065-48-3	Demeton					4.0E−05	I					1	0.1	
103-23-1	Di(2-ethylhexyl)adipate	1.2E−03	I			6.0E−01	I					1	0.1	
2303-16-4	Diallate	6.1E−02	H									1	0.1	
333-41-5	Diazinon					7.0E−04	A					1	0.1	
96-12-8	Dibromo-3-chloropropane, 1,2-	8.0E−01	P	6.0E−03	P	2.0E−04	P	2.0E−04	I	V	M	1		9.79E+02
106-37-6	Dibromobenzene, 1,4-					1.0E−02	I					1	0.1	
124-48-1	Dibromochloromethane	8.4E−02	I	2.7E−05	C	2.0E−02	I			V		1	0.1	8.02E+02
106-93-4	Dibromoethane, 1,2-	2.0E+00	I	6.0E−04	I	9.0E−03	I	9.0E−03	I	V		1		1.34E+03
74-95-3	Dibromomethane (Methylene Bromide)					1.0E−02	H	4.0E−03	X	V		1		2.82E+03
84-74-2	Dibutyl Phthalate					1.0E−01	I					1	0.1	
NA	Dibutyltin Compounds					3.0E−04	P					1	0.1	
1918-00-9	Dicamba					3.0E−02	I					1	0.1	
764-41-0	Dichloro-2-butene, 1,4-			4.2E−03	P					V		1		5.19E+02
1476-11-5	Dichloro-2-butene, cis-1,4-			4.2E−03	P					V		1	0.1	5.19E+02
110-57-6	Dichloro-2-butene, trans-1,4-			4.2E−03	P					V		1	0.1	7.60E+02
79-43-6	Dichloroacetic Acid	5.0E−02	I			4.0E−03	I					1	0.1	
95-50-1	Dichlorobenzene, 1,2-					9.0E−02	I	2.0E−01	H	V		1		3.76E+02
106-46-7	Dichlorobenzene, 1,4-	5.4E−03	C	1.1E−05	C	7.0E−02	A	8.0E−01	I	V		1		
91-94-1	Dichlorobenzidine, 3,3′-	4.5E−01	I	3.4E−04	C							1	0.1	
90-98-2	Dichlorobenzophenone, 4,4′-					9.0E−03	X					1	0.1	
75-71-8	Dichlorodifluoromethane					2.0E−01	I	2.0E−01	H	V		1		8.45E+02
75-34-3	Dichloroethane, 1,1-	5.7E−03	C	1.6E−06	C	2.0E−01	P			V		1		1.69E+03
107-06-2	Dichloroethane, 1,2-	9.1E−02	I	2.6E−05	I	2.0E−02	P	2.4E+00	A	V		1		2.98E+03
75-35-4	Dichloroethylene, 1,1-					5.0E−02	I	2.0E−01	I	V		1		1.19E+03
540-59-0	Dichloroethylene, 1,2- (Mixed Isomers)					9.0E−03	H			V		1		1.29E+03
156-59-2	Dichloroethylene, 1,2-cis-					1.0E−02	P			V		1		2.37E+03
156-60-5	Dichloroethylene, 1,2-trans-					2.0E−02	I	6.0E−02	P	V		1		1.67E+03

120-83-2	Dichlorophenol, 2,4-					3.0E−03	I				1	0.1	
94-75-7	Dichlorophenoxy Acetic Acid, 2,4-					1.0E−02	I				1	0.05	
94-82-6	Dichlorophenoxy)butyric Acid, 4-(2,4-					8.0E−03	I				1	0.1	
78-87-5	Dichloropropane, 1,2-	3.6E−02	C	1.0E−05	C	9.0E−02	A	4.0E−03	I	V	1		1.36E+03
142-28-9	Dichloropropane, 1,3-					2.0E−02	P			V	1		1.49E+03
616-23-9	Dichloropropanol, 2,3-					3.0E−03	I				1	0.1	
542-75-6	Dichloropropene, 1,3-	1.0E−01	I	4.0E−06	I	3.0E−02	I	2.0E−02	I	V	1		1.57E+03
62-73-7	Dichlorvos	2.9E−01	I	8.3E−05	C	5.0E−04	I	5.0E−04	I		1	0.1	
77-73-6	Dicyclopentadiene					8.0E−03	P	7.0E−03	P	V	1		1.32E+02
60-57-1	Dieldrin	1.6E+01	I	4.6E−03	I	5.0E−05	I				1	0.1	
NA	Diesel Engine Exhaust			3.0E−04	C			5.0E−03	I		1	0.1	
111-42-2	Diethanolamine							3.0E−03	C				
84-66-2	Diethyl Phthalate					8.0E−01	I				1	0.1	
112-34-5	Diethylene Glycol Monobutyl Ether					3.0E−02	P	1.0E−04	P		1	0.1	
111-90-0	Diethylene Glycol Monoethyl Ether					6.0E−02	P	3.0E−04	P		1	0.1	
617-84-5	Diethylformamide					1.0E−03	P				1	0.1	
56-53-1	Diethylstilbestrol	3.5E+02	C	1.0E−01	C						1	0.1	
43222-48-6	Difenzoquat					8.0E−02	I				1	0.1	
35367-38-5	Diflubenzuron					2.0E−02	I				1	0.1	
75-37-6	Difluoroethane, 1,1-							4.0E+01	I	V	1		1.43E+03
94-58-6	Dihydrosafrole	4.4E−02	C	1.3E−05	C								
108-20-3	Diisopropyl Ether							4.0E−01	P	V	1		2.26E+03
1445-75-6	Diisopropyl Methylphosphonate					8.0E−02	I			V	1		5.30E+02
55290-64-7	Dimethipin					2.0E−02	I				1	0.1	
60-51-5	Dimethoate					2.0E−04	I				1	0.1	
119-90-4	Dimethoxybenzidine, 3,3′-	1.4E−02	H								1	0.1	
756-79-6	Dimethyl methylphosphonate	1.7E−03	P			6.0E−02	P				1	0.1	
60-11-7	Dimethylamino Azobenzene [p-]	4.6E+00	C	1.3E−03	C						1	0.1	
21436-96-4	Dimethylaniline HCl, 2,4-	5.8E−01	H								1	0.1	
95-68-1	Dimethylaniline, 2,4-	7.5E−01	H								1	0.1	
121-69-7	Dimethylaniline, N,N-					2.0E−03	I			V	1		8.30E+02
119-93-7	Dimethylbenzidine, 3,3′-	1.1E+01	P								1	0.1	

(Continued)

Appendix K (Continued)

CAS No.	Chemical Name	SFO (mg/kg-day)$^{-1}$		IUR (μg/m^3)$^{-1}$		RfD$_o$ (mg/kg-day)		RfC$_i$ (mg/m^3)		VOC	Muta-gen	GIABS	ABS	C$_{sat}$ (mg/kg)
68-12-2	Dimethylformamide					1.0E−01	P	3.0E−02	I			1	0.1	
57-14-7	Dimethylhydrazine, 1,1-					1.0E−04	X	2.0E−06	X			1	0.1	
540-73-8	Dimethylhydrazine, 1,2-	5.5E+02	C	1.6E−01	C							1	0.1	
105-67-9	Dimethylphenol, 2,4-					2.0E−02	I					1	0.1	
576-26-1	Dimethylphenol, 2,6-					6.0E−04	I					1	0.1	
95-65-8	Dimethylphenol, 3,4-					1.0E−03	I					1	0.1	
120-61-6	Dimethylterephthalate					1.0E−01	I			V		1		5.45E+00
513-37-1	Dimethylvinylchloride	4.5E−02	C	1.3E−05	C									
534-52-1	Dinitro-o-cresol, 4,6-					8.0E−05	X					1	0.1	
131-89-5	Dinitro-o-cyclohexyl Phenol, 4,6-					2.0E−03	I					1	0.1	
528-29-0	Dinitrobenzene, 1,2-					1.0E−04	P					1	0.1	
99-65-0	Dinitrobenzene, 1,3-					1.0E−04	I					1	0.1	
100-25-4	Dinitrobenzene, 1,4-					1.0E−04	P					1	0.1	
51-28-5	Dinitrophenol, 2,4-					2.0E−03	I					1	0.1	
25321-14-6	Dinitrotoluene Mixture, 2,4/2,6-	6.8E−01	I									1	0.1	
121-14-2	Dinitrotoluene, 2,4-	3.1E−01	C	8.9E−05	C	2.0E−03	I					1	0.102	
606-20-2	Dinitrotoluene, 2,6-					1.0E−03	P					1	0.099	
35572-78-2	Dinitrotoluene, 2-Amino-4,6-					2.0E−03	S					1	0.006	
19406-51-0	Dinitrotoluene, 4-Amino-2,6-					2.0E−03	S					1	0.009	
88-85-7	Dinoseb					1.0E−03	I					1	0.1	
123-91-1	Dioxane, 1,4-	1.1E−02	I	7.7E−06	C	1.0E−01	A	3.6E+00	A			1	0.1	
	Dioxins													
NA	~Hexachlorodibenzo-p-dioxin, Mixture	6.2E+03	I	1.3E+00	I							1	0.03	
1746-01-6	~TCDD, 2,3,7,8-	1.3E+05	C	3.8E+01	C	1.0E−09	A	4.0E−08	C			1	0.03	
957-51-7	Diphenamid					3.0E−02	I					1	0.1	
127-63-9	Diphenyl Sulfone					8.0E−04	X					1	0.1	
122-39-4	Diphenylamine					2.5E−02	I					1	0.1	
122-66-7	Diphenylhydrazine, 1,2-	8.0E−01	I	2.2E−04	I							1	0.1	
85-00-7	Diquat					2.2E−03	I					1	0.1	

1937-37-7	Direct Black 38	7.4E+00	C	2.1E−03	C						1	0.1	
2602-46-2	Direct Blue 6	7.4E+00	C	2.1E−03	C						1	0.1	
16071-86-6	Direct Brown 95	6.7E+00	C	1.9E−03	C						1	0.1	
298-04-4	Disulfoton					4.0E−05	I				1	0.1	
505-29-3	Dithiane, 1,4-					1.0E−02	I				1	0.1	
330-54-1	Diuron					2.0E−03	I				1	0.1	
2439-10-3	Dodine					4.0E−03	I				1	0.1	
759-94-4	EPTC					2.5E−02	I			V	1		4.07E+02
115-29-7	Endosulfan					6.0E−03	I				1	0.1	
145-73-3	Endothall					2.0E−02	I				1	0.1	
72 20 8	Endrin					3.0E−04	I				1	0.1	
106-89-8	Epichlorohydrin	9.9E−03	I	1.2E−06	I	6.0E−03	P	1.0E−03	I	V	1		1.05E+04
106-88-7	Epoxybutane, 1,2-							2.0E−02	I	V	1		1.53E+04
16672-87-0	Ethephon					5.0E−03	I				1	0.1	
563-12-2	Ethion					5.0E−04	I				1	0.1	
111-15-9	Ethoxyethanol Acetate, 2-					3.0E−01	H	3.0E−01	C		1	0.1	
110-80-5	Ethoxyethanol, 2-					4.0E−01	H	2.0E−01	I		1	0.1	
141-78-6	Ethyl Acetate					9.0E−01	I			V	1		1.08E+04
140-88-5	Ethyl Acrylate	4.8E−02	H							V	1		2.50E+03
75-00-3	Ethyl Chloride							1.0E+01	I	V	1		2.12E+03
60-29-7	Ethyl Ether					2.0E−01	I			V	1		1.01E+04
97-63-2	Ethyl Methacrylate					9.0E−02	H			V	1		1.10E+03
2104-64-5	Ethyl-p-nitrophenyl Phosphonate					1.0E−05	I				1	0.1	
100-41-4	Ethylbenzene	1.1E−02	C	2.5E−06	C	1.0E−01	I	1.0E+00	I	V	1		4.80E+02
109-78-4	Ethylene Cyanohydrin					3.0E−02	P				1	0.1	
107-15-3	Ethylene Diamine					9.0E−02	P				1	0.1	
107-21-1	Ethylene Glycol					2.0E+00	I	4.0E−01	C		1	0.1	
111-76-2	Ethylene Glycol Monobutyl Ether					1.0E−01	I	1.6E+00	I		1	0.1	
75-21-8	Ethylene Oxide	3.1E−01	C	8.8E−05	C			3.0E−02	C	V	1		1.21E+05
96-45-7	Ethylene Thiourea	4.5E−02	C	1.3E−05	C	8.0E−05	I				1	0.1	
151-56-4	Ethyleneimine	6.5E+01	C	1.9E−02	C								
84-72-0	Ethylphthalyl Ethyl Glycolate					3.0E+00	I				1	0.1	
101200-48-0	Express					8.0E−03	I				1	0.1	

(Continued)

Appendix K (Continued)

CAS No.	Chemical Name	SFO (mg/kg-day)$^{-1}$		IUR (µg/m^3)$^{-1}$		RfD$_o$ (mg/kg-day)		RfC$_i$ (mg/m^3)		VOC	Muta-gen	GIABS	ABS	C$_{sat}$ (mg/kg)
22224-92-6	Fenamiphos					2.5E−04	I					1	0.1	
39515-41-8	Fenpropathrin					2.5E−02	I					1	0.1	
2164-17-2	Fluometuron					1.3E−02	I					1	0.1	
16984-48-8	Fluoride					4.0E−02	C	1.3E−02	C					
7782-41-4	Fluorine (Soluble Fluoride)					6.0E−02	I	1.3E−02	C			1		
59756-60-4	Fluridone					8.0E−02	I					1	0.1	
56425-91-3	Flurprimidol					2.0E−02	I					1	0.1	
66332-96-5	Flutolanil					6.0E−02	I					1	0.1	
69409-94-5	Fluvalinate					1.0E−02	I					1	0.1	
133-07-3	Folpet	3.5E−03	I			1.0E−01	I					1	0.1	
72178-02-0	Fomesafen	1.9E−01	I									1	0.1	
944-22-9	Fonofos					2.0E−03	I					1	0.1	
50-00-0	Formaldehyde			1.3E−05	I	2.0E−01	I	9.8E−03	A			1	0.1	
64-18-6	Formic Acid					2.0E+00	H	3.0E−03	P			1	0.1	
39148-24-8	Fosetyl-AL					3.0E+00	I					1	0.1	
	Furans													
132-64-9	~Dibenzofuran					1.0E−03	X			V		1		1.71E+02
110-00-9	~Furan					1.0E−03	I			V		1		6.22E+03
67-45-8	Furazolidone	3.8E+00	H									1	0.1	
98-01-1	Furfural					3.0E−03	I	5.0E−02	H			1	0.1	
531-82-8	Furium	1.5E+00	C	4.3E−04	C							1	0.1	
60568-05-0	Furmecyclox	3.0E−02	I	8.6E−06	C							1	0.1	
77182-82-2	Glufosinate, Ammonium					4.0E−04	I					1	0.1	
111-30-8	Glutaraldehyde							8.0E−05	C					
765-34-4	Glycidyl					4.0E−04	I	1.0E−03	H			1	0.1	
1071-83-6	Glyphosate					1.0E−01	I					1	0.1	
42874-03-3	Goal					3.0E−03	I					1	0.1	
86-50-0	Guthion					3.0E−03	A	1.0E−02	A			1	0.1	
69806-40-2	Haloxyfop, Methyl					5.0E−05	I					1	0.1	
79277-27-3	Harmony					1.3E−02	I					1	0.1	

76-44-8	Heptachlor	4.5E+00	I	1.3E−03	I	5.0E−04	I				1	0.1	
1024-57-3	Heptachlor Epoxide	9.1E+00	I	2.6E−03	I	1.3E−05	I				1	0.1	
87-82-1	Hexabromobenzene					2.0E−03	I				1	0.1	
68631-49-2	Hexabromodiphenyl Ether, 2,2′,4,4′,5,5′-(BDE−153)					2.0E−04	I						
118-74-1	Hexachlorobenzene	1.6E+00	I	4.6E−04	I	8.0E−04	I				1	0.1	
87-68-3	Hexachlorobutadiene	7.8E−02	I	2.2E−05	I	1.0E−03	P				1	0.1	
319-84-6	Hexachlorocyclohexane, Alpha-	6.3E+00	I	1.8E−03	I	8.0E−03	A				1	0.1	
319-85-7	Hexachlorocyclohexane, Beta-	1.8E+00	I	5.3E−04	I						1	0.1	
58-89-9	Hexachlorocyclohexane, Gamma-(Lindane)	1.1E+00	C	3.1E−04	C	3.0E−04	I				1	0.04	
608-73-1	Hexachlorocyclohexane, Technical	1.8E+00	I	5.1E−04	I						1	0.1	
77-47-4	Hexachlorocyclopentadiene					6.0E−03	I	2.0E−04	I		1	0.1	
67-72-1	Hexachloroethane	1.4E−02	I	4.0E−06	I	1.0E−03	I				1	0.1	
70-30-4	Hexachlorophene					3.0E−04	I				1	0.1	
121-82-4	Hexahydro-1,3,5-trinitro-1,3,5-triazine (RDX)	1.1E−01	I			3.0E−03	I				1	0.015	
822-06-0	Hexamethylene Diisocyanate, 1,6-							1.0E−05	I	V	1		5.19E+03
110-54-3	Hexane, N-					6.0E−02	H	7.0E−01	I	V	1		1.41E+02
124-04-9	Hexanedioic Acid					2.0E+00	P				1	0.1	
591-78-6	Hexanone, 2-					5.0E−03	I	3.0E−02	I	V	1		3.28E+03
51235-04-2	Hexazinone					3.3E−02	I				1	0.1	
302-01-2	Hydrazine	3.0E+00	I	4.9E−03	I			3.0E−05	P		1		
10034-93-2	Hydrazine Sulfate	3.0E+00	I	4.9E−03	I						1		
7647-01-0	Hydrogen Chloride							2.0E−02	I		1		
7664-39-3	Hydrogen Fluoride					4.0E−02	C	1.4E−02	C		1		
7783-06-4	Hydrogen Sulfide							2.0E−03	I		1		
123-31-9	Hydroquinone	6.0E−02	P			4.0E−02	P				1	0.1	
35554-44-0	Imazalil					1.3E−02	I				1	0.1	
81335-37-7	Imazaquin					2.5E−01	I				1	0.1	
7553-56-2	Iodine					1.0E−02	A						
36734-19-7	Iprodione					4.0E−02	I				1	0.1	
7439-89-6	Iron					7.0E−01	P				1		

(Continued)

Appendix K (Continued)

CAS No.	Chemical Name	SFO (mg/kg-day)$^{-1}$		IUR (μg/m^3)$^{-1}$		RfD$_o$ (mg/kg-day)		RfC$_i$ (mg/m^3)		VOC	Muta-gen	GIABS	ABS	C$_{sat}$ (mg/kg)
78-83-1	Isobutyl Alcohol					3.0E−01	I			V		1		1.00E+04
78-59-1	Isophorone	9.5E−04	I			2.0E−01	I	2.0E+00	C			1	0.1	
33820-53-0	Isopropalin					1.5E−02	I					1	0.1	
67-63-0	Isopropanol							7.0E+00	C			1	0.1	
1832-54-8	Isopropyl Methyl Phosphonic Acid					1.0E−01	I					1	0.1	
82558-50-7	Isoxaben					5.0E−02	I					1	0.1	
NA	JP-7							3.0E−01	A	V		1		
23950-58-5	Kerb					7.5E−02	I					1	0.1	
77501-63-4	Lactofen					2.0E−03	I					1	0.1	
	Lead Compounds													
301-04-2	~Lead Acetate	2.8E−01	C	8.0E−05	C									
7439-92-1	~Lead and Compounds											1		
1335-32-6	~Lead Subacetate	3.8E−02	C	1.1E−05	C									
78-00-2	~Tetraethyl Lead					1.0E−07	I					1	0.1	
330-55-2	Linuron					2.0E−03	I					1	0.1	
7439-93-2	Lithium					2.0E−03	P					1		
7791-03-9	Lithium Perchlorate					7.0E−04	I					1		
83055-99-6	Londax					2.0E−01	I					1	0.1	
94-74-6	MCPA					5.0E−04	I					1	0.1	
94-81-5	MCPB					1.0E−02	I					1	0.1	
93-65-2	MCPP					1.0E−03	I					1	0.1	
121-75-5	Malathion					2.0E−02	I					1	0.1	
108-31-6	Maleic Anhydride					1.0E−01	I	7.0E−04	C			1	0.1	
123-33-1	Maleic Hydrazide					5.0E−01	I					1	0.1	
109-77-3	Malononitrile					1.0E−04	P					1	0.1	
8018-01-7	Mancozeb					3.0E−02	H					1	0.1	
12427-38-2	Maneb					5.0E−03	I					1	0.1	
7439-96-5	Manganese (Diet)					1.4E−01	I	5.0E−05	I			1		
7439-96-5	Manganese (Water)					2.4E−02	I	5.0E−05	I			0.04		
950-10-7	Mephosfolan					9.0E−05	H					1	0.1	

24307-26-4	Mepiquat Chloride					3.0E−02	I				1	0.1	
	Mercury Compounds												
7487-94-7	~Mercuric Chloride					3.0E−04	I	3.0E−05	C		0.07		
1344-48-5	~Mercuric Sulfide					3.0E−04	S				1		
7439-97-6	~Mercury (elemental)					1.6E−04	C	3.0E−04	I	V	1		3.13E+00
NA	~Mercury, Inorganic Salts					3.0E−04	S				0.07		
22967-92-6	~Methyl Mercury					1.0E−04	I				1		
62-38-4	~Phenylmercuric Acetate					8.0E−05	I				1	0.1	
150-50-5	Merphos					3.0E−05	I				1	0.1	
78-48-8	Merphos Oxide					3.0E−05	I				1	0.1	
57837-19-1	Metalaxyl					6.0E−02	I				1	0.1	
126-98-7	Methacrylonitrile					1.0E−04	I	7.0E−04	H	V	1		4.58E+03
10265-92-6	Methamidophos					5.0E−05	I				1	0.1	
67-56-1	Methanol					5.0E−01	I	4.0E+00	C		1	0.1	
950-37-8	Methidathion					1.0E−03	I				1	0.1	
16752-77-5	Methomyl					2.5E−02	I				1	0.1	
99-59-2	Methoxy-5-nitroaniline, 2-	4.9E−02	C	1.4E−05	C						1	0.1	
72-43-5	Methoxychlor					5.0E−03	I				1	0.1	
110-49-6	Methoxyethanol Acetate, 2-					2.0E−03	H	9.0E−02	C		1	0.1	
109-86-4	Methoxyethanol, 2-					3.0E−03	P	2.0E−02	I		1	0.1	
79-20-9	Methyl Acetate					1.0E+00	H			V	1		2.90E+04
96-33-3	Methyl Acrylate					3.0E−02	H			V	1		6.75E+03
78-93-3	Methyl Ethyl Ketone (2-Butanone)					6.0E−01	I	5.0E+00	I	V	1		2.84E+04
108-10-1	Methyl Isobutyl Ketone (4-methyl-2-pentanone)					8.0E−02	H	3.0E+00	I	V	1		3.36E+03
624-83-9	Methyl Isocyanate							1.0E−03	C				
80-62-6	Methyl Methacrylate					1.4E+00	I	7.0E−01	I	V	1		2.36E+03
298-00-0	Methyl Parathion					2.5E−04	I				1	0.1	
993-13-5	Methyl Phosphonic Acid					6.0E−02	X				1	0.1	
25013-15-4	Methyl Styrene (Mixed Isomers)					6.0E−03	H	4.0E−02	H	V	1		3.83E+02
66-27-3	Methyl Methanesulfonate	9.9E−02	C	2.8E−05	C						1	0.1	
1634-04-4	Methyl tert-Butyl Ether (MTBE)	1.8E−03	C	2.6E−07	C			3.0E+00	I	V	1		8.87E+03
99-55-8	Methyl-5-Nitroaniline, 2-	3.3E−02	H								1	0.1	

(Continued)

Appendix K (Continued)

CAS No.	Chemical Name	SFO (mg/kg-day)$^{-1}$		IUR (μg/m^3)$^{-1}$		RfD$_o$ (mg/kg-day)		RfC$_i$ (mg/m^3)		VOC	Muta-gen	GIABS	ABS	C$_{sat}$ (mg/kg)
70-25-7	Methyl-N-nitro-N-nitrosoguanidine, N-	8.3E+00	C	2.4E−03	C									
636-21-5	Methylaniline Hydrochloride, 2-	1.3E−01	C	3.7E−05	C							1	0.1	
124-58-3	Methylarsonic Acid					1.0E−02	A					1	0.1	
56-49-5	Methylcholanthrene, 3-	2.2E+01	C	6.3E−03	C							1	0.1	
75-09-2	Methylene Chloride	7.5E−03	I	4.7E−07	I	6.0E−02	I	1.0E+00	A	V		1		3.32E+03
101-14-4	Methylene-bis(2-chloroaniline), 4,4′-	1.0E−01	P	4.3E−04	C	2.0E−03	P				M	1	0.1	
101-61-1	Methylene-bis(N,N-dimethyl) Aniline, 4,4′-	4.6E−02	I	1.3E−05	C							1	0.1	
101-77-9	Methylenebisbenzenamine, 4,4′-	1.6E+00	C	4.6E−04	C			2.0E−02	C			1	0.1	
101-68-8	Methylenediphenyl Diisocyanate							6.0E−04	I			1	0.1	
98-83-9	Methylstyrene, Alpha-					7.0E−02	H			V		1		5.00E+02
51218-45-2	Metolachlor					1.5E−01	I					1	0.1	
21087-64-9	Metribuzin					2.5E−02	I					1	0.1	
8012-95-1	Mineral Oils					3.0E+00	P							
2385-85-5	Mirex	1.8E+01	C	5.1E−03	C	2.0E−04	I					1	0.1	
2212-67-1	Molinate					2.0E−03	I					1	0.1	
7439-98-7	Molybdenum					5.0E−03	I					1		
10599-90-3	Monochloramine					1.0E−01	I					1		
100-61-8	Monomethylaniline					2.0E−03	P					1	0.1	
74-31-7	N,N′-Diphenyl-1,4-benzenediamine					3.0E−04	X					1	0.1	
300-76-5	Naled					2.0E−03	I					1	0.1	
64724-95-6	Naphtha, High Flash Aromatic (HFAN)					3.0E−02	X	1.0E−01	P	V				
91-59-8	Naphthylamine, 2-	1.8E+00	C	0.0E+00	C							1	0.1	
15299-99-7	Napropamide					1.0E−01	I					1	0.1	
13463-39-3	Nickel Carbonyl					5.0E−02	C	5.0E−05	C			0.04		
1313-99-1	Nickel Oxide					5.0E−02	C	1.0E−04	C					
NA	Nickel Refinery Dust			2.4E−04	I	5.0E−02	C	5.0E−05	C			0.04		
7440-02-0	Nickel Soluble Salts			2.6E−04	C	2.0E−02	I	9.0E−05	A			0.04		
12035-72-2	Nickel Subsulfide	1.7E+00	C	4.8E−04	I	5.0E−02	C	5.0E−05	C			0.04		
14797-55-8	Nitrate					1.6E+00	I					1		

14797-65-0	Nitrite					1.0E−01	I					1		
88-74-4	Nitroaniline, 2-					1.0E−02	X	5.0E−05	X			1	0.1	
100-01-6	Nitroaniline, 4-	2.0E−02	P			4.0E−03	P	6.0E−03	P			1	0.1	
98-95-3	Nitrobenzene			4.0E−05	I	2.0E−03	I	9.0E−03	I	V		1		3.05E+03
9004-70-0	Nitrocellulose					3.0E+03	P							
67-20-9	Nitrofurantoin					7.0E−02	H					1	0.1	
59-87-0	Nitrofurazone	1.3E+00	C	3.7E−04	C							1	0.1	
55-63-0	Nitroglycerin	1.7E−02	P			1.0E−04	P					1	0.1	
556-88-7	Nitroguanidine					1.0E−01	I					1	0.1	
75-52-5	Nitromethane			9.0E−06	P			2.0E−02	P	V		1		1.80E+04
79-46-9	Nitropropane, 2-			2.7E−03	H			2.0E−02	I	V		1		4.86E+03
759-73-9	Nitroso-N-ethylurea, N-	2.7E+01	C	7.7E−03	C							1	0.1	
684-93-5	Nitroso-N-methylurea, N-	1.2E+02	C	3.4E−02	C							1	0.1	
924-16-3	Nitroso-di-N-butylamine, N-	5.4E+00	I	1.6E−03	I					V		1		7.10E+03
621-64-7	Nitroso-di-N-propylamine, N-	7.0E+00	I	2.0E−03	C							1	0.1	
1116-54-7	Nitrosodiethanolamine, N-	2.8E+00	I	8.0E−04	C							1	0.1	
55-18-5	Nitrosodiethylamine, N-	1.5E+02	I	4.3E−02	I						M	1	0.1	
62-75-9	Nitrosodimethylamine, N-	5.1E+01	I	1.4E−02	I	8.0E−06	P	4.0E−05	X		M	1	0.1	
86-30-6	Nitrosodiphenylamine, N-	4.9E−03	I	2.6E−06	C							1	0.1	
10595-95-6	Nitrosomethylethylamine, N-	2.2E+01	I	6.3E−03	C							1	0.1	
59-89-2	Nitrosomorpholine [N-]	6.7E+00	C	1.9E−03	C							1	0.1	
100-75-4	Nitrosopiperidine [N-]	9.4E+00	C	2.7E−03	C							1	0.1	
930-55-2	Nitrosopyrrolidine, N-	2.1E+00	I	6.1E−04	I							1	0.1	
99-08-1	Nitrotoluene, m-					1.0E−04	X					1	0.1	
88-72-2	Nitrotoluene, o-	2.2E−01	P			9.0E−04	P			V		1		1.51E+03
99-99-0	Nitrotoluene, p-	1.6E−02	P			4.0E−03	P					1	0.1	
111-84-2	Nonane, n-					3.0E−04	X	2.0E−01	P	V				6.86E+00
27314-13-2	Norflurazon					4.0E−02	I					1	0.1	
85509-19-9	Nustar					7.0E−04	I					1	0.1	
32536-52-0	Octabromodiphenyl Ether					3.0E−03	I					1	0.1	
2691-41-0	Octahydro-1,3,5,7-tetranitro-1,3,5,7-tetra (HMX)					5.0E−02	I					1	0.006	
152-16-9	Octamethylpyrophosphoramide					2.0E−03	H					1	0.1	

(*Continued*)

Appendix K (Continued)

CAS No.	Chemical Name	SFO $(mg/kg\text{-}day)^{-1}$		IUR $(\mu g/m^3)^{-1}$		RfD_o (mg/kg-day)		RfC_i (mg/m^3)		VOC	Muta-gen	GIABS	ABS	C_{sat} (mg/kg)
19044-88-3	Oryzalin					5.0E−02	I					1	0.1	
19666-30-9	Oxadiazon					5.0E−03	I					1	0.1	
23135-22-0	Oxamyl					2.5E−02	I					1	0.1	
76738-62-0	Paclobutrazol					1.3E−02	I					1	0.1	
1910-42-5	Paraquat Dichloride					4.5E−03	I					1	0.1	
56-38-2	Parathion					6.0E−03	H					1	0.1	
1114-71-2	Pebulate					5.0E−02	H					1	0.1	
40487-42-1	Pendimethalin					4.0E−02	I					1	0.1	
32534-81-9	Pentabromodiphenyl Ether					2.0E−03	I					1	0.1	
60348-60-9	Pentabromodiphenyl Ether, 2,2′,4,4′,5- (BDE−99)					1.0E−04	I							
608-93-5	Pentachlorobenzene					8.0E−04	I					1	0.1	
76-01-7	Pentachloroethane	9.0E−02	P									1	0.1	
82-68-8	Pentachloronitrobenzene	2.6E−01	H			3.0E−03	I					1	0.1	
87-86-5	Pentachlorophenol	1.2E−01	I	5.1E−06	C	3.0E−02	I					1	0.25	
109-66-0	Pentane, n-							1.0E+00	P	V				3.88E+02
14797-73-0	Perchlorate and Perchlorate Salts					7.0E−04	I					1		
52645-53-1	Permethrin					5.0E−02	I					1	0.1	
62-44-2	Phenacetin	2.2E−03	C	6.3E−07	C							1	0.1	
13684-63-4	Phenmedipham					2.5E−01	I					1	0.1	
108-95-2	Phenol					3.0E−01	I	2.0E−01	C			1	0.1	
108-45-2	Phenylenediamine, m-					6.0E−03	I					1	0.1	
95-54-5	Phenylenediamine, o-	4.7E−02	H									1	0.1	
106-50-3	Phenylenediamine, p-					1.9E−01	H					1	0.1	
90-43-7	Phenylphenol, 2-	1.9E−03	H									1	0.1	
298-02-2	Phorate					2.0E−04	H					1	0.1	
75-44-5	Phosgene							3.0E−04	I	V		1		1.61E+03
732-11-6	Phosmet					2.0E−02	I					1	0.1	
7803-51-2	Phosphine					3.0E−04	I	3.0E−04	I			1		

7664-38-2	Phosphoric Acid							1.0E−02	I		1		
7723-14-0	Phosphorus, White					2.0E−05	I				1		
100-21-0	Phthalic Acid, P-					1.0E+00	H				1	0.1	
85-44-9	Phthalic Anhydride					2.0E+00	I	2.0E−02	C		1	0.1	
1918-02-1	Picloram					7.0E−02	I				1	0.1	
96-91-3	Picramic Acid (2-Amino-4,6-dinitrophenol)					1.0E−04	X				1	0.1	
29232-93-7	Pirimiphos, Methyl					1.0E−02	I				1	0.1	
59536-65-1	Polybrominated Biphenyls	3.0E+01	C	8.6E−03	C	7.0E−06	H				1	0.1	
	Polychlorinated Biphenyls (PCBs)												
12674-11-2	~Aroclor 1016	7.0E−02	I	2.0E−05	I	7.0E−05	I				1	0.14	
11104-28-2	~Aroclor 1221	2.0E+00	I	5.7E−04	I					V	1	0.14	7.57E+02
11141-16-5	~Aroclor 1232	2.0E+00	I	5.7E−04	I					V	1	0.14	7.32E+01
53469-21-9	~Aroclor 1242	2.0E+00	I	5.7E−04	I						1	0.14	
12672-29-6	~Aroclor 1248	2.0E+00	I	5.7E−04	I						1	0.14	
11097-69-1	~Aroclor 1254	2.0E+00	I	5.7E−04	I	2.0E−05	I				1	0.14	
11096-82-5	~Aroclor 1260	2.0E+00	I	5.7E−04	I						1	0.14	
39635-31-9	~Heptachlorobiphenyl, 2,3,3′,4,4′,5,5′- (PCB 189)	1.3E+01	C	3.8E−03	C						1	0.14	
52663-72-6	~Hexachlorobiphenyl, 2,3′,4,4′,5,5′- (PCB 167)	1.3E+00	C	3.8E−04	C						1	0.14	
69782-90-7	~Hexachlorobiphenyl, 2,3,3′,4,4′,5′- (PCB 157)	6.5E+01	C	1.9E−02	C						1	0.14	
38380-08-4	~Hexachlorobiphenyl, 2,3,3′,4,4′,5- (PCB 156)	6.5E+01	C	1.9E−02	C						1	0.14	
32774-16-6	~Hexachlorobiphenyl, 3,3′,4,4′,5,5′- (PCB 169)	1.3E+03	C	3.8E−01	C						1	0.14	
65510-44-3	~Pentachlorobiphenyl, 2′,3,4,4′,5- (PCB 123)	1.3E+01	C	3.8E−03	C						1	0.14	
31508-00-6	~Pentachlorobiphenyl, 2,3′,4,4′,5- (PCB 118)	1.3E+01	C	3.8E−03	C						1	0.14	
32598-14-4	~Pentachlorobiphenyl, 2,3,3′,4,4′- (PCB 105)	1.3E+01	C	3.8E−03	C						1	0.14	

(Continued)

Appendix K (Continued)

CAS No.	Chemical Name	SFO (mg/kg-day)$^{-1}$		IUR (μg/m^3)$^{-1}$		RfD$_o$ (mg/kg-day)		RfC$_i$ (mg/m^3)		VOC	Muta-gen	GIABS	ABS	C_{sat} (mg/kg)
74472-37-0	~Pentachlorobiphenyl, 2,3,4,4′,5- (PCB 114)	6.5E+02	C	1.9E−02	C							1	0.14	
57465-28-8	~Pentachlorobiphenyl, 3,3′,4,4′,5- (PCB 126)	1.3E+04	C	3.8E+00	C							1	0.14	
1336-36-3	~Polychlorinated Biphenyls (high risk)	2.0E+00	I	5.7E−04	C							1	0.14	
1336-36-3	~Polychlorinated Biphenyls (low risk)	4.0E−01	I	1.0E−04	I							1	0.14	
1336-36-3	~Polychlorinated Biphenyls (lowest risk)	7.0E−02	I	2.0E−05	C							1	0.14	
32598-13-3	~Tetrachlorobiphenyl, 3,3′,4,4′- (PCB 77)	1.3E+01	C	3.8E−03	C							1	0.14	
70362-50-4	~Tetrachlorobiphenyl, 3,4,4′,5- (PCB 81)	1.3E+01	C	3.8E−03	C							1	0.14	
9016-87-9	Polymeric Methylene Diphenyl Diisocyanate (PMDI)							6.0E−04	I			1	0.1	
	Polynuclear Aromatic Hydrocarbons (PAHs)													
83-32-9	~Acenaphthene					6.0E−02	I			V		1	0.13	
120-12-7	~Anthracene					3.0E−01	I			V		1	0.13	
56-55-3	~Benz[a]anthracene	7.3E−01	E	1.1E−04	C						M	1	0.13	
205-82-3	~Benzo(j)fluoranthene	1.2E+00	C	1.1E−04	C									
50-32-8	~Benzo[a]pyrene	7.3E+00	I	1.1E−03	C						M	1	0.13	
205-99-2	~Benzo[b]fluoranthene	7.3E−01	E	1.1E−04	C						M	1	0.13	
207-08-9	~Benzo[k]fluoranthene	7.3E−02	E	1.1E−04	C						M	1	0.13	
218-01-9	~Chrysene	7.3E−03	E	1.1E−05	C						M	1	0.13	
53-70-3	~Dibenz[a,h]anthracene	7.3E+00	E	1.2E−03	C						M	1	0.13	
192-65-4	~Dibenzo(a,e)pyrene	1.2E+01	C	1.1E−03	C									
57-97-6	~Dimethylbenz(a)anthracene, 7,12-	2.5E+02	C	7.1E−02	C							1	0.13	
206-44-0	~Fluoranthene					4.0E−02	I					1	0.13	
86-73-7	~Fluorene					4.0E−02	I			V		1	0.13	
193-39-5	~Indeno[1,2,3-cd]pyrene	7.3E−01	E	1.1E−04	C						M	1	0.13	
90-12-0	~Methylnaphthalene, 1-	2.9E−02	P			7.0E−02	A			V		1		3.94E+02
91-57-6	~Methylnaphthalene, 2-					4.0E−03	I			V		1		3.68E+02
91-20-3	~Naphthalene			3.4E−05	C	2.0E−02	I	3.0E−03	I	V		1	0.13	

57835-92-4	~Nitropyrene, 4-	1.2E+00	C	1.1E−04	C								
129-00-0	~Pyrene					3.0E−02	I			V	1	0.13	
7778-74-7	Potassium Perchlorate					7.0E−04	I				1		
67747-09-5	Prochloraz	1.5E−01	I			9.0E−03	I				1	0.1	
26399-36-0	Profluralin					6.0E−03	H				1	0.1	
1610-18-0	Prometon					1.5E−02	I				1	0.1	
7287-19-6	Prometryn					4.0E−03	I				1	0.1	
1918-16-7	Propachlor					1.3E−02	I				1	0.1	
709-98-8	Propanil					5.0E−03	I				1	0.1	
2312-35-8	Propargite					2.0E−02	I				1	0.1	
107-19-7	Propargyl Alcohol					2.0E−03	I				1	0.1	
139-40-2	Propazine					2.0E−02	I				1	0.1	
122-42-9	Propham					2.0E−02	I				1	0.1	
60207-90-1	Propiconazole					1.3E−02	I				1	0.1	
123-38-6	Propionaldehyde							8.0E−03	I	V			3.26E+04
103-65-1	Propyl benzene					1.0E−01	X	1.0E+00	X	V	1	0.1	2.64E+02
115-07-1	Propylene							3.0E+00	C				
57-55-6	Propylene Glycol					2.0E+01	P				1	0.1	
6423-43-4	Propylene Glycol Dinitrate							2.7E−04	A	V	1		1.51E+03
1569-02-4	Propylene Glycol Monoethyl Ether					7.0E−01	H				1	0.1	
107-98-2	Propylene Glycol Monomethyl Ether					7.0E−01	H	2.0E+00	I		1	0.1	
75-56-9	Propylene Oxide	2.4E−01	I	3.7E−06	I			3.0E−02	I	V	1		7.77E+04
81335-77-5	Pursuit					2.5E−01	I				1	0.1	
51630-58-1	Pydrin					2.5E−02	I				1	0.1	
110-86-1	Pyridine					1.0E−03	I			V	1		5.30E+05
13593-03-8	Quinalphos					5.0E−04	I				1	0.1	
91-22-5	Quinoline	3.0E+00	I								1	0.1	
NA	Refractory Ceramic Fibers							3.0E−02	A		1		
10453-86-8	Resmethrin					3.0E−02	I				1	0.1	
299-84-3	Ronnel					5.0E−02	H				1	0.1	
83-79-4	Rotenone					4.0E−03	I				1	0.1	
94-59-7	Safrole	2.2E−01	C	6.3E−05	C						1	0.1	
78587-05-0	Savey					2.5E−02	I				1	0.1	

(Continued)

Appendix K (Continued)

CAS No.	Chemical Name	SFO (mg/kg-day)$^{-1}$		IUR (μg/m^3)$^{-1}$		RfD$_o$ (mg/kg-day)		RfC$_i$ (mg/m^3)		VOC	Mutagen	GIABS	ABS	C$_{sat}$ (mg/kg)
7783-00-8	Selenious Acid					5.0E−03	I					1		
7782-49-2	Selenium					5.0E−03	I	2.0E−02	C			1		
7446-34-6	Selenium Sulfide					5.0E−03	C	2.0E−02	C			1		
74051-80-2	Sethoxydim					9.0E−02	I					1	0.1	
7631-86-9	Silica (crystalline, respirable)							3.0E−03	C					
7440-22-4	Silver					5.0E−03	I					0.04		
122-34-9	Simazine	1.2E−01	H			5.0E−03	I					1	0.1	
62476-59-9	Sodium Acifluorfen					1.3E−02	I					1	0.1	
26628-22-8	Sodium Azide					4.0E−03	I					1		
148-18-5	Sodium Diethyldithiocarbamate	2.7E−01	H			3.0E−02	I					1	0.1	
7681-49-4	Sodium Fluoride					5.0E−02	A	1.3E−02	C			1		
62-74-8	Sodium Fluoroacetate					2.0E−05	I					1	0.1	
13718-26-8	Sodium Metavanadate					1.0E−03	H					1		
7601-89-0	Sodium Perchlorate					7.0E−04	I					1		
961-11-5	Stirofos (Tetrachlorovinphos)	2.4E−02	H			3.0E−02	I					1	0.1	
7440-24-6	Strontium, Stable					6.0E−01	I					1		
57-24-9	Strychnine					3.0E−04	I					1	0.1	
100-42-5	Styrene					2.0E−01	I	1.0E+00	I	V		1		8.67E+02
80-07-9	Sulfonylbis(4-chlorobenzene), 1,1′-					8.0E−04	P					1	0.1	
7664-93-9	Sulfuric Acid							1.0E−03	C					
88671-89-0	Systhane					2.5E−02	I					1	0.1	
21564-17-0	TCMTB					3.0E−02	H					1	0.1	
34014-18-1	Tebuthiuron					7.0E−02	I					1	0.1	
3383-96-8	Temephos					2.0E−02	H					1	0.1	
5902-51-2	Terbacil					1.3E−02	I					1	0.1	
13071-79-9	Terbufos					2.5E−05	H					1	0.1	
886-50-0	Terbutryn					1.0E−03	I					1	0.1	
5436-43-1	Tetrabromodiphenyl ether, 2,2′,4,4′- (BDE−47)					1.0E−04	I							
95-94-3	Tetrachlorobenzene, 1,2,4,5-					3.0E−04	I					1	0.1	

630-20-6	Tetrachloroethane, 1,1,1,2-	2.6E−02	I	7.4E−06	I	3.0E−02	I			V	1		6.80E+02
79-34-5	Tetrachloroethane, 1,1,2,2-	2.0E−01	I	5.8E−05	I	4.0E−03	P			V	1		1.90E+03
127-18-4	Tetrachloroethylene	5.4E−01	C	5.9E−06	C	1.0E−02	I	2.7E−01	A	V	1		1.66E+02
58-90-2	Tetrachlorophenol, 2,3,4,6-					3.0E−02	I				1	0.1	
5216-25-1	Tetrachlorotoluene, p- alpha, alpha, alpha-	2.0E+01	H								1	0.1	
3689-24-5	Tetraethyl Dithiopyrophosphate					5.0E−04	I				1	0.1	
811-97-2	Tetrafluoroethane, 1,1,1,2-							8.0E+01	I	V	1		1.09E+03
479-45-8	Tetryl (Trinitrophenylmethylnitramine)					4.0E−03	P				1	0.1	
7440-28-0	Thallium (Soluble Salts)										1		
28249-77-6	Thiobencarb					1.0E−02	I				1	0.1	
111-48-8	Thiodiglycol					7.0E−02	X				1	0.0075	
39196-18-4	Thiofanox					3.0E−04	H				1	0.1	
23564-05-8	Thiophanate, Methyl					8.0E−02	I				1	0.1	
137-26-8	Thiram					5.0E−03	I				1	0.1	
7440-31-5	Tin					6.0E−01	H				1		
7550-45-0	Titanium Tetrachloride							1.0E−04	A		1		
108-88-3	Toluene					8.0E−02	I	5.0E+00	I	V	1		8.18E+02
106-49-0	Toluidine, p-	1.9E−01	H								1	0.1	
8001-35-2	Toxaphene	1.1E+00	I	3.2E−04	I						1	0.1	
66841-25-6	Tralomethrin					7.5E−03	I				1	0.1	
688-73-3	Tri-n-butyltin					3.0E−04	A				1	0.1	
2303-17-5	Triallate					1.3E−02	I				1	0.1	
82097-50-5	Triasulfuron					1.0E−02	I				1	0.1	
615-54-3	Tribromobenzene, 1,2,4-					5.0E−03	I				1	0.1	
126-73-8	Tributyl Phosphate	9.2E−03	P			2.0E−01	P				1	0.1	
NA	Tributyltin Compounds					3.0E−04	P				1	0.1	
56-35-9	Tributyltin Oxide					3.0E−04	I				1	0.1	
76-13-1	Trichloro-1,2,2-trifluoroethane, 1,1,2-					3.0E+01	I	3.0E+01	H	V	1		9.10E+02
76-03-9	Trichloroacetic Acid										1	0.1	
33663-50-2	Trichloroaniline HCl, 2,4,6-	2.9E−02	H								1	0.1	
634-93-5	Trichloroaniline, 2,4,6-	3.4E−02	H								1	0.1	
87-61-6	Trichlorobenzene, 1,2,3-					8.0E−04	X			V	1	0.1	1.51E+02

(Continued)

Appendix K (Continued)

CAS No.	Chemical Name	SFO (mg/kg-day)$^{-1}$		IUR (μg/m^3)$^{-1}$		RfD$_o$ (mg/kg-day)		RfC$_i$ (mg/m^3)		VOC	Mutagen	GIABS	ABS	C$_{sat}$ (mg/kg)
120-82-1	Trichlorobenzene, 1,2,4-	2.9E−02	P			1.0E−02	I	2.0E−03	P	V		1		4.04E+02
71-55-6	Trichloroethane, 1,1,1-					2.0E+00	I	5.0E+00	I	V		1		6.40E+02
79-00-5	Trichloroethane, 1,1,2-	5.7E−02	I	1.6E−05	I	4.0E−03	I			V		1		2.16E+03
79-01-6	Trichloroethylene	5.9E−03	C	2.0E−06	C					V		1		6.92E+02
75-69-4	Trichlorofluoromethane					3.0E−01	I	7.0E−01	H	V		1		1.23E+03
95-95-4	Trichlorophenol, 2,4,5-					1.0E−01	I					1	0.1	
88-06-2	Trichlorophenol, 2,4,6-	1.1E−02	I	3.1E−06	I	1.0E−03	P					1	0.1	
93-76-5	Trichlorophenoxyacetic Acid, 2,4,5-					1.0E−02	I					1	0.1	
93-72-1	Trichlorophenoxypropionic Acid, 2,4,5-					8.0E−03	I					1	0.1	
598-77-6	Trichloropropane, 1,1,2-					5.0E−03	I			V		1		1.28E+03
96-18-4	Trichloropropane, 1,2,3-	3.0E+01	I			4.0E−03	I	3.0E−04	I	V	M	1		1.40E+03
96-19-5	Trichloropropene, 1,2,3-					3.0E−03	X	3.0E−04	P	V		1		4.51E+02
58138-08-2	Tridiphane					3.0E−03	I					1	0.1	
121-44-8	Triethylamine							7.0E−03	I	V		1		2.79E+04
1582-09-8	Trifluralin	7.7E−03	I			7.5E−03	I					1	0.1	
512-56-1	Trimethyl Phosphate	3.7E−02	H									1	0.1	
95-63-6	Trimethylbenzene, 1,2,4-							7.0E−03	P	V		1		2.19E+02
108-67-8	Trimethylbenzene, 1,3,5-					1.0E−02	X			V		1		1.82E+02
99-35-4	Trinitrobenzene, 1,3,5-					3.0E−02	I					1	0.019	
118-96-7	Trinitrotoluene, 2,4,6-	3.0E−02	I			5.0E−04	I					1	0.032	
791-28-6	Triphenylphosphine Oxide					2.0E−02	P					1	0.1	
13674-87-8	Tris(1,3-Dichloro-2-propyl) Phosphate					2.0E−02	A					1	0.1	
115-96-8	Tris(2-chloroethyl)phosphate	2.0E−02	P			7.0E−03	P					1	0.1	
78-42-2	Tris(2-ethylhexyl)phosphate	3.2E−03	P			1.0E−01	P					1	0.1	
NA	Uranium (Soluble Salts)					3.0E−03	I	3.0E−04	A			1		
51-79-6	Urethane	1.0E+00	C	2.9E−04	C									
1314-62-1	Vanadium Pentoxide			8.3E−03	P	9.0E−03	I	7.0E−06	P			0.026		
36907-42-3	Vanadium Sulfate					2.0E−02	H					0.026		
NA	Vanadium and Compounds					5.0E−03	S					1		
7440-62-2	Vanadium, Metallic					7.0E−05	P	1.0E−04	A			0.026		

1929-77-7	Vernolate					1.0E−03	I					1	0.1	
50471-44-8	Vinclozolin					2.5E−02	I					1	0.1	
108-05-4	Vinyl Acetate					1.0E+00	H	2.0E−01	I	V		1		2.75E+03
593-60-2	Vinyl Bromide			3.2E−05	H			3.0E−03	I	V		1		3.37E+03
75-01-4	Vinyl Chloride	7.2E−01	I	4.4E−06	I	3.0E−03	I	1.0E−01	I	V	M	1		3.92E+03
81-81-2	Warfarin					3.0E−04	I					1	0.1	
1330-20-7	Xylene, Mixture					2.0E−01	I	1.0E−01	I	V		1		2.58E+02
106-42-3	Xylene, p-					2.0E−01	I	7.0E−01	C	V		1		3.90E+02
108-38-3	Xylene, m-					2.0E−01	I	7.0E−01	C	V		1		3.88E+02
95-47-6	Xylene, o-					2.0E−01	I	7.0E−01	C	V		1		4.34E+02
7440-66-6	Zinc (Metallic)					3.0E−01	I					1		
1314-84-7	Zinc Phosphide					3.0E−04	I					1		
12122-67-7	Zineb					5.0E−02	I					1	0.1	

Key: I = IRIS; P = PPRTV; A = ATSDR; C = Cal EPA; X = PPRTV Appendix; H = HEAST; J = New Jersey; E = Environmental Criteria and Assessment Office; S = see user guide Section 5; L = see user guide on lead; M = mutagen; V = volatile; F = see FAQ #29; c = cancer; * = where: n SL < 100X c SL; ** = where n SL < 10X c SL; n = noncancer; m = concentration may exceed ceiling limit (see user guide); s = concentration may exceed Csat (see user guide); SSL values are based on DAF = 1.

Reference: Regional Screening Levels (Formerly PRGs) Version 9 May 2010 Stanford J. Smucker Ph.D., EPA Region IX Technical Support Section (SFD-8-4) 75 Hawthorne Street, San Francisco, CA 94105-3901.

Appendix L: Industry Chemical Matrix

CAS No.	Chemical Name	Wood Preservation Industry	Organophosphate and Agro-Chemicals Industry Sectors	The Carbon Black Industry	Oil and Gas Industry	Organophosphate and Agro-Chemicals Industry Sectors	Iron and Steel Industry	Food and Dairy Industry	Textile Industry	Construction Industry	Power Utilities Industry
75-07-0	Acetaldehyde								x		x
67-64-1	Acetone	x			x					x	
75-86-5	Acetone cyanohydrin						x				
75-05-8	Acetonitrile									x	
74-86-2	Acetylene			x							
107-02-8	Acrolein										x
79-10-7	Acrylic acid										x
107-13-1	Acrylonitrile										x
107-18-6	Allyl alcohol									x	
7429-90-5	Aluminum				x					x	x
1344-28-1	Aluminum oxide				x		x			x	
	Aluminum (pyro powders and welding fumes, as Al)						x			x	x
7664-41-7	Ammonia	x	x		x		x		x	x	x
62-53-3	Aniline (and homologs)										x
62-53-3	Aniline oil									x	
65996-93-2	Anthracene & benzo(a)pyrene	x			x		x			x	x
7440-36-0	Antimony	x	x		x		x		x	x	x
7803-52-3	Antimony hydride				x						
7803-52-3	Antimony trihydride				x						
7647-01-0	Aqueous hydrogen chloride (i.e., Hydrochloric acid, Muriatic acid)	x	x		x				x	x	
7664-39-3	Aqueous hydrogen fluoride (i.e., Hydrofluoric acid)				x						
7440-38-2	Arsenic (inorganic compounds, as As)	x	x		x				x	x	
7440-38-2	Arsenic metal: Arsenia	x							x		x

	Arsenic, organic compounds (as As)	x	x		x		x	x	
1332-21-4	Asbestos				x			x	x
1912-24-9	Atrazine		x						
17804-35-2	Benomyl		x						
71-43-2	Benzene				x	x		x	x
92-87-5	Benzidine				x				
100-44-7	Benzyl chloride								x
7440-41-7	Beryllium & beryllium compounds (as Be)				x				x
314-40-9	Bromacil		x						
7726-95-6	Bromine		x						x
106-99-0	1,3-Butadiene		x		x	x			
106-97-8	n-Butane			x					
111-76-2	2-Butoxyethanol								x
141-32-2	Butyl acrylate							x	x
71-36-3	n-Butyl alcohol	x	x		x	x	x	x	x
78-92-2	sec-Butyl alcohol					x		x	x
75-65-0	tert-Butyl alcohol		x		x			x	x
7440-43-9	Cadmium dust (as Cd)		x		x		x		
1306-19-0	Cadmium fume (as Cd)		x		x		x		
156-62-7	Calcium cyanamide					x			
133-06-2	Captan		x						
63-25-2	Carbaryl		x						
1563-66-2	Carbofuran		x						
75-15-0	Carbon disulfide		x	x	x	x			
630-08-0	Carbon monoxide			x					
56-23-5	Carbon tetrachloride		x	x					
7782-50-5	Chlorine	x	x		x	x	x	x	x
10049-04-4	Chlorine dioxide				x		x		
108-90-7	Chlorobenzene		x		x			x	
67-66-3	Chloroform							x	
76-06-2	Chloropicrin		x						
	Chromium(II) compounds (as Cr)	x	x		x	x	x	x	
	Chromium(III) compounds (as Cr)	x	x		x	x	x	x	x

(Continued)

Appendix L (Continued)

CAS No.	Chemical Name	Wood Preservation Industry	Organophosphate and Agro-Chemicals Industry Sectors	The Carbon Black Industry	Oil and Gas Industry	Organophosphate and Agro-Chemicals Industry Sectors	Iron and Steel Industry	Food and Dairy Industry	Textile Industry	Construction Industry	Power Utilities Industry
7440-47-3	Chromium metal	x	x		x				x	x	x
10210-68-1	Cobalt carbonyl (as Co)				x						
16842-03-8	Cobalt hydrocarbonyl (as Co)				x						
7440-48-4	Cobalt metal dust and fume (as Co)		x		x		x		x		
7440-50-8	Copper (dusts and mists, as Cu)	x	x		x		x		x	x	
1317-38-0	Copper fume (as Cu)	x			x		x		x	x	
108-39-4	m-Cresol				x		x			x	
95-48-7	o-Cresol				x		x			x	
106-44-5	p-Cresol				x		x			x	
98-82-8	Cumene		x		x					x	
110-82-7	Cyclohexane				x						
108-93-0	Cyclohexanol		x								
333-41-5	Diazinon®		x								
84-74-2	Dibutyl phthalate	x							x	x	
95-50-1	o-Dichlorobenzene									x	
106-46-7	p-Dichlorobenzene									x	
75-71-8	Dichlorodifluoromethane		x		x						
62-73-7	Dichlorvos		x								
77-73-6	Dicyclopentadiene		x								
111-42-2	Diethanolamine		x		x		x		x		
84-66-2	Diethyl phthalate									x	
124-40-3	Dimethylamine		x								
68-12-2	Dimethylformamide		x								x
77-78-1	Dimethyl sulfate		x								
34590-94-8	Dipropylene glycol methyl ether				x						

330-54-1	Diuron		x						
110-80-5	2-Ethoxyethanol						x	x	
140-88-5	Ethyl acrylate							x	
100-41-4	Ethyl benzene				x				
107-21-1	Ethylene glycol	x	x		x	x	x	x	
75-21-8	Ethylene oxide				x			x	
50-00-0	Formaldehyde	x	x		x		x	x	
64-18-6	Formic acid		x				x		
77-47-4	Hexachlorocyclopentadiene		x						
822-06-0	Hexamethylene diisocyanate		x						
110-54-3	n Hexane		x				x		x
302-01-2	Hydrazine		x		x				x
74-90-8	Hydrogen cyanide				x	x	x		
7664-39-3	Hydrogen fluoride		x						
2148878	Hydrogen sulfide			x					x
123-31-9	Hydroquinone		x						
75-28-5	Isobutane			x					
4098-71-9	Isophorone diisocyanate		x						
67-63-0	Isopropyl alcohol		x		x		x	x	
7439-92-1	Lead		x		x	x	x	x	x
58-89-9	Lindane		x						
121-75-5	Malathion		x						
108-31-6	Maleic anhydride		x				x		
7439-96-5	Manganese compounds and fume (as Mn)		x		x	x		x	
7439-97-6	Mercury compounds [except (organo) alkyls] (as Hg)				x				
	Mercury (organo) alkyl compounds (as Hg)				x				
72-43-5	Methoxychlor		x						
96-33-3	Methyl acrylate							x	
101-77-9	4,4′-Methylenedianiline								x
624-83-9	Methyl isocyanate		x						
80-62-6	Methyl methacrylate	x			x		x	x	x
298-00-0	Methyl parathion		x						

(Continued)

Appendix L (Continued)

CAS No.	Chemical Name	Wood Preservation Industry	Organophosphate and Agro-Chemicals Industry Sectors	The Carbon Black Industry	Oil and Gas Industry	Organophosphate and Agro-Chemicals Industry Sectors	Iron and Steel Industry	Food and Dairy Industry	Textile Industry	Construction Industry	Power Utilities Industry
21087-64-9	Metribuzin		x								
91-20-3	Naphthalene	x	x		x		x		x	x	x
3173-72-6	Naphthalene diisocyanate		x								
7440-02-0	Nickel metal and other compounds (as Ni)		x		x		x		x	x	x
7697-37-2	Nitric acid	x	x		x		x			x	x
10102-43-9	Nitric oxide										x
98-95-3	Nitrobenzene									x	x
10102-44-0	Nitrogen dioxide			x							x
1910-42-5	Paraquat (Paraquat dichloride)		x								
56-38-2	Parathion		x								
	Particulates not otherwise regulated			x							
87-86-5	Pentachlorophenol	x									
109-66-0	n-Pentane			x							
594-42-3	Perchloromethyl mercaptan		x								
108-95-2	Phenol	x	x		x		x		x	x	x
7803-51-2	Phosphine		x								
7664-38-2	Phosphoric acid	x	x		x		x		x	x	
7723-14-0	Phosphorus (yellow)										x
85-44-9	Phthalic anhydride										x
74-98-6	Propane			x							
114-26-1	Propoxur		x								
627-13-4	n-Propyl nitrate										x
110-86-1	Pyridine		x				x			x	
	Rosin core solder, pyrolysis products (as formaldehyde)				x						

7782-49-2	Selenium				x			x	x
7440-22-4	Silver (metal dust and soluble compounds, as Ag)				x	x			x
100-42-5	Styrene	x			x	x	x	x	x
7446-09-5	Sulfur dioxide			x					
7664-93-9	Sulfuric acid	x	x	x		x	x	x	x
35400-43-2	Sulprofos		x						
3383-96-8	Temephos		x						
127-18-4	Tetrachloroethylene	x			x	x	x	x	x
78-00-2	Tetraethyl lead (as Pb)				x				
75-74-1	Tetramethyl lead (as Pb)				x				
137-26-8	Thiram		x						
7440-31-5	Tin				x				
108-88-3	Toluene	x	x		x	x	x	x	x
584-84-9	Toluene-2, 4-diisocyanate	x	x				x		
120-82-1	1,2,4-Trichlorobenzene		x				x		
79-00-5	1,1,2-Trichloroethane				x				x
79-01-6	Trichloroethylene					x	x	x	x
121-44-8	Triethylamine		x						
95-63-6	1,2,4-Trimethylbenzene		x		x	x	x	x	x
1314-62-1	Vanadium dust		x		x	x			x
1314-62-1	Vanadium fume		x		x	x			x
108-05-4	Vinyl acetate				x			x	x
75-01-4	Vinyl chloride		x				x		x
108-38-3	m-Xylene	x	x		x	x	x	x	x
95-47-6	o-Xylene	x	x		x	x	x	x	x
106-42-3	p-Xylene	x	x		x	x	x	x	x
1477-55-0	m-Xylene-alpha, alpha′-diamine	x	x				x		
7440-65-5	Yttrium				x				
7646-85-7	Zinc chloride fume				x	x			x
1314-13-2	Zinc oxide				x				
557-05-1	Zinc stearate				x				

Index

F

Kolibri 360 GmbH
Ferdinand-Jühlke-Straße 7
99095 Erfurt, Deutschland
produktsicherheit@kolibri360.de